D1755087

H. U. Blaser, E. Schmidt (Eds.)

Asymmetric Catalysis on Industrial Scale

Further Reading from Wiley-VCH

Aehle, W.

Enzymes in Industry, 2. Ed.

2003. ISBN 3-527-29592-5

Weissermel, K., Arpe H.-J.

Industrial Organic Chemistry, 4. Ed.

2003. ISBN 3-527-30578-5

Bamfield, P.

Research & Development Management in the Chemical and Pharmaceutical Industry, 2. Ed.

2003. ISBN 3-527-30667-6

Cornils, B., Herrmann, W. A.

Applied Homogeneous Catalysis with Organometallic Compounds, 3 Vols., 2. Ed.

2002. ISBN 3-527-30434-7

Asymmetric Catalysis on Industrial Scale

Challenges, Approaches and Solutions

Edited by
H. U. Blaser, E. Schmidt

WILEY-VCH

WILEY-VCH Verlag GmbH & Co. KGaA

Editors

Dr. Hans-Ulrich Blaser
Solvias AG
P.O. Box
CH-4002 Basel
Switzerland
hans-ulrich.blaser@solvias.com

Dr. Elke Schmidt
Syngenta Crop Protection AG
CH-4058 Basel
Switzerland
elke.schmidt@syngenta.com

■ This book was carefully produced. Nevertheless, editors, authors and publisher do not warrant the information contained therein to be free of errors. Readers are advised to keep in mind that statements, data, illustrations, procedural details or other items may inadvertently be inaccurate.

Library of Congress Card No.: applied for

British Library Cataloguing-in-Publication Data
A catalogue record for this book is available from the British Library.

Bibliographic information published by Die Deutsche Bibliothek
Die Deutsche Bibliothek lists this publication in the Deutsche Nationalbibliografie; detailed bibliographic data is available in the Internet at <http://dnb.ddb.de>

© 2004 WILEY-VCH Verlag GmbH & Co. KGaA, Weinheim, Germany

All rights reserved (including those of translation in other languages). No part of this book may be reproduced in any form – by photoprinting, microfilm, or any other means – nor transmitted or translated into machine language without written permission from the publishers. Registered names, trademarks, etc. used in this book, even when not specifically marked as such, are not to be considered unprotected by law.

Printed in the Federal Republic of Germany
Printed on acid-free paper

Composition K+V Fotosatz GmbH, Beerfelden
Printing betz-druck GmbH, Darmstadt
Bookbinding J. Schäffer GmbH & Co. KG, Grünstadt

ISBN 3-527-30631-5

Contents

List of Contributors *XXI*

Introduction *1*
Hans-Ulrich Blaser and Elke Schmidt
1 Background and Motivation *1*
2 Goals and Concept *1*
3 Our (the Editors) Assessment of the Resulting Monograph *2*
4 The Organization of the Book
 (Types of Intermediates – Catalysts – Situations) *3*
4.1 Category: New Processes for Existing Active Compounds *5*
4.2 Category II: New Catalysts and/or Processes for Important Building Blocks *5*
4.3 Category III: Adaptation of Existing Catalysts for Important Building Blocks *6*
4.4 Category IV: Processes for New Chemical Entities (NCE) *6*
5 Missing Processes *6*
6 Some Important Messages *9*
6.1 Transformations and Catalyst Types *10*
6.2 Scale, Development Stage *13*
6.3 Synthesis Planning and Time Lines *13*
6.4 Major Technical Problems *14*
6.5 Technology and Patents *14*
7 Final Comments and Conclusions *15*
8 Glossary *16*
9 References *18*

I	**New Processes for Existing Active Compounds** *21*
1	**Asymmetric Hydrogenations – The Monsanto L-Dopa Process** *23*
	William S. Knowles
1.1	The Development of Chiral Phosphane Ligands *23*
1.2	Synthesis and Properties of the Phosphanes *31*
1.3	Mechanism of the Asymmetric Catalysis *35*
1.4	Concluding Comments *37*
1.5	Acknowledgements *38*
1.6	References *38*
2	**The Other L-Dopa Process** *39*
	Rüdiger Selke
2.1	Introduction *39*
2.2	Choice of the Substrate *42*
2.3	The Catalyst *44*
2.4	Improving the Hydrogenation Reaction *45*
2.5	Effect of Solvent and Anions *46*
2.6	Immobilization of the Catalyst *49*
2.7	Production Process *49*
2.8	Acknowledgements *52*
2.9	References *52*
3	**The Chiral Switch of Metolachlor: The Development of a Large-Scale Enantioselective Catalytic Process** *55*
	Hans-Ulrich Blaser, Reinhard Hanreich, Hans-Dieter Schneider, Felix Spindler, and Beat Steinacher
3.1	Introduction and Problem Statement *55*
3.2	Route Selection *56*
3.2.1	Enamide Hydrogenation *56*
3.2.2	Nucleophilic Substitution of an (R)-Methoxyisopropanol Derivative *57*
3.2.3	Hydrogenation of MEA Imine *57*
3.2.4	Catalytic Alkylation with Racemic Methoxyisopropanol *58*
3.2.5	Assessment and Screening of the Proposed Routes *58*
3.3	Imine Hydrogenation: Laboratory Process *59*
3.3.1	Finding the Right Metal-Ligand Combination *59*
3.3.1.1	Screening of Rh-Diphosphine Complexes *59*
3.3.1.2	Screening of Ir-Diphosphine Complexes *60*
3.3.1.3	Synthesis and Screening of a New Ligand Class *60*
3.3.2	Optimization of Reaction Medium and Conditions *61*
3.3.3	Ligand Fine Tuning *61*
3.4	Imine Hydrogenation: Technical Process *62*
3.4.1	Strategy for Process Development *62*
3.4.2	The Production of the MEA Imine at the Required Quality *62*
3.4.3	Scale-up of the Ligand Synthesis *63*

3.4.4	Fine Optimization of the Ir-Catalyst Formulation 65	
3.4.5	Choice of Reactor Technology 65	
3.4.6	Scale-Up to the Production Autoclave 66	
3.4.7	Work-up, Separation of the Catalyst from the Product 67	
3.5	Summary and Conclusions 68	
3.6	Acknowledgements 69	
3.7	References 69	
4	**Enantioselective Hydrogenation:** **Towards a Large-Scale Total Synthesis of (R,R,R)-α-Tocopherol** 71 *Thomas Netscher, Michelangelo Scalone, and Rudolf Schmid*	
4.1	Introduction: Vitamin E as the Target 71	
4.2	Routes to (2R,4′-R,8′R)-α-Tocopherol by Total Synthesis 74	
4.3	Synthesis of Prochiral Allylic Alcohols 76	
4.4	Asymmetric Hydrogenation of Prochiral Allylic Alcohols 78	
4.5	Synthesis of the Diphosphane Ligands 81	
4.6	Procedures for Stereochemical Analysis 84	
4.7	Concluding Remarks 86	
4.8	Acknowledgements 87	
4.9	References 87	
5	**Comparison of Four Technical Syntheses** **of Ethyl (R)-2-Hydroxy-4-Phenylbutyrate** 91 *Hans-Ulrich Blaser, Marco Eissen, Pierre F. Fauquex, Konrad Hungerbühler, Elke Schmidt, Gottfried Sedelmeier, and Martin Studer*	
5.1	Introduction 91	
5.2	Synthetic Pathways to HPB Ester 92	
5.2.1	Route A: Synthesis and Enantioselective Reduction of Keto Acid 3 with Immobilized *Proteus vulgaris* Followed by Esterification 94	
5.2.2	Route B: Enantioselective Reduction of 3 with D-LDH in a Membrane Reactor 94	
5.2.3	Route C. Synthesis and Enantioselective Hydrogenation of Keto Ester 5 95	
5.2.4	Route D: Synthesis and Enantioselective Hydrogenation of Diketo Ester 6, Followed by Hydrogenolysis 95	
5.3	Comparison of Routes A–D with Respect to Mass Consumption, Environmental, Health and Safety Aspects 96	
5.3.1	Definitions 96	
5.3.2	Overall Material Masses Consumed and Produced for Routes A–D 97	
5.3.3	Mass Consumption of the Reduction Systems A–D 98	
5.3.4	Problematic Chemicals: Environmental, Health and Safety Aspects (EHS) 99	
5.3.4.1	Safety Aspects 99	
5.3.4.2	Health/Toxicology 100	

5.3.4.3 Environment 100
5.4 Overall Comparison of Routes A–D and General Conclusions 100
5.4.1 Conclusions for the Reduction Steps 100
5.4.2 Conclusions for the Overall Syntheses 101
5.5 References 103

6 Biocatalytic Approaches for the Large-Scale Production of Asymmetric Synthons 105
Nicholas M. Shaw, Karen T. Robins, and Andreas Kiener
6.1 Introduction 105
6.2 Asymmetric Biocatalysis 106
6.2.1 L-Carnitine 106
6.2.2 Asymmetric Reduction: (R)-Ethyl-4,4,4-trifluorohydroxybutanoate 107
6.3 Resolution of Racemic Mixtures 108
6.3.1 (R)- and (S)-3,3,3-Trifluoro-2-hydroxy-2-methylpropionic Acid 108
6.3.2 (S)-2,2-Dimethylcyclopropane Carboxamide 111
6.3.3 CBZ-D-proline [(R)-N-CBZ-proline] 113
6.3.4 (1R,4S)-1-Amino-4-hydroxymethyl-cyclopent-2-ene 114
6.4 Summary 114
6.5 References 115

7 7-Aminocephalosporanic Acid – Chemical Versus Enzymatic Production Process 117
Thomas Bayer
7.1 Introduction 117
7.2 Synthesis of 7-ACA 119
7.2.1 The Chemical 7-ACA Route 120
7.2.2 Single Step Biocatalytic 7-ACA Synthesis 121
7.2.3 Two-step Biocatalytic 7-ACA Synthesis 122
7.2.4 Two-Step Biocatalytic Process 125
7.2.4.1 Oxidation 126
7.2.4.2 Deacylation 127
7.3 Comparison of Chemical and Two-Step Biocatalytic 7-ACA Process 127
7.4 Conclusion 129
7.4 References 130

8 Methods for the Enantioselective Biocatalytic Production of L-Amino Acids on an Industrial Scale 131
Harald Gröger and Karlheinz Drauz
8.1 Introduction 131
8.2 L-Amino Acids via Enzymatic Resolutions 133
8.2.1 Processes Using L-Amino Acylases 133
8.2.2 Processes Using L-Amidases 135
8.2.3 Processes Using L-Hydantoinases 137

8.2.4	Processes via Lactam Hydrolysis *139*
8.3	L-Amino Acids via Asymmetric Biocatalysis *140*
8.3.1	Processes via Reductive Amination *140*
8.3.2	Processes via Transamination *142*
8.3.3	Processes via Addition of Ammonia to α,β-Unsaturated Acids *143*
8.4	Conclusion *145*
8.5	References *145*

| II | **New Catalysts for Existing Active Compounds** *149* |

1	**The Large-Scale Biocatalytic Synthesis of Enantiopure Cyanohydrins** *151*
	Peter Poechlauer, Wolfgang Skranc, and Marcel Wubbolts
1.1	Introduction *151*
1.2	Chiral Cyanohydrins as Building Blocks in the Synthesis of Fine Chemicals *151*
1.3	Synthesis of Enantiopure Cyanohydrins *152*
1.3.1	General Methods *152*
1.3.1.1	Sources of HCN *152*
1.3.1.2	The Sources of Chirality of Cyanohydrins *153*
1.3.2	Syntheses via Chiral Metal Catalysts *154*
1.3.3	Cyanohydrin Synthesis via Cyclic Dipeptides *154*
1.3.4	The Cyanohydrin Synthesis via Lipases *156*
1.3.5	Synthesis of Chiral Cyanohydrins Using Hydroxy Nitrile Lyases (HNLs) *157*
1.4	Large-Scale Cyanohydrin Production *158*
1.4.1	General Considerations *158*
1.4.2	Scale-up of HNL-Catalyzed Cyanohydrin Formation *159*
1.5	Examples of Large-Scale Production of Cyanohydrins *161*
1.5.1	Production of (S)-3-Phenoxybenzaldehyde Cyanohydrin (SCMB) *162*
1.5.2	Production of (R)- and (S)-Mandelic Acid Derivatives *163*
1.6	Summary *163*
1.7	References *163*

2	**Industrialization Studies of the Jacobsen Hydrolytic Kinetic Resolution of Epichlorohydrin** *165*
	Larbi Aouni, Karl E. Hemberger, Serge Jasmin, Hocine Kabir, Jay F. Larrow, Isidore Le-Fur, Philippe Morel, and Thierry Schlama
2.1	Background *165*
2.2	The HKR Catalyst *166*
2.2.1	Preparation and Isolation of Active Catalyst *167*
2.2.2	Activation Step Characterization *169*
2.2.2.1	Quantitative Analysis of Co(II) and Co(III) Species *169*
2.2.2.2	Kinetic Limitation of the Co(II) to Co(III) Oxidation Reaction *169*
2.2.3	Scale-up Considerations for the (Salen)-Co(III)-OAc Process *174*
2.3	Kinetic Modelling and Simulation of the HKR Reaction *176*

2.3.1	Objectives and Approach 176
2.3.2	Development of a Global Reaction Scheme 177
2.3.3	Kinetic Modelling 179
2.3.3.1	Kinetic Rate Equations and Assumptions 179
2.3.3.2	Experimentation and Kinetic Model for a Fed-Batch Vessel 180
2.3.3.3	Parameter Deviation Study 188
2.4	HKR Process Optimization and Scale-up 190
2.4.1	Process Description 190
2.4.2	Process Optimization 190
2.4.2.1	HKR Stage Optimization 191
2.4.2.1.1	Shortcut Optimization Method 191
2.4.2.1.2	Process Temperature Control 193
2.4.2.2	Isolation of Resolved Epichlorohydrin 196
2.4.2.2.1	Reaction Medium Stabilization 196
2.4.2.2.2	Azeotropic Drying and Epichlorohydrin Isolation 197
2.5	Conclusions 198
2.6	References 199
3	**Scale-up Studies in Asymmetric Transfer Hydrogenation** 201
	John Blacker and Juliette Martin
3.1	Background 201
3.2	The Catalytic System 202
3.2.1	Catalyst 202
3.2.2	Hydrogen Donor 206
3.2.3	Substrate 209
3.3	Process 210
3.3.1	Temperature 211
3.3.2	Reactant Concentration 211
3.3.3	Reaction Control 212
3.4	Case Studies 212
3.4.1	Example 1: (*R*)-1-Tetralol 212
3.4.2	Example 2: (*S*)-4-Fluorophenylethanol 213
3.4.3	Example 3: (*R*)-*N*-Diphenylphosphinyl-1-methylnaphthylamine 213
3.5	Conclusions 214
3.6	Acknowledgements 215
3.7	References 216
4	**Practical Applications of Biocatalysis for the Manufacture of Chiral Alcohols such as (*R*)-1,3-Butanediol by Stereospecific Oxidoreduction** 217
	Akinobu Matsuyama and Hiroaki Yamamoto
4.1	Introduction 217
4.2	Screening of Microorganisms Producing Optically Active 1,3-BDO from 4-Hydroxy-2-butanone (4H2B) by Asymmetric Reduction 219
4.3	Screening of Microorganisms Producing Optically Active 1,3-BDO from the Racemate 221

4.4	Preparation of (R)- and (S)-1,3-BDO by the Same Strain or from the Same Material *222*	
4.5	Large-Scale Preparation of (R)-1,3-BDO from the Racemate Using *Candida parapsilosis* IFO 1396 *223*	
4.6	Purification and Characterization of (S)-1,3-Butanediol Dehydrogenase from *Candida parapsilosis* IFO 1396 *225*	
4.7	Cloning and Expression of a Gene Coding for a Secondary Alcohol Dehydrogenase from *Candida parapsilosis* IFO 1396 in *Eschericha coli* *226*	
4.8	Preparation of Ethyl (R)-4-Chloro-3-hydroxybutanoate (ECHB) by Recombinant *E. coli* Cells Expressing CpSADH *230*	
4.9	Conclusion and Outlook *231*	
4.10	References *231*	

5 Production of Chiral C3 and C4 Units via Microbial Resolution of 2,3-Dichloro-1-propanol, 3-Chloro-1,2-propanediol and Related Halohydrins *233*
N. Kasai and T. Suzuki

5.1	Introduction *233*
5.2	C3 Chiral Synthetic Units *235*
5.2.1	(R)- and (S)-2,3-dichloro-1-propanol (DCP) *235*
5.2.1.1	(R)-2,3-Dichloro-1-propanol (DCP) Assimilating Bacterium for (S)-2,3-Dichloro-1-propanol (DCP) and (R)-epichlorohydrin (EP) Production *235*
5.2.1.2	(S)-Epichlorohydrin (EP) Production and Isolation of a DCP Assimilating Bacterium *236*
5.2.2	Large-Scale Trials Using Immobilized Cells *236*
5.2.2.1	Production on a 100 L Scale *236*
5.2.2.2	Production on a 5000 L Scale *237*
5.2.3	Production of (R)- and (S)-3-Chloro-1,2-propanediol (CPD) and (R)- and (S)-Glycidol (GLD) *238*
5.2.3.1	Screening of (R)- and (S)-3-Chloro-1,2-propanediol (CPD) Assimilating Bacteria *238*
5.2.4	Scale-up of the Production of C3 Chiral Synthetic Units *239*
5.2.4.1	Fed-Batch Fermentation and Control *239*
5.2.4.2	Control Logic System *240*
5.3	Production of C4 Chiral Synthetic Units *245*
5.3.1	Production of 4-Chloro-3-acetoxybutyronitrile (BNOAc) by Ester-Degrading Enzymes *246*
5.3.2	Production of (R)-4-Chloro-3-hydroxy-butyrate (CHB) and (S)-3-hydroxy-γ-butyrolactone (HL) by *Enterobacter* sp. *247*
5.3.2.1	Cultivation and Preparation of *Enterobacter* sp. Resting Cells with High Degradation Activity *248*

5.3.2.2	Ton-Scale Production of (R)-4-Chloro-3-hydroxy-butyrate (CHB) and (S)-3-Hydroxy-γ-butyrolactone (HL) by the Resting Cells of *Enterobacter* sp.	251
5.3.2.3	Effect of Base on the Stereoselectivity of *Enterobacter* sp.	252
5.3.2.4	Intelligent Production of (R)-Methyl 4-chloro-3-hydroxybutyrate (CHBM) and (S)-3-Hydroxy-γ-butyrolactone (HL) with a Fed-Batch System of Substrate	253
5.4	Purification of Chiral Units	253
5.5	Racemization	255
5.6	References	256

III	**Adaptation of Existing Catalysts for Important Building Blocks**	**257**
1	**Synthesis of Unnatural Amino Acids**	**259**
	David J. Ager and Scott A. Laneman	
1.1	Introduction	259
1.2	Scope of Enamide Substrates	260
1.3	Preparation of Enamides	263
1.4	Mechanism	264
1.5	Ligand Synthesis	265
1.6	Other Substrates	266
1.7	Summary	267
1.8	References	268
2	**The Application of DuPHOS Rhodium(I) Catalysts for Commercial Scale Asymmetric Hydrogenation**	**269**
	Christopher J. Cobley, Nicholas B. Johnson, Ian C. Lennon, Raymond McCague, James A. Ramsden, and Antonio Zanotti-Gerosa	
2.1	Introduction	269
2.2	Large-Scale Manufacture of α-Amino Acids	271
2.2.1	N-Boc-(S)-3-Fluorophenylalanine	272
2.2.2	Experimental for N-t-Butoxycarbonyl-(S)-3-fluorophenylalanine	273
2.2.2.1	N-Acetyl Dehydro-3-fluorophenylalanine 8 via an Azlactone Intermediate	273
2.2.2.2	N-Acetyl-3-fluorophenylalanine 9	274
2.2.2.3	N-BOC-(S)-3-fluorophenylalanine 10	274
2.3	Practical and Economic Considerations for Manufacture	274
2.3.1	Manufacture and Choice of Precatalysts	275
2.3.2	Substrate Synthesis, Purity and Catalyst Loading	276
2.3.3	Catalyst Charging	278
2.3.4	Catalyst Removal	279
2.4	Conclusions	280
2.5	References	281

3	Liberties and Constraints in the Development of Asymmetric Hydrogenations on a Technical Scale 283
	John F. McGarrity, Walter Brieden, Rudolf Fuchs, Hans-Peter Mettler, Beat Schmidt, and Oleg Werbitzky
3.1	Introduction 283
3.2	Asymmetric Hydrogenation in the Lonza Biotin Process 284
3.2.1	Historical Development 284
3.2.2	Laboratory Screening Experiments 286
3.2.2.1	Ligand Screening 286
3.2.2.2	Substrate Screening 287
3.2.2.3	Parameter Screening and Optimization 290
3.2.3	Scale-up and Production 291
3.2.4	Preparation of Josiphos (R)-(S)-PPF-P(^{t}Bu)$_2$ on a Technical Scale 291
3.2.5	Conclusions 292
3.3	The Lonza Dextrometorphan Process 293
3.3.1	Introduction 293
3.3.2	Imine Hydrogenation 294
3.3.2.1	Background 294
3.3.2.2	Screening Experiments 295
3.3.2.3	Optimization Phase: Development of an Economic Process 296
3.3.3	Scale-up 297
3.3.4	Conclusion 298
3.4	4-Boc-piperazine-2(S)-N-^{t}butylcarboxamide 299
3.4.1	Introduction 299
3.4.2	Initial Studies 299
3.4.3	Scale-up 301
3.4.4	Further Improvements in Synthetic Strategy 302
3.4.5	Conclusions 303
3.5	Intermediates for SB-214857 (Lotrafiban) 303
3.5.1	Introduction 303
3.5.2	Initial Screening of Catalysts 304
3.5.3	Statistically Evaluated Screening of Experimental Conditions 306
3.6	References 308
4	Large-Scale Applications of Biocatalysis in the Asymmetric Synthesis of Laboratory Chemicals 309
	Roland Wohlgemuth
4.1	Introduction 309
4.2	History of Applied Biocatalysis at Fluka 310
4.2.1	Resolution of Racemic Mixtures 311
4.2.1.1	Resolution of Racemic Amino Acids 311
4.2.1.2	Resolution of Racemic Alcohols 311
4.2.2	Decomposition Reactions of Educts and Side Products 311
4.2.3	Introduction and Removal of Protecting Groups 311
4.2.4	Development and Production of Biocatalysts 312

4.2.5	Complex Reaction Paths to Natural Products *312*
4.2.6	Regio- and Stereoselective Synthesis *313*
4.3	Reasons for Large-Scale Application of Biocatalysts for the Synthesis of Laboratory Chemicals *314*
4.4	Examples of Large-Scale Application of Biocatalysis *314*
4.4.1	Matrix Approach: Product Groups Produced with Specific Enzyme Classes *314*
4.4.2	(S)-2-Octanol *315*
4.4.3	(R)-2-Octanol *316*
4.4.4	Comparison of Classical Resolution with the Biocatalytic Procedure *316*
4.4.5	2-Oxabicyclo[3.3.0]-oct-6-en-3-one *316*
4.4.6	3-Oxabicyclo[3.3.0]-oct-6-en-2-one *317*
4.4.7	Comparison of the Classical Procedure with the Biocatalytic Baeyer-Villiger Reaction *317*
4.5	Discussion and Outlook *318*
4.6	References *318*

IV Processes for New Chemical Entities (NCE) *321*

1 Development of an Efficient Synthesis of Chiral 2-Hydroxy Acids *323*
Junhua Tao and Kevin McGee

1.1	Introduction *323*
1.2	Results and Discussion *326*
1.3	Conclusion *330*
1.4	Experimental Section *331*
1.4.1	General Remarks *331*
1.4.2	Sodium 3-(4-Fluorophenyl)-2-oxo-propionate (6) *331*
1.4.3	(R)-3-(4-Fluorophenyl)-2-hydroxy propionic acid (1) *331*
1.4.4	Reactor Set-up and Preparation *332*
1.4.5	Enzyme Loading and Replenishment *333*
1.5	Acknowledgment *333*
1.6	References *333*

2 Factors Influencing the Application of Literature Methods Toward the Preparation of a Chiral *trans*-Cyclopropane Carboxylic Acid Intermediate During Development of a Melatonin Agonist *335*
Ambarish K. Singh, J. Siva Prasad, and Edward J. Delaney

2.1	Introduction *335*
2.2	Chemical Approach Employed During Preclinical Research and Development (Route A) *336*
2.3	Chemical Approaches Employed to Support Early Clinical Development (Route B *338*
2.4	Development of Process Technology to Support Phase II/III Clinical Studies and Future Commercialization (Routes C and D) *339*

2.4.1	Enzymatic Approaches to Convert 5 into 19	*341*
2.4.2	Chemical Approaches to Convert 5 into 19	*341*
2.5	Definition of Process Technology for the Optimal Route (Route D)	*343*
2.6	Summary and Conclusions	*345*
2.7	Acknowledgement	*347*
2.8	References	*347*

3 **Hetero Diels-Alder-Biocatalysis Approach for the Synthesis of (S)-3-[2-{(Methylsulfonyl)oxy}ethoxy]-4-(triphenylmethoxy)-1-butanol Methanesulfonate: Successful Application of an Enzyme Resolution Process** *349*
Jean-Claude Caille, Jim Lalonde, Yiming Yao, and C.K. Govindan

3.1	Introduction	*349*
3.2	The Hetero Diels-Alder-Biocatalysis Strategy	*350*
3.3	Hetero Diels-Alder Reaction: Synthesis of 2-Ethoxycarbonyl-3,6-dihydro-2H-pyran, (R,S)-4	*352*
3.4	Enzymatic Resolution of (R,S)-4	*353*
3.4.1	Secondary Screen and E Determination	*353*
3.4.2	Protease Screen	*354*
3.4.3	Optimization of the Resolution of (R,S)-4	*355*
3.4.3.1	Buffer pH and Concentration	*355*
3.4.3.2	Optimization of Enzyme Loading and Other Parameters	*356*
3.5	Pilot Plant Trials	*358*
3.6	Attempted Resolution of 3,6-Dihydro-2H-pyran-2-ylmethanol, (R,S)-5	*358*
3.7	Experimental	*359*
3.7.1	Hydrolase Library Screening – Hydrolysis of (R,S)-4	*359*
3.7.2	Gram Scale Runs	*359*
3.7.3	Butyl Ester	*360*
3.7.4	Determination of Enantiomeric Purity by Gas Chromatography	*360*
3.7.5	Resolution of (R,S)-4	*360*
3.7.6	Reduction of (S)-4	*361*
3.7.7	Tritylation of 5	*361*
3.7.8	Reductive Ozonolysis of (S)-2-Trityloxymethyl-3,6-dihydro-2H-pyran, 6	*361*
3.7.9	Preparation of 1	*362*
3.8	Summary	*362*
3.9	References	*363*

4		**Multi-Kilo Resolution of XU305, a Key Intermediate to the Platelet Glycoprotein IIb/IIIa Receptor Antagonist Roxifiban via Kinetic and Dynamic Enzymatic Resolution** *365*
		Jaan A. Pesti and Luigi Anzalone
4.1		Introduction *365*
4.2		Scale-up of the Kinetic Enzymatic Resolution of 9 to 1 *367*
4.2.1		Development of the Resolution of (R,S)-9 *368*
4.2.2		Racemization of 9 *369*
4.2.2.1		Mechanism of Isoxazoline (9) Racemization *370*
4.3		Process Development of the Dynamic Enzymatic Resolution of Thioester 10b *372*
4.3.1		Identification of Efficient Reaction Conditions for Dynamic Enzymatic Resolution *373*
4.3.2		Preparation of 1-Propyl Thioester (10b) *377*
4.3.3		Scale-up of the Dynamic Enzymatic Resolution Chemistry into the Plant *379*
4.4		Conclusions *380*
4.5		Acknowledgements *381*
4.6		References *381*
5		**Protease-Catalyzed Preparation of (S)-2-[(tert-Butylsulfonyl)methyl]-hydrocinnamic Acid for Renin Inhibitor RO0425892** *385*
		Beat Wirz, Stephan Doswald, Ernst Kupfer, Wolfgang Wostl, Thomas Weisbrod, and Heinrich Estermann
5.1		Introduction *385*
5.2		The Process Research Synthesis *387*
5.2.1		The Synthetic Concept *387*
5.2.2		Alternative Concepts *387*
5.2.3		Enzymatic Approaches *388*
5.2.3.1		Racemic Resolution of Ethyl Sulfopropionate 2 with α-Chymotrypsin *388*
5.2.3.2		Aminolysis of Ethyl Sulfopropionate 2 with Histidine Methylester 4 *388*
5.3		Enzymatic Resolution of Ethyl Sulfopropionate 2 with Subtilisin Carlsberg *389*
5.3.1		Process Research *389*
5.3.1.1		Enzyme Screening *389*
5.3.1.2		Parameter Optimization *390*
5.3.1.3		Work-up *390*
5.3.2		Process Development *391*
5.3.2.1		Process Parameters *391*
5.3.2.2		Improved Enzyme Preparation *392*
5.3.2.3		Work-up *392*
5.3.2.3.1		Equipment *393*
5.3.2.3.2		Isolation of Sulfopropionic Acid (S)-3 *393*

5.3.3	Pilot Production	*394*
5.3.3.1	Equipment	*394*
5.3.3.2	Enzyme Reaction	*395*
5.3.3.3	Work-up	*395*
5.4	Discussion	*395*
5.5	Acknowlegements	*396*
5.6	References	*397*

6 Protease-Catalyzed Preparation of Chiral 2-Isobutyl Succinic Acid Derivatives for Collagenase Inhibitor RO0319790 *399*

Beat Wirz, Milan Soukup, Thomas Weisbrod, Florian Stäbler, and Rolf Birk

6.1	Introduction	*399*
6.2	The Process Research Synthesis	*401*
6.2.1	The Synthetic Concept	*401*
6.2.2	Alternative Concepts	*401*
6.2.3	Enzymatic Approaches	*401*
6.2.3.1	Racemic Resolution of Nitrile Ester 10	*401*
6.2.3.2	Enantio- and Regioselective Monohydrolysis of Diester 9	*402*
6.2.3.3	Aminolysis in Organic Solvent Systems	*403*
6.2.3.4	Coupling of (R)-2a with Amine 13 in Aqueous Milieu	*403*
6.3	Enzymatic Resolution of Diethyl 2-Isobutyl Succinate 9	*403*
6.3.1	Process Research	*403*
6.3.1.1	Selection of the Enzyme	*403*
6.3.1.2	Parameter Optimization	*404*
6.3.1.3	Substrate Engineering	*405*
6.3.1.4	Work-up	*406*
6.3.1.5	Racemization of the Antipodal Diester (S)-9	*406*
6.3.2	Process Development	*406*
6.3.2.1	Process Parameters	*406*
6.3.2.2	Work-up	*407*
6.3.2.3	Racemization	*408*
6.3.3	Pilot Production	*408*
6.3.3.1	Equipment	*408*
6.3.3.2	Enzyme Reaction	*408*
6.3.3.3	Work-up	*409*
6.4	Discussion	*409*
6.5	Acknowledgments	*410*
6.6	References	*410*

7	An Innovative Asymmetric Sulfide Oxidation: The Process Development History Behind the New Antiulcer Agent Esomeprazole *413*
	Hans-Jürgen Federsel and Magnus Larsson
7.1	Major Events at a Glance: An Overview of Achievements from 1979–2000 *414*
7.2	Introduction: Omeprazole (Losec®) as the Starting Point for a Challenging Project *414*
7.3	Early Attempts: Turning Unfavorable Odds into Success at Last *416*
7.4	The Way Forward: Options and Development Strategies *420*
7.5	The Breakthrough: What a Difference a Day Made! *423*
7.5.1	The Base *423*
7.5.2	The Solvent *424*
7.5.3	The Chiral Auxiliary *424*
7.5.4	The Amount of H_2O *424*
7.5.5	The Effect of Oxidation Time *425*
7.6	Going from a Bench-scale Synthesis to a Fully Fledged Process *425*
7.6.1	Solvent *426*
7.6.2	Base *426*
7.6.3	Composition of Titanium Complex *426*
7.6.4	Equilibration of Titanium Complex *427*
7.6.5	Equivalents of Ti-Complex *427*
7.6.6	Layout of Synthesis *428*
7.7	The Big Test: Going to Plant Scale *428*
7.8	Reaching the Final Target: A Robust Commercial Process *431*
7.8.1	Water *431*
7.8.2	Diethyl Tartrate *431*
7.8.3	Hünigs Base *432*
7.8.4	Cumene Hydroperoxide *432*
7.8.5	Stability of the Catalytic Complex *432*
7.8.6	Reaching the Goal – A Robust Process *433*
7.9	Points to Learn and Conclusions *434*
7.10	Acknowledgements *435*
7.11	References *435*
8	Development of a Biocatalytic Process for the Resolution of (R)- and (S)-Ethyl-3-amino-4-pentynoate Isomers Using Enzyme Penicillin G Amidohydrolase *437*
	Ravindra S. Topgi
8.1	Introduction *437*
8.2	Enzymatic Approach *437*
8.2.1	Deacylation *438*
8.2.2	Acylation *440*
8.3	Optimization of Reaction Conditions *441*
8.3.1	The Enzyme Penicillin G Amidohydrolase *441*
8.3.2	Reaction Monitoring *441*

8.3.3	Acyl Group Specificities	*442*
8.3.4	Organic Co-solvent	*442*
8.3.5	Phosphate Buffer	*443*
8.3.6	pH of the Reaction Medium	*443*
8.3.7	Optimum Amount of Acylating Agent	*443*
8.3.8	Optimum Amount of Enzyme	*444*
8.3.9	Optimum Amount of Substrate	*445*
8.3.10	Optimum Reaction Volume	*445*
8.3.11	Enzyme Activity	*446*
8.3.12	Design of Experiment Study	*446*
8.4	Conclusion	*447*
8.5	Acknowledgments	*448*
8.6	References	*448*

Subject Index *451*

List of Contributors

D. Ager
DSM Pharmaceuticals Inc., 5900 NW
Greenville Blvd.
Greenville, NC 27834
USA

L. Anzalone
Johnson & Johnson
Pharmaceutical Research
& Development, LLC
Spring House, PA 19477
USA

L. Aouni
Rhodia, Centre de Recherches de Lyon
85 Avenue des Frères Perret
F-69192 St. Fons Cedex
France

T. Bayer
Siemens AG
Industriepark Höchst, G 811
D-65926 Frankfurt am Main
Germany

R. Birk
F. Hoffmann-La Roche Ltd
Grenzacherstrasse 124
Postfach
CH-4070 Basel
Switzerland

J. Blacker
Avecia Process Technology
Leeds Road, P.O. Box 521
Huddersfield, HD2 5AG
UK

H.-U. Blaser
Solvias AG
P.O. Box
CH-4002 Basel
Switzerland

W. Brieden
Lonza AG
Process R&D
CH-3930 Visp
Switzerland

J. C. Caille
PPG-Sipsy, Sipsy Chimie Fine
Z.I. la Croix Cadeau B.P. 79
F-49242 Avrillé Cedex
France

C. J. Cobley
Dowpharma
Chirotech Technology Ltd
A Subsidiary of
The Dow Chemical Company
Unit 321 Cambridge Science Park
Milton Road
Cambridge CB4 0WG
UK

E. Delaney
Bristol-Myers Squibb
Process R&D
Pharmaceutical Institute
P.O. Box 191
New Brunswick, NJ 08903
USA

S. Doswald
F. Hoffmann-La Roche Ltd
Grenzacherstrasse 124
Postfach
CH-4070 Basel
Switzerland

K. Drauz
Degussa AG
Fine Chemicals
Rodenbacher Chaussee 4
D-63457 Hanau-Wolfgang
Germany

M. Eissen
Laboratory of Chemical Engineering
Swiss Federal Institute of Technology
Zürich
CH-8093 Zürich
Switzerland

H. Estermann
F. Hoffmann-La Roche Ltd
Grenzacherstrasse 124
Postfach
CH-4070 Basel
Switzerland

P.F. Fauquex
Novartis Pharma AG
Klybeckstrasse 141
CH-4057 Basel
Switzerland

H.J. Federsel
Astra Zeneca
Process R&D
S-15185 Södertalje
Sweden

R. Fuchs
Lonza AG
Process R&D
CH-3930 Visp
Switzerland

C.K. Govindan
PPG-Sipsy, Sipsy Chimie Fine
Z.I. la Croix Cadeau B.P. 79
F-49242 Avrillé Cedex
France

H. Gröger
Degussa AG
Project House Biotechnology
Rodenbacher Chaussee 4
D-63457 Hanau-Wolfgang
Germany

R. Hanreich
Syngenta Crop Protection
Münchwilen AG
CH-4333 Münchwilen
Switzerland

K.E. Hemberger
Rhodia ChiRex
56 Roland Street
Boston, MA 02129
USA

K. Hungerbühler
Laboratory of Chemical Engineering
Swiss Federal Institute of Technology
Zürich
CH-8093 Zürich
Switzerland

S. Jasmin
Rhodia ChiRex
56 Roland Street
Boston, MA 02129
USA

N. B. Johnson
Dowpharma
Chirotech Technology Ltd
A Subsidiary of
The Dow Chemical Company
Unit 321 Cambridge Science Park
Milton Road
Cambridge CB4 0WG
UK

H. Kabir
Rhodia, Centre de Recherches de Lyon
85 Avenue des Frères Perret
F-69192 St. Fons Cedex
France

N. Kasai
Osaka Prefecture University
Laboratory of Food Chemistry
Division of Applied Biological
Chemistry
Graduate School of Agriculture
& Biological Sciences
Gakuen-cho, Sakai, Osaka 599-8531
Japan

A. Kiener
Lonza AG
Department of Biotechnology
CH-3930 Visp
Switzerland

W. S. Knowles
661 East Monroe Avenue
St. Louis, MO63122
USA

E. Kupfer
F. Hoffmann-La Roche Ltd
Grenzacherstrasse 124
Postfach
CH-4070 Basel
Switzerland

J. Lalonde
PPG-Sipsy, Sipsy Chimie Fine
Z.I. la Croix Cadeau B.P. 79
F-49242 Avrillé Cedex
France

S. A. Lanemann
Albany Molecular Research Inc
601 East Kensington Road
Mount Prospect IL 60056
USA

J. F. Larrow
Rhodia ChiRex
56 Roland Street
Boston, MA 02129
USA

M. Larsson
Astra Zeneca
Process R&D
S-15185 Södertalje
Sweden

I. Le-Fur
Rhodia, Centre de Recherches de Lyon
85 Avenue des Frères Perret
F-69192 St. Fons Cedex
France

I. C. Lennon
Dowpharma
Chirotech Technology Ltd
A Subsidiary of
The Dow Chemical Company
Unit 321 Cambridge Science Park
Milton Road
Cambridge CB4 0WG
UK

J. Martin
Avecia Process Technology
Leeds Road, P.O. Box 521
Huddersfield, HD2 5AG
UK

List of Contributors

A. MATSUYAMA
Tsukuba Research Center
Daicel Chemical Industries Ltd
Tsukuba, Ibaraki
Japan

R. MCCAGUE
Dowpharma
Chirotech Technology Ltd
A Subsidiary of
The Dow Chemical Company
Unit 321 Cambridge Science Park
Milton Road
Cambridge CB4 0WG
UK

J. F. MCGARRITY
Lonza AG
Process R&D
CH-3930 Visp
Switzerland

K. MCGEE
Pfizer Inc.
Global Research and Development
La Jolla, San Diego, California
USA

H. P. METTLER
Lonza AG
Process R&D
CH-3930 Visp
Switzerland

P. MOREL
Rhodia, Centre de Recherches de Lyon
85 Avenue des Frères Perret
F-69192 St. Fons Cedex
France

T. NETSCHER
Roche Vitamins Ltd
Research and Development
Bldg 214/27
CH-4070 Basel
Switzerland

J. PESTI
Bristol-Myers Squibb
Process R&D
Pharmaceutical Research Institute
B.O. Box
Princeton, NJ 08543-4000
USA

P. POECHLAUER
DSM Fine Chemicals Austria
R&D Center
Linz
Austria

J. S. PRASAD
Bristol-Myers Squibb
Process R&D
Pharmaceutical Institute
P.O. Box 191
New Brunswick, NJ 08903
USA

J. A. RAMSDEN
Dowpharma
Chirotech Technology Ltd
A Subsidiary of
The Dow Chemical Company
Unit 321 Cambridge Science Park
Milton Road
Cambridge CB4 0WG
UK

K. T. ROBINS
Lonza AG
Department of Biotechnology
CH-3930 Visp
Switzerland

M. SCALONE
F. Hoffmann-La Roche Ltd
Pharmaceuticals Division
Process Research & Catalysis
CH-4070 Basel
Switzerland

T. SCHLAMA
Rhodia, Centre de Recherches de Lyon
85 Avenue des Frères Perret
F-69192 St. Fons Cedex
France

R. SCHMID
F. Hoffmann-La Roche Ltd
Pharmaceuticals Division
Process Research & Catalysis
CH-4070 Basel
Switzerland

B. SCHMIDT
Lonza AG
Process R&D
CH-3930 Visp
Switzerland

E. SCHMIDT
Syngenta Crop Protection AG
CH-4058 Basel
Switzerland

H.-D. SCHNEIDER
Syngenta Crop Protection
Münchwilen AG
CH-4333 Münchwilen
Switzerland

G. SEDELMEIER
Novartis Pharma AG
Klybeckstrasse 141
CH-4057 Basel
Switzerland

R. SELKE
Leibniz-Institut für Organische Katalyse an der Universität Rostock e.V.
Buchbinderstraße 5–6
D-18055 Rostock
Germany

N.M. SHAW
Lonza AG
Department of Biotechnology
CH-3930 Visp
Switzerland

A.K. SINGH
Bristol-Myers Squibb
Process R&D
Pharmaceutical Institute
P.O. Box 191
New Brunswick, NJ 08903
USA

W. SKRANC
DSM Fine Chemicals Austria
R&D Center
Linz
Austria

M. SOUKUP
Efeuweg 5
CH-4103 Bottmingen
Switzerland

F. SPINDLER
Solvias AG
P.O. Box
CH-4002 Basel
Switzerland

F. STÄBLER
F. Hoffmann-La Roche Ltd
Grenzacherstrasse 124
Postfach
CH-4070 Basel
Switzerland

B. STEINACHER
Syngenta Crop Protection
Münchwilen AG
CH-4333 Münchwilen
Switzerland

M. STUDER
Solvias AG
P.O. Box
CH-4002 Basel
Switzerland

T. Suzuki
Research Laboratories
of DAISO Co. Ltd
9 Otakasu-Cho, Amagasaki
Hyogo 660-0842
Japan

J. Tao
Pfizer Inc.
Global Research and Development
La Jolla, San Diego, California
USA

R. S. Topgi
Pfizer Inc.
Research and Development
Chemical Sciences
Skokie, IL 60077
USA

T. Weisbrod
European Patent Office
Bayerstr. 35
D-80335 München
Germany

O. Werbitzky
Lonza AG
Process R&D
CH-3930 Visp
Switzerland

B. Wirz
F. Hoffmann-La Roche Ltd
Grenzacherstrasse 124
Postfach
CH-4070 Basel
Switzerland

R. Wohlgemuth
Fluka Group
Industriestrasse 25
CH-9470 Buchs
Switzerland

W. Wostl
F. Hoffmann-La Roche Ltd
Grenzacherstrasse 124
Postfach
CH-4070 Basel
Switzerland

M. Wubbolts
DSM Fine Chemicals
P.O. Box 43
6130 AA Sittard
The Netherlands

Y. Yao
PPG-Sipsy, Sipsy Chimie Fine
Z.I. la Croix Cadeau B.P. 79
F-49242 Avrillé Cedex
France

H. Yamamoto
Tsukuba Research Center
Daicel Chemical Industries Ltd
Tsukuba, Ibaraki
Japan

A. Zanotti-Gerosa
Dowpharma
Chirotech Technology Ltd
A Subsidiary of
The Dow Chemical Company
Unit 321 Cambridge Science Park
Milton Road
Cambridge CB4 0WG
UK

Introduction
Hans-Ulrich Blaser and Elke Schmidt

1
Background and Motivation

After some soul searching we agreed to collaborate with Wiley-VCH on this book project on the technical application of enantioselective chemo- and biocatalysis. Some of the reasons for our positive decision were the following:

- While there are quite a number of recent books and monographs on the science of enantioselective catalysis using homogeneous, heterogeneous or biocatalysts, no good reference book exists focusing on relevant aspects of the large-scale application of these technologies.
- It is generally very difficult to obtain reliable information on industrial processes, on the one hand due to secrecy concerns and on the other hand because writing publications is not a central aspect of industrial work. In addition, the existing reports are often found in congress proceedings or scattered in monographs and thus not easily accessible.
- An additional incentive was of course the awarding of the Nobel Prize 2001 to W. S. Knowles, R. Noyori, and K. B. Sharpless for their work in the area of enantioselective catalysis. From the point of view of the industrial chemist, it was especially gratifying that the development of a technical process for L-dopa was the basis for the award to W.S. Knowles, a very rare event indeed! And to our great satisfaction, he agreed to let us include his Nobel lecture in our book.

2
Goals and Concept

When we contacted prospective authors for the planned monograph, we defined the central goal as follows: *"To show the organic chemist working in process development that enantioselective catalysis is not just an academic toy but is really a suitable tool for large-scale production of enantioenriched intermediates. To serve as a source of information and maybe also inspiration for academic research and last but not least strengthen the position of the industrial catalyst specialists working in the exciting but sometimes frustrating field of enantioselective catalysis"*.

Asymmetric Catalysis on Industrial Scale: Challenges, Approaches and Solutions
Edited by H. U. Blaser, E. Schmidt
Copyright © 2004 Wiley-VCH Verlag GmbH & Co. KGaA, Weinheim
ISBN: 3-527-30631-5

For this purpose we decided to collect case studies on the development of industrial scale enantioselective processes written exclusively by the specialists who were very closely involved with the work described in their contribution. In this context, technical-scale can be anything from a few hundred kilograms of a complex chiral intermediate for a pharmaceutical to several thousand tons for a herbicide or an amino acid. Because up to now there were no similar books available, it was less important whether a particular process had been described previously but rather that interesting and useful information be collected and discussed in a context where it can be found easily.

We asked the authors to illustrate important aspects of development work, such as:

- the environment and situation for carrying out process development in industry such as time pressures, the fit of the catalytic step into the over-all synthesis or the competition with other synthetic approaches and so on;
- the typical problems that are encountered in the various phases of the development of a technical enantioselective process such as finding/developing the catalyst, optimization of the process or choice of the equipment etc.;
- the successful (and also the unsuccessful!) approaches to solve the problem(s) at hand.

In addition, and we realized that due to problems of confidentiality this would not always be possible, we suggested that the authors also address frequently asked questions such as

- which is the preferred catalyst (homogeneous, heterogeneous, biocatalyst)
- how to separate soluble catalysts from the reaction mixture
- how to handle the sometimes very sensitive catalysts (impurities, air sensitivity, temperature sensitivity, etc.)
- the need for catalyst recycling (is it feasible or to be avoided?)
- where to get the enzyme/chiral catalyst/ligand/auxiliary in commercial quantities
- what are the critical parameters to optimize an enantioselective catalytic reaction
- questions of special equipment (pressure, high/low temperature)
- which were the success factors and which the critical factors during process development.

3
Our (the Editors) Assessment of the Resulting Monograph

Originally, we contacted 41 potential authors we knew to be involved in the technical application of asymmetric catalysis, 27 agreed to write a chapter, a surprisingly high "yield" of almost 70%, and in fact we can present 25 contributions. Major reasons for declining the invitation were on the one hand confidentiality, secrecy concerns or lack of time and on the other hand because the contact person was no longer in process development or had even changed company (or the company had changed hands). In particular, the latter occurred more often than we expected, reflecting the dynamic nature of today's business.

Quantity is one aspect – how about the quality of the case studies? As editors, we can say that our efforts have paid off and that we are satisfied that each of the reports contributes to the high information content of the monograph. What we find especially important is the fact that each study has its own character, its own way of telling the story behind the results presented. Each contribution clearly fulfills our major requirements described above, discussing the issues and problems with the strongest effect on the course and the outcome of the work.

The most important question is of course: Are these studies relevant, or in other words, are these processes actually typical and are they applied for commercial production? Also here, the answer is an almost unqualified YES. As discussed below, we are sure to have a representative mix of different types of catalysts, transformations and intended use of the chiral products. There are two caveats. Firstly, for various reasons, a number of important technically mature enantioselective processes are not described in our book. In order to at least partially remedy this gap we have listed selected "missing processes" and some relevant references for further reading. Secondly, even though more than half of the processes are either already applied on a technical scale or their introduction is scheduled, not all processes are or will in fact be used commercially (sometimes this was actually the reason, why clearance was given to publish the results!). In any case, we have made sure that only processes are included in our collection where the technical feasibility has been demonstrated at least in a pilot setting.

Unquestionably there are some drawbacks associated with the format of our book. For one, the terminology used by the various authors can differ significantly (for some help in this respect see Glossary) and of course the proficiency in English of the different authors from at least ten nations is also variable. It was beyond the capacity of the two editors as well as of the Wiley-VCH team to assure an integral "unite de doctrine". Also quite variable is the depth of the provided information; some authors chose not to focus on a single process, but rather present an overview on practiced technology, others give a very detailed description and assessment of a specific technology. We have decided to accept this difference in approaches in the hope that the sum of the various articles would eventually give a balanced and, even more important, a relevant picture of the present situation regarding large scale asymmetric catalysis.

4
The Organization of the Book (Types of Intermediates – Catalysts – Situations)

There are many ways of organizing a monograph on industrial asymmetric catalysis. Rather than using the catalyst type (chemo- or biocatalyst) or the type of transformation (hydrogenation, oxidation, etc.), or the type or use of the target product (pharma, agro, etc.) we decided to apply the *nature of the task* of the development chemist as the major classification criterion. Even though some of the classifications might be debatable, we think that this subdivision will best serve the reader who would like to find how his colleagues have dealt with a particular situation. Accordingly, we have arranged the following groups of applications:

I. New processes for existing active compounds
II. New catalysts and/or processes for important building blocks
III. Adaptation of existing catalysts for important building blocks
IV. Processes for new chemical entities (NCE)

For each contribution we have indicated the issues which have been emphasized by the authors and are therefore discussed in some depth.

Tab. 1 New processes for existing active compounds (category I)

Authors	Title	Emphasized topics
W. S. Knowles	Asymmetric hydrogenation: The Monsanto L-dopa process	New process, search for new ligand, ligand synthesis
R. Selke	The other L-dopa process	New ligand, reaction parameters, technology transfer
H. U. Blaser, R. Hanreich, H.-D. Schneider, F. Spindler, B. Steinacher	The chiral switch of Metolachlor: the development of a large-scale enantioselective catalytic process	Search for new catalyst and ligand, process development, technical ligand synthesis, integration into existing process
T. Netscher, M. Scalone, R. Schmid	Enantioselective hydrogenation: towards a large-scale total synthesis of (R,R,R)-α-tocopherol	Route selection, substrate synthesis, process optimization, technical ligand synthesis
H. U. Blaser, M. Eissen, P. F. Fauquex, K. Hungerbühler, E. Schmidt, G. Sedelmeier, M. Studer	Comparison of four industrial syntheses of ethyl (R)-2-hydroxy-4-phenylbutyrate	Case studies, biocatalysis vs. heterogeneous catalysis, systematic comparison of overall processes and of catalytic steps, ecological parameters
N. M. Shaw, K. T. Robins, A. Kiener	Biocatalytic approaches for the large-scale production of asymmetric synthons	Combination of chemical and biocatalytic steps, toolbox of biocatalysts, back integration
T. Bayer	7-Aminocephalosporanic acid – chemical vs. enzymatic production process	Recombinant enzyme, process development, process comparison concerning waste
H. Gröger, K. Drauz	Methods for the enantioselective biocatalytic production of L-amino acids on an industrial scale	Overview of various production methods, types of biocatalysts

4.1
Category I: New Processes for Existing Active Compounds (Tab. 1)

Described are either the development of what is sometimes called a 2nd generation process (i.e., replacing an unsatisfactory existing process) or a new process to allow a chiral switch (to replace a racemic product and process). This is a problem-driven approach and the emphasis of most contribution is on the development of the catalyst, the catalytic process and also how to fit the new chemistry into the existing production facilities. Originators who want to defend an "old" product or companies who want to move into an existing market such as generics producers are concerned.

4.2
Category II: New Catalysts and/or Processes for Important Building Blocks (Tab. 2)

This strategy is mostly technology-driven and the emphasis is on the development of the catalyst and new catalytic process technology; usually the domain of fine chemicals companies who want to apply their own technology and who can choose the molecules for which to develop a new process. Important issues are most often the search for the catalyst, development of the process and the technical catalyst synthesis.

Tab. 2 New catalysts and/or processes for important building blocks (category II)

Authors	Title	Emphasized topics
P. Poechlauer, W. Skranc, M. Wubbolts	The large-scale biocatalytic synthesis of enantiopure cyanohydrins	Overview of methods, process development, recombinant enzyme, equipment
L. Aouni, K. E. Hemberger, S. Jasmin, H. Kabir, J. F. Larrow, I. Le-Fur, P. Morel, T. Schlama	Industrialization studies of the Jacobsen hydrolytic kinetic resolution of racemic epichlorhydrin	Process development, catalyst preparation, quantitative kinetic and reaction modeling (engineering approach)
J. Blacker, J. Martin	Scale-up studies in asymmetric transfer hydrogenation	Catalyst optimization, application scope, reaction parameters, case studies
A. Matsuyama, H. Yamamoto	Practical applications of biocatalysis for the manufacture of chiral alcohols such as (R)-1,3-butanediol by stereospecific oxidoreduction	Microorganism screening, recombinant biocatalyst, reaction parameters, product isolation
N. Kasai, T. Suzuki	Production of chiral C3 and C4 units via microbial resolution of 2,3-dichloro-1-propanol, 3-chloro-1,2-propanediol and related halohydrins	Microorganism screening and production, process control logic, reaction parameters, coupled production

Tab. 3 Adaptation of existing catalysts for important building blocks (category III)

Authors	Title	Emphasized topics
D. J. Ager, S. A. Laneman	Synthesis of unnatural amino acids	Ligand and substrate synthesis, reaction parameters, enantio-enrichment
C. J. Cobley, N. B. Johnson, I. C. Lennon, R. McCague, J. A. Ramsden, A. Zanotti-Gerosa	The application of DuPHOS rhodium(I) catalysts for commercial scale asymmetric hydrogenation	Substrate synthesis, reaction parameters, enantioenrichment via biocatalysis, catalyst handling, economic considerations
J. F. McGarrity, W. Brieden, R. Fuchs, H.-P. Mettler, B. Schmidt, O. Werbitzky	Master the scale-up of catalytic asymmetric hydrogenations for the manufacture of fine chemicals	Catalyst screening, substrate design, process optimization, application of Josiphos ligands, ligand synthesis, scale-up
R. Wohlgemuth	Large-scale applications of biocatalysis in the asymmetric synthesis of laboratory chemicals	Production considerations, comparison with classical methods

4.3
Category III: Adaptation of Existing Catalysts for Important Building Blocks (Tab. 3)

This is also a technology-driven strategy where the emphasis is on the adaptation of known catalysts and the development of new processes. Usually the domain of fine chemicals companies who want to apply their in-house technology to a selected molecule (family) where substrate synthesis and backintegration are important considerations.

4.4
Category IV: Processes for New Chemical Entities (NCE) (Tab. 4)

Here, emphasis is on testing of synthetic alternatives, route selection and the integration of the catalytic step into the over-all process often with a very limited time schedule; obviously an entirely problem-driven approach. A situation most often encountered by integrated originator companies, in recent years also by technology companies (custom manufacturers) working for originators with missing competence or capacity.

5
Missing Processes

In a recent review [1], entitled *Enantioselective Catalysis in Fine Chemicals Production*, we have summarized and critically discussed the known state of the art of the industrial application of chiral chemical catalysts. For the industrial applica-

Tab. 4 Processes for new chemical entities (category **IV**)

Authors	Title	Emphasized topics
J. Tao, K. McGee	Development of an efficient synthesis of chiral 2-hydroxy acids	Discussion of methods, substrate synthesis, continuous process, membrane reactor, enzyme deactivation
A. K. Singh, J. S. Prasad, E. J. Delaney	Factors influencing the application of literature methods toward the preparation of a chiral *trans*-cyclopropane carboxylic acid intermediate during development of a melatonin agonist	Development phases, route comparison, test of many alternative methods, technical catalyst preparation, process optimization, ee enrichment via crystallization, industrial assessment
J.-C. Caille, J. Lalonde, Y. Yao, C. K. Govindan	Hetero Diels-Alder-Biocatalysis approach for the synthesis of (S)-3-{2-[(methyl-sulfonyl)oxy]ethoxy}-4-(triphenylmethoxy)-1-butanol methanesulfonate: successful application of an enzyme resolution process	Route selection, alternative catalytic methods, enzyme screening, process optimization, experimental details
J. A. Pesti, L. Anzalone	Multi-kilo resolution of XU305, a key intermediate to the platelet glycoprotein IIb/IIIa receptor antagonist roxifiban via kinetic and dynamic enzymatic resolution	Background information, fit into overall synthesis, racemization of unwanted enantiomer, dynamic kinetic resolution as 2nd generation process
B. Wirz, S. Doswald, E. Kupfer, W. Wostl, T. Weisbrod, H. Estermann	Protease-catalyzed preparation of (S)-2-[(*tert*-butylsulfonyl)methyl]-hydrocinnamic acid for renin inhibitor RO0425892	Route selection, parameter and process optimization, work-up of emulsion, racemization
B. Wirz, M. Soukup, T. Weisbrod, F. Stäbler, R. Birk	Protease-catalyzed preparation of chiral 2-isobutyl succinic acid derivatives for collagenase inhibitor RO0319790	Route selection, alternative approaches, parameter and process optimization, racemization
H.-J. Federsel, M. Larsson	An innovative asymmetric sulfide oxidation: the process development history behind the new antiulcer agent Esomeprazole	Change from racemate to single enantiomer, adaptation of known catalyst system, integration into existing process, time line pressure
R. S. Topgi	Development of a biocatalytic process for the resolution of (R)- and (S)-ethyl 3-amino-4-pentynoate isomers using enzyme penicillin G amidohydrolase	pH-dependent acylation/deacetylation, parameter optimization, design of experiment approach with statistical analysis

Tab. 5 Production scale processes for existing active compounds (category I)

Product	Reaction and catalyst	Scale	Company [Ref.]
L-Menthol, hydroxy-dihydro-citronellal, D- and L-citronellol	Isomerization of allyl amine with Rh-binap	>1000 t/y	Takasago [5–7]
Aspartame	Dipeptide formation from D,L-phenylalanine with thermolysin	kilo-tons	Holland Sweetener Company [8]
Vitamin E	Hydrogenation of allyl alcohol with Ru-binap	300 t/y	Takasago [5, 6]
L-Dopa	Addition of ammonia with L-tyrosine phenol lyase from *Erwinia herbicolad*	250 t/y	Ajinomoto [9]
Carbapenem	Hydrogenation of α-substituted β-keto ester Ru-tolbinap	50–120 t/y	Takasago [5, 6, 10]
(S)-Oxafloxazin	Hydrogenation of α-hydroxy ketone with Ru-tolbinap	50 t/y	Takasago [5]
(+)-*cis*-Methyl dihydrojasmonate	Hydrogenation of α,β-unsaturated ester with Ru-josiphos	multi t/y	Firmenich [11]
Glycidol	Sharpless epoxidation with Os-cinchona	multi t/y	PPG-Sipsy [12]
Aspartame	Hydrogenation of enamide with Rh-eniphos	15 t	Enichem/Anic [13]
(S)-Naproxen	Kinetic resolution by carboxylic ester hydrolysis (recombinant esterase)	Development stage	Chirotech & Shasun Chemicals, India [14]

tion of biocatalysts, we have relied on selected monographs [2] and reviews [3, 4]. Tables 5–7 list processes which have been described in reasonable detail elsewhere and which have bearing on the topic of this monograph. In our analyses presented below we have included information gathered in these surveys.

The most impressive homogeneous missing process in category I is certainly the isomerization of an allyl amine, the key step of Takasago's L-menthol synthesis, but also several hydrogenation processes by Takasago and Firmenich. The glycidol and the Enichem/Anic aspartame processes are no longer in operation, the latter obviously could not compete successfully with the biocatalytic variant. In addition, there were further reports on three pilot processes by Roche (2) and Ciba-Geigy/Solvias (1) as well as seven bench-scale processes by Roche (3), Takasago (1), Sumitomo (1), Degussa Hüls (1), and Monsanto (1) [1].

In categories II and III, the most important processes with volumes of up to 2000 t/y are carried out with biocatalysts, whereas only small scale or pilot processes were reported with homogeneous metal complexes. In addition, there were further reports on four bench-scale processes by Takasago (3) and HoechstMarion-Roussel/Aventis (1) [1].

Tab. 6 Production and pilot processes for chiral intermediates/building blocks (categories **II** and **III**)

Reaction and catalyst	Scale	Company [Ref.]
C–Cl hydrolysis for 2-chloropropionic acid with whole cells of *Pseudomonas putida* (dehalogenase)	2000 t/y	Zeneca Life Science Molecules [3]
C–O bond formation, L-malic acid from fumaric acid with fumarase	2000 t/y	Amino GmbH, Tanabe Seiyaku [15]
Transesterification for the production of chiral 1-phenylethyl-amines with lipases	>100 t/y	BASF [16]
Redox reaction for producing *cis*-1,2-dihydroxycatechol with benzoate dioxygenase from *Pseudomonas putida*	multi t/y	ICI [17]
CBS reduction of α-chloro ketone	"commercial production"	PPG-Sipsy (Callery) [18]
Epoxide ring opening with Mn-salen	"small scale production"	Merck/ChiRex [19]
Hydrogenation of β-keto ester with Ru-tmbtp	multi 100 kg	Chemi [20]
Sulfide oxidation with Ti-tartrate	<100 kg	Lonza [21]
Sharpless dihydroxylation with Os-cinchona	multi 10 kg	ChiRex [22]

Because most examples in category IV described in this monograph use biocatalysts, it could be inferred that chemical catalysis is not suitable for the synthesis of NCEs. Just to show that this is not correct but rather due to the choice of our authors, Tab. 7 shows an extensive list of examples illustrating this point. In addition to these production and pilot processes, there were further reports on 19 bench-scale processes using homogeneous catalysts by Roche (7), HoechstMarion-Roussel/Aventis (5), SmithKline Beecham (2), Sepracor (2), Pharmacia/Upjohn (1), Eli Lilly (1) and Merck (1) [1].

6
Some Important Messages

In order to identify the major problems and issues when developing enantioselective catalysis for large-scale application in a somewhat more structured way, we devised a questionnaire and sent this to all authors as well as to some additional contact people. We received 26 answers to our questionnaire, allowing some statistically relevant conclusions to be drawn (summarized in Tab. 8) and furthermore several additional comments describing critical points. In the following paragraphs we will discuss selected issues and will try to draw some general conclusions.

Tab. 7 Processes for NCEs on production and pilot scale (category **IV**)

Product	Reaction and catalyst	Scale	Company [Ref.]
Tipranavir	Hydrogenation of C=C bond (E/Z mixture) with Rh-duphos	"production scale"	Chirotech (for Pharmacia & Upjohn) [23]
Trusopt	Redox reaction with suspended whole cells of *Neurospora crassa* (alcohol dehydrogenase)	multi-ton	Zeneca Life Science Molecules/Merck & Co. [24]
Cilastatin	Cyclopropanation with Cu complex	"small scale"	Sumitomo [25]
Orlistat	Hydrogenation of β-keto ester with Ru-biphep or Raney-Ni-tartrate[a]	ton	Roche [26, 27]
Trocade	ene reaction with Ti-binol	multi 100 kg	Roche [26, 28]
HMR 2906	Hydrogenation of enamide with Rh-bpm	multi 10 kg	HoechstMarion-Roussel/Aventis [1]
Mibefradil	Hydrogenation of α,β-unsaturated acid with Ru-biphep	multi 10 kg	Roche [26, 29, 30]
Cilazapril	Hydrogenation of cyclic enamide with Ru-biphep	multi 10 kg	Roche [26, 29]
HMG-CoA reductase inhibitor	Hydrogenation of β-keto ester with Ru-binap	10 kg	HoechstMarion-Roussel/Aventis [31]
SK&F 107647	Hydrogenation of di-enamide with Rh-duphos	multi kg	Chirotech (for Nycomed) [32]
Zeaxanthin	Hydrogenation of cyclic enol acetate with Rh-duphos	multi kg	Roche [29]

[a] Modified heterogeneous catalyst.

6.1
Transformations and Catalyst Types

An analysis of Tab. 9 shows that hydrogenation/reduction and hydrolysis reactions are by far the predominant transformations (31 out of 38) that have successfully been developed to industrial processes. The most important reason for this fact is the broad scope of these reaction types. For the catalytic hydrogenation it could also be attributed to the early success of Knowles with the L-dopa process, because for many years most academic and industrial research was focused on this transformation. Concerning hydrolysis, important points might be the rather good substrate tolerance (low substrate specificity) and the availability of a number of commercial enzymes.

The most prevalent catalysts are homogeneous metal complexes mostly with chiral diphosphine ligands, isolated enzymes and whole cell preparations, whereas

Tab. 8 Answers to the questionnaire grouped according to categories I–IV (number of answers)

	I (8)	II (4)	III (6)	IV (7)	Total (25)
Type of catalyst used					
enzyme, microbial catalyst	3	2	1	5	11
homogeneous complex	6	2	6	3	17
heterogeneous catalyst	1	–	–	–	1
commercial catalyst	2	1	4	2	9
new development	6	2	4	1	12
Type of reaction					
reduction/hydrogenation	6	2	6	1	15
oxidation	1	1	–	1	3
hydrolysis	2	1	1	4	8
C–C coupling	1	1	–	1	3
kinetic resolution	1	2	1	5	9
other	–	–	–	2	0
Intended scale					
<1 t/y	–	1	–	1	2
1–10 t/y	3	1	3	3	10
>10 t/y	6	4	2	2	15
Realized development stage					
pilot	3	1	3	5	12
production (realized/planned)	6	3	4	2	17
abandoned	2	1	1	4	8
X-step synthesis?					
yes	3	–	–	6	11
if yes, number of steps	3–11	–	–	6–12	
Other approaches studied?					
yes	8	1	6	5	21
if yes, catalytic	5	1	3	4	13
if yes, stoichiometric	5	1	5	2	13
Schedule/time line					
was critical	5	2	4	6	17
no problem	4	2	2	1	9
Development time for catalysis					
<6 months	–	1	2	5	8
<1 year	2	–	4	4	10
>1 year	7	3	–	–	11

continued on next page

modified heterogeneous metal catalysts are seldom used. An inspection of which catalyst type is used for which transformation category listed in Tab. 9 shows that biocatalysts and chemical catalysts are actually complementary rather than competing. With the exception of the reduction of C=O groups which is feasible with both catalyst classes, all other transformations have a preferred catalyst type (at least for large scale applications).

Tab. 8 (cont.)

	I (8)	II (4)	III (6)	IV (7)	Total (25)
Major problem(s)					
finding suitable catalyst	4	1	1	3	9
enantioselectivity	4	–	2	4	10
(bio-)catalyst activity	3	–	1	2	6
catalyst productivity	5	–	2	3	10
scale-up of (bio-)catalyst	1	2	1	3	6
large-scale catalyst	2	2	2	3	9
(bio)-catalyst separation	1	2	–	3	5
catalyst recycling	2	–	–	3	5
process development	4	4	4	3	15
impurities starting material	7	–	–	1	10
reproducibility	1	–	1	2	5
special equipment	3	1	1	1	6
Why a particular technology					
in-house know how	6	2	5	4	17
technology driven	1	3	3	3	10
best alternative	5	4	6	6	19
only choice	2	–	–	–	2
Patent issues					
own patents filed	9	4	4	6	23
license needed	1	1	1	–	3
none	–	–	2	3	5

Tab. 9 Transformations and catalysts types described in the individual contributions

Reaction type	Substrates/comment	E	WC	H	Het	
C–C coupling	Cyanohydrin, cyclopropanation	1	–	1	–	2
Hydrogenation C=C	Dehydroacylaminoacid, allylic alcohol, tetrasubstituted C=C, enamine	–	–	8	–	8
Hydrogenation/ reduction C=O, C=N	α-Keto acid derivatives, var. ketones and imines	2	3	3	2	10
Oxidation	Alcohol, sulfide	–	1	1	–	2
Hydrolysis, acetylation	Amide, hydantoin, ester, thioester, nitrile, carbamate, epoxide	8	4	1	–	13
Various	Assimilation, carnitine synth.	–	3	–	–	3
Total		11	11	14	2	38

E enzyme, **WC** whole cell, **H** homogeneous metal complex, **Het** heterogeneous catalyst.

6.2
Scale, Development Stage

With very few exceptions, the envisaged or actual production scale is between 1–10 t/y (11 cases) or >10 t/y (15 cases). The largest processes in operation are described for (S)-metolachlor (>10000 t/y, Blaser et al., p. 55) and several amino acids (up to 5–10000 t/y, Gröger and Drauz, p. 131). Also 7-ACA (>400 t/y, Bayer), which is not actually an enantioselective process but the stereoselective transformation of a chiral substrate, and several chiral C3 and C4 units (Kasai and Suzuki, p. 233) are produced on very large scale.

All processes described in the monograph have been developed either to the pilot or production stage. It has to be pointed out that at least 7 processes have been abandoned, mostly because the product was either not commercialized or is no longer produced.

6.3
Synthesis Planning and Time Lines

As a rule, the asymmetric catalytic reaction is part of a more extensive multi-step synthesis. This is particularly pronounced for the cases where the active substance is the goal of the development work (categories A and D), but also for the more simple intermediates described in B and C. This means that the catalytic step has to be integrated into the overall synthesis and therefore, the route selection is a very important phase of process development. Very detailed discussions of this aspect can be found, e.g., in the contributions of Wirz et al. (p. 385, 399), Netscher et al. (p. 71) or Caille et al. (p. 349). It is important to realize that the effectiveness of the catalytic step is only one, albeit often an important, factor but that it is the cost of the overall synthesis which is decisive for the final choice as to which route will be chosen. The comparison of competing routes is not always easy and different approaches can be found in the contributions of Blaser et al. (p. 91), Pesti and Anzalone (p. 365), or Singh et al. (p. 335). In some cases, the overall synthesis is actually designed around an effective enantioselective transformation as for example described for the metolachlor process by Blaser et al. (p. 55). This situation will become rarer when more catalysts with well described scope and limitations will be commercially accessible.

The time line can be a problem, especially when the optimal catalyst has yet to be developed or when no commercial catalyst is available for a particular substrate (substrate specificity) and/or when not much is known about the desired catalytic transformation (technological maturity). When developing a process for a new chemical entity (NCE) in the pharmaceutical or agrochemical industry, time restraints can be severe (see Fig. 1). In these cases it is more important to find a competitive process on time than an optimal process too late. For second generation processes, chiral switches (for existing products sold as racemates) or other products, the time factor is often not so important but here, the process must be the most cost effective.

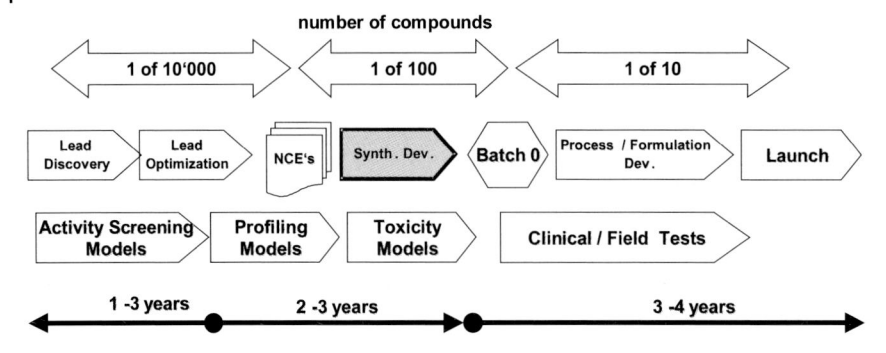

Fig. 1 Development process for a new chemical entity in the pharmaceutical industry.

6.4
Major Technical Problems

In Tab. 8, the most frequent technical problems are (not surprisingly) related to the catalyst performance and to the development of the process. Finding the right catalyst (with satisfactory enantioselectivity and productivity) was problematic especially when making active compounds (categories I and IV). Process development was a major issue in about two-thirds of all studies and in some cases, very extensive work was carried out as very well illustrated by Aouni et al. (p. 165) or by Kasai and Suzuki (p. 233). Scale-up and/or large-scale availability of the catalyst were also frequent problems, whereas catalyst separation and recycling seem to be relatively unproblematic. Finally, substrate purity (leading to bad reproducibility) and the need for special equipment were also mentioned to be of concern especially for second generation processes.

6.5
Technology and Patents

The answers to the question, why a particular technology was chosen and implemented successfully are quite straightforward: In most cases, in-house know how was already available before the particular project was tackled and in most cases, the catalytic process was the best alternative. Purely technological driven approaches were less common and in only two instances was the catalytic route the only feasible solution to the problem.

Patent issues can be very important when deciding whether to implement a particular technology for commercial production. In particular, the pharmaceutical industry is extremely reluctant to pay royalties for using a particular technology and in many instances, this might actually kill an otherwise successful route. This is a bit different for producing intermediates or chiral building blocks and indeed it is for such applications where license were actually needed. On the other hand, if ever possible, the process developed for a specific process was patented in order to protect the R&D investment and in many cases also to prevent generics producers or other competitors from using the advanced technology.

7
Final Comments and Conclusions

The collection of case studies makes a convincing argument that enantioselective catalysis is not just an academic exercise but that it can compete in an industrial context with classical stoichiometric approaches. Most of the described applications are for the production of biologically active compounds or their intermediates. Particularly impressive is the wide scope of successful solutions concerning the type of catalysts and reactions, and also concerning the scale and value of the manufactured products, ranging from relatively cheap, high volume intermediates to small scale high value active compounds. In almost all cases, other synthesis variants were investigated but turned out to be less cost effective.

The success factors most often mentioned were a broad technology expertise and practical know how in the field of synthesis and especially catalysis, either with biocatalysts or homogeneous catalysts. The most successful teams usually work in firms where this ability is part of the company culture and is supported by senior management. We see a trend that some of this development work is outsourced to custom manufacturers or technology companies with a high relevant technology expertise.

Not surprisingly, the critical issues most often mentioned are related to the selection of the overall synthesis, the catalyst selection and performance, the process optimization and the development time. It has to be stressed, that the importance of these problems can be quite different for the four categories of tasks we have distinguished. As a rule, catalyst performance (and cost) and process optimization are more important for the manufacture of relatively cheap chiral building blocks whereas overall synthesis and time issues are dominant when developing active compounds with very high added value.

There is little doubt in our mind that the industrial use of asymmetric catalysis will grow significantly in the coming years. Some important growth factors are the rapidly increasing number of commercially available biocatalysts and chiral ligands and complexes with well known scope and limitation, the impressive progress made by academic research with new catalysts and transformations with very high enantioselectivity, and last but not least the increasing economical and ecological pressure to find the best process for a commercial product – which will be quite often a catalytic process.

8
Glossary

Absolute configuration	arrangement of the four substituents around a stereogenic center according to defined rules; the two forms are distinguished with the prefix (*R*) or (*S*); for some classes of natural products D or L is used
Asymmetric synthesis	see enantioselective synthesis
Catalyst activity	measure of the efficiency of a catalyst for a specific reaction, usually expressed as rate/g catalyst or as turnover frequency
Catalyst productivity	mol product/g catalyst or turnover number
Chiral auxiliary	chiral molecules or groups used to induce enantioselective transformations
Chiral center	should not be used any more, see stereogenic center
Chiral compound	molecules where image and mirror image are not superimposable
Chiral ligand	chiral auxiliary used to render homogeneous metal complex enantioselective (see important chiral ligands, Fig. 2)
Conversion (%)	fraction of starting material reacted (converted)
Dynamic kinetic resolution	kinetic resolution with concomitant racemization of the less reactive enantiomer
EC number	specification of an enzyme, developed by the *Enzyme Commission*; each enzyme is assigned a four-digit EC number EC (i), (ii), (iii), (iv).
Enantiomeric excess (ee)	also used: optical yield; % major enantiomer – % minor enantiomer
Enantioselective homogeneous catalyst	complex consisting of a central metal atom and at least one chiral ligand
Enantioselective synthesis/catalysis	Synthesis of enantiomerically enriched chiral product starting from non-chiral substrate, also called asymmetric
Enzyme activity	expressed mostly in Units, 1 Unit = 1 µmol of substrate transformed per minute
Enzyme classes	1. oxidoreductases, 2. transferases, 3. hydrolases, 4. lyases, 5. isomerases, 6. ligases
Feed	reactant(s) in a continuous process
Hydrolase	catalysis of the hydrolysis or formation of esters, amides, lactones, lactams, epoxides, nitriles, anhydrides, glycosides
Isomerase	catalysis of isomerizations (e.g., racemizations, epimerizations)

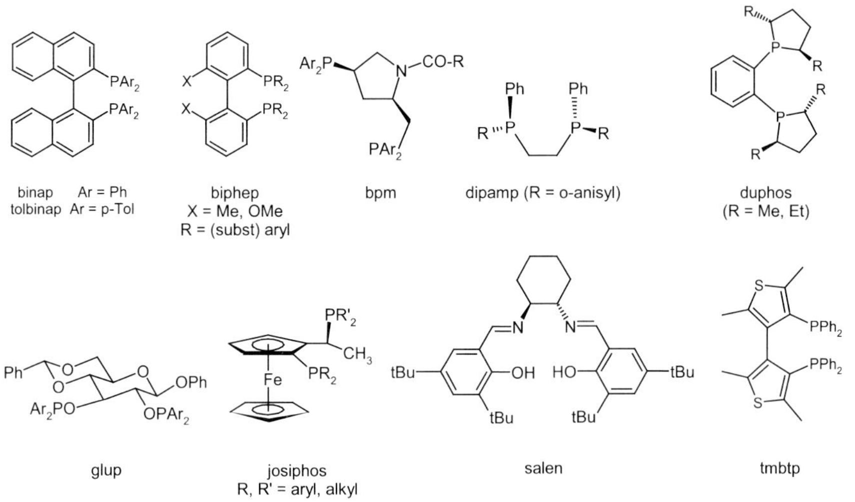

Fig. 2 Structures of selected chiral ligands.

Kinetic resolution	preferential reaction of one enantiomer, leading to enriched product and starting material
Lyase	catalysis of the elimination of a leaving group (formation of double bonds), addition of small molecules on C=O, C=N, C=C bonds
Ligase	catalysis of the bond formation (or cleavage) coupled with a triphosphate cleavage
Optical yield	see enantiomeric excess
Oxidoreductase	catalysis of oxidation-reduction reactions: oxygenation of C–H, C–C, C=C bonds, overall removal or addition of hydrogen atom equivalents
Selectivity (%)	ratio of desired product/total products
Space-time yield	Product formed per volume and time unit, usually expressed in (g product)/(L reactor volume x hour)
Stereogenic center	atom with four different substituents
Substrate	reactant or starting material in a catalytic process
tof, turnover frequency (1/time)	activity scale, moles of substrate reacted/moles catalyst/unit of time
ton, turnover number	mol product/mol catalyst
Transferase	catalysis of the transfer of functional groups such as acyl, sugar, phosphoryl, methyl, aldehydic or ketonic groups

9
References

1   H. U. Blaser, F. Spindler, M. Studer, Appl. Catal. A: General 2001, 221, 119.
2   A. Liese, K. Seelbach, C. Wandrey, Industrial Biotransformations, Wiley-VCH, Weinheim, 2000. S. J. C. Taylor, K. W. Holt, R. C. Brown, P. A. Keene, I. N. Taylor in Stereoselective Biocatalysis (R. N. Patel, Ed.), Marcel Dekker, New York. 2000, p. 397.
3   P. S. J. Cheetham in Applied Biocatalysis (J. M. S, Cabral, D. Best, L. Boross, J. Tramper, Eds.), Harwood Academic Publishers, Chur, 1994, p. 70.
4   A. J. J. Straathof, S. Panke, A. Schmid, Curr. Opin. Biotechnol. 2002, 13, 548. J. P. Pachlatko, Chimia, 1999, 53, 577.
5   H. Kumobayashi, Recl. Trav. Chim. Pays-Bas 1996, 115, 201.
6   S. Akutagawa, Appl. Catal. 1995, 128, 171. S. Akutagawa, Takasago, personal communication.
7   S. Akutagawa, K. Tani in Catalytic Asymmetric Synthesis (I. Ojima Ed.), VCH Publishers, N.Y., 1993, p. 41.
8   A. F. V. Wagner, T. Sonke, P. J. L. M. Quadflieg, EP 1013663 A1 (2000) and T. Harada, S. Irino, Y. Kunisawa, K. Oyama, EP 768384 A1 (1997) both assigned to Holland Sweetener Co. K. Nakanishi, A. Takeuchi, R. Matsuno, Appl. Microbiol. Biotechnol. 1990, 32, 633.
9   T. Kotani, K. Lizumi, M. Takeuchi, EP 636695 A1 – (1995) and T. Tsuchida, Y. Nishimoto, T. Kotani, Takuya, K. Iiizumi, JP 05123177 A2, 1993 both assigned to Ajinomoto Co., Inc.
10 R. Noyori, M. Tokunago, M. Kitamura Bull. Chem. Soc. Jpn. 1995, 68, 36.
11 D. A. Dobbs, K. P. M. Vanhessche, E. Brazi, V. Rautenstrauch, J.-Y. Lenoir, J.-P. Genet, J. Wiles, S. H. Bergens, Angew. Chem., Int. Ed. Engl. 2000, 39, 1992. D. Dobbs, K. Vanhessche, V. Rautenstrauch, WO 98/52,687 (1997) assigned to Firmenich.
12 W. P. Shum, M. J. Cannarsa in Chirality in Industry II (A. N. Collins, G. N. Sheldrake, J. Crosby, Eds.), John Wiley, 1997, p. 363.
13 M. Fiorini, M. Giongo, M. Riocci, EP 077'099 (1983), assigned to Anic SPA. Described in I. Ojima, N. Clos, C. Bastos, Tetrahedron 1989, 45, 6901.
14 J. P. Rasor, E. Voss, Appl. Catal. A: General 2001, 221, 145.
15 H. J. Danneel, M. Busse, R. Faurie, Mededelingen – Faculteit Landbouwkundige en Toegepaste Biologische Wetenschappen, Universiteit Gent (Forum for Applied Biotechnology, Part 1) 1995, 60, 2093. I. Takata, T. Tosa, Bioprocess Technology (Ind. Appl. Immobilized Biocatal.) 1993, 16, 53. H. J. Danneel, R. Geiger, DE 4424664 C1 (1995), assigned to Amino GmbH.
16 F. Balkenhohl, K. Ditrich, B. Hauer, W. Ladner, J. Prakt. Chem./Chem.-Ztg., 1997, 339, 381.
17 S. C. Taylor, M. D. Turnbull, EP 253485 A2 (1988), assigned to ICI PLC.
18 J. C. Caille, M. Bulliard, B. Laboue in Chirality in Industry II (A. N. Collins, G. N. Sheldrake, J. Crosby, Eds.), John Wiley, 1997, p. 391.
19 C. H. Senanayake, E. N. Jacobsen in Process Chemistry in the Pharmaceutical Industry (K. G. Gadamasetti, Ed.), Marcel Dekker Inc, New York, 1999, p. 347. I. W. Davies, P. J. Reider, Chem. Eng. News, 3 June 1996, p. 413. D. L. Hughes, G. B. Smith, J. Liu, G. G. Dezeny, C. H. Senanayake, R. D. Larsen, T. R. Verhoeven, P. J. Reider, J. Org. Chem. 1997, 62, 2222.
20 T. Benincori, S. Rizzo, F. Sannicolo, O. Piccolo, Proceedings of the Chira-Source 2000 Symposium, The Catalyst Group, Spring House, USA, 2000.
21 W. Brieden in Proceedings of the Chira-Source '99 Symposium, The Catalyst Group, Spring House, USA, 1999.
22 A. A. Smith, Proceedings of the Chira-Tech'96 Symposium, The Catalyst Group, Spring House, USA, 1996. R. Pettman, Specialty Chemicals, 1994, 12.
23 B. D. Hewitt, Pharmacia & Upjohn, Chem. Eng. News, 1 November 1999, p. 35.

24 R. A. Holt, Chim. Oggi 1996, 14, 17.
25 T. Aratani in Comprehensive Asymmetric Catalysis (E. N. Jacobsen, H. Yamamoto, A. Pfaltz, Eds.), Springer, Berlin, 1999, p. 1451. T. Aratani, Pure Appl. Chem. 1985, 57, 1839.
26 R. Schmid, M. Scalone in Comprehensive Asymmetric Catalysis (E.N. Jacobsen, H. Yamamoto, A. Pfaltz, Eds.), Springer, Berlin, 1999, p. 1439.
27 a) R. Schmid, E. A. Broger, Proceedings of the ChiralEurope '94 Symposium, Spring Innovation, Stockport, UK, 1994, p. 79. b) K. Püntener, M. Scalone, R. Schmid, unpublished results.
28 P. Brown, H. Hilpert EP 96110814 (1996) assigned to Hoffmann-La Roche.
29 M. Scalone, R. Schmid, E. A. Broger, W. Burkart, M. Cereghetti, Y. Crameri, J. Foricher, M. Henning, F. Kienzle, F. Montavon, G. Schoettel, D. Tesauro, S. Wang, R. Zell, U. Zutter in Proceedings of the ChiraTech '97 Symposium, The Catalyst Group, Spring House, USA, 1997.
30 Y. Crameri, J. Foricher, U. Hengartner, C. J. Jenni, F. Kienzle, H. Ramuz, M. Scalone, M. Schlageter, R. Schmid, S. Wang, Chimia 1997, 51, 303.
31 H. Jendralla, Proceedings of the ChiraTech '97 Symposium, The Catalyst Group, Spring House, USA,1997. G. Beck, Proceedings of the ChiraTech '96 Symposium, The Catalyst Group, Spring House, USA, 1996.
32 J. Hiebl, H. Kollmann, F. Rovenszky, K. Winkler, J. Org. Chem. 1999, 64, 1947.

I
New Processes for Existing Active Compounds

1
Asymmetric Hydrogenations – The Monsanto L-Dopa Process
WILLIAM S. KNOWLES

Abstract

The development of catalysts for asymmetric hydrogenation began with the concept of replacing the triphenylphosphane ligand of the Wilkinson catalyst with a chiral ligand. With these new catalysts, it should be possible to hydrogenate prochiral olefins.

Knowles and his co-workers were convinced that the phosphorus atom played a central role in this selectivity, as only chiral phosphorus ligands such as (R,R)-DIPAMP, the stereogenic center of which lies directly on the phosphorus atom, lead to high enantiomeric excesses when used as catalysts in asymmetric hydrogenation reactions. This hypothesis was shown to be wrong by the development of ligands with chiral carbon backbones. Although the exact mechanism of action of the phosphane ligands has not been incontrovertibly determined to this day, they provide an easy point of access to a large number of chiral compounds.

1.1
The Development of Chiral Phosphane Ligands [1]

This particular story is actually the account of the genesis of an invention. The inventive process is not clearly understood, but one factor that seems to be important is the necessity for a high degree of naivety. That is why, so frequently, it is not the experts that do the inventing, but they are the ones who, once a lead has been established, come in and exploit the area. Our work is an excellent illustration of this phenomenon.

In the study of any of the life sciences, chiral compounds are important. In the past when chiral compounds were required, chemists had to use biochemical processes or prepare racemic mixtures followed by laborious resolutions. In industry the problem is particularly severe, since resolution, with its numerous recycling loops and fractional crystallizations, is an inherently expensive process.

Thus, large-volume products such as monosodium L-glutamate, L-lysine, and L-menthol have traditionally been prepared through biochemical routes, even though efficient procedures are available to produce their racemic forms.

Asymmetric Catalysis on Industrial Scale: Challenges, Approaches and Solutions
Edited by H. U. Blaser, E. Schmidt
Copyright © 2004 Wiley-VCH Verlag GmbH & Co. KGaA, Weinheim
ISBN: 3-527-30631-5

In the early 1960s we became aware of this problem when we prepared a paper on the evaluation of a monosodium glutamate process. The racemic mixture was easy to obtain, but by the time resolution had been implemented, the projected costs had doubled, even though we racemized and recycled the unwanted D-isomer. It looked as though, if one wanted to beat the bug, it would be necessary to have a catalyst that would direct the reaction in order to give a predominance of the desired isomer, when an asymmetric center was formed. For this purpose, 100% efficiency of the enzymes would not be necessary to produce something of real value.

At this point in time I was aware of the extensive studies by Akabori, which began in the mid-1950s, in which heterogeneous catalysts such as Raney nickel and palladium were modified with a chiral agent. The asymmetric bias was always too small to be of preparative interest. However, all these thoughts remained dormant for several years.

In the interim, I became part of a program for carrying out exploratory research. I was given a new Ph.D. student to train for a year before going on to more pressing things. Industrial labs always wrestle with the problem of how much undirected research they should do, but this is just one of many ways to achieve a goal. I had been working with several new employees on a number of projects, when I became aware of Professor Wilkinson's discovery of chlorotris(triphenylphosphane) rhodium, [RhCl(PPh$_3$)$_3$], and its amazing properties as a soluble hydrogenation catalyst for unhindered olefins. Homogeneous catalysts had been reported before, but this was the first one where the rates were comparable with the well-known heterogeneous counterparts.

A second development in the mid-1960s was in the area of methods for making chiral phosphanes, by Mislow and also by Horner. Phosphorus, like carbon, is tetrahedral and, when four different substituents are attached, can exist in D- and L-forms. In the case of phosphanes, the lone pair of electrons counts as a substituent. Earlier, it was thought that phosphanes might invert pyramidally like their nitrogen analogues, but Mislow and also Horner showed that they were stable at room temperature. They turned out to have a half-life of a couple of hours at 115 °C. For our contemplated hydrogenations, this stability would be quite adequate. The basic strategy was then to replace the triphenylphosphane of Wilkinson's catalyst with a chiral counterpart and hydrogenate a prochiral olefin. This experiment was performed on α-phenylacrylic acid using the known chiral methylpropylphenylphosphane, and gave an enantiomeric excess (ee) value of 15% (Fig. 1).

This modest result, of course, was of no preparative value, but it did establish that the hydrogenation technique gave a definite asymmetric bias. In order to achieve this bias, the hydrogen atom, the ligand, and the substrate all had to be on the metal at the same time. Furthermore, we established that the hydrogenation was accomplished in solution and not from some extraneous rhodium plating out in our reactor. The inherent generality of the method offered almost unlimited opportunities for matching substrate and catalyst, moving towards the goal of achieving efficient results.

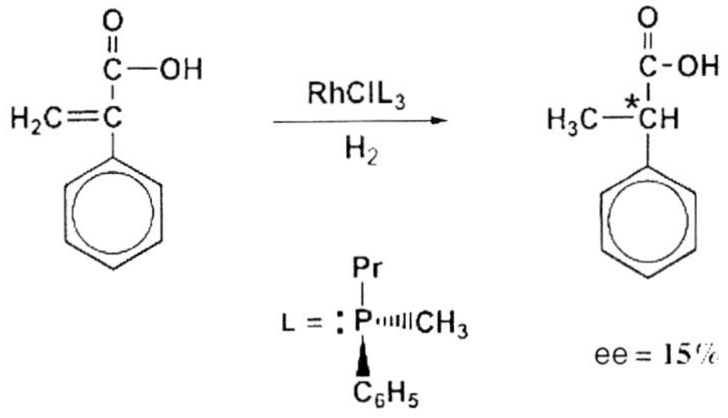

Fig. 1 First use of a chiral ligand in the hydrogenation of an olefin.

$$H_2C=C(C_6H_5)(COOH) \xrightarrow[H_2]{RhClL_3} H_3C-*CH(C_6H_5)(COOH)$$

$$L = :P(Pr)(CH_3)(C_6H_5)$$

ee = 15%

We were not alone in having this idea, but were the first to report on it. I believe it was discussed in the question session after Wilkinson's lecture on his soluble hydrogenation catalyst at a Welch Foundation conference. Horner, shortly after our paper appeared, reported even more modest results with the same methylpropylphenylphosphane on a substituted styrene. Other workers were using other phosphanes with uninteresting results. Seemingly, we may have been the only ones naive enough to pursue this lead to any great depth. In fact, there was definitely nothing in the literature to encourage us to proceed further. A mechanistic study showed that just two ligands were all that were needed and not the three, as in Wilkinson's structure. α-Phenylacrylic acid worked better as a triethylamine salt, but even so we never obtained good results with this system.

While delving around in this area, another apparently unrelated development appeared, which played an important role in our project. This was the discovery that a fairly massive dose of L-dopa was useful in treating Parkinson's disease, which created a sizable demand for this rare amino acid. Because of Monsanto's position in the production of vanillin, which provided the 3,4-dihydroxyphenyl moiety, we learned that they were custom-manufacturing a racemic intermediate, which Hoffman-LaRoche resolved and deblocked to give L-dopa. The synthesis, which closely followed the Erlenmeyer azlactone procedure described in typical organic syntheses, went by way of a prochiral enamide that was hydrogenated to give blocked D,L-dopa (Fig. 2). This enamide offered a golden opportunity for commercializing this burgeoning technology. We soon found out that these prochiral enamide precursors of α-amino acids hydrogenated much faster than one would expect for such a highly substituted olefin. Even so, the chiral results were only 28%, but the stage was then nicely set to carry out a structure versus activity study. We had a good test reaction, as shown in Fig. 3, in which we used the simple phenylalanine intermediate.

We also had a good test for efficiency, since all we had to do was run rotation measurements on an appropriately diluted reaction mixture. Our job was to find a phosphane of the appropriate structure. Early on we tried phosphanes with a chiral alkyl side chain, and the asymmetric bias was barely detectable. We felt strongly that, if one wanted to obtain high ee values, the asymmetry would have to be directly on the phosphorus. That is where the action is.

Initially, we varied the alkyl groups on the phosphorus, by converting the normal propyl to the more hindered isopropyl or cyclohexyl, but ee values still re-

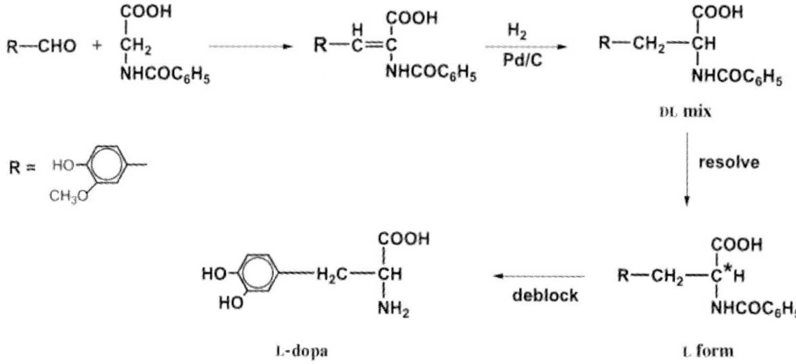

1 Asymmetric Hydrogenations – The Monsanto L-Dopa Process

$$R\text{—CHO} + \underset{\underset{NHCOC_6H_5}{|}}{\overset{\overset{COOH}{|}}{CH_2}} \longrightarrow R\text{—}\underset{\underset{NHCOC_6H_5}{|}}{\overset{\overset{COOH}{|}}{\underset{}{C}}}=\overset{H}{C} \xrightarrow{H_2, Pd/C} R\text{—}CH_2\text{—}\underset{\underset{NHCOC_6H_5}{|}}{\overset{\overset{COOH}{|}}{CH}}$$

R = HO—⟨⟩—, CH₃O

DL mix ↓ resolve

L-dopa ←—deblock—— L form

Fig. 2 The Hoffman-LaRoche L-dopa process.

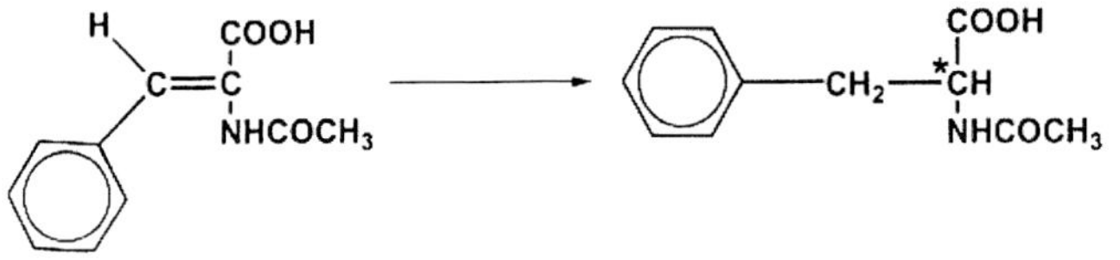

(Fig. 3 scheme: PhCH=C(NHCOCH₃)COOH → PhCH₂—*CH(NHCOCH₃)COOH)

Fig. 3 Test reaction for the structure-activity-relationship study.

mained in the range of 28–32%. Our first real variation was to introduce the o-anisyl group. This should provide some steric hindrance as well as a possible hydrogen-bonding site. Furthermore, the ether linkage would be stable enough to survive the rigors of a phosphane synthesis. In those days, our small group was in constant contact and what we decided to do was arrived at by informal consensus. I hate to admit it, but it is much easier to invent when working in a small, underfunded group. Being lean and hungry is conducive to invention.

We prepared methylphenyl-o-anisylphosphane (PAMP) and obtained ee values, after playing with the hydrogenation conditions, of up to 58%. Further modification of this molecule gave us methylcyclohexyl-o-anisylphosphane (CAMP); this change gave up to 88%. These results are summarized in Fig. 4.

It all seemed too easy and simple, but this was the first time ever that anyone had obtained enzyme-like selectivity with a man-made catalyst! Never in our wildest imagination did we think a structure versus activity study would converge so quickly to a product with commercial potential. CAMP was our sixth candidate. As I look back from this perspective, I do not think that we were actually emotionally equipped to realize what we had done. Here, with this simplest of molecules (CAMP), we had solved one of the toughest synthetic problems. For the last hundred years, it had been almost axiomatic among chemists that only nature's enzymes could ever do this job.

Our patent department always considered our invention was the use of chiral phosphanes with rhodium but, of course, without finding PAMP and CAMP, we would have only had a new way of doing what had been done before. The lawyers

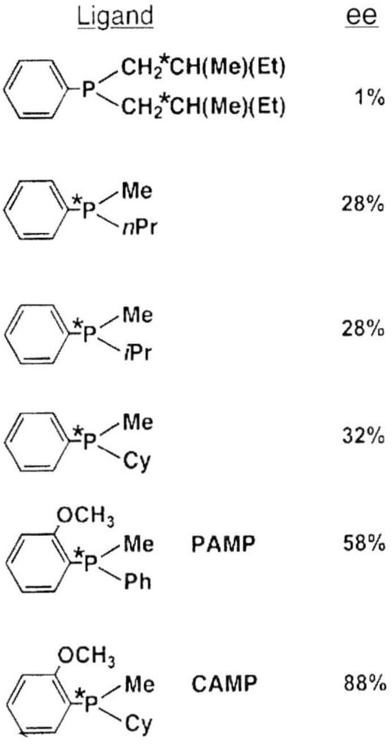

Fig. 4 Phosphane ligands for asymmetric hydrogenation.

Ligand		ee
Ph-P(CH₂*CH(Me)(Et))₂		1%
Ph-*P(Me)(nPr)		28%
Ph-*P(Me)(iPr)		28%
Ph-*P(Me)(Cy)		32%
(o-OCH₃-C₆H₄)-*P(Me)(Ph)	PAMP	58%
(o-OCH₃-C₆H₄)-*P(Me)(Cy)	CAMP	88%

felt that this first result was not much, but that we could very rapidly come up with an improved embodiment, and in this they were unusually prescient.

We have called these catalysts man-made but this is not strictly true. We have not violated the general principal that, if you want chiral molecules, you will have to get them with the assistance of previously formed natural products. Our asymmetry was obtained from the (–)-menthol used in the chiral phosphane synthesis, but being a catalyst, a small amount of (–)-menthol could lead to a large amount of chiral product. CAMP worked equally well for the L-dopa precursor (Fig. 2), and it made no difference whether the amine-blocking group was benzoyl or acetyl.

At this point, we were strongly motivated to develop a commercial L-dopa process. It is a rare thing that the emergence of a substantial demand for a chemical is so closely timed with an invention for a new way of making it. Our management reluctantly increased our manpower but did not really believe we could do it until the hydrogenation was achieved on a 50 gallon scale without incident.

Since CAMP was clearly good enough, we stopped exploring phosphanes and concentrated on converting this unique hydrogenation into large-scale production. This process was helped when another fortuitous event occurred. Monsanto decided to get out of manufacturing its first product, saccharin, and an idle plant was now available for this type of fine-chemical manufacture. These things were then brought together to give our simplified L-dopa process, which started with vanillin (Fig. 5).

The chiral hydrogenation was the simplest step in the sequence. We started with a slurry of prochiral olefin in an alcohol-water mix and ended up with a slur-

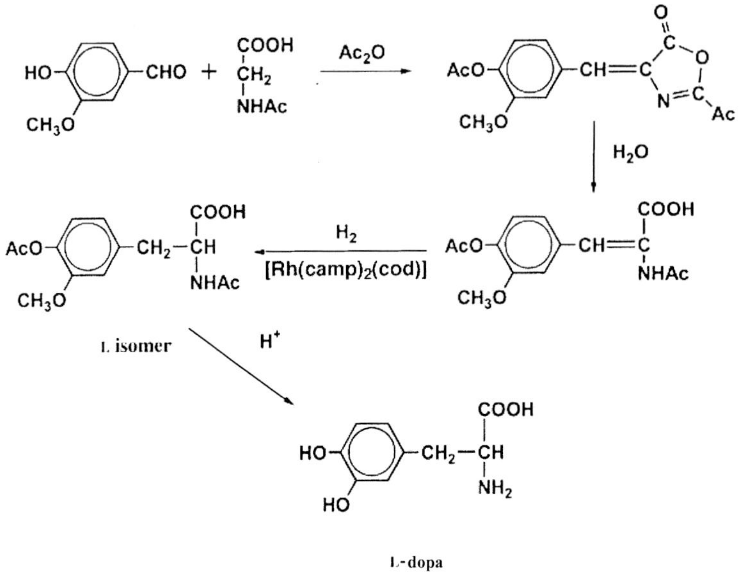

Fig. 5 Monsanto L-dopa process.

ry of chiral product which could be filtered, to leave the catalyst and residual racemate in the mother liquor. We could use an *in situ* prepared catalyst, but it was more convenient to use a solid air-stable complex of the type $[Rh(1,5\text{-cod})L_2]^+BF_4^-$ (cod = cyclooctadiene). These catalysts were fast, so that mole ratios of substrate:catalyst were about 20000:1. Thus, even this super-expensive complex was used at close to throwaway levels.

Even in the best case, some racemic product is produced and must be separated out. This separation is easy or hard, depending on the nature of the racemate. If the racemic modification has a different crystalline form to that of the pure D or L, then separation of the pure excess enantiomer will be inefficient. If one achieves a 90% ee value, then it is quite possible to get out only 75–80% pure enantiomer. With lower ee values, the losses become prohibitive. For such a system, a catalyst of very high efficiency must be used. Unfortunately, most compounds are of this type; their racemic modifications do not crystallize as pure D- or L-forms. If, on the other hand, the racemic modification is a conglomerate or an equal mix of D- and L-crystals, then recovery of the excess the L-form can be achieved with no losses. Since the L- and D,L-forms are not independently soluble, a 90% ee value easily gives a 90% recovery of pure isomer. In our L-dopa process, the intermediate is just such a conglomerate and separations are efficient. This lucky break was most welcome. If one thinks back, ours was the same luck that Pasteur encountered in his classical tartaric acid separations, 150 years ago.

At the time of our initial commercialization, we learned of a new, efficient ligand discovered by Kagan et al., which they called DIOP. This was a chelating bisphosphane ligand prepared from tartaric acid with chirality on the carbon back-

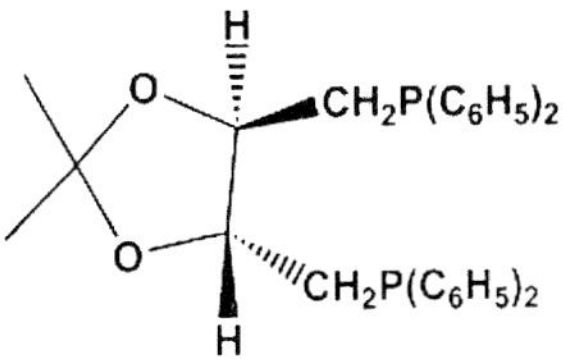

Fig. 6

83% ee

Fig. 7

(R,R)-DIPAMP

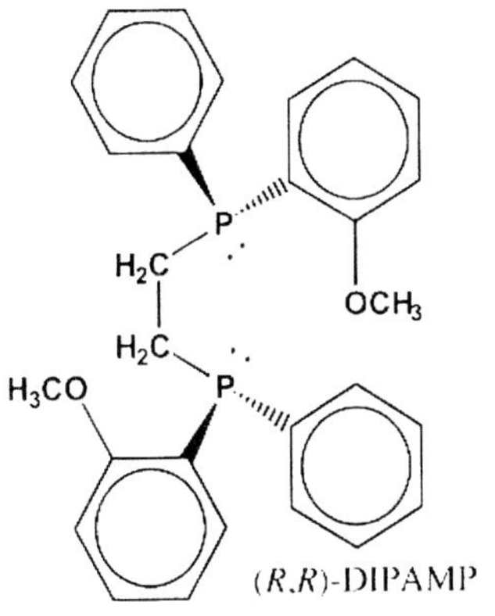

95% ee

bone (Fig. 6) and it gave results comparable to CAMP. We had hypothesized that, to obtain good results, one needed chirality directly on the phosphorus atom. It made sense, but Kagan showed us to be totally wrong. It is most appropriate that this invention using tartaric acid should have come from a Frenchman in the land of Louis Pasteur, who, of course, was the one who got it all started. Kagan's discovery was the wave of the future for a whole series of bisphosphane ligands with asymmetry on the chiral backbone (Fig. 6) [2].

Shortly afterwards, we came up with our own chelating bisphosphane ligand, by dimerizing PAMP through another Mislow procedure. We called it DIPAMP, and chirality resided on the phosphorus atom. DIPAMP worked at about 95% ee in our L-dopa system and we quickly converted our commercial process to be able to use it. Part of our motivation to make a quick change was that DIPAMP was easier to make than CAMP, and, in addition, it was a nice, crystalline air-stable solid (Fig. 7).

When we started this work we expected these man-made systems to have a highly specific match between substrate and ligand, just like enzymes. Generally, in our experience and that of those that followed us, a good candidate is useful for quite a range of applications. This feature has substantially enhanced their value in synthesis. It turned out that these chiral hydrogenations, as applied to enamides, were entirely general, especially with DIPAMP. It should be pointed out here that these prochiral enamides can exist in both *E* and *Z* forms. The *Z* form hydrogenates efficiently whereas the *E* form hydrogenates less so. Both give the same product. Fortunately for us, the base condensation used in their preparation gives us only the desirable *Z* form (Fig. 8) [3]. The *R* group can be just about any-

Fig. 8 Generality for Z enamides.

Fig. 9

1 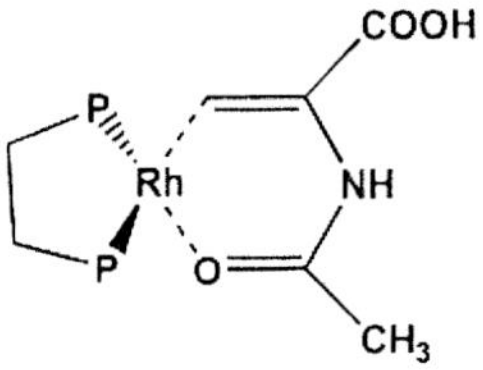 enamides
2 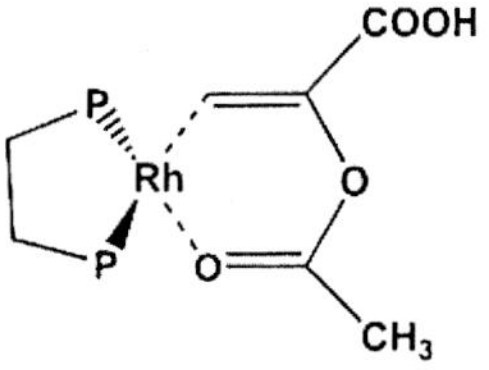 enol esters
3 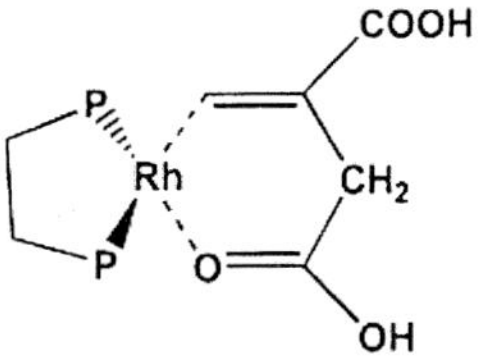 carboxylic acids

thing except –COOH. It is easy to see how our rhodium catalyst could become confused with two carboxy groups so close together. Thus, aspartic acid is best made by an enzymatic process.

However, almost all the known familiar α-amino acids can be prepared in this way since, at least in principle, an enamide precursor is possible. Evidently the polar carboxy and amide groups overwhelm any variation in the R group. Also, the carboxy- and the nitrogen-blocking groups can be varied extensively. Once again lady luck was with us, since if we had a choice where the catalyst would be useful, we could not have selected a more important area than the α-amino acids, the building blocks of the proteins.

A few of the more important ones are listed in Tab. 1. Our colleagues at Hoffmann-LaRoche have added about a dozen more nonaromatic members to this list using DIPAMP. This generality can be extended to a variety of enol esters and itaconic acid derivatives. Evidently what is required is the ability to chelate with the metal. Thus, the nitrogen atom can be replaced with an oxygen atom or a methylene group (1–3, Fig. 9) [3].

Tab. 1 Important amino acids produced by asymmetric catalysis.

Product	ee Value (%)
L-dopa	94
L-phenylalanine	96
L-tryptophan	93
L-alanine	90
L-lysine	85

One compound which did not work well in our system was our original model, α-phenylacrylic acid. A number of these aryl propionic acids are valuable as nonsteroidal antiarthritics. Here, as is typically the case, only one enantiomer is active and thus a process to prepare one isomer directly was needed. We tried hard to solve this problem, even using ruthenium-ligand systems, but without success. It took Professor Noyori with his BINAP-ruthenium complex to solve this problem [4]. This is just another example in the history of invention. The one who makes the first discovery seldom makes the second. On a grander stage, this may explain why there are so few double Nobel Laureates.

Soon after the appearance of DIOP and DIPAMP, a considerable number of bisphosphane ligands with chiral carbon backbones were found. All of these worked well with the same enamides and on related oxygen analogues. A few of these are shown in Fig. 10 [5].

It is interesting that, over the years, we made a lot of chiral phosphanes but never managed to prepare a good one without our beloved o-anisyl group. Others have used it in connection with their bisphosphanes but it gives them no particular advantage. Thus, the choice of suitable structures is still pretty much guesswork.

In our hands, DIPAMP was by far the most versatile ligand, and remained supreme for enamides for many years. Later, in the 1990s, an improved bisphosphane ligand was reported by Burk, then at DuPont, to which he gave the name DuPHOS (Fig. 11) [6].

This bisphosphane ligand, complexed with rhodium, gave fast hydrogenations of enamides with efficiencies of 99%. Once again, the next invention was made by someone else. These high ee values can be important where the racemate is not a conglomerate.

1.2
Synthesis and Properties of the Phosphanes

The key to asymmetric hydrogenation is the structure of the chiral ligand. The phosphanes are prepared by a multistep route and are quite expensive, but fortunately one mole of catalyst will make many thousands of moles of product. Even so, the ligand must be made from cheap starting materials. Some economy of scale is achieved by making a ten year supply in a few plant-sized batches. At

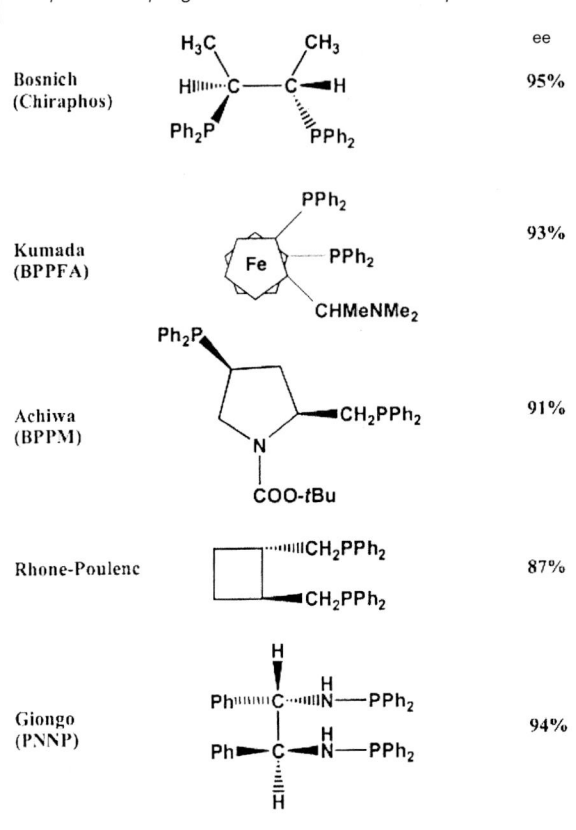

		ee
Bosnich (Chiraphos)	[structure: H₃C-C(H)(PPh₂)-C(H)(CH₃)(PPh₂)]	95%
Kumada (BPPFA)	[ferrocene with PPh₂, PPh₂, CHMeNMe₂]	93%
Achiwa (BPPM)	[pyrrolidine with Ph₂P, CH₂PPh₂, N-COO-tBu]	91%
Rhone-Poulenc	[cyclobutane with two CH₂PPh₂]	87%
Giongo (PNNP)	[Ph-C(H)-N(H)-PPh₂ / Ph-C(H)-N(H)-PPh₂]	94%

Fig. 10 Phosphane catalysts used in the catalytic hydrogenation of Z-α-acetamidocinnamic acid.

[Structure of bis(phospholane) ligand with R groups, R = Me or Et] — 99% ee

Fig. 11

first, CAMP was prepared from phenyldichlorophosphane via Mislow's menthyl ester, by introducing the o-anisyl group last. Unfortunately, the desired isomer was produced in minor amounts and, to correct this situation, it was necessary to reverse the order of addition of the aryl groups. The sequence starting with trimethylphosphite is outlined in Figs. 12 and 13.

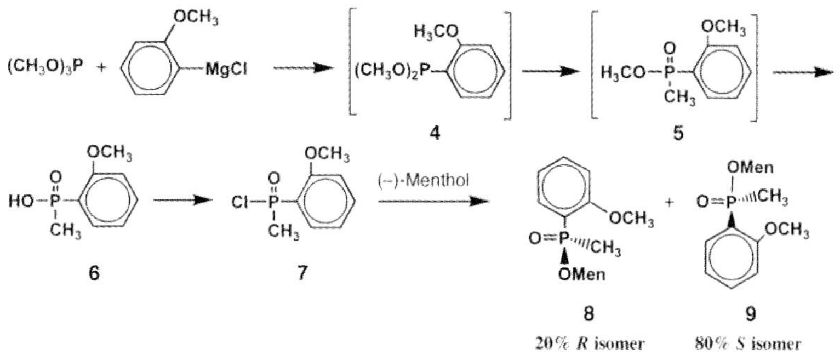

Fig. 12 Synthesis and resolution of the menthyl ester; men = menthyl.

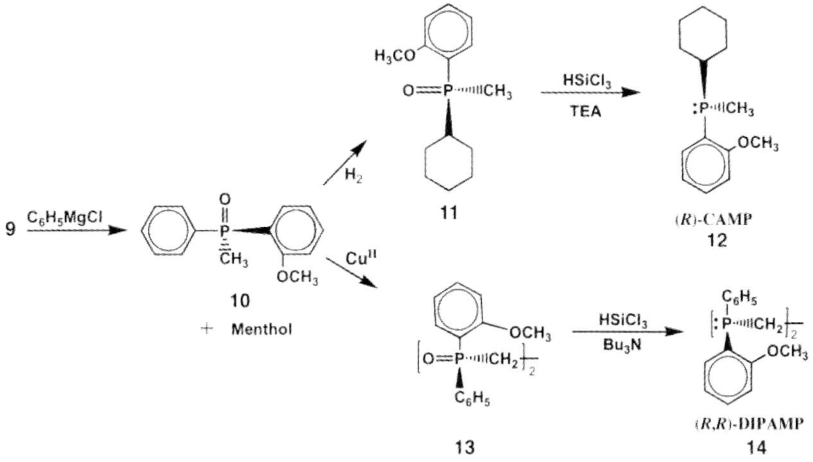

Fig. 13 Improved synthetic procedure for CAMP and DIPAMP.

A large excess of trimethylphosphite was required to achieve good yields of monosubstituted product **4**. In the sequence **4 → 6**, only the nicely crystalline phosphinic acid **6** was isolated. The fact that the acid chloride **7** can be converted into an 80:20 mix of (S) and (R) isomers means that the menthol preferentially reacts with one form while the other isomer rapidly racemizes. Thus, the catalyst preparation was greatly facilitated by an asymmetric synthesis of its own directed by (–)-menthol.

Another advantage of the sequence in Figs. 12 and 13 was that CAMP and DIPAMP were prepared from a common intermediate (**10**) and no new resolution procedure needed to be worked out. Thus, the change to an improved ligand could be done with minimum dislocation, both at the synthesis and the utilization end. It is a clear advantage of catalytic processes that it is often easy to shift from the old to the new.

CAMP was prepared by a selective hydrogenation reaction of **10** (Fig. 13) using a heterogeneous rhodium-on-carbon catalyst. It was important to monitor the reaction closely and to stop before the anisyl ring started to hydrogenate. Reduction with trichlorosilane and triethylamine (TEA) gave (R)-CAMP **12**, with inversion. The (R)-menthyl ester **8** could also have been used in this sequence if the last step was run with pure HSiCl$_3$; this modification results in retention of configuration.

In the case of DIPAMP the copper-coupling step, run with lithium diisopropyl amide and CuCl$_2$, did not affect the stereochemistry. However, only the base-promoted reaction with trichlorosilane to give a double inversion was applicable. In this case, an empirical study showed that use of tributylamine minimized formation of the *meso* product.

In principle, the menthol recovered in Fig. 13 could be recycled, which makes the use of the chiral agent derived from nature truly minimal, but in practice it has not been worth the effort. More useful is the recovery by hydrolysis of the phosphinic acid from the (R)-menthyl ester (Fig. 12).

In contrast to CAMP, DIPAMP is a stable solid that melts at 102 °C. Heated at 100 °C, it has a half-life of 3–5 h. This racemization was somewhat faster than Mislow's phosphanes, which did not invert appreciably until 10–15 °C higher. The rate was reasonable, if one considers that inversion at either end destroys chirality. DIPAMP complexed to rhodium must invert much more slowly because efficient, asymmetric hydrogenations have been obtained at 95–100 °C. For the sake of convenience, particularly on a large scale, a solid complex was made by reacting two equivalents of phosphane with one equivalent of [Rh(cod)Cl]$_2$ in alcohol. This air-stable orange solid [Rh(bisligand)(cod)]$^+$BF$_4^-$ made a most suitable catalyst precursor.

We have used the resolved menthyl ester **9** to make a variety of phosphane ligands. The first and most obvious use is to convert DIPAMP into DICAMP. You will recall that, in the monophosphane series, the exchange of a phenyl group for a cyclohexyl group gave an enormous increase in selectivity. Not so with DICAMP, which gave only 60–65% ee in our enamide systems. It was, however, our best candidate for preparing the more hindered amino acid, valine, for which the other systems were very poor (Fig. 14).

In the monophosphane series, we only found one ligand that was marginally better than CAMP (Fig. 15). This scarcity of good monophosphanes shows how lucky we were to find an efficient one on almost the first try. We never found a good candidate that did not have the *o*-anisyl group. This is in contrast to all our colleagues in other labs who never found much benefit from it.

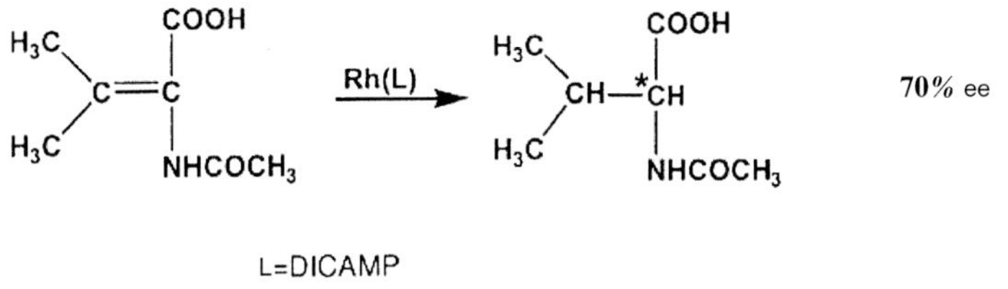

Fig. 14 Synthesis of valine by use of the DICAMP ligand.

Fig. 15

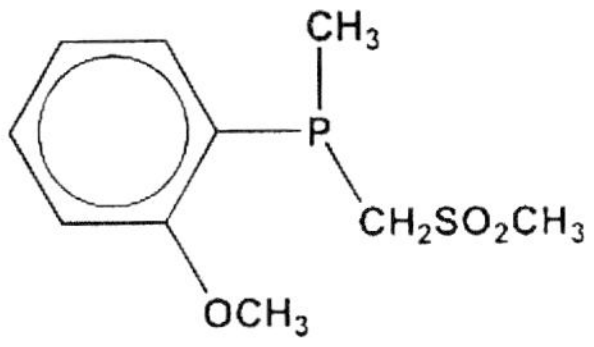

[Structure: o-anisyl group with P bonded to CH₃ and CH₂SO₂CH₃] 92% ee

We could sulfonate DIPAMP and make it water soluble. It worked fine but gave only 85% ee which, by current standards, is too low. I winced when I came in one morning and saw our valuable DIPAMPO being treated with concentrated sulfuric acid, but it worked. This exploratory effort suffices to show that, as one might expect, the catalysis continues to be a very sensitive function of ligand structure and that our ability to predict or proceed in a rational manner is severely limited.

1.3
Mechanism of the Asymmetric Catalysis

Now that we have these catalysts and have the ability to use them commercially, we would like to know how they work. When we look at energy calculations and realize that, to get 90% ee, we are talking about only a 2 kcal difference, and this is just about the same as the rotation barrier in ethane. Thus, the asymmetric bias may be caused by very subtle effects.

Using the ball-and-stick models in Fig. 16 to illustrate a typical prochiral olefin, we can see that attack at the *si* face gives the D-isomer and at the *re* face the L, which correspond to *R* and *S* isomers in more modern nomenclature. These of course are mirror images and our catalyst must discriminate between them.

We examined the X-ray crystal structure of the catalyst precursor [Rh(cod)(dipamp)]$^+$BF$_4^-$, and we noticed that this system presented an array of four aryl groups arranged in an edge-face manner. The phenyl groups present an edge and the o-anisyl group a face. This is depicted in Fig. 17 where, for the sake of clarity, we have omitted the cyclooctadiene ligand and the counterion, as well as oversimplifying the structure. In this picture, we are looking along the phosphorus-rhodium-phosphorus plane. One could speculate that an approaching substrate might prefer to lie on the flat face rather than on the hindered edge. We can more easily show this by a quadrant diagram (Fig. 18), in which the shaded quadrants represent the edge or hindered side. We speculated that a prochiral olefin might prefer to lie in the unhindered quadrant.

You will note that all the other bisphosphanes in Fig. 10 also present a similar array of four phenyl groups in their X-ray crystal structures, though not quite as convincingly, but there was always a face-exposed ring next to the skewed methylene group.

It makes no difference whether one attributes the bias to the edge-face configuration, as I prefer, or to the skewed methylene group. By using this quadrant interpretation, we could predict what the chirality of the product would be, from the

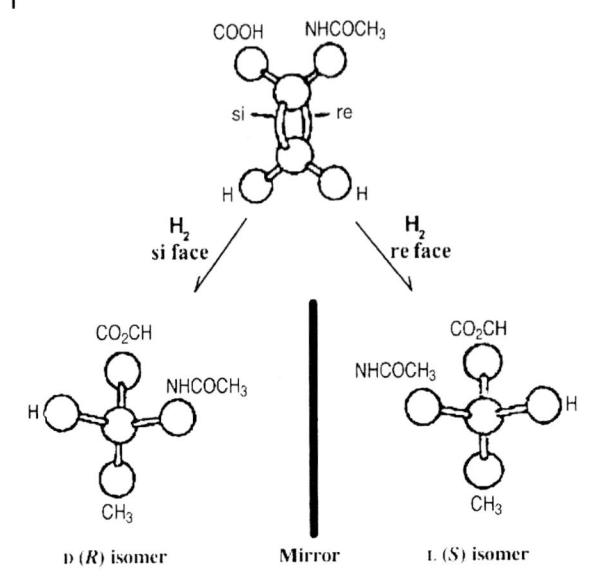

Fig. 16 Ball-and-stick models of the prochiral olefin and the resulting isomers.

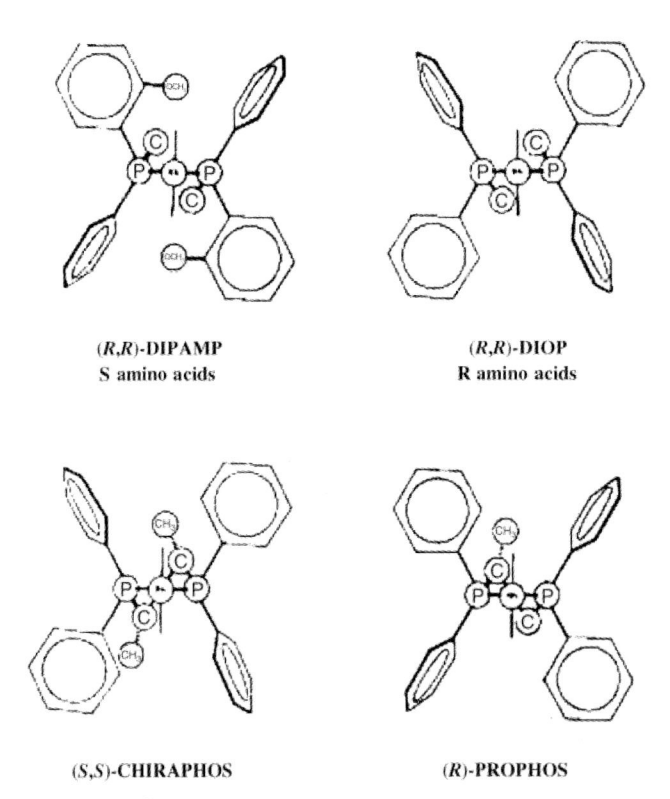

Fig. 17 Edge-face diagrams.

Fig. 18 Quadrant diagrams of positioning of the prochiral olefin.

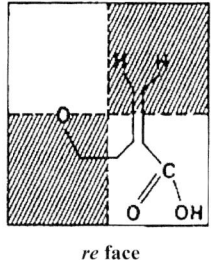

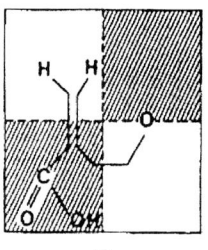

re face   *si* face

chirality of the phosphane. For any single case where there is a 50% chance of being right, such a prediction has no significance, but having predicted correctly for five cases where X-ray crystal structures were available, one gains credibility. We felt pretty good about how things were fitting into place.

Then along came Halpern's studies [7]. He had been able to isolate a more advanced intermediate, in which the enamide substrate actually formed a complex with the metal-ligand system. He obtained it in crystalline form, and it was with considerable eagerness that we awaited the X-ray crystallographic analysis results. It turned out that the enamide was lying nicely in the hindered quadrant.

So much for our theory. As so often happens in science, one comes up with an explanation in which everything seems to be fitting together nicely, and someone else then shows that your whole interpretation may well be wrong. We were stranded with the argument that, at the square-planar stage, Halpern reported that these steric factors may not be important. However, to get asymmetric bias, we know that the hydrogen atom, the ligand, and the substrate must all be on the metal center at the same time. Such a configuration requires an octahedral structure. Perhaps then these quadrant constraints are important. So far as I know, there is no evidence either to support or reject this contention. Our theory, though possibly wrong, does predict correctly.

All of this thinking does not explain CAMP, unless we argue that, during the hydrogenation step, this monodentate ligand prefers to occupy adjacent sites on the metal center and acts as a bidentate species.

Whatever the case, this unique catalysis has enabled chemists to study mechanistic details that could not previously be studied. When one thinks of it, it is quite remarkable that we are even in a position to debate such subtle features.

1.4
Concluding Comments

These soluble hydrogenation catalysts have begun a new era in catalytic processes. Since we are now dealing with pure complexes, we can design something to do just the job we want. This catalysis will continue to find many uses in industry, whenever an efficient route to the unsaturated precursor is available. These catalytic processes can be a good alternative but will by no means replace biochemical processes. Here, the problems with dilute solutions and difficult isolations are often less than the problems involved in a multistep synthesis. One area where these catalysts will reign supreme is in the preparation of d-amino acids or other non-natural isomers, where biochemical alternatives will not be available.

Perhaps the most important use of these catalysts will be to provide an easy way of making a large number of chiral compounds. In the past, research chemists have been reluctant to run laborious resolutions and have done so only when necessary. Now they can obtain chiral compounds for their life-sciences research with very little effort. We can look on these catalysts as a labor-saving device for the laboratory. For this, they will have an impact for as long as chemists run reactions.

For an invention to succeed, Paul Ehrlich, the father of chemotherapy, stated that four Gs are required: Geist, Geld, Geduld and Glück. The first of these is axiomatic; you have got to have a good idea. The second is essential; one needs financial support, but I would suggest a proper balance, too much or too little is inhibitory. For the third, you must have patience. Things never move as fast as you would have them. Finally, luck is all-important. I suspect that no invention has ever been made without some fortuitous help.

1.5
Acknowledgements

I have pointed out, and will continue to do so, that ours has been very much a joint effort. It would not have been possible without my associates, Jerry Sabacky and Billy Vineyard.

1.6
References

1 W. S. Knowles, Acc. Chem. Res. 1983, 16, 106, and references therein.
2 I. V. Komarov, A. Börner, Angew. Chem. 2001, 113, 1237; Angew. Chem. Int. Ed. 2001, 40, 1197.
3 K. E. Koenig in Asymmetric Synthesis, Vol. 5, Chiral Catalysis (J. D. Morrison, Ed.), Academic Press, New York, 1985, chap. 3: The Applicability of Asymmetric Homogeneous Catalytic Hydrogenation.
4 R. Noyori, Angew. Chem. 2002, 114, 2108; Angew. Chem. Int. Ed. 2002, 41, 2008.
5 H. B. Kagan in Asymmetric Synthesis, Vol. 5, Chiral Catalysis (J. D. Morrison, Ed.), Academic Press, New York, 1985, chap. 1: Chiral Ligands for Asymmetric Catalysis.
6 M. J. Burk, J. E. Feaster, W. A. Nugent, R. L. Harlow, J. Am. Chem. Soc. 1993, 115, 10125; M. J. Burk, M. F. Gross, J. P. Martinez, J. Am. Chem. Soc. 1995, 117, 9375.
7 J. Halpern in Asymmetric Synthesis, Vol. 5, Chiral Catalysis (J. D. Morrison, Ed.), Academic Press, New York, 1985, chap. 2: Asymmetric Catalytic Hydrogenation: Mechanism and Origin of Enantioselection.

2
The Other L-Dopa Process
RÜDIGER SELKE

Abstract

In January 1986 Isicom, a prescription drug used to treat the Parkinson syndrome, went on sale in the former German Democratic Republic. Isicom is a combination of L-dopa [L-dopa = (S)-3-(3,4-dihydroxyphenyl)-alanine] and L-carbidopa. For both components, new synthetic procedures were developed independently at the former Academy of Sciences of the GDR and the production of the active substances was realized at the former VEB ISIS-Chemie Zwickau. In this contribution the history and details of the development of the L-dopa process are recounted. For the commercial application of enantioselective catalysis, this is of some importance, since it was the first industrial process utilizing asymmetric complex catalysis to be realized in the former socialistic countries as well as in Europe.

2.1
Introduction

The roots for the activity in the field of preparation of enantiopure amino acids in the "Leibniz-Institut für Organische Katalyse an der Universität Rostock e.V." (formerly known as "Bereich Komplexkatalyse" which was a part of the "Zentralinstitut für Organische Chemie der Akademie der Wissenschaften der DDR") were planted at the end of the 1960s by Horst Pracejus, who was its director at that time (Fig. 1). In the 1950s Pracejus had previously worked on asymmetric catalysis and published outstanding results on the reaction of nucleophiles with ketenes catalyzed by chiral bases and developed a fundamental understanding of the mechanism of such enantioselective processes controlled by opposed entropic and enthalpic parts of the free activation enthalpy [1].

Pracejus was fascinated by the idea of functionalizing cellulose as the cheapest chiral material and to use it in this form as a carrier for monovalent rhodium for asymmetric hydrogenation. Rh was shown by Wilkinson to be useful as a catalytically active central metal in phosphane complexes. However, the catalytic activities of the new cellulose immobilized complexes in the hydrogenation of unsaturated amino acid precursors were low and the enantioselectivities did not exceed 35%

Asymmetric Catalysis on Industrial Scale: Challenges, Approaches and Solutions
Edited by H. U. Blaser, E. Schmidt
Copyright © 2004 Wiley-VCH Verlag GmbH & Co. KGaA, Weinheim
ISBN: 3-527-30631-5

Fig. 1 Horst Pracejus, 1927–1987.

ee [2, 3]. This was not much in comparison with the good results of Knowles who had already developed an industrial L-dopa process [4] and of Kagan who realized up to 80% ee with his well-known Rh-diop catalyst [5].

To improve the accessibility of the catalytic active centers the author was asked by Pracejus to prepare crosslinked polysaccharides, to functionalize them with chlorodiphenylphosphine and to load the resulting polyphosphinite with simple rhodium complexes. This was a difficult job and for approximately three years (1971–1974) neither the activity nor the enantioselectivity could be decisively enhanced.

Thus we decided, with the consent of Pracejus, to first find out the steric prerequisites for highly enantioselective derivatives of monomeric carbohydrates substituted by phosphorus. However, the purification of these newly prepared carbohydrate-phosphanes was tedious and the enantioselectivities of its rhodium chelates did not even reach 50% ee (after two further years of research).

The synthesis of carbohydrate-phosphinites was, however, unexpectedly successful. The preparation was much easier than that of the phosphanes. The products could be obtained in crystalline state and were less sensitive to oxidation by air. Their thermal stability was high and drying could be done in vacuum up to 100 °C without decomposition. The rhodium(I) chelate of methyl 4,6-O-benzylidene-2,3-bis(O-diphenylphosphino)-α-D-glucopyranoside led to 75% ee (S)-N-acetylphenylalanine **2a** on hydrogenation of (Z)-2-acetamidocinnamic acid **1a** (Fig. 2) [6].

The hydrogenation activity which was very low for the μ-chloro-bridged neutral rhodium(I) complexes **3** could be enhanced tremendously by reaction with silver tetrafluoroborate according to Fig. 3, transforming them into cationic species **4** possessing two additional free coordination sites to bind both the substrates, olefin and hydrogen, in the transition state during the catalytic reaction.

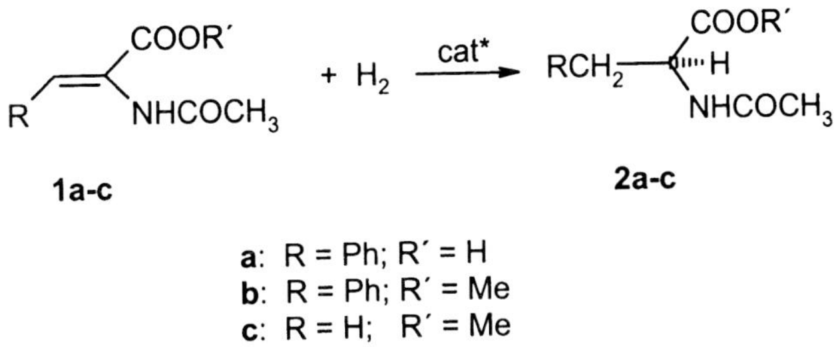

2.1 Introduction

Fig. 2

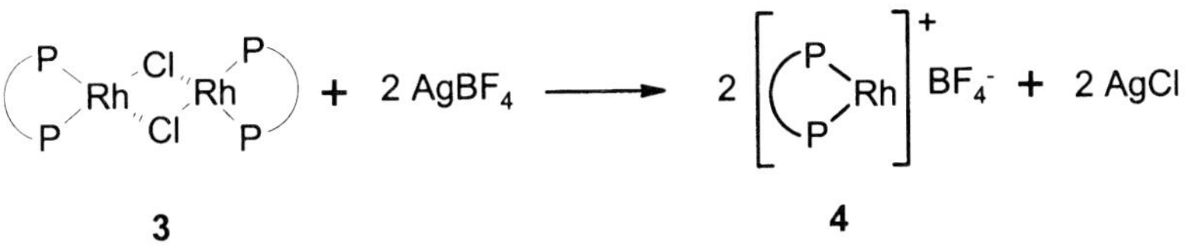

$$\text{1a-c} + H_2 \xrightarrow{\text{cat*}} \text{2a-c}$$

a: R = Ph; R' = H
b: R = Ph; R' = Me
c: R = H; R' = Me

Fig. 3

$$3 + 2\,\text{AgBF}_4 \longrightarrow 2\,[4]^+ \text{BF}_4^- + 2\,\text{AgCl}$$

The preparation of the cationic complexes was already known [7], however, in contrast to bisphosphane complexes its effect on the activity turned out to be particularly high for rhodium(I)-bisphosphinites. This enhancement of the activity was essential for an industrial application [8]. It could be shown that the usually sensitive P–O bonds of the ligand were well protected against solvolysis by complexation with rhodium(I).

In 1978 we tried to establish contacts with an industrial partner, which were realized by talks with Dr. Lohmann, at that time head of the research department of the VEB ISIS-Chemie, Zwickau. Earlier it was a "Kommanditgesellschaft" with government participation but was later converted into a nationalized state-run plant with the status of a "VEB" ("Volkseigener Betrieb", i.e., a firm owned by the people), a typical state-owned plant in the former GDR. The company was interested in producing L-dopa and we saw a good opportunity for an industrial process using the asymmetric hydrogenation of O-functionalized derivatives of (Z)-2-acetamidocinnamic acid **1a** with our catalysts as shown in Fig. 4.

For this application we chose phenyl 4,6-O-(R)-benzylidene-2,3-bis(O-diphenylphosphino)-β-D-glucopyranoside **9** as the chiral ligand, which we abbreviated to Ph-β-glup (Fig. 5). It was advantageous that phenyl β-D-glucopyranoside **7** with an aryl group as aglycon could for instance be prepared in large amounts much more easily by the method of Helferich [9] as a precursor for Ph-β-glup **8**, than the analogous methyl β-D-glucopyranoside. Particularly useful was the superior enantioselectivity of rhodium(I)-chelates of Ph-β-glup, which later could be increased in some cases up to 99% ee.

Meanwhile Cullen and Sugi [10] as well as Jackson and Thompson [11] had also published work in this field and we were in a hurry to develop the process. However, further publication activity was prohibited by our industrial partner.

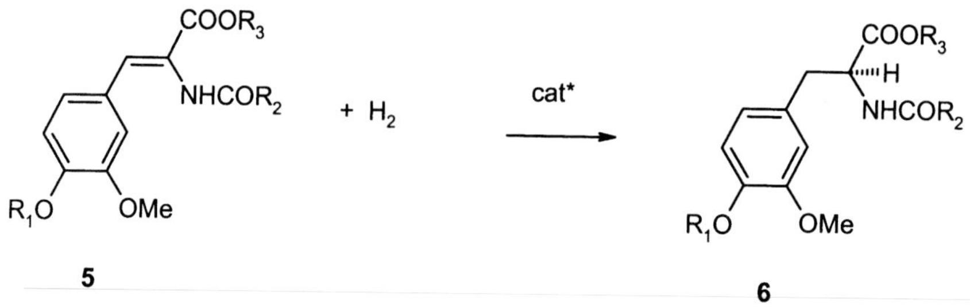

Fig. 4

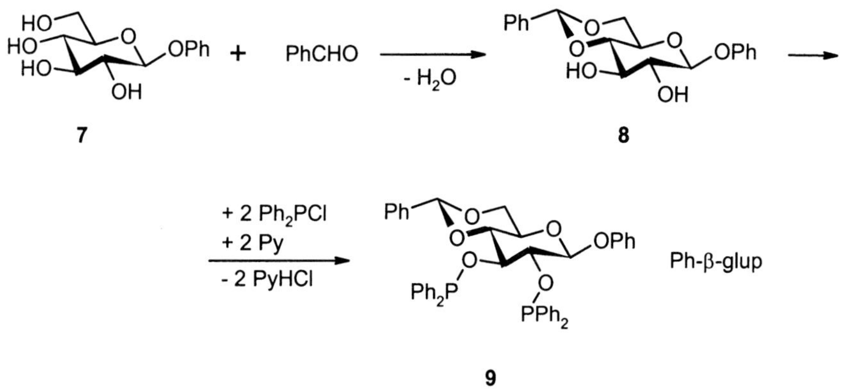

Fig. 5

2.2
Choice of the Substrate

Analogous with (Z)-2-acetamidocinnamic acid **1a**, their derivatives **5**, O-functionalized at the 3,4-position of the phenyl ring, could be readily synthesized by the Erlenmeyer condensation (Fig. 6).

Azlactones **12** were obtained from hippuric acid (**11**, R_2=phenyl) in higher purity and in better yield than those from N-acetylglycine (**11**, R_2=methyl). The question as to which of the aldehydes, vanillin (**10**, R_1=acetyl) or veratrumaldehyde (**10**, R_1=methyl), should be chosen was decided by the availability of the latter at a relatively low price from Italy. By asymmetric hydrogenation applying [Rh(Ph-β-glup)(cod)]BF$_4$ (see formula **15**, cod=cis,cis-cycloocta-1,5-diene, A$^-$=tetrafluoroborate) all acid substrates led to higher enantioselectivities than their esters (Tab. 1) [12].

Thus, deviating from Knowles' starting material, (Z)-3,4-dimethoxyphenyl-2-N-acetyl-acrylic acid (**5d**, R_1=Me, R_2=Ph, R_3=H) became the favorite substrate for the process in Zwickau, a useful synthesis for which had already been published by Kropp and Decker [13]. As can be seen from Tab. 1, at the beginning of the hydrogenation the substrates are not completely soluble under the applied conditions. However, we soon found that the hydrogenation reactions ran particularly well in suspension and with the amount of solvent reduced in such a way that good stirring, essential for the hydrogen uptake, was still possible.

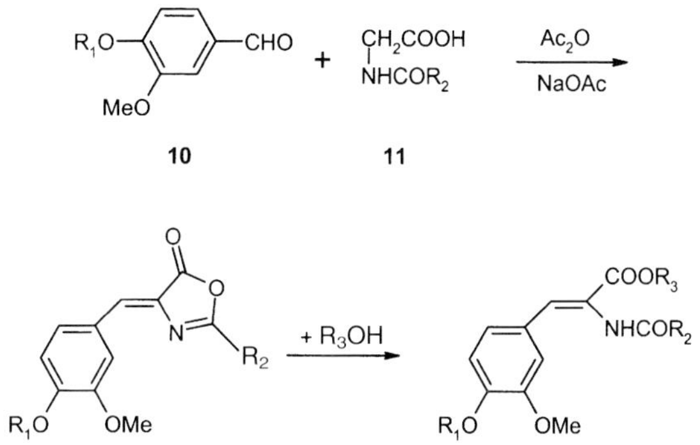

Fig. 6 **12** **5**

Tab. 1 Hydrogenation of various cinnamic acid derivatives.

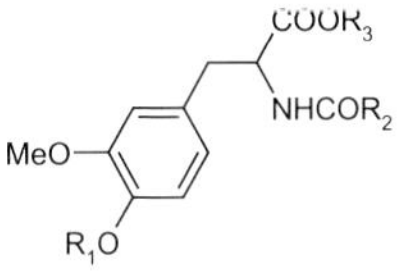				*Hydrogenation results*		
				Dissolved substrate at the start (%)	*t/2 (min)*	*% ee (S)-6*
No.	**R_1**	**R_2**	**R_3**			
5a	Me	Me	**H**	37	67	**96**
5b	Me	Me	Me	2	105	87
5c	Me	Me	Et	20	58	90
5d	Me	Ph	**H**	49	60	**95**
5e	Me	Ph	Me	29	56	83
5f	Me	Ph	Et	96	62	88
5g	Me	Ph	i-Pr	–	230	94
5h	Ac	Me	**H**	8	55	**94**
5i	Ac	Me	Me	23	61	91
5k	Ac	Me	Et	29	57	87
5l	Ac	Ph	Et	45	65	90
5m	H	Ph	Me	92	78	85
5n	H	Ph	**H**	98	57	**93**

Hydrogenation conditions: 20 mmol substrate **5** in 250 mL ethanol, 0.08 mmol catalyst [Rh(Ph-β-glup)(cod)]BF_4, 25 °C, 0.1 MPa.

2.3
The Catalyst

The superiority of the rhodium(I) chelates with the ligand Ph-β-glup (possessing only equatorial oriented substituents on the hexopyranoside ring), as compared with other carbohydrate bisphosphinites as catalysts for asymmetric hydrogenation of N-acyldehydroamino acids, has been documented by us [14–17] and others [18] in a number of papers, which will not be discussed in detail here. The negative influence of axially oriented hexopyranoside substituents on the enantioselectivity can be seen in Fig. 10 (Section 2.5).

In order to realize an industrial process it was very important that cationic rhodium(I)-bisphosphinites could be obtained without the use of silver salts. For this purpose neutral rhodium(I)-bisphosphinite-cyclooctadiene-acetylacetonate was prepared according to Fig. 7 and transformed into a cationic species by reaction with acids, particularly sulfuric acid (Fig. 8, A = HSO_4). This could also be achieved as a one pot reaction [19]. It should be noted that in addition we improved the existing procedures for the preparation of the precursor complexes $[Rh(cod)Cl]_2$, Rh(cod)acac [20] and $[Rh(cod)_2]BF_4$.

What was astonishing and really unexpected was the high stability of the P-O bonds, which are normally very sensitive to hydrolysis. Because of the protecting properties of the coordinated rhodium, the cleavage of the P-O bonds is inhibited even under strongly acid conditions. High-grade and even concentrated sulfuric acid are tolerated. However, strong acid leads to a time-dependent solvolysis of the

Rh(cod)acac + Ph-β-glup ⟶ Rh(Ph-β-glup)(cod)acac

13   14

Fig. 7

Rh(Ph-β-glup)(cod)acac $\xrightarrow[-\text{Hacac}]{+\text{HA}}$ [Rh(Ph-β-glup)(cod)]$^+$A$^-$

14   15

Fig. 8

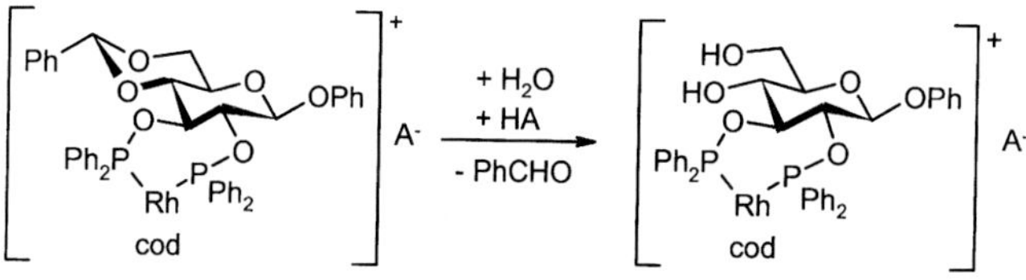

Fig. 9

benzylidene group and gives rise to a new precatalyst **16** (Fig. 9) [21] with increased catalytic activity and selectivity. The enantioselectivity was particularly improved for ester substrates (94–95% ee), which on hydrogenation by precatalyst **15** led in most cases only to 90–91% ee [22, 23]. Even though we applied the non-solvolyzed catalyst **15** in the technical hydrogenation runs, we believe that under the action of the sulfuric acid that is always added, large parts of the catalyst act in the form of **16**, carrying two hydroxy groups.

2.4
Improving the Hydrogenation Reaction

One of the main challenges for the introduction of a new catalytic process into industrial production is the systematic enhancement of the turnover number (TON, the number of substrate molecules that can be transformed into the product by one molecule of catalyst) and to establish the reason for the inhibition or destruction of the catalyst. From Tab. 2, third column, one can see that the TON for a quantitative conversion (in practice at least 98%) of methyl 3,4-dimethoxy-cinnamate **5e** increased considerably with the scale of the reaction.

This was of course encouraging for an industrial application and an indication that oxygen was responsible for catalyst deactivation. Indeed the turnover number could be enhanced by adding an oxygen-consuming additive such as acetaldehyde, however, this is not practicable in an industrial process.

With the substrate (Z)-2-benzamido-(3,4-dimethoxy)-cinnamic acid **5d**, as used in the technical process, we could realize turnover numbers of 10 000–12 000 in our laboratory under very careful oxygen exclusion. However, when substrate suspensions were used, this good result could be obtained only with the colorless methanol solvate **5d**×MeOH. It is not clear why the methanol-free, yellow colored samples of the substrate with the otherwise same chemical purity led to a much poorer result and with a TON of only 2 000. This yellow solvent-free modification can be obtained

Tab. 2 Dependence of the maximum TON on the reaction scale and on added acetaldehyde for the hydrogenation of **5e**.

Without acetaldehyde			In presence of acetaldehyde		
mmol substrate 5e	mmol catalyst 15 $A^- = BF_4^-$	TON	mmol CH_3CHO	mmol catalyst 15 $A^- = BF_4^-$	TON
1	0.005	200	1	0.002	500
20	0.04	500	4	0.02	1000
100	0.10	1000			
1000	0.67	1500			

Conditions: 15 mL methanol per 1 mmol substrate, 25 °C, 0.1 MPa.

by removing the solvating methanol under vacuum, or as well-formed yellow crystals by recrystallization of the substrate from some solvent mixtures, or even from pure methanol or ethanol, by seeding with yellow crystals of **5d**. This effect cannot be due to a contamination with the intermediate azlactone **12**, which is also yellow, because addition of some azlactone had no influence on the reactivity despite the fact that azlactones themselves could not be hydrogenated with our catalysts. As yet, we have no explanation for the remarkable effect described above.

2.5
Effect of Solvent and Anions

For the hydrogenation with [Rh(Ph-β-glup)(cod)]BF$_4$ all of the solvents investigated led to high enantioselectivities of generally more than 80% ee for the ester and >90% ee for the acid substrates. In contrast, using the α-glucosidic analogues in benzene led for acid substrates to D-amino acid derivatives with the opposite absolute configuration [15]. The cheap methanol was particularly attractive for the technical process due to its low evaporation enthalpy. Halogen-containing solvents allowed only low TON because they soon deactivated the catalyst.

In water the hydrogenation rate and the enantioselectivities were considerably lower. However, the addition of amphiphiles as proposed by G. Oehme mediates the dispersion of the relatively hydrophobic reactants and effects an enormous increase in hydrogenation activity and selectivity [17, 24, 25]. Fig. 10 depicts the results of the model hydrogenation reaction of methyl 2-acetamidoacrylate **1c** (see

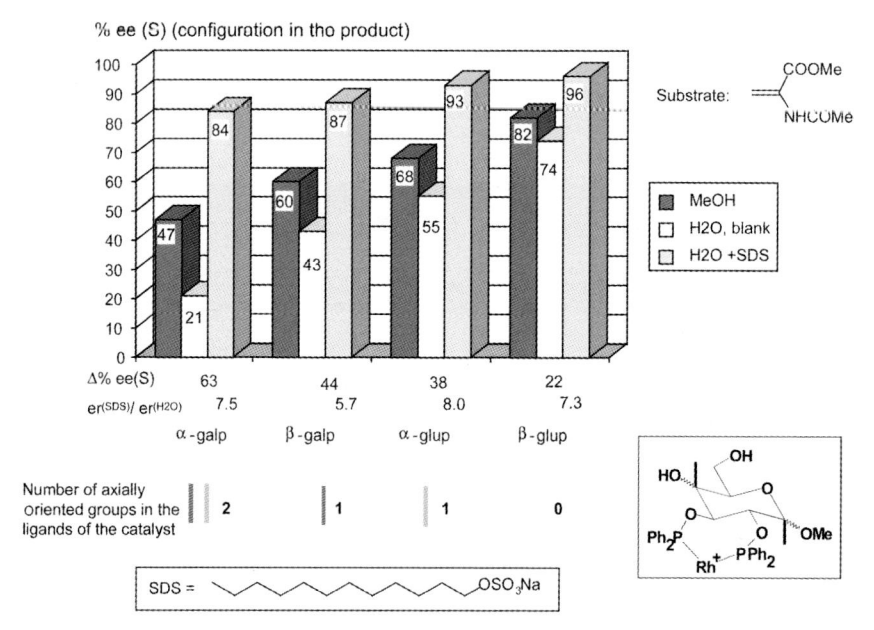

Fig. 10 Optimization of the enantioselectivity in water.

Fig. 2) with four different methyl D-hexopyranoside-2,3-bisphosphinite catalysts prepared from methyl α- or β-galacto-, respectively, α- or β-gluco-pyranoside.

It can be seen that the best selectivities are obtained with all equatorially oriented hexopyranoside substituents in β-D-glucopyranoside, for all solvents investigated. The addition of sodium dodecyl sulfate generally increases the low enantioselectivity observed in pure water and the values rise above the selectivities achievable in methanol. This is generally valid not only for water-soluble substrates such as **1c** but also for a great number of L-dopa precursors **5** [17, 25].

It has now been confirmed that the micelle forming amphiphiles act by incorporation of the catalyst as well as the substrate into the micelles [26], in which the hydrogenation rate is enhanced by a higher local concentration of the reagents and the enantioselectivity is often higher than in methanol [27]. For the latter improvement it seems important that catalysts possess sufficient flexibility, as with our seven-membered rhodium(I) chelates, to be able to adopt the optimal conformation for the environment of low polarity [28]. When using five- or six-membered ring chelates the effect of the amphiphiles was very small in our experience [26].

By optimizing the temperature as well as the concentration of substrate, catalyst and amphiphile, it was possible to adjust the conditions in such a way that at the end of the hydrogenation 75% of the product precipitated as absolutely enantiopure product (100±0.2% ee L-dopa precursor **6i**, Fig. 11). For the industrial syn-

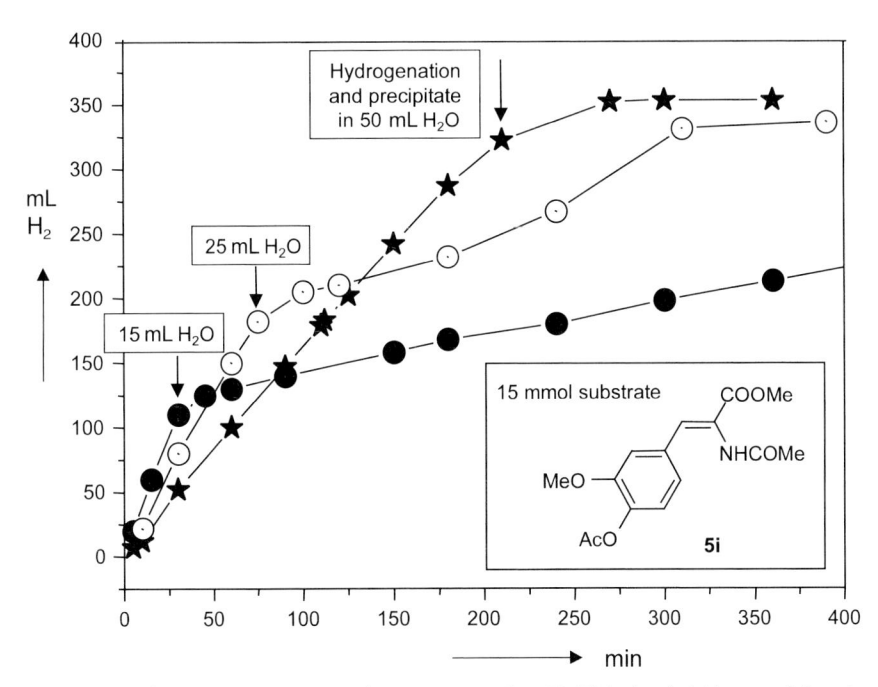

Fig. 11 Optical pure (100±0.2% ee) L-dopa precursor **6i** in 75.4% isolated yield as precipitated from 50 mL water at 40°C and 0.1 MPa.

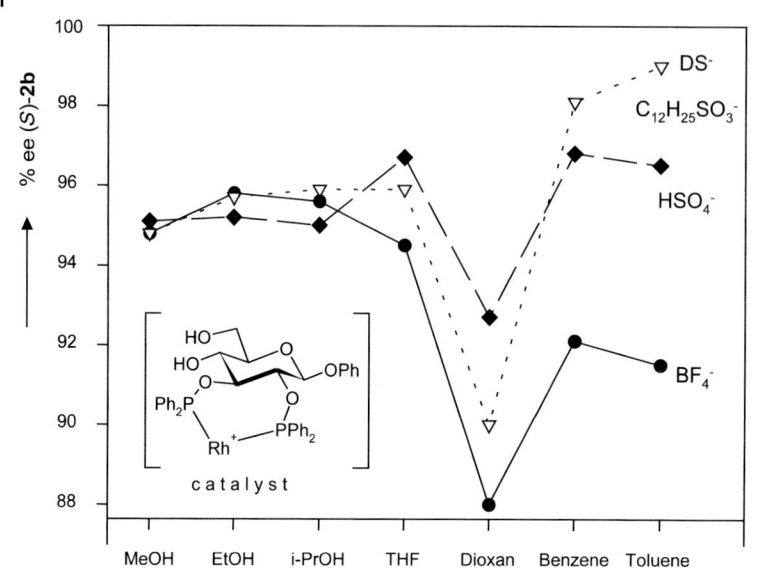

Fig. 12 Influence of different anions A⁻ on the enantioselectivity of catalyst **16** as a dependence on the solvent polarity in the hydrogenation of **1b**.

thesis of L-dopa this was not applicable, because the TON was less than 1 000 and this could not compete with the reaction in methanol. Moreover, the working up of methanol solutions is cheaper.

The influence of the anion A⁻ in the precatalyst **15** is relatively small in polar solvents such as methanol [96.3±0.2% ee for (Z)-2-acetamidocinnamic acid **1a** as substrate] due the complete dissociation of the ion pairs [15]. However, in solvents of lower polarity, such as for instance benzene, we observed a larger variation of enantioselectivity. This indicates that in the non-dissociated ion pairs (spatial proximity of the both cation and anion) the type of anion A⁻ influences the preferred conformation of the rhodium(I) chelate catalyst and thereby the ratio of the hydrogenation routes of diastereomeric re- and si-catalyst-substrate complexes [29]. For the precatalyst **16** containing hydroxy groups, the increasing influence on the ee of the nature of the anions A⁻ in solvents with decreasing polarity is shown in Fig. 12 for the hydrogenation of (Z)-methyl 2-acetamidocinnamate **1b** as a model substrate. The best results, methyl (S)-N-acetylphenylalaninate **2b** with 99% ee, were obtained with dodecylsulfate (DS⁻) as well as dodecylsulfonate as the anions with toluene as the solvent.

2.6
Immobilization of the Catalyst

For the preparation of immobilized catalysts we initially used polystyrene crosslinked by varying amounts of divinylbenzene and carrying aldehyde groups. Acetalization of glucosides analogous to Fig. 5 and further reaction with ClPPh$_2$ to the polymer bound ligand or even direct acetalization of the polystyrene aldehyde groups with the hydroxy groups of precatalyst [Rh(Ph-β-glup-OH)(cod)]BF$_4$ **16** (A$^-$=BF$_4^-$) could be realized with good yields. However, the activity of the polymer catalysts was disappointingly low. It seems that the accessibility of the catalytic centers for the substrates was too low.

Only the immobilization of the catalyst in the prehydrogenated [Rh(Ph-β-glup)(solv)$_n$]BF$_4$ complexes on cation exchangers of sulfonated polystyrene crosslinked with less than 2% divinylbenzene [30, 31] or on functionalized silica (SiO$_2$)-O-Si(CH$_2$)$_3$-O-C$_6$H$_4$SO$_3$H [32] was successful as regards the activity of the immobilized catalysts. The enantioselectivity for ester substrates even increased from 90 to 94% ee because fast hydrolysis of the catalyst by the action of the acid ion exchangers soon led to the more enantioselective [Rh(Ph-β-glup-OH)(solv)$_n$]BF$_4$. We believe that the high activity of these immobilized catalysts can be explained by the high mobility of the catalytically active cations. Thus it is mainly the free cations that are probably active in the hydrogenation. Despite this high mobility the leaching per hydrogenation step was less than 0.1% of the applied rhodium. Unprecedented at that time was the high reproducibility, with 94±1% ee for 15 to 20 runs with the same catalyst, which was washed with methanol three times after each run. This high selectivity could be kept up through to the last run when the catalytic activity almost ceased. This must be due to the fact that the rhodium residue of the decomposed catalyst has no further catalytic activity in our case. This is in contrast to similar rhodium(I) catalysts with the aminophosphine-phosphinite propraphos [33] which showed a gradually decreasing ee with repeated use [31]. It should be mentioned here that this ligand prepared by Krause [33] by reacting (+)- or (–)-propranolol with chlorodiphenylphosphine was in close competition with [Rh(Ph-β-glup)(solv)$_n$]BF$_4$ for use in the L-dopa process in Zwickau. In the end propraphos was unsuccessful due to the high cost of the separation of the enantiomers of the propranolol, despite the advantage of the availability of both enantiomers.

Even though the results with the immobilized catalyst in hydrogenation experiments with a substrate/catalyst ratio of 100:1 were good, their industrial application failed because with a substrate/catalyst ratio of 1000:1 re-use of the immobilized catalyst for the second run was impossible.

2.7
Production Process

The process was introduced by the VEB ISIS-Chemie and the production of L-dopa began in 1985. Accounts of this were first published by its staff in 1987, having partially used our reports and whom the author is indepted for the good cooperation [34]. Vocke was the process engineer assisted by Hänel, and Flöther was the head of the research department. Also instrumental to the success of the process were contributions from König in connection with the regeneration of the applied rhodium and from Foken and Grüner for the choice and preparation of the substrate as well as for the scale-up of the catalyst preparation and its precursors. The pressure of 0.01 MPa communicated in that paper is a misprint. The hydrogenation was performed under normal pressure of 0.1 MPa at 40 °C. Medium pressure could presumably be advantageous and might facilitate an increase in the turnover number. The enantioselectivity proved to be independent up to a pressure 10 MPa [14]. However, at that time VEB ISIS-Chemie was not able to work at higher pressures.

The hydrogenation reaction was carried out only in 80 L glass vessels (in German known as Hängegefäß). Despite of the fact that a large-scale process could be much more economical (increase in the hydrogenation rate and turnover number, decrease in the personnel expenditure), one did not dare to use a larger vessel fearing the loss of the valuable substrate if one batch should fail. The hydrogenation process was carried out at 40 °C with 32 mol of prochiral substrate per batch. Care had to be taken of course for efficient removal of the reaction enthalpy of ca. 110 kJ/mol because an increase in the reaction temperature would have led to a decrease in the enantioselectivity [14].

The technical process ran with a substrate/catalyst ratio of 2000 in methanol with more than 90% ee. Hydrogen uptake was 240 L/h and on average the hydrogenation ended after three hours. The application of less catalyst was possible, but this resulted in an unwanted prolongation of the reaction time and a decrease in the enantioselectivity. However, further development of the process showed that a considerable increase in the substrate/catalyst ratio up to 20 000 was possible, without loss of yield and selectivity, by increasing the temperature to 50–60 °C at the end of the hydrogenation. Under these conditions the crystallization of the product during the hydrogenation could be avoided.

Of particular importance was the careful exclusion of traces of oxygen. For this purpose both the hydrogen and the inert gas nitrogen were purified over special contacts from Leuna (Leuna-652 or Leuna-7748-Katalysator [34]) and the residual oxygen was continuously controlled with a "Permolyt Ex" instrument from Junkalor, Dessau. A flow diagram of the technical hydrogenation plant and the purification of nitrogen and hydrogen was published by Vocke et al. [34].

At the end of the process the hydrogenation product was hydrolyzed with HBr (Fig. 13) and benzoic acid was recycled. The rhodium was recycled by adsorption on a cation exchanger KP2 from Wolfen (sulfonated polystyrene crosslinked by 2% divinylbenzene) [35]. 80% of the rhodium could be removed, the residual rhodium accumulated in the final mother liquor which contained only a very small amount of L-dopa. This mother liquor and the loaded ion exchanger were ashed and the resulting rhodium reacted with chlorine under heating to redness to give rhodium(III)-chloride. This was used directly for the synthesis of the dimer rhodium(I)-cyclooctadiene-chloride from which rhodium(I)-cyclooctadiene-acetylacetonate could easily be obtained.

$$\underset{\textbf{6d}}{\text{MeO}\text{-}C_6H_3(\text{OMe})\text{-}CH_2\text{-}\overset{\text{COOH}}{\underset{\text{NHCOPh}}{CH}}} \xrightarrow[\substack{- \text{PhCOOH} \\ - 2\,\text{MeBr}}]{\substack{+ 2\,\text{HBr} \\ + \text{H}_2\text{O}}} \underset{\text{L-dopa}}{\text{MeO}\text{-}C_6H_3(\text{OMe})\text{-}CH_2\text{-}\overset{\text{COOH}}{\underset{\text{NH}_2}{CH}}}$$

Fig. 13

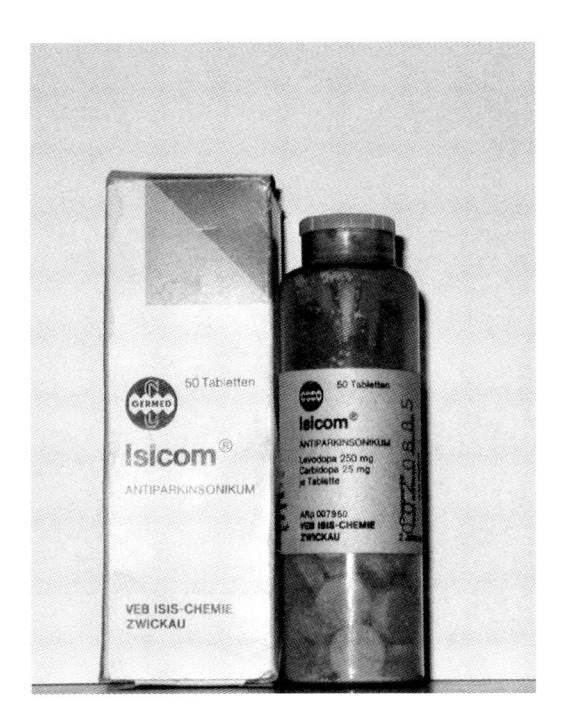

Fig. 14 The anti-Parkinsonicum Isicom.

Using the described process, roughly 1 ton per year of L-dopa was produced between 1985 and 1990. The first tablets were put on sale under the name of *Isicom* (Fig. 14) in January 1986 and were a combination of L-dopa and L-carbidopa [L-carbidopa = (S)-2-hydrazino-2-(3,4-dihydroxybenzyl)-propionic acid monohydrate] also prepared in Zwickau by a new process developed by Schmitz and Andreae [36, 37]. The presence of small amounts of carbidopa is very important, allowing a considerable minimization of the amount of L-dopa required for the effective treatment of the symptoms of the Morbus Parkinson.

An enlargement of the plant was envisioned because export into other socialistic countries was planned. However, production of L-dopa ended in 1990, one year after the collapse of the socialist system and one month after the management had actually decided to increase production capacity for L-dopa. The particular circumstances led to a general halt of all chemical manufacture by the successor companies of the former VEB ISIS-Chemie Zwickau!

2.8 Acknowledgements

I would like to express my deeply felt thanks to Prof. H. Pracejus[†] for development of the field of asymmetric catalysis and to the staff of the research department of the former VEB ISIS-Chemie in Zwickau, particularly Dr. W. Vocke and Dr. F.-U. Flöther, for the elaboration of the industrial L-dopa process. For practical assistance I particularly thank Mrs. R. Pinske, Mrs. H. Burneleit, DC H. Foken, Dr. H. Grüner, Dr. R. Hänel and DC G. König. I gratefully acknowledge the support of Prof. G. Oehme, Dr. C. Fischer and Dr. H. Dreyer[†] particularly with respect to analytical problems. I am indebted to Prof. H.-W. Krause for useful advice in many instances.

2.9 References

1 H. Pracejus, Fortschritte Chem. Forsch. 1967, 8, 493–553.
2 H. Pracejus, M. Bursian, DD 92 031 (1972).
3 H. Walzel, Dissertation, University Rostock, 1976.
4 W. S. Knowles, M. J. Sabacky, B. D. Vineyard, Chemtech. 1972, 2, 590–591. For the development of the L-dopa process, Knowles was honoured in 2001 with a share of the Nobel Prize for Chemistry; see Nobel lecture: W. S. Knowles, Angew. Chem. 2002, 114, 2096–2107; Angew. Chem. Int. Ed. 2002, 41, 1998–2007.
5 H. B. Kagan, T. P. Dang, J. Am. Chem. Soc. 1972, 6429–6433.
6 R. Selke, H. Pracejus, DD 140 036 (1978).
7 R. R. Schrock, J. A. Osborn, J. Am. Chem. Soc. 1976, 98, 2134–2143; 2143–2147.
8 R. Selke, React. Kinet. Catal. Lett. 1979, 10, 135–138.
9 G. H. Coleman, Meth. Carbohydr. Chem. 1963, 2, 397–399.
10 W. R. Cullen, Y. Sugi, Tetrahedron Lett. 1978, 1635–1636.
11 R. Jackson, D. J. Thompson, J. Organomet. Chem. 1979, 159, C29–C31.
12 R. Selke, Dissertation B, University Rostock 1986, pp. 50–52.
13 W. Kropp, H. Decker, Chem. Ber. 1909, 42, 1184–1192.
14 R. Selke, H. Pracejus, J. Mol. Catal. 1986, 37, 213–225.
15 R. Selke, J. Prakt. Chem. 1987, 329, 717–724.
16 R. Selke, M. Schwarze, H. Baudisch, I. Grassert, M. Michalik, G. Oehme, N. Stoll, B. Costisella, J. Mol. Catal. 1993, 84, 223–237.
17 R. Selke, M. Ohff, A. Riepe, Tetrahedron 1996, 52, 15079–15101.
18 T. V. RajanBabu, T. A. Ayers, A. L. Casalnuovo, J. Am. Chem. Soc. 1994, 116, 4101–4102.
19 R. Selke, J. Organomet. Chem. 1989, 370, 241–248.
20 R. Selke, E. Paetzold, H. Grüner, DD 243 189 (1981).
21 R. Selke, F.-U. Flöther, G. König, DD 275 623 (1988).
22 R. Selke, H. Foken, P. Winkler, DD 275 671 (1988).
23 R. Selke, J. Organomet. Chem. 1989, 370, 249–256.
24 G. Oehme, E. Paetzold, R. Selke, J. Mol. Catal. 1992, 71, L1–L5.
25 A. Kumar, G. Oehme, J. P. Roque, M. Schwarze, R. Selke, Angew. Chem. 1994, 106, 2272–2275; Angew. Chem., Int. Ed. Engl. 1994, 33, 2197–2199.
26 M. Ludwig, R. Kadyrov, H. Fiedler, K. Haage, R. Selke, Chem. Eur. J. 2001, 7, 3298–3304.
27 R. Selke, Enantiomers 1997, 2, 415–419.

28 R. Kadyrov, A. Börner, R. Selke, Eur. J. Inorg. Chem. 1999, pp. 705–711.

29 V. Fehring, R. Kadyrov, M. Ludwig, J. Holz, K. Haage, R. Selke, J. Organomet. Chem. 2001, 621, 120–129.

30 R. Selke, J. Mol. Catal. 1986, 37, 227–234.

31 R. Selke, K. Häupke, H. W. Krause, J. Mol. Catal. 1989, 56, 315–328.

32 R. Selke, M. Čapka, J. Mol. Catal. 1990, 63, 319–334.

33 H. W. Krause, F. Foken, H. Pracejus, New J. Chem. 1989, 13, 615–620.

34 W. Vocke, R. Hänel, F.-U. Flöther, Chem. Techn. 1987, 39, 123–125.

35 G. König, J. Börner, S. Vicenda, DD 298 231 (1981).

36 S. Andreae, E. Schmitz, S. Schramm, F.-M. Albert, D. Lohmann, DD 230 865 (1985).

37 S. Andreae, F.-M. Albert, F.-U. Flöther, E. Schmitz, Mitteilungsblatt Chem. Ges. DDR 1989, 36, 31–34; CA 1989, 110, 212 218s.

3
The Chiral Switch of Metolachlor: The Development of a Large-Scale Enantioselective Catalytic Process[1]

Hans-Ulrich Blaser, Reinhard Hanreich, Hans-Dieter Schneider, Felix Spindler, and Beat Steinacher

Abstract

The development of an enantioselective catalytic process for the large-scale production of (S)-metolachlor (trade name Dual Magnum®) is described. Four synthetic routes have been assessed and investigated. The key step for the final technical process is the asymmetric hydrogenation of an imine intermediate made possible by a novel Ir-ferrocenyl diphosphine catalyst. The various phases of the process development are described in some detail. Important issues were purity requirements for the starting materials, catalyst formulation, technical ligand synthesis, work-up procedure, separation of catalyst and reactor design. The resulting catalytic process has an exceptionally high efficiency: at a hydrogen pressure of 80 bar and a reaction temperature of 50 °C, more than 1 000 000 turnovers can be accomplished. In addition, with a capacity of >10 000 t/y, this is it at the moment the largest-scale enantioselective catalytic process.

3.1
Introduction and Problem Statement

Metolachlor is the active ingredient of Dual®, one of the most important grass herbicides for use with maize and a number of other crops. It is an N-chloroacetylated, N-alkoxyalkylated ortho-disubstituted aniline. The unusual functionalization pattern renders the amino function extremely sterically hindered. Metolachlor has two chiral elements: a chiral axis (atropisomerism, due to hindered rotation around the C_{Ar}-N axis) and a stereogenic center, leading to four stereoisomers (Fig. 1). Dual® was brought to the market in 1976, and contained a mixture of all four metolachlor stereoisomers produced via the Pt-catalyzed reductive alkylation of 2-methyl-5-ethyl-aniline (MEA) with aqueous methoxyacetone in the presence of traces of sulfuric acid followed by chloroacetylation (see Fig. 2) [1]. By 1982 it had already been shown that about 95% of the herbicidal activity of metolachlor resides in the two (1'S)-diastereomers [2]. In 1997, after years of intensive re-

---

[1] This is an amended version of a paper first published in Chimia 1999, 53, 275.

Asymmetric Catalysis on Industrial Scale: Challenges, Approaches and Solutions
Edited by H. U. Blaser, E. Schmidt
Copyright © 2004 Wiley-VCH Verlag GmbH & Co. KGaA, Weinheim
ISBN: 3-527-30631-5

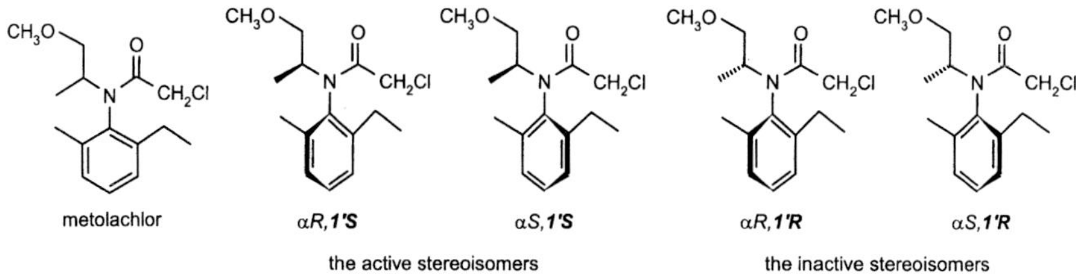

Fig. 1 Structure of metolachlor and its individual stereoisomers.

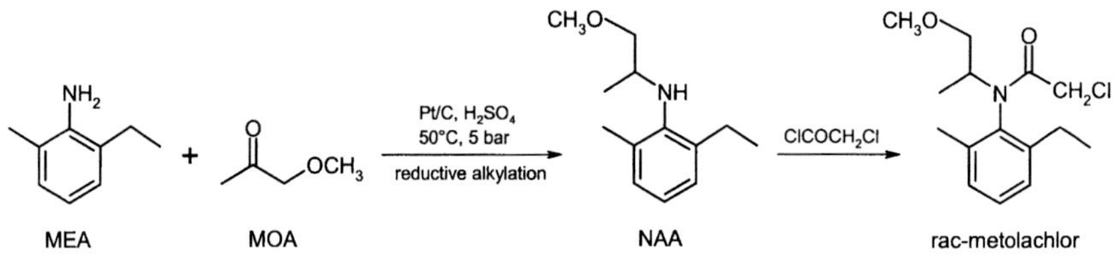

Fig. 2 The process for the industrial production of racemic metolachlor.

search [3], Dual Magnum® with a (1′S)-diastereomers content of approximately 90% and with the same biological effect at about 65% of the use rate of Dual® was introduced in the USA. To make this "chiral switch" possible, a new technical process had to be found for the economical production of the enantiomerically enriched precursor of metolachlor. This account describes what problems were encountered and how these were solved in the course of the development work.

3.2
Route Selection

Even though many possibilities exist for the enantioselective preparation of enriched (S)-metolachlor, it was clear from the beginning that because of the relatively low price and the large volume (>10 000 t/y) of the racemic product, only a catalytic route would be feasible. The following four synthetic routes were studied in some detail.

3.2.1
Enamide Hydrogenation

This idea clearly was inspired by the successful L-dopa process of Monsanto [4]. At that time, little was known about the effects of the substituents at the C=C bond and the amide nitrogen. A selective synthesis of one of the three possible enamide isomers depicted in Fig. 3 looked difficult.

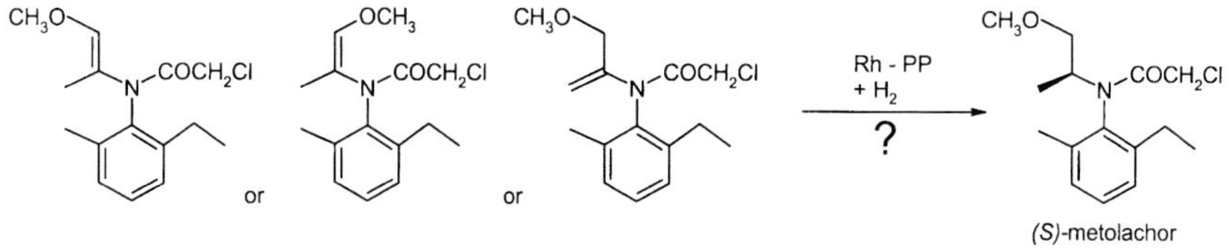

Fig. 3 Enamide hydrogenation: structures of tested enamides.

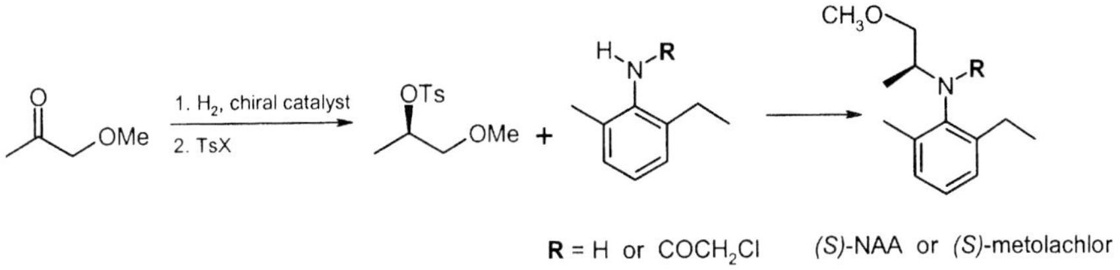

Fig. 4 Enantioselective hydrogenation of methoxyacetone and nucleophilic substitution with an MEA derivative.

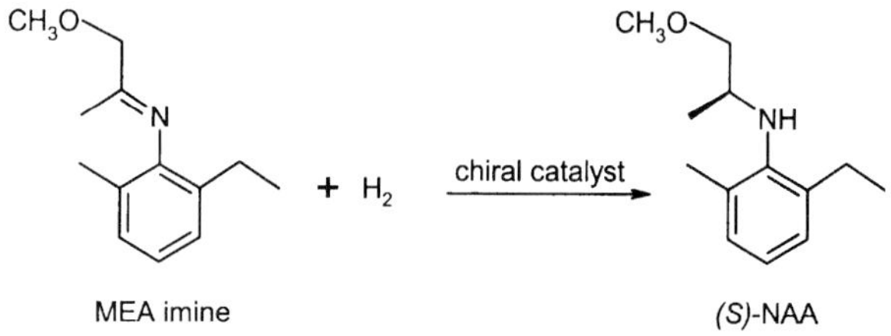

Fig. 5 Imine hydrogenation: structures of MEA imine and (S)-N-alkylated aniline.

3.2.2
Nucleophilic Substitution of an (R)-Methoxyisopropanol Derivative

Here, the key step was the enantioselective hydrogenation of methoxyacetone (Fig. 4), in analogy with the Pt-cinchona-catalyzed hydrogenation of α-ketoesters [5] (the Ru-binap system was not known at that time). It was anticipated that the nucleophilic substitution with a clean inversion would be difficult.

3.2.3
Hydrogenation of MEA Imine

Because the racemic metolachlor was produced via a reductive alkylation, it was obvious that hydrogenating the MEA imine intermediate should be tried (Fig. 5), either isolated or formed *in situ*. Unfortunately, at that time only one single imine hydrogenation had been described in the literature with an ee of only 22% [6].

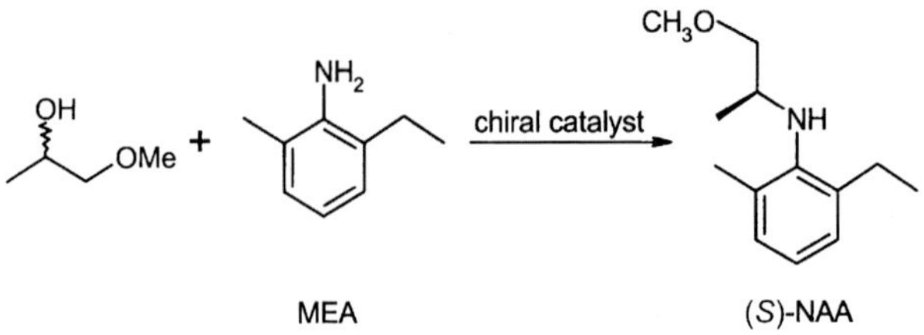

Fig. 6 Alkylation of MEA with methoxyisopropanol.

Tab. 1 Comparison of possible routes for the synthesis of (S)-metolachlor.

Route	Catalytic step	Other steps	Cost	Ecology	Priority
Enamide	close analogy ee >90%	enamide synthesis difficult	high	medium	1
Substitution	weak analogy ee >80%	substitution difficult	high	bad	2
Imine	weak analogy ee <30%	as in current process	medium	good	3
Direct alkylation	no precedent	as in current process	low	very good	4

3.2.4
Catalytic Alkylation with Racemic Methoxyisopropanol

This idea (Fig. 6) was based on an alternative process developed for the racemic product with heterogeneous catalysts in the gas phase [7] and some results of the N-alkylation of aliphatic amines with primary alcohols using homogeneous Ru phosphine catalysts [8].

3.2.5
Assessment and Screening of the Proposed Routes

In order to assess the potential synthetic routes, the following criteria were considered to be important:

- chances of success for the catalytic step according to precedents, i.e., closely related, efficient catalytic transformations
- number and perceived difficulty of the non-catalytic steps
- first approximations for costs and ecology of the over-all synthesis

In Tab. 1 the four proposed routes are classified according to these criteria. The over-all ranking was used for setting the priorities to carry out experimental work. Because the enantioselective catalysis is usually considered to be the most difficult step, its chances of success very often dominate the decision-making process and accordingly, the enamide and the substitution route were tested first.

Enamide Route
The preparation of the three MEA enamides proved to be rather difficult. Disappointingly, we did not succeed in hydrogenating any of the three isomers using seven different Rh-diphosphine complexes at normal pressure and temperatures up to 50 °C.

Substitution Route
The hydrogenation of methoxyacetone was somewhat more successful: using a Pt/C catalyst modified with cinchonidine as described by Orito et al. [5] (R)-methoxyisopropanol was obtained in good yields, but the ee values were never higher than 12%.

Direct Alkylation
This was not tested experimentally, because chances for success were considered to be too low.

Conclusion
The results of the route screening left the hydrogenation of the MEA imine as the only realistic possibility.

3.3
Imine Hydrogenation: Laboratory Process

3.3.1
Finding the Right Metal-Ligand Combination

The history of the development of a technically feasible catalyst for the enantioselective hydrogenation of MEA imine has been described [3]. Collaborations, initially with a research team from the University of British Columbia at Vancouver and later with the group of J. A. Osborn of the University of Strasbourg were very important.

3.3.1.1 Screening of Rh-Diphosphine Complexes
The first positive results were obtained by trying to adapt Rh-diphosphine catalysts that had originally been developed for the hydrogenation of olefins. An extensive ligand screening led to [Rh(nbd)Cl]$_2$/cycphos (ligand structures see Fig. 7) as

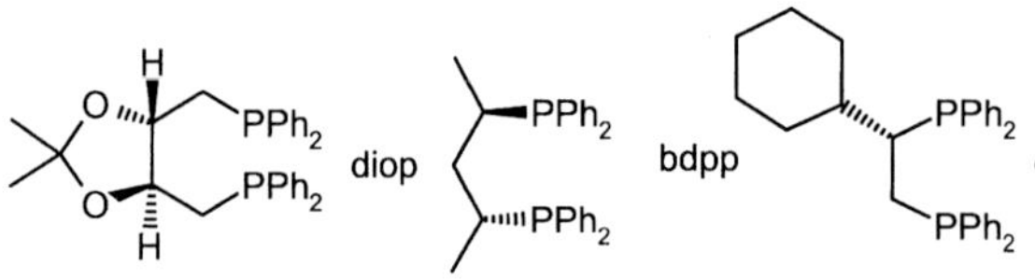

Fig. 7 Imine hydrogenation: structure of important ligands.

the best catalyst: 69% ee was achieved at –25 °C, the best turnover frequency (tof) was 15/h at 65 bar and room temperature, far too low for any industrial application [9]. Nevertheless, these results represented remarkable progress in the enantioselective hydrogenation of N-aryl imines.

3.3.1.2 Screening of Ir-Diphosphine Complexes

The next breakthrough was obtained when iridium was used instead of rhodium. This idea was inspired by results from Crabtree who had described an extraordinarily active Ir-tricyclohexylphosphine-pyridine catalyst that was able to hydrogenate even *tetra*-substituted C=C bonds. For the MEA imine hydrogenation very good ee values were obtained with an Ir-bdpp catalyst in the presence of iodide ions (ee 84% at 0 °C) but the activity was disappointing. Turnover numbers (ton) of up to 10000 and tof numbers of 250/h (100 bar and 25 °C) but somewhat lower ee values were obtained with Ir-diop-iodide catalysts [10, 11]. A major problem with these new Ir-diphosphine catalysts was an irreversible catalyst deactivation.

These results, in particular the good enantioselectivities, were very promising and represented by far the best catalyst performance for the enantioselective hydrogenation of imines at that time. Nevertheless, it was also clear that the ambitious goals could probably not be reached using Ir complexes with "classical" diphosphine ligands. Even though Ir-diop and Ir-bdpp catalysts showed much higher activities than the best Rh complexes for MEA imine, they were still far below the requirements. A new approach was clearly required.

3.3.1.3 Synthesis and Screening of a New Ligand Class

As a consequence, new ligand types were tested, among others the novel ferrocenyl diphosphines (PPF) developed by Togni and Spindler [12]. Their mode of preparation (see Fig. 8) allows an efficient fine tuning of the electronic and steric properties of the two phosphino groups, something that is often difficult with other ligand classes. Indeed, the Ir complexes of such diphosphines proved to be very efficient. In particular PPF-P[3,5-$(CH_3)_2C_6H_3]_2$ (R=Ph, R'=3,5-xylyl), known as xyliphos, turned out to give an exceptionally active catalyst and, even more importantly, it did not deactivate!

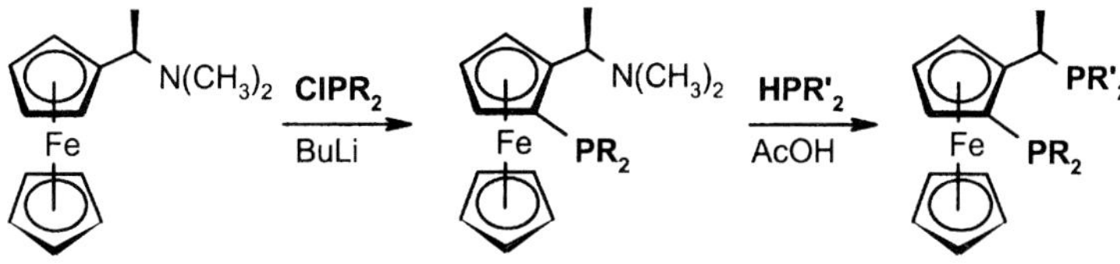

Fig. 8 Preparation and structure of ferrocenyl diphosphine ligands.

Tab. 2 MEA imine hydrogenation with selected Ir-ferrocenyl diphosphine complexes (formulas see Fig. 8)

R	R'	Ton	tof (per h)	ee	Comments
Ph	3,5-xylyl	1 000 000	>200 000	79	production process
p-CF$_3$C$_6$H$_4$	3,5-xylyl	800	400	82	ligand screening
Ph	4-^{t}Bu-C$_6$H$_4$	5 000	80	87	low temperature
Ph	4-(^{n}Pr)$_2$N-3,5-xyl	100 000	28 000	83	optimized conditions

3.3.2
Optimization of Reaction Medium and Conditions

Using xyliphos as the ligand, a screening of solvents and additives as well as an optimization of the reaction conditions was carried out. Most remarkable was the effect observed when 30% acetic acid was added to the reaction mixture of MEA imine and Ir-xyliphos-NBu$_4$I: a rate increase by a factor of 5 was observed while the time for 100% conversion was more than 20 times shorter than without additives. The effects of pressure and temperature were investigated in the presence of acid and iodide. The reaction rate was approximately proportional to the hydrogen pressure and also increased with temperature, ee values decreased from 81% at –10 °C to 76% at 60 °C but were not affected by changing the hydrogen pressure.

3.3.3
Ligand Fine Tuning

As described above, the Ir-xyliphos catalysts showed extremely high catalyst activities and productivities. On the other hand, the enantioselectivity to the desired *S*-enantiomer just barely met the requirements. Therefore, we tried to improve the ee values by tuning of the electronic and steric properties of the new ferrocenyl ligands. As shown in Tab. 2, it was indeed possible to increase the selectivity of the catalyst, however, as observed earlier with other ligands, any gain in selectivity was always offset by a loss in catalyst activity and often productivity. In the end, xyliphos was the best compromise as regards activity and selectivity for a technical process.

3.4
Imine Hydrogenation: Technical Process

Once a catalyst system with the required performance had been found and confirmed, the attention turned to finding a technically feasible overall process. The technical preparation of methoxyacetone and 2-methyl-6-ethyl-aniline as well as the chloroacetylation step had already been established in the existing process for racemic metolachlor.

3.4.1
Strategy for Process Development

For the production of enriched (S)-NAA the reductive alkylation step in the original process had to be replaced by a condensation reaction, followed by isolation and purification of the imine and a subsequent homogeneous asymmetric hydrogenation at high pressure (80 bar and 50 °C). As outlined above, a catalyst system had been developed that was able to fulfill the minimal requirements to make a process commercially feasible: substrate to catalyst ratio (s/c) >100 000, reaction time <8 h for >99% conversion and enantioselectivity ≥80%. The selected catalyst system was a mixture of four components: a dimeric iridium cyclooctadiene complex $[Ir(cod)Cl]_2$, the xyliphos ligand, tetrabutyl ammonium iodide as the iodide source and acetic acid as the preferred acid at this stage. During the process development, an optimization led to a simpler catalyst system with only three components by using HI as both the iodide and proton source. Because of the limited time available for the development of a definitive process, it was decided to change as few parameters as possible and to focus development activities on the following topics: purity requirements of the starting materials, catalyst formulation, ligand synthesis, work-up procedure, separation of the catalyst and reactor design.

3.4.2
The Production of the MEA Imine at the Required Quality

Surprisingly, the seemingly simple condensation of MEA with methoxy acetone (Fig. 9) turned out to be quite tricky: significant side product formation was observed when trying to force the conversion of the reaction to 100%. When different qualities of MEA-imine were tested, the reproducibility of the results was very poor. Sometimes some crude samples gave surprisingly good results when hydrogenated, while distillation did not always lead to an improvement of catalyst activity. In the end it was concluded that depending on the composition of the side product spectrum and as a result of the thermal instability of the imine and its sensitivity to air and moisture, a significant deactivation of the catalyst could result. With an efficient separation of the water formed during condensation of methoxyacetone (see Fig. 10) and MEA and a rather complicated multi-step continuous distillation process for the purification of MEA imine, recovery of solvent

3.4 Imine Hydrogenation: Technical Process

Fig. 9 Condensation of MOA with MEA.

Fig. 10 The two water separators for the removal of water during the condensation reaction.

and non-reacted starting materials, an excellent imine quality was provided for the subsequent enantioselective hydrogenation step.

3.4.3
Scale-up of the Ligand Synthesis

At the present stage of the application of enantioselective homogeneous catalysis, very few chiral ligands are commercially available on a kilogram scale or larger. This means that the development of a technical ligand synthesis must be part of the over-all development process. In the case of the xyliphos ligand this meant that a 6-step synthesis (see Fig. 11) had to be scaled-up from a laboratory process for making gram amounts to a commercial process producing hundreds of kilo-

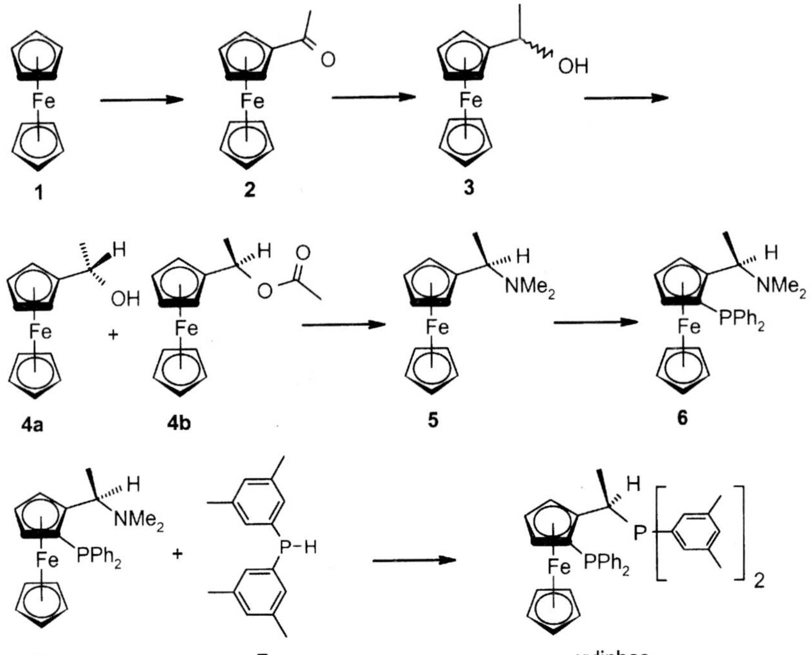

Fig. 11 The technical 6-step synthesis of xyliphos.

grams of ligand in a reproducible form and quality. Again the starting point was the original synthesis and the experience that had been gained a few years earlier for the synthesis of a similar ferrocenyl diphosphine, bppfoh, by a development team of the Pharma division of Ciba-Geigy [13].

The challenge for the ligand synthesis team was on the one hand to find an economical and technically feasible process giving a high quality of xyliphos, while at the same time providing xyliphos of a constant quality during all phases of the development of the MEA imine hydrogenation. The synthetic strategy was similar to one developed for the synthesis of the bppfoh ligand [13]. The starting material for the synthesis of xyliphos was ferrocene **1**. Acetylation of ferrocene **1** gave racemic acetylferrocene **2** which was reduced to the racemic alcohol **3**. This alcohol needed a careful work-up to obtain the appropriate quality for the subsequent enzymatic kinetic resolution via a lipase-catalyzed acetylation reaction [14]. For the conversion of **3** to **4b** fine optimization of reaction temperature, quantity of enzyme and reaction time was crucial to achieve the optimum selectivity. Lipases from different suppliers showed similar selectivities but had significant differences in activity. During work-up, special care had to be taken because of the thermal instability of **4b**. The enriched dimethylamino compound **5** was obtained by reacting a mixture of **4a** and **4b** with dimethylamine, whereby the acetate group was replaced with retention of configuration. Compound **5** was converted into intermediate **6** by reaction with *n*-butyl lithium and chloro-diphenylphosphine. After crystallization of **6**, ee values of >99.5% were obtained. Reaction of **6** with bis(3,5-dimethylphenyl)-phosphine **7**, which is now commercially available, gave the final product xyliphos in high purity.

This synthesis was carried out in reactors of up to 2500 L and is feasible for the preparation of xyliphos in quantities of hundreds of kilograms. In order to run an economical process it was crucial to define the most important parameters, to optimize these and to have them under good control at the production scale.

3.4.4
Fine Optimization of the Ir-Catalyst Formulation

The use of a solid multi-component catalyst mixture was challenging, especially because catalyst addition to a high-pressure autoclave is usually time consuming and slows down the cycle time of the process. An advantage of the solid catalyst used at the beginning was the slow release of the catalyst activity due to the low dissolution rate of the components and, as a consequence, it was easy to control the exothermicity of the hydrogenation reaction. However, in cases of incomplete conversion, e.g., because of catalyst deactivation, new catalyst had to be added and that was difficult with a solid catalyst. Therefore, our attention was focused on the development of a liquid catalyst formulation, which would allow easy addition to the reaction vessel whenever necessary. Many attempts to work with catalyst solutions failed due to the instability of the dissolved catalyst: only freshly prepared solutions could be used. In the end, a liquid, highly active catalyst formulation was developed which was stable over several months. Now it was possible to feed the catalyst safely and easily into the hydrogenation reactor at any stage of the reaction. After catalyst addition full activity was available immediately so that cycle time and catalyst amount could be further optimized.

3.4.5
Choice of Reactor Technology

Laboratory experiments had shown that the enantioselectivity in the hydrogenation of the MEA imine was mainly influenced by the temperature, whereas hydrogen pressure only had a significant effect on the reaction rate. In the pilot trials it was confirmed that rate and selectivity of the reaction reach their optimum at 50 °C and 80 bar. Because under these conditions more than 70% of the reaction takes place within the first hour, control of reaction temperature could only be achieved using large external heat exchangers. For optimal mass and heat transfer a loop reactor (see Fig. 12) was therefore the best choice. In this technology, the reaction mixture is pumped via a heat exchanger through a nozzle where hydrogen is fed into the reaction solution, allowing both very good mixing and the use of the appropriate exchange surface.

Fig. 12 Bottom part of loop reactor.

3.4.6
Scale-Up to the Production Autoclave

The scale-up factor of the reaction from laboratory to production was >10 000. Laboratory experiments for screening and optimization were run in 50 mL up to 1 L high pressure autoclaves. Owing to the small amount of catalyst necessary and the high sensitivity of the hydrogenation to impurities in the starting materials, reproducibility of experimental results was a critical factor and a big challenge for the experimental skills of the technicians. For the design of the new production unit, valuable experience was gained during the pilot trials. In some cases results obtained in the pilot plant were much better than those from the laboratory; the discovery of the high performing liquid catalyst system would have been very unlikely without these trials. Under optimized conditions it was possible to reduce the catalyst amount significantly to a molar s/c ratio of 2 000 000. The new, ready to use catalyst solution proved its outstanding performance and pushed enantioselective hydrogenation into new dimensions. During these investigations, use of on-line NIR (near infra red) and polarimetry was very helpful for monitoring the conversion and selectivity of the enantioselective reaction.

Fig. 13 Bottom part of thin film evaporator.

3.4.7
Work-up, Separation of the Catalyst from the Product

The following three separation methods of product from the Ir-catalyst were evaluated: distillation, extraction and filtration. For the last two options the preparation of new modified extractable or immobilized xyliphos ligands was necessary. However, lower activity and selectivity of these xyliphos derivatives and the additional development work that would have been required led to the decision to stay with the already well optimized soluble xyliphos system. After the hydrogenation step, a continuous aqueous extraction is performed to neutralize and eliminate the acid from the crude product. After flash distillation to remove residual water the catalyst is separated from (S)-NAA in a subsequent distillation on a thin film evaporator (see Fig. 13). From the organic distillation residue, Ir could in principle be recovered whereas the chiral ligand decomposes. Owing to the very low catalyst concentrations, Ir recovery is not economical.

Tab. 3 Milestones in the history of (S)-metolachlor.

1970	Discovery of the biological activity of rac-metolachlor (patent for product and synthesis)
1978	Full-scale plant for the production of rac-metolachlor in operation (capacity >10 000 t/y)
1982	Synthesis and biological tests of the four stereoisomers of metolachlor
1983	First unsuccessful attempts to synthesize (S)-metolachlor via enantioselective catalysis
1985	Rhodium-cycphos catalyst gives 69% ee for the imine hydrogenation (UBC Vancouver)
1986	Discovery of new iridium diphosphine catalysts that are more active and selective than Rh catalysts for the hydrogenation of MEA imine
1993	Ir-ferrocenyl diphosphine catalysts and acid effect are discovered. Process development starts
1993/4	Patents for rac-metolachlor expire
1995/6	Pilot results for (S)-metolachlor: ee 79%, ton 1 000 000, tof >200 000/h, first 300 t produced
1996	Full-scale plant for production of >10 000 t/y (S)-metolachlor starts operation
2003	Since the start of production >50 000 tons of (S)-NAA have been produced without major problems and without the need for further optimization

3.5
Summary and Conclusions

Tab. 3 gives an overview of the timetable and the milestones for the development of a technical process for the production of enriched (S)-metolachlor. It took so many years to reach the ambitious goal because the efforts to find a suitable catalyst system for the enantioselective imine hydrogenation had to start almost from zero. The final "result" of our efforts can be seen in Fig. 14 which depicts a partial view of the production unit.

The case of (S)-metolachlor allows some generalized conclusions.

- The chiral switch from the racemate to an enriched form is attractive not only for pharmaceuticals but also for agrochemicals, and enantioselective hydrogenation is a particularly suitable and commercially feasible technology to allow this.
- The activity of the catalyst and not necessarily its enantioselectivity was the major problem to be solved, and an appreciable amount of patience and intuition of the chemists involved, as well as some luck, were necessary to reach the challenging goal.
- The selection of the catalytic system was especially difficult because the required catalyst performance was very ambitious and very little was known about enantioselective imine hydrogenation.

Fig. 14 View of the production unit with hydrogen generator and 200 bar storage in the foreground.

3.6
Acknowledgements

Obviously, the success in finding and implementing a technically feasible process for the commercial production of (S)-metolachlor is due to a large number of people. We acknowledge all of their contributions and achievements and thank all of them for their commitment to the project.

3.7
References

1 R. R. Bader, H. U. Blaser, Stud. Surf. Sci. Catal. 1997, 108, 17. R. R. Bader, P. Flatt, P. Radimerski, EP 605363-A1 (Ciba-Geigy AG, 1992).
2 H. Moser, G. Ryhs, H. P. Sauter, Z. Naturforsch. 1982, 37b, 451.
3 For a case history of the discovery of the new catalyst system see F. Spindler, B. Pugin, H. P. Jalett, H. P. Buser, U. Pittelkow, H. U. Blaser, Chem. Ind. (Dekker) 1996, 68, 153.
4 D. Vineyard, W. Knowles, M. Sabacky, G. Bachmann, D. Weinkauf, J. Am. Chem. Soc. 1977, 99, 5946.
5 Y. Orito, S. Ima, S. Niwa, J. Chem. Soc. Jpn. 1979, 1118; 1980, 670 and 1982, 137.
6 A. Levi, G. Modena, G. Scorrano, J. Chem. Soc., Chem. Commun. 1975, 6.
7 M. Rusek, Stud. Surf. Sci. Catal. 1991, 59, 359.
8 Y. Watanabe, Y. Tsuji, H. Ige, Y. Ohsugi, T. Ohta, J. Org. Chem. 1984, 49, 3359 and references cited therein.
9 W. R. Cullen, M. D. Fryzuk, B. R. James, J. P. Kutney, G.-J. Kang, G. Herb, I. S. Thorburn, R. Spogliarich, J. Mol. Cat. al. 1990, 62, 243.
10 Y. Ng Cheong Chan, J. A. Osborn, J. Am. Chem. Soc. 1990, 112, 9400.
11 F. Spindler, B. Pugin, H. U. Blaser, Angew. Chem., Int. Ed. Engl. 1990, 29, 558.

12 A. Togni, C. Breutel, A. Schnyder, F. Spindler, H. Landert, A. Tijani, J. Am. Chem. Soc. 1994, 116, 4062.

13 H. U. Blaser, R. Gamboni, G. Sedelmeier, B. Schaub, E. Schmidt, B. Schmitz, F. Spindler, Hj. Wetter in Approaches to Pharmaceutical Process Development (K.G. Gadamasetti, Ed.), 1999, p. 189.

14 A. Wickli, E. Schmidt, J.R. Bourne, in Biocatalysis in Non-Conventional Media (J. Tramper, M. Vermüe, H. Beeftink, U. von Stockar, Eds.), Elsevier, Amsterdam, 1992, p. 577.

4
Enantioselective Hydrogenation:
Towards a Large-Scale Total Synthesis of (R,R,R)-α-Tocopherol

Thomas Netscher, Michelangelo Scalone and Rudolf Schmid

Abstract

(2R,4'R,8'R)-α-Tocopherol, the compound possessing highest vitamin E activity, is an economically important product due to its specific biological and antioxidant properties. For the generation of the two chiral centers in the isoprenoid sidechain, homogeneous asymmetric hydrogenation of allylic alcohols by ruthenium diphosphane catalysts was used as the key technology. Based on the fundamental findings of Takaya and Noyori et al., very efficient processes applicable on a production scale of the chiral building blocks on the C_{10} and C_{15} stages were established. Excellent results with regard to chemo- and enantioselectivity (yields and ee/de (diastereo-isomeric excess) values >98%) were obtained with BIPHEP-type ligands with very high substrate-to-catalyst ratios (up to 150 000). Special emphasis was placed on the careful consideration of the requirements for technical feasibility of the process. In addition, details for the large-scale preparation of the most successful chiral ligands and the determination of the optical purity of products are given. Impediments concerning the realization of a complete total synthesis of (R,R,R)-α-tocopherol are also discussed.

4.1
Introduction: Vitamin E as the Target

Homogeneous asymmetric catalysis has become a key technology for the efficient preparation of isoprenoid building blocks which are of importance for possible large-scale syntheses of various chiral lipid-soluble natural products [1]. A prominent representative is (2E,7R,11R)-phytol (**1**, Fig. 1), ubiquitously occurring in nature as a constituent of chlorophyll in green plants. The C_{20}-moiety possessing the same stereochemistry is found in vitamin K_1 (**2**) [2, 3] and in the components of vitamin E, such as the tocopherols **3–6** and the tocotrienols **7–10**. (2R,4'R,8'R)-α-Tocopherol (**3**), the main component of vitamin E discovered in the 1920s [4–6], is a product of particular value. Therefore, some introductory remarks concerning definitions, stereochemistry, biological role, and commercial significance of vitamin E [7–9] will be given.

Asymmetric Catalysis on Industrial Scale: Challenges, Approaches and Solutions
Edited by H.U. Blaser, E. Schmidt
Copyright © 2004 Wiley-VCH Verlag GmbH & Co. KGaA, Weinheim
ISBN: 3-527-30631-5

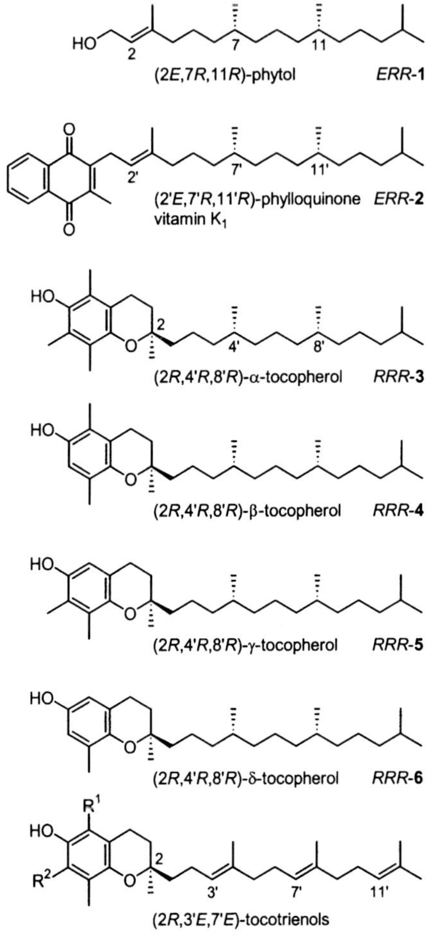

Fig. 1 Structures of natural phytol, vitamin K₁, tocopherols, and tocotrienols.

(2E,7R,11R)-phytol — ERR-1

(2′E,7′R,11′R)-phylloquinone ERR-2
vitamin K₁

(2R,4′R,8′R)-α-tocopherol — RRR-3

(2R,4′R,8′R)-β-tocopherol — RRR-4

(2R,4′R,8′R)-γ-tocopherol — RRR-5

(2R,4′R,8′R)-δ-tocopherol — RRR-6

(2R,3′E,7′E)-tocotrienols

R^1	R^2	prefix	
CH₃	CH₃	α-	REE-7
CH₃	H	β-	REE-8
H	CH₃	γ-	REE-9
H	H	δ-	REE-10

The great economic importance of vitamin E as an essential food ingredient [10–12] is due to its specific biological activity and antioxidant properties [13–15]. While the term vitamin E covers all tocol and tocotrienol compounds and derivatives (in whatever stereoisomeric form) exhibiting qualitatively the biological activity of α-tocopherol [16], the naturally occurring components (tocopherols **3–6** and tocotrienols **7–10**) are single-isomer products as specified in Fig. 1. The determination of the biological activity by the fetal resorption-gestation test in rats has shown that (2R,4′R,8′R)-α-tocopherol (*RRR*-**3**) has the highest value of the eight naturally occurring compounds *RRR*-**3** to *RRR*-**6** and *REE*-**7** to *REE*-**10**, and of the eight possible stereoisomers of α-tocopherol [17–20].

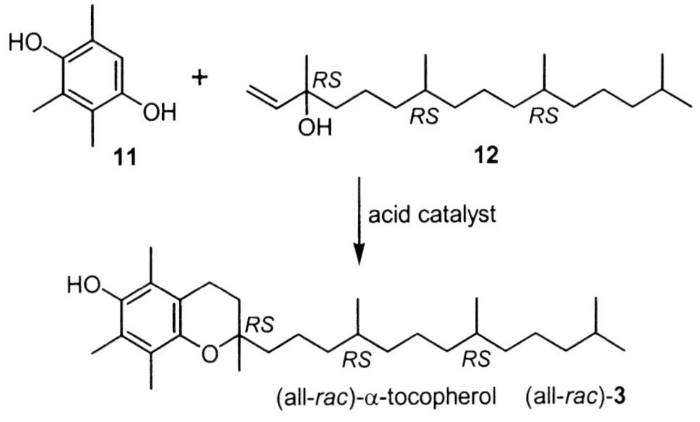

Fig. 2 Industrial synthesis of (all-*rac*)-α-tocopherol.

Based on these biological data, two commercial forms of α-tocopherol (or their more stable acetate derivatives) are currently being produced by independent approaches [7, 8]. Totally synthetic vitamin E, which is an equimolar mixture of all eight stereoisomers of α-tocopherol, is produced at a rate of over 25 000 tons per year for the application in feed, food, and the pharma industry. The large-scale industrial synthesis of (all-*rac*)-α-tocopherol uses 2,3,5-trimethylhydroquinone (**11**) as the aromatic building block and the C_{20} compound isophytol (**12**). The acid-catalyzed condensation reaction in the last step delivers (all-*rac*)-**3** (Fig. 2) [21–25].

About one tenth of the total amount of vitamin E produced is semi-synthetic, isomerically pure (2*R*,4′*R*,8′*R*)-α-tocopherol (*RRR*-**3**) which is used almost exclusively in human applications (mainly pharma). This product, originating from natural sources, is obtained from soya deodorizer distillates (SDD). Vegetable oils refined on a large scale are the major sources of vitamin E compounds [26–28]. The deodorizer distillate, originally a waste stream, contains considerable amounts (up to 10%) of α-, β-, γ- and δ-tocopherols (*RRR*-**3** to *RRR*-**6**) which are isolated by several separation methods. To increase the value of the vitamin E concentrate of mixed tocopherols obtained from SDD, the lower β-, γ- and δ-homologues (*RRR*-**4** to *RRR*-**6**, content ca. 90%) have to be transformed subsequently into the biologically more active α-tocopherol (*RRR*-**3**, only ca. 5% in the original mixture) by permethylation reactions. Permethylations are performed by chloro-, amino-, or hydroxymethylation reactions to provide functionalized alkylated intermediates, which are reductively converted into *RRR*-**3** (Fig. 3) [29].

When our company planned to enter the market of vitamin E from natural sources, two major issues had to be addressed. First of all, large amounts of SDDs had to be made available as the starting material for the process. This was realized by a collaboration with Cargill Inc. [30]. Secondly, the permethylation procedures known at that time (that is in the early 1990s) had several serious drawbacks. After intensive research and development work we now perform such processes on an industrial scale in high yield and selectivity and with recyclable reagents [31–33].

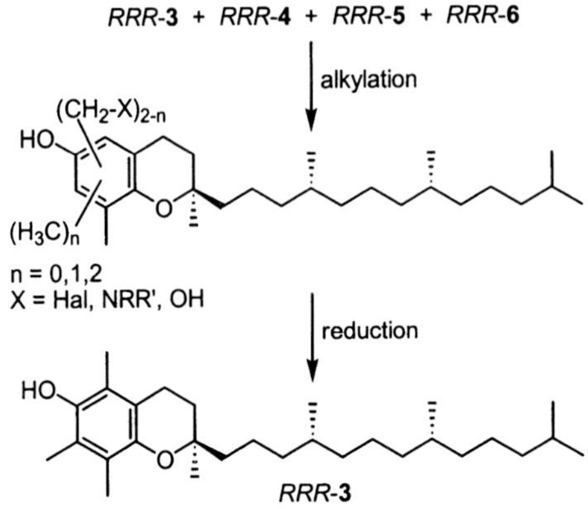

Fig. 3 Industrial synthesis of *RRR*-3 from natural-source material by permethylation.

4.2
Routes to (2*R*,4′*R*,8′*R*)-α-Tocopherol by Total Synthesis

A general problem remains when a semi-synthetic approach for the synthesis of α-tocopherol is followed: owing to the given limited availability of starting material from natural sources, an increasing demand for (2*R*,4′*R*,8′*R*)-α-tocopherol can only be satisfied to a certain level. Additional amounts of this product could be made available by an economical total synthesis. Therefore, over the last three decades considerable efforts have been directed towards the development of stereoselective syntheses of *RRR*-3 and of the corresponding building blocks. General routes entailed classical optical resolution, biocatalysis (by microorganisms and isolated enzymes), the use of chiral-pool starting materials, the application of stoichiometric and catalytic amounts of chiral auxiliaries, as well as asymmetric catalysis. For large-scale applications, many of these methods (for reviews cf. [34, 35]) suffer from their complexity, limited space-time yield, and formation of excessive amounts of waste material.

Economical as well as ecological factors determine the success of processes in the competitive field of vitamins and fine chemicals. In this regard, asymmetric homogeneous catalysis with metal complexes has, in general, a unique potential to fulfill the requirements of highly selective and efficient transformations [36]. In recognition of this fact, the 2001 Nobel Prize in Chemistry was awarded to W. S. Knowles [37], R. Noyori [38], and K. B. Sharpless [39] for their seminal contributions [40–42]. In particular, enantioselective catalytic isomerization (for example allylamine → enamine [43–48]) and hydrogenation reactions [49–56] have been shown to be key technologies enabling exceptionally promising strategies for the vitamin and fragrance and flavor, as well as the pharma industry. As is already the case with the Monsanto L-dopa process [37, 57] and the large-scale Rh(I)-catalyzed synthesis of citronellal and menthol [1 b, 52], the reactions mediated by ruthenium diphosphane catalysts represented a breakthrough in the synthesis of chiral terpenoid compounds. Selected transformations for the construction of the chiral isoprenoid side-chain of *RRR*-3, already realized with high efficiency, are illustrated in Fig. 4.

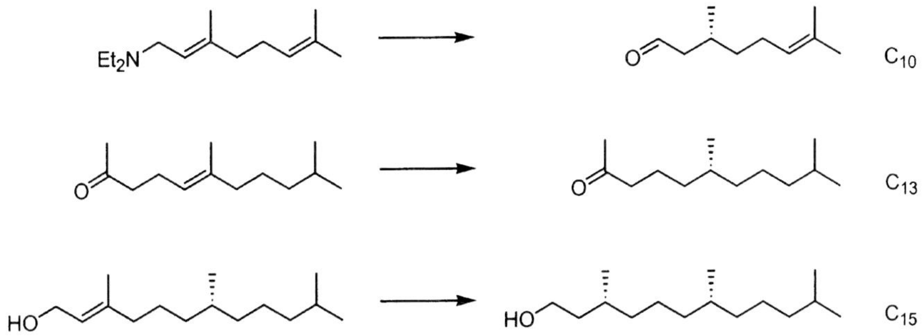

Fig. 4 Selected examples of catalytic asymmetric reactions as part of the preparation of *RRR*-**3**.

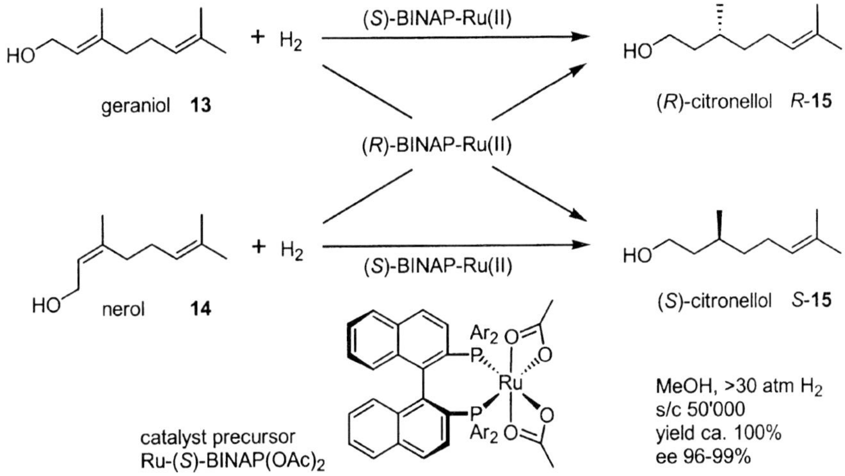

Fig. 5 Ru-BINAP enantioselective hydrogenation of geraniol/nerol by Takaya and Noyori et al. [58, 59].

Important features for industrial applicability of such procedures are high reaction rates and the possibility to work with a high substrate concentration in organic solvents and high molar substrate-to-catalyst ratios (s/c ratios). Besides the suitable combination of reaction conditions, the selection of the chiral organometallic complex is a key factor for developing high-performance processes. The prototype diphosphane ligand BINAP showed exceptionally high chemo- and enantioselectivity in the ruthenium-catalyzed preparation of both enantiomers of citronellol (**15**) from geraniol (**13**) and nerol (**14**) (Fig. 5) [58, 59]. In our work a large variety of C_2-symmetric diphosphanes of the biphenyl series (BIPHEP ligands) were tested in the allylic alcohol hydrogenations. Among them, the BIPHEMP [46] and MeOBIPHEP [60] ligands (cf. Fig. 8) were found to be at least as efficient as the original BINAP ligand [1 b, 61–63].

4.3
Synthesis of Prochiral Allylic Alcohols

The two chiral centers (4′R and 8′R) of the side-chain of (2R,4′R,8′R)-α-tocopherol (RRR-3) are both generated by asymmetric hydrogenation. Good access to prochiral primary (E)-allylic alcohols as starting materials for these reactions is one requirement for a technically feasible process. Excellent chemical and also isomeric purities of the substrates are prerequisites for achieving high (>98%) ee values. The preparation of such high-quality prochiral substrates, in particular of the C_{15} building block (2E,7R)-tetrahydrofarnesol (E-19) on the way to (3R,7R)-3,7,11-trimethyldodecanol (20), is a task that cannot be considered negligible in the overall synthesis of RRR-3.

A typical lab-scale sequence for a C_5 homologation [16 (C_{10}) → 19 (C_{15})] is depicted in Fig. 6. This requires several steps with a variety of reagents in at least equimolar amounts. E/Z-separation was performed at the stage of the α,β-unsaturated ester 17 by column chromatography on silica gel to arrive finally at isomerically pure (2E,7R)-6,7,10,11-tetrahydrofarnesol (E-19). This homologation was performed on a several-kg scale. However, for a preparation on a larger scale, steps causing extensive waste formation had to be avoided. Accordingly, a homologation sequence based entirely on catalytic transformations was developed (Fig. 7). It eliminated (or at least substantially reduced) the use of halides, excess of strong bases, complex hydrides, and, consequently, salt formation. For the start of this homologation sequence (Fig. 7), catalytic gas phase dehydrogenation was used to

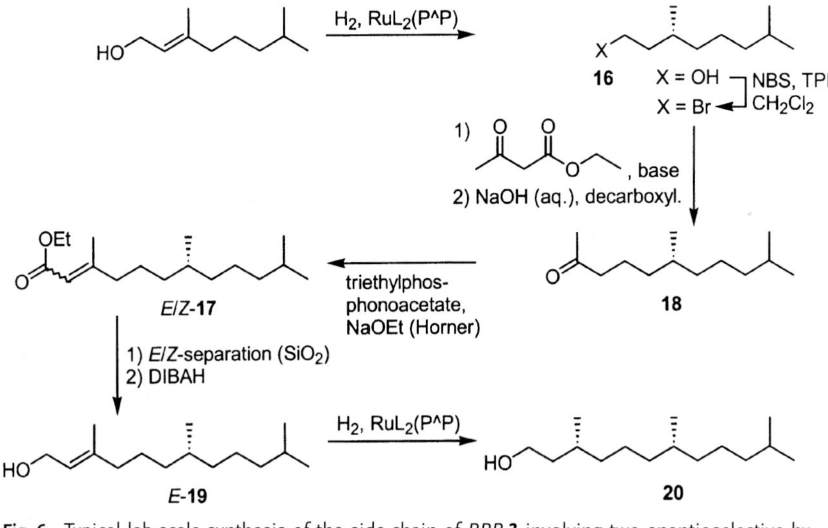

Fig. 6 Typical lab-scale synthesis of the side-chain of RRR-3 involving two enantioselective hydrogenation steps.

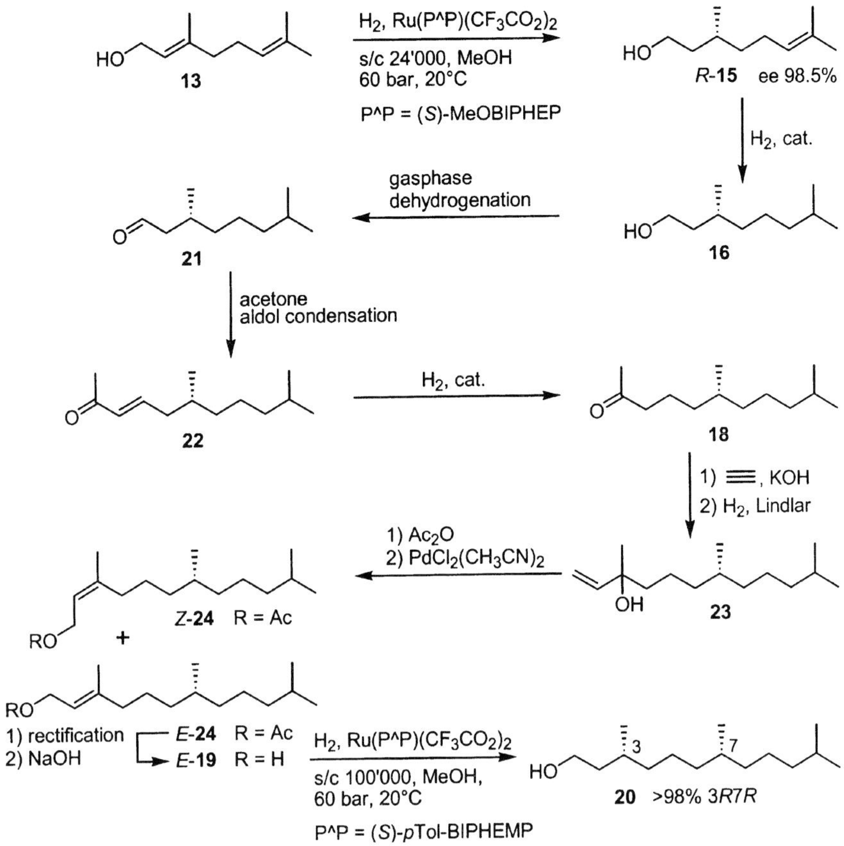

Fig. 7 Large-scale preparation of **20**, C$_{15}$ side-chain building block in the route to *RRR*-**3**.

activate (*R*)-dihydrocitronellol (**16**). This reaction, performed on a 50-kg scale, was run in the presence of a brass catalyst at ca. 470 °C and 100 to 200 mbar with a selectivity of ca. 95% at a conversion of around 85%. It is important to note that the ee value of isolated aldehyde **21** (purity > 99.5%) and of starting material **16** recovered for recycling after continuous rectification decreased only slightly from 98.6 to 98.4 ± 0.1% (for the challenging determination of ee values see Section 4.6). In contrast, dehydrogenation of (*R*)-citronell (*R*-**15**) at ca. 440 °C did not work very selectively, due to the formation of various side products by hydrogen transfer reactions. The next steps of the sequence were a sodium hydroxide-catalyzed aldol condensation with acetone and a subsequent heterogeneous olefin hydrogenation. During scale up of the latter reaction **22** → **18**, problems arose from partial racemization when using palladium-based catalysts, for example Pd on CaCO$_3$. Racemization was absent when applying Raney-nickel or platinum catalysts. Subsequent addition of acetylene in liquid ammonia with catalytic amounts of powdered potassium hydroxide afforded the ethinyl alcohol. Further on in the sequence,

Lindlar semi-hydrogenation to C_{15} vinyl alcohol **23**, subsequent acetylation, and Pd-mediated rearrangement of the tertiary allylic acetate [64] provided access to **24** as a ca. 85:15 (E/Z)-mixture. Notably, full conversion in the preceding hydrogenation reaction is important, since the acetylenic acetate is a strong catalyst poison for the allylic rearrangement.

While long-term experience exists in utilizing such catalytic methodology in other processes on a multi-ton scale, two serious problems were encountered in the industrial realization of the overall sequence illustrated in Fig. 7. Although the yields and selectivities of the individual steps are generally very good, the large number of transformations of the lengthy process caused numerous unit operations, which is hardly compatible with cost effectiveness. Secondly, the separation of E/Z-isomers on the C_{15} stage is a bottleneck that is difficult to overcome technically. Rectification of allylic alcohols (E/Z)-**19** is not a possibility, due to a strong tendency to decomposition induced by thermal stress. Alternatively, and despite the small difference in boiling points (Δbp ca. 2 °C), the acetate E-**24** could be purified from a ca. 4:1 mixture with Z-**24** by fractional distillation on a miniplant scale (2.2 kg) up to a content of >99.5%. However, it was clear that adaptation of this purification procedure to a large-scale process would be highly demanding.

4.4
Asymmetric Hydrogenation of Prochiral Allylic Alcohols

Optimization of the enantioselective catalytic key steps calls for careful experimental investigation of many reaction parameters. Besides temperature, concentration of substrate, solvent effects, pressure, and conversion rate, a defined robustness of the process towards impurities, for example contained in reagents, as well as its sensitivity towards air (oxygen) or moisture at various temperatures are important aspects. In particular, the purity of prochiral substrates is of utmost importance for the success of asymmetric hydrogenation experiments. As a consequence, considerable attention had to be paid to even the smallest differences in the impurity profile of substrates, which may be due to different preparation and/or purification procedures at lab, pilot, or production scale.

In view of efficient research and development activities within the limited time frames allocated to projects today, broad ligand and catalyst libraries are indispensable tools for rapid evaluation of the technical feasibility of the processes. For production purposes, short reaction times (full conversion in less than 10 hours) at high s/c ratios (preferably >50 000) and high ee values have to be reached. A very high enantioselectivity (preferably >98% ee) is indeed essential, since the optical purity of (liquid) chiral isoprenoids cannot be improved in an economical manner, such as by crystallization.

Another important aspect is the selection of the organometallic catalyst precursor responsible for generation of the active catalytic species. The suitable combination of metal complex and chiral ligand determines the efficiency of the process, in particular the conversion rate as well as the yield and optical purity of products. The enantioselective hydrogenation of geraniol (13) (cf. Figs. 5 and 7), already achieved in an exceptionally efficient manner by Takaya and Noyori et al. [58, 59], has been further investigated by us. A reliable procedure for the *in situ* preparation of catalysts from a Ru-precursor and the chiral ligand has been worked out [65]. On screening of novel ligands, with the aim of yielding an enantiomeric excess as high as possible, (S)-MeOBIPHEP (S-29) turned out to be the optimum ligand affording 0.5–1% higher ee values than (S)-BINAP (S-25), (S)-pTol-BINAP (S-26), and (S)-BIPHEMP (S-27). Results obtained with other C_2-symmetric ligands (S-30 to S-32, Fig. 8) [65] are compiled in Tab. 1. The purity of the starting material indeed had a strong influence on the reaction rate, due to possible deactivation of the catalyst. Therefore, only high-purity geraniol enabled the hydrogenation to be performed with s/c ratios in the order of 50 000.

Reproducible results were obtained by batch hydrogenations in a 20 L autoclave starting from a 40% (w/w) solution of 5 kg geraniol (13) in technical grade methanol. Use of [Ru((S)-MeOBIPHEP)(CF$_3$CO$_2$)$_2$] (s/c 24 000) at 60 bar hydrogen and 20 °C gave 100% conversion after 24 hours, yielding (R)-citronellol (R-15) in 98% with an ee of 98.5%. For subsequent heterogeneous hydrogenation of the 6,7-double bond, R-15 (which can still contain the homogeneous Ru-catalyst) was treated in 10 kg batches with Raney-nickel at 20 °C and 80 bar hydrogen to afford full conversion after 24–36 hours. (R)-Dihydrocitronellol (16) was obtained in >99% yield without detectable loss of enantiomeric purity.

Optimization work on the asymmetric hydrogenation of (2E,7R)-tetrahydrofarnesol (E-19) (Figs. 6 and 7) gave excellent results. When starting from high-purity batches of E-19, high ee values at C-3 (cf. the remark in Tab. 2) and very short reaction times were obtained with very high s/c ratios of 100 000, well above the originally reported ones [58]. As can be seen from the selection in Tab. 2, comparison of

Tab. 1 Enantioselective hydrogenation of geraniol (13) with chiral Ru(tfa)$_2$-complexes.

Diphosphane ligand		s/c	Conversion (%) after				% ee (R)
			0.25 h	1 h	5 h	22 h	
(S)-MeOBIPHEP	S-29	12 000	100				99.2
(S)-MeOBIPHEP	S-29	30 000		82	96	99	99.0
(S)-pTol-MeOBIPHEP	S-30	12 000	100				99.2
(S)-pTol-MeOBIPHEP	S-30	50 000	42	91	100		99.3
(S)-HO-BIPHEP	S-31	12 000	100				96.0
(S)-HO-BIPHEP	S-31	50 000	46	95	100		96.2
(S)-Tri-MeOBIPHEP	S-32	12 000		100			98.8
(S)-Tri-MeOBIPHEP	S-32	50 000	15	46	86	97	98.6
(S)-BIPHEMP	S-27	12 000		100			98.2
(S)-pTol-BINAP	S-26	12 000		100			97.7

Conditions: MeOH, c = 20%, 20 °C, 60 bar.
tfa = Trifluoroacetate, s/c = substrate-to-catalyst molar ratio.

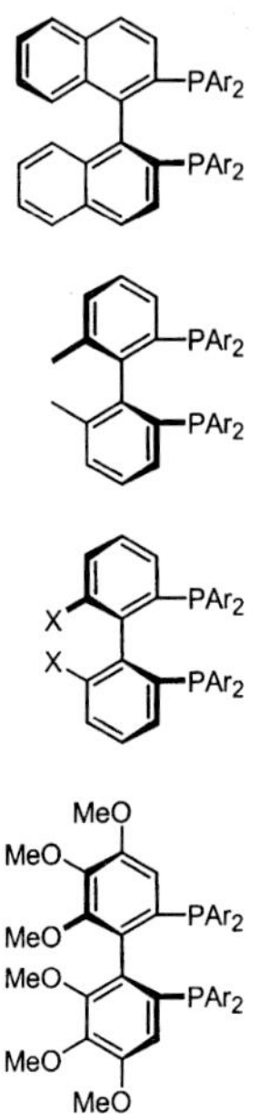

Fig. 8 Diphosphane ligands used in asymmetric hydrogenation reactions.

S-25 (S)-BINAP   Ar = Ph
S-26 (S)-pTol-BINAPP Ar = pTol

S-27 (S)-BIPHEMP   Ar = Ph
S-28 (S)-pTol-BIPHEMP   Ar = pTol

S-29 (S)-MeOBIPHEP   X = OCH$_3$ Ar = Ph
S-30 (S)-pTol-MeOBIPHEP X = OCH$_3$ Ar = pTol
S-31 (S)-HO-BIPHEP   X = OH   Ar = Ph

S-32 (S)-Tri-MeOBIPHEP   Ar = Ph

ligands S-25 to S-28 revealed that (S)-pTol-BIPHEMP (S-28) gave the best results. In general, BIPHEMP-type ligands (27, 28) form the most stereoselective catalysts, generating 0.4–0.7% higher ee values than BINAP-type diphosphanes (25, 26). Even at 35 bar, full conversion in less than 3 hours was obtained with these ligands, although with a slight decrease (0.3–0.7%) of the ee compared with reactions performed at a pressure of 60 bar. In all these hydrogenations strict control of temperature is extremely important, since even minor temperature increases caused by an inefficient removal of the reaction heat bring about lower ee values.

On a kg scale, (3R,7R)-3,7,11-trimethyldodecanol (20) was obtained in high quality under conditions similar to those given in Tab. 2. With an s/c ratio of 150 000 and ligand S-28, total conversion of E-19 in less than 12 hours at a substrate concentration of 40% and a technically convenient pressure of 35 bar hydrogen gave a >98% isolated yield of pure 20 after flash distillation. The high chemical purity (99.9 area-% by gas chromatography, no detectable Ru by XRF spectroscopy) and a 98.0% ee at C-3 certainly represent a satisfactory result.

All hydrogenations described in this chapter were carried out in commercial steel autoclaves. The steel type (DIN 1.4435, DIN 1.4980, Hastelloy C4) had no influence on the results, whereas efficient stirring and temperature control were definitely important. Since the catalysts are air sensitive, exclusion of air was guaranteed by charging the autoclaves with a volume up to 500 mL in a glove box under argon with <2 ppm oxygen. On a larger scale, the results were reproduced only if oxygen was again carefully excluded from the reaction. Therefore, the catalyst was added only after the autoclave containing the substrate and solvent had been carefully purged with argon.

Tab. 2 Enantioselective hydrogenation of (2E,7R)-tetrahydrofarnesol (E-19) with chiral Ru(tfa)$_2$-complexes

Diphosphane ligand		s/c	educt conc. (%)	p H$_2$ (bar)	Conversion after (%)		ee (%) (3R)[a)]
					0.25 h	1 h	
(S)-pTol-BIPHEMP	S-28	100 000	40	60	79	99	98.4
(S)-pTol-BIPHEMP	S-28	100 000	40	35	50	95	98.1
(S)-pTol-BIPHEMP	S-28	150 000	60	60	27	89	98.6
(S)-pTol-BIPHEMP	S-28	150 000	60	35		70	97.9
(S)-BIPHEMP	S-27	100 000	40	60	36	96	98.3
(S)-BIPHEMP	S-27	100 000	40	35	12	87	98.0
(S)-pTol-BINAP	S-26	100 000	40	60	83	99	97.7
(S)-pTol-BINAP	S-26	100 000	40	35	19	93	97.3
(S)-BINAP	S-25	100 000	40	60	16	86	97.9

a) Although it is formally not correct, this value is calculated for quantifying the extent of optical purity at C-3, omitting stereochemical information on C-7: ee(3R) = [(3R7R + 3R7S) minus (3S7R + 3S7S)] divided by [sum of all four stereoisomers (3R7R + 3R7S + 3S7R + 3S7S)].
Conditions: MeOH, 20 °C.
tfa = Trifluoroacetate, s/c = substrate-to-catalyst molar ratio.
"conc. educt (%)" means concentration of educt (starting material E–19) dissolved in MeOH; e.g. 40% means 40 g of E–19 in 100 g solution.

Finally, the recovery and/or recycling of the catalysts was not investigated. On one hand, the situation was ideal, since the products were isolated by a simple distillation, whereas the catalyst was left in the residue. On the other hand, the amounts of catalyst were too tiny even on the largest scale applied to permit any reasonably accurate investigation. Nevertheless, the clarification of this issue would certainly become necessary, if the asymmetric hydrogenation of **13** and **19** were to be further developed and scaled up. In fact, in a multi-ton process the transition metal and the chiral diphosphane would add up to significant amounts. Thus, recovery and recycling of the chiral diphosphanes, which are costly to produce, would very probably be worth pursuing. In contrast, the relatively low price of ruthenium coupled with its high reprocessing costs would render the recovery questionable. Nevertheless, an exact control of its fate would be dictated by environmental reasons.

4.5
Synthesis of the Diphosphane Ligands

The investigation and development of the enantioselective catalytic hydrogenations of the allylic alcohols required the ready availability of a number of BIPHEP type diphosphanes for screening and optimization purposes as well the availability of the most interesting ligands, i.e., of (S)-MeOBIPHEP (S-29) and (S)-pTol-BI-

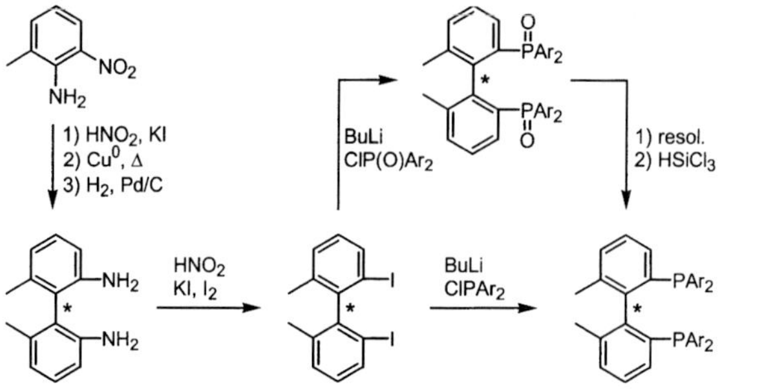

Fig. 9 Synthesis of BIPHEMP ligands.

PHEMP (*S*-28) in larger amounts for technical process development and scale-up. The availability on a larger scale was particularly important for ligand *S*-29 employed in the geraniol hydrogenation where it proved difficult to reach s/c ratios higher than 50 000. Accordingly, chemistry towards BIPHEP ligands was developed providing both the flexibility to prepare many ligand analogues and the potential for scale-up for individual ligands.

The synthesis of BIPHEMP ligands (Fig. 9) was initially realized via resolution of the readily available 6,6'-dimethyl-2,2-diaminobiphenyl followed by a sequence in the optically active series of double diazotization/iodination, iodine/lithium exchange and phosphanylation [46, 66]. This approach, although offering flexibility in terms of the synthesis of various analogues (cf. [56]), suffered from shortcomings, in particular from the rather ineffective resolution at the diamine stage. Therefore, for the preparation of specific ligands such as e.g., **27** and **28**, a route was developed involving the preparation of the racemic bis(phosphane oxide) from the racemic diiodide, resolution of the bis(phosphane oxide), and final reduction to the diphosphane [67]. Even if the resolution step is carried out late in this route, this proves more favorable due to efficient resolutions of the bis(phosphane oxides) with resolving agents such as *O,O'*-dibenzoyltartaric acid (DBTA), 2,3-bis(phenylaminocarbonyloxy)succinic acid or 2-phenylaminocarbonyloxy-propionic acid [67b].

A similar approach involving resolution at the bis(phosphane oxide) stage was used to prepare MeOBIPHEP type diphosphanes, e.g., **29** and **30** (Fig. 10) [60]. The bis(phosphane oxides) were synthesized starting from 3-bromoanisole *via* a phosphinoylation, *ortho*-lithiation/iodination, Ullmann reaction sequence. Such Ullmann reactions of phosphinoyl-substituted iodobenzenes proceeded smoothly and with high yields in DMF at surprisingly low temperatures of 130–150 °C. Resolutions were performed in good yields with DBTA. Alternatively, MeOBIPHEP-type ligands were obtained by a route involving a common optically active intermediate, i.e., an optically active bis(diethyl phosphonate), itself accessible from 3-bromoanisole by a sequence of phosphorylation, *ortho*-lithiation/iodination, Ullmann reaction and resolution [68]. The enantiomerically pure (*R*)- and (*S*)-bis(phosphonates) represent optimal intermediates for the rapid preparation of a diversity of MeOBI-

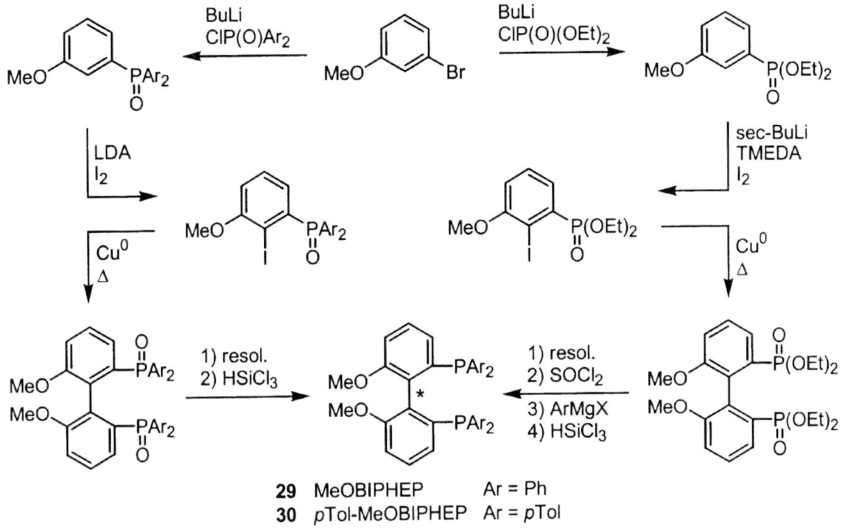

4.5 Synthesis of the Diphosphane Ligands

Fig. 10 Synthesis of MeOBIPHEP ligands.

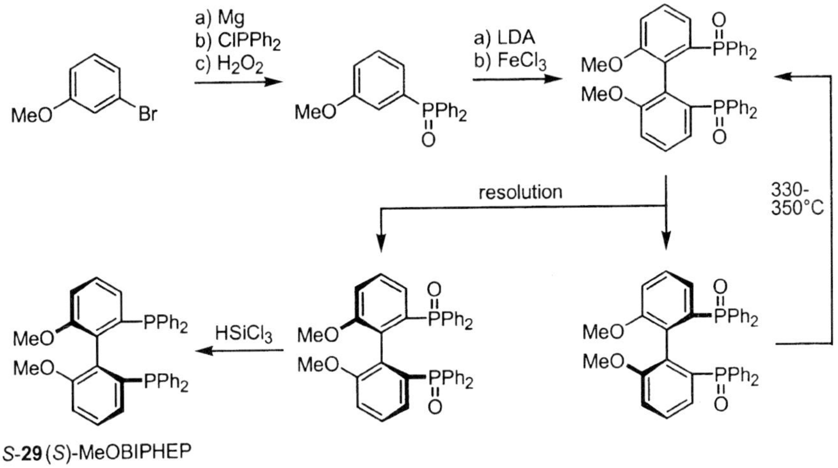

29 MeOBIPHEP   Ar = Ph
30 pTol-MeOBIPHEP Ar = pTol

Fig. 11 Large-scale synthesis of (S)-MeOBIPHEP (S-**29**).

PHEP-type ligands, since, through activation to the bis(phosphonic dichlorides) they react with a large variety of aryl, heteroaryl or alkyl Grignard reagents to provide the bis(phosphane oxides), which then are reduced to the diphosphanes (cf. [56] and literature cited therein).

The bis(phosphane oxide) resolution route to (S)-MeOBIPHEP (S-**29**), the ligand for the geraniol hydrogenation, was successfully developed to a larger scale process (Fig. 11). In this process, phosphinoylation of 3-bromoanisole was realized via Grignard formation, reaction with chlorodiphenylphosphane (cheaper and more readily available than the corresponding P(V) reagent) and hydrogen peroxide oxidation in a one-pot process. Most decisively, the previous two-step *ortho*-lithiation/iodination/Ullmann reaction sequence to construct the biphenyl system was replaced by an expedient one-pot *ortho*-lithiation/oxidative coupling process to provide the racemic bis(phosphane oxide) in high yield [69]. Resolution with DBTA followed by reduction then afforded *S*-**29** and its corresponding (*R*)-enantiomer each in 33% (that is each in 66% of the theoretical) overall yield over the four-step process. This route was developed to a 100 kg scale and the process proved reliable and robust. In view of the possible use of one of the MeOBIPHEP enantiomers only, a racemization/recycle protocol was also established [70]. It consists in the thermal racemization at 330–350 °C of the undesired bis(phosphane oxide) which, very remarkably, proceeds without significant decomposition and provides the racemate in yields in the order of 90%. Such racemization/recycle, pending the successful technical development and scale-up, has the potential for further improvement of the economics of the synthesis of the (*S*)-MeOBIPHEP ligand *S*-**29** or its enantiomer.

4.6
Procedures for Stereochemical Analysis

The reliable determination of the stereoisomeric purity of products resulting from enantio- or diastereoselective reactions is a prerequisite for the development of new methods and procedures. This applies in particular to the optimization of catalytic systems. In the case of the enantioselective hydrogenation of allylic alcohols, fine-tuning of metal complexes and reaction conditions required the non-trivial accurate measurement of ee (de) values of routinely above 95% of several primary alcohols, aldehydes, and ketones bearing unfunctionalized hydrocarbon moieties. For the analysis of chiral intermediates involved in a synthetic sequence to *RRR*-**3** (cf. Section 4.3, Figs. 6 and 7), three methods have been developed in our laboratories, which show clear advantages over earlier protocols. The gas chromatographic separation of diastereoisomeric esters and acetals, derived from C_{10} to C_{18} isoprenoids possessing one or two chiral centers, is the basis for two procedures. The third method is applied to the end product **3**. This set of protocols allows the complete stereochemical analysis on practically all chain length levels of vitamin E syntheses, in particular of highly (*R*)- or (*R*,*R*)-enriched samples of the key compounds **16**, **18**, and **20** in the sequence of Fig. 12. In addition, it should be mentioned that the availability of several independent methods allows the verification of even very small differences in the stereoisomeric purity of products.

(*S*)-Trolox^{TM} methyl ether has been introduced as a chiral reagent (commercially available for several years [71]), being superior to others for the determination of the optical purity of aliphatic alcohols via their diastereoisomeric esters **33** by GC or SFC (supercritical fluid chromatography) [72–75]. The first successful use of this reagent for the determination of the stereoisomer composition of citro-

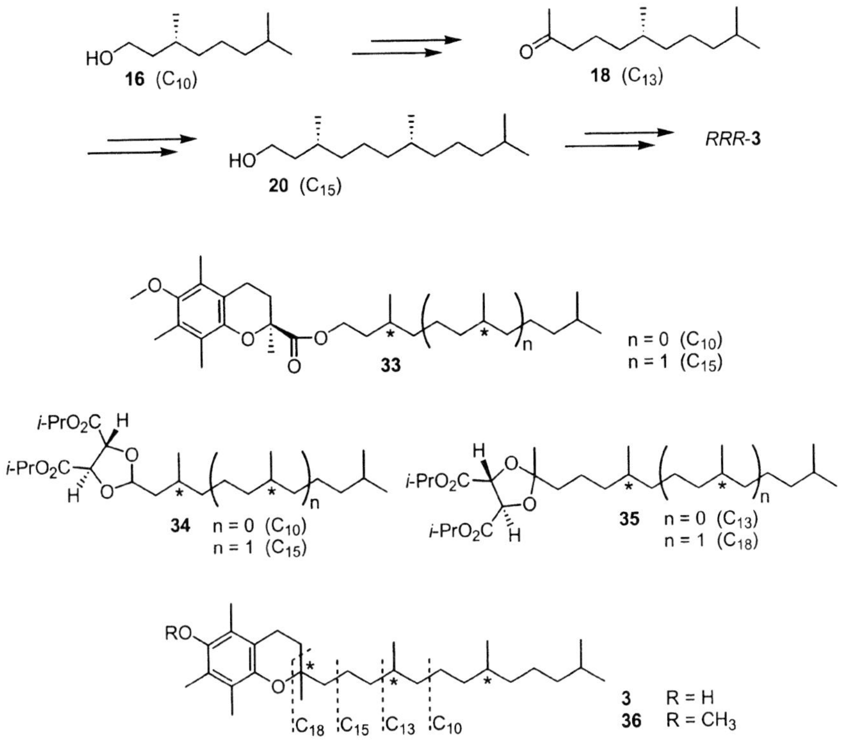

Fig. 12 Chiral side-chain building blocks and derivatives for stereochemical analysis.

nellol (**15**, cf. Fig. 5) was a historically very important example [58] during the early development stages of enantioselective hydrogenation procedures.

Even two chiral centers (in C_{15} or C_{18} methyl-branched carbonyl compounds, cf. Fig. 12) can be determined simultaneously with high accuracy ($\pm 0.3\%$) by GC analysis of acetals **34** and **35** ($n=1$) derived from C_2-symmetric tartrate esters. Laboratory-made 80- to 100-meter capillary columns were used in the original work [76, 77]. Recently we have shown [78] that this methodology can be applied routinely by using commercially available reagents [79] and GC columns [80]. It has to be pointed out that up to now this protocol is the only one that is capable of accomplishing such an analysis.

At the end of the synthesis, the outcome of enantioselective reactions can be checked by the full stereochemical characterization of the final product **3**. The chromatographic separation of all eight stereoisomers of **3**, that is the concurrent determination of all three chiral centers [81, 82], is in principle based on the separation of enantiomers by HPLC on a chiral phase and separation of diastereoisomers on an achiral GC column. While also in this case special laboratory-made columns were used in the original procedure [82] for both HPLC and GC separation steps, analyses are done today routinely on the stable and easy-to-prepare methyl ether derivative **36** with the use of commercial columns [80, 83].

4.7
Concluding Remarks

The following chemical aspects of the synthesis of $(2R,4'R,8'R)$-α-tocopherol (RRR-3) may be summarized: The individual catalytic asymmetric hydrogenation steps can be performed in high yield and excellent optical induction (>98% ee/de of aliphatic alcohols obtained). High substrate concentration, a reaction temperature around room temperature, relatively low pressure, short reaction times, and extremely high s/c ratios are almost ideal conditions for technically feasible processes. Chemical waste and hazardous reagents are not actual major problems. Well-elaborated procedures for the synthesis of optimum chiral ligands and catalyst precursors applicable to the production scale are available. Consequently, and in contrast to many other examples, catalytic asymmetric hydrogenation steps do not contribute to a major part of the production cost in this case. All these results indeed meet the requirements for technical realization. Why then is an economical total synthesis of nature-identical $(2R,4'R,8'R)$-α-tocopherol (RRR-3) so difficult to achieve? Mainly three reasons have so far prevented the realization of this project.

First of all, the access to high-purity C_{15} allylic alcohol still remains a technical challenge, mainly due to the difficult separation of the E/Z-isomers. Secondly, the efficient generation of the chiral tertiary center of the chroman unit, for example by enantio/diastereoselective ring closure, and its subsequent coupling to the side-chain, represents a highly demanding task. Although considerable progress has been made in this area, an economical solution is still lacking. In parallel to these chemistry- and process-related reasons, management and marketing considerations, also being part of any such project, developed in the course of the investigation. Eventually it was decided to discontinue the total synthesis approach in favor of the semi-synthetic one. In order to establish a medium-term solution to satisfy customer's needs, manpower had to be concentrated on the alternative route, which uses a natural-source mixture of vitamin E compounds as the starting material and its subsequent chemical permethylation to $(2R,4'R,8'R)$-α-tocopherol (RRR-3). As an outcome of this strategy, our efforts resulted in the successful market introduction of the highly pure acetate derivative valuable for various human applications.

In conclusion, although the large-scale total synthesis of $(2R,4'R,8'R)$-α-tocopherol (RRR-3) is (so far) not a success story within the area of industrial total syntheses of natural products, the incontestably high potential and usefulness of catalytic methods again became evident. If there ever will be a chemical solution, the key technology contributing decisively to this would very probably have to be labeled as the title of this book: asymmetric catalysis on industrial scale.

4.8 Acknowledgements

The achievements described in this chapter would not have been possible without the expert work of many synthetic and analytical laboratories. In particular, we gratefully acknowledge the contributions of Dres. E.A. Broger, M. Cereghetti, Y. Crameri, Mr. J. Foricher, Dres. B. Heiser, R.K. Müller, M. Vecchi, W. Vetter, W. Walther, and their co-workers.

4.9 References

1 a) R. Noyori, Asymmetric Catalysis in Organic Synthesis, John Wiley & Sons, New York, 1994, and references cited therein; b) S. Akutagawa, Appl. Catal. A: General 1995, 128, 171–207, and references cited therein.
2 A. Rüttimann, Chimia 1986, 40, 290–306.
3 R. Schmid, S. Antoulas, A. Rüttimann, M. Schmid, M. Vecchi, H. Weiser, Helv. Chim. Acta 1990, 73, 1276–1299.
4 H.M. Evans, K.S. Bishop, Science 1922, 56, 650–651.
5 H.M. Evans, O.H. Emerson, G.A. Emerson, J. Biol. Chem. 1936, 113, 319–332.
6 E. Fernholz, J. Am. Chem. Soc. 1938, 60, 700–705.
7 K.U. Baldenius, L. von dem Bussche-Hünnefeld, E. Hilgemann, P. Hoppe, R. Stürmer, in Ullmann's Encyclopedia of Industrial Chemistry, VCH Verlagsgesellschaft, Weinheim, 1996, Vol. A27, pp. 478–488.
8 H.G. Ernst, in Ullmanns Encyclopädie der technischen Chemie, Verlag Chemie, Weinheim, 1983, Vol. 23, pp. 643–649.
9 Chemical Market Reporter, November 10, 1997, pp. 5, 12.
10 P.J. McLaughlin, J.L. Weihrauch, J. Am. Diet. Ass. 1979, 75, 647–665.
11 H. Crawley, in The Technology of Vitamins in Food (P.B. Ottaway, Ed.), Blackie Academic & Professional, Chapman & Hall Inc., Glasgow, 1993, chapter 2, p. 19.
12 A.J. Sheppard, J.A.T. Pennington, J.L. Weihrauch, in Vitamin E in Health and Disease (L. Packer, J. Fuchs, Eds.), Marcel Dekker, Inc., New York, 1993, chapter I, A 2, pp. 9–31.
13 G.W. Burton, K.U. Ingold, Acc. Chem. Res. 1986, 19, 194–201.
14 G. Pongratz, H. Weiser, D. Matzinger, Fat. Sci. Technol. 1995, 97, 90–1041.
15 A. Kamal-Eldin, L.-A. Åppelqvist, Lipids 1996, 31, 671–701 (remark: this review contains several errors of nomenclature and wrong structures drawn from literature).
16 IUPAC-IUB, Eur. J. Biochem. 1982, 123, 473–475; Pure Appl. Chem. 1982, 54, 1507–1510.
17 H. Weiser, M. Vecchi, Internat. J. Vit. Nutr. Res. 1982, 52, 351–370.
18 B.J. Weimann, H. Weiser, Am. J. Clin. Nutr. 1991, 53, 1056S–1060S.
19 H. Weiser, G. Riss, A.W. Kormann, J. Nutr. 1996, 126, 2539–2549.
20 W. Cohn, Eur. J. Clin. Nutr. 1997, 51, Suppl. 1, S80–S85.
21 P. Karrer, H. Fritzsche, B.H. Ringier, H. Salomon, Helv. Chim. Acta 1938, 21, 820–825.
22 M. Matsui, N. Karibe, K. Hayashi, H. Yamamoto, Bull. Chem. Soc. Jpn. 1995, 68, 3569–3571.
23 K. Ishihara, M. Kubota, H. Yamamoto, Synlett 1996, pp. 1045–1046.
24 M. Matsui, J. Jpn. Oil Chem. Soc. 1996, 45, 821–831; Chem. Abstr. 1996, 125, 301241.
25 W. Bonrath, A. Haas, E. Hoppmann, Th. Netscher, H. Pauling, F. Schager, A. Wildermann, Adv. Synth. Catal. 2002, 344, 37–39, and lit. cited therein.
26 F.V.K. Young, C. Poot, E. Biernoth, N. Krog, L.A. O'Neill, N.G.J. Davidson,

27 M. K. Gupta, INFORM 1993, 4, no. 11 (November), 1267–1272.
28 L. Walsh, R. L. Winters, R. G. Gonzalez, INFORM 1998, 9, no. 1 (January), 78–83.
29 Th. Netscher, in Lipid Synthesis and Manufacture (F. D. Gunstone, Ed.), Sheffield Academic Press Ltd, Sheffield, UK, 1999, pp. 250–267.
30 Chemical Market Reporter, 13 June 1994.
31 R. K. Mueller, H. Schneider (F. Hoffmann-La Roche AG), EP 0735035 A1 (21.03.1996, prior. date 28.03.1995); US 6066731 (Roche Vitamins Inc., 23.05.2000).
32 K. Brüggemann, J. R. Herguijuela, Th. Netscher, J. Riegl (F. Hoffmann-La Roche AG), EP 0769497 A1 (09.10.1996, prior. date 18.10.1995); US 5892058 (Roche Vitamins Inc., 06.04.1999).
33 M. Breuninger (F. Hoffmann-La Roche AG), EP 0882722 A1 (28.05.1998, prior. date 06.06.1997); US 5932748 (Roche Vitamins Inc., 03.08.1999).
34 G. Saucy, N. Cohen, in New Synthetic Methodology and Biologically Active Substances (Z. Yoshida, Ed.), Elsevier, Amsterdam, 1981, pp. 155–175.
35 a) D. L. Coffen, N. Cohen, A. M. Pico, R. Schmid, M. J. Sebastian, F. Wong, Heterocycles 1994, 39, 527–552, and cit. lit.; b) Th. Netscher, Chimia 1996, 50, 563–567, and references cited therein.
36 a) H. U. Blaser, F. Spindler, M. Studer, Appl. Catal. A: General 2001, 221, 119–143; b) R. Schmid, Chimia 1996, 50, 110–113; c) R. Schmid, E. A. Broger, Proceedings of the Chiral Europe '94, 19–20 Sept. 1994, Nice, France, pp. 79–86.
37 W. S. Knowles, Angew. Chem. Int. Ed. 2002, 41, 1998–2007.
38 R. Noyori, Angew. Chem. Int. Ed. 2002, 41, 2008–2022.
39 K. B. Sharpless, Angew. Chem. Int. Ed. 2002, 41, 2024–2032.
40 S. L. Buchwald, M. J. Burk, E. N. Jacobsen, Sh. Kobayashi, A. Pfaltz, Ch.-H. Wong, Adv. Synth. Catal. 2001, 343, 753.
41 J. M. Brown, Adv. Synth. Catal. 2001, 343, 755–756.
42 I. Markó, Adv. Synth. Catal. 2001, 343, 757–758.
43 K. Tani, T. Yamagata, S. Akutagawa, H. Kumobayashi, T. Taketomi, H. Takaya, A. Miyashita, R. Noyori, S. Otsuka, J. Am. Chem. Soc. 1984, 106, 5208–5217.
44 K. Tani, Pure Appl. Chem. 1985, 57, 1845–1854.
45 K. Tani, T. Yamagata, S. Otsuka, H. Kumobayashi, S. Akutagawa, Org. Synth. 1989, 67, 33–43.
46 R. Schmid, M. Cereghetti, B. Heiser, P. Schönholzer, H.-J. Hansen, Helv. Chim. Acta 1988, 71, 897–929.
47 R. Schmid, H.-J. Hansen, Helv. Chim. Acta 1990, 73, 1258–1275.
48 R. Noyori, Asymmetric Catalysis in Organic Synthesis, John Wiley & Sons, New York, 1994, chapter 3, pp. 95–121.
49 I. Ojima, N. Clos, C. Bastos, Tetrahedron 1989, 45, 6901–6939.
50 R. Noyori, M. Kitamura, in Modern Synthetic Methods, Vol. 5 (R. Scheffold, Ed.), Springer, Berlin, 1989, pp. 115–198.
51 H. Takaya, T. Ohta, R. Noyori, in Catalytic Asymmetric Synthesis (I. Ojima, Ed.), VCH Publishers Inc., New York, 1993, chapter 1, pp. 1–39.
52 R. Noyori, Asymmetric Catalysis in Organic Synthesis, John Wiley & Sons, New York, 1994, chapter 2, pp. 16–94.
53 A. Pfaltz, J. M. Brown, in Houben-Weyl, Methods of Organic Chemistry, Vol. E21d, Stereoselective Synthesis, (G. Helmchen, R. W. Hoffmann, J. Mulzer, E. Schaumann, Eds.), Georg Thieme, Stuttgart, 1995, chapter 2.5.1.2., pp. 4334–4359.
54 T. Ohkuma, M. Kitamura, R. Noyori, in Catalytic Asymmetric Synthesis, 2nd edn (I. Ojima, Ed.), Wiley-VCH, New York, 2000, chapter 1, pp. 1–110.
55 G.-Q. Lin, Y.-M. Li, A. S. C. Chan, in Principles and Applications of Asymmetric Synthesis, Wiley-Interscience, New York, 2001, chapter 6, pp. 331–396.
56 R. Schmid, E. A. Broger, M. Cereghetti, Y. Crameri, J. Foricher, M. Lalonde, R. K. Müller, M. Scalone, G. Schoettel, U. Zutter, Pure Appl. Chem. 1996, 68, 131–138.

57 W. S. Knowles, Acc. Chem. Res. 1983, 16, 106–112.
58 H. Takaya, T. Ohta, N. Sayo, H. Kumobayashi, S. Akutagawa, S. Inoue, I. Kasahara, R. Noyori, J. Am. Chem. Soc. 1987, 109, 1596–1597.
59 H. Takaya, T. Ohta, S. Inoue, M. Tokunaga, M. Kitamura, R. Noyori, Org. Synth. 1995, 72, 74–85.
60 R. Schmid, J. Foricher, M. Cereghetti, P. Schönholzer, Helv. Chim. Acta 1991, 74, 370–389.
61 R. Noyori, H. Takaya, Acc. Chem. Res. 1990, 23, 345–350.
62 T. Ohta, H. Takaya, R. Noyori, Inorg. Chem. 1988, 27, 566–569.
63 H. Takaya, T. Ohta, K. Mashima, R. Noyori, Pure Appl. Chem. 1990, 62, 1135–1138.
64 K. Meyer (F. Hoffmann-La Roche & Co AG), Ger. Offen. DE 2513198 A1, 16.10.1974, prior. date 3.04.1974.
65 B. Heiser, E. A. Broger, Y. Crameri, Tetrahedron: Asymmetry 1991, 2, 51–62.
66 M. Cereghetti, R. Schmid, P. Schönholzer, A. Rageot, Tetrahedron Lett. 1996, 37, 5343–5346.
67 a) M. Cereghetti, A. Rageot (F. Hoffmann-La Roche AG), unpublished results; b) M. Cereghetti, A. Rageot (F. Hoffmann-La Roche AG), EP 0511558 B1 (16.04.1992, prior. date 29.04.1991); US 5334766 (02.08.1992).
68 J. Foricher, B. Heiser, R. Schmid (F. Hoffmann-La Roche AG), EP 0530335 B1 (12.03.1992, prior. date 15.03.1991); US 5,302,738 (12.04.1994).
69 J. Foricher, R. Schmid (F. Hoffmann-La Roche AG), EP 0926152 (17.12.1998, prior. date 23.12.1997); US 6162929 (19.12.2000).
70 S. Wang, R. Schmid, F. Kienzle, M. Lalonde (F. Hoffmann-La Roche AG), EP Appl. 00114219.9 (05.08.2000, prior. date 09.08.1999); US 6288280 (11.09.2001).
71 Trolox^{TM} methyl ether is available in both enantiomeric forms from Fluka Chemie GmbH, Buchs, Switzerland.
72 W. Walther, W. Vetter, M. Vecchi, H. Schneider, R. K. Müller, Th. Netscher, Chimia 1991, 45, 121–123.
73 W. Walther, W. Vetter, Th. Netscher, J. Microcol. Separat. 1992, 4, 45–49.
74 Th. Netscher, I. Gautschi, Liebigs Ann. Chem. 1992, 543–546.
75 W. Walther, Th. Netscher, Chirality 1996, 8, 397–401.
76 A. Knierzinger, W. Walther, B. Weber, Th. Netscher, Chimia 1989, 43, 163–164.
77 A. Knierzinger, W. Walther, B. Weber, R. K. Müller, Th. Netscher, Helv. Chim. Acta 1990, 73, 1087–1107.
78 W. Walther, A. Staempfli, S. Brogly, B. Wüstenberg, A. Pfaltz, Th. Netscher, unpublished results.
79 Diisopropyl tartrate and diisopropyl-O,O'-bis-(trimethylsilyl)-tartrate are available in both enantiomeric forms from Fluka Chemie GmbH, Buchs, Switzerland.
80 A 50-m×0.25 mm capillary column CP-Sil-88 is available from Chrompack, The Netherlands.
81 N. Cohen, C. G. Scott, Ch. Neukom, R. J. Lopresti, G. Weber, G. Saucy, Helv. Chim. Acta 1981, 64, 1158–1173.
82 M. Vecchi, W. Walther, E. Glinz, Th. Netscher, R. Schmid, M. Lalonde, W. Vetter, Helv. Chim. Acta 1990, 73, 782–789.
83 Chiracel OD column, 250×4.6 mm from Daicel Chemical Ind. Ltd.

5
Comparison of Four Technical Syntheses of Ethyl (R)-2-Hydroxy-4-Phenylbutyrate

Hans-Ulrich Blaser, Marco Eissen, Pierre F. Fauquex, Konrad Hungerbühler, Elke Schmidt, Gottfried Sedelmeier, and Martin Studer

Abstract

Four technically feasible routes for the large-scale preparation of the valuable intermediate (R)-2-hydroxy-4-phenylbutyrate (HPB ester) are compared. In all cases, the stereogenic center is introduced via enantioselective reduction, in two cases via biochemical and in two cases via chemical methods. Starting from the relatively simple materials diethyl oxalate and acetophenone or ethyl hydrocinnamate, the mass consumption of the 3–4 step routes varies between 40 and 105 kg per kg of enantiomerically pure HPB ester. In all cases, the enantioselective reduction is the most difficult step using sophisticated catalysts and either high pressure equipment or dedicated biochemical reactors. The shortest route, which includes the enantioselective hydrogenation of an α,γ-diketo ester with a cinchona-modified Pt-alumina catalyst has the lowest overall mass consumption, even though ee and yield for the reduction step are the lowest. This is compensated for by fewer steps, higher atom efficiency, and lower solvent consumption for the synthesis and extraction. The same route is presumably also the economically most attractive one, because in addition to the lower mass consumption the lower priced acetophenone instead of ethyl hydrocinnamate is used.

5.1
Introduction

A variety of commercially important Angiotensin Converting Enzyme (ACE) inhibitors contain an (S)-homophenylalanine moiety which can be introduced starting from various building blocks [1]. One of the most useful is ethyl (R)-2-hydroxy-4-phenylbutyrate (HPB ester) **1** which after introduction of a leaving group can be coupled with the respective amino moiety with inversion of the stereogenic center (Fig. 1). Since the patents for several ACE inhibitors have already expired or will soon do so, the production costs will become very important. This calls for the development of more efficient syntheses both for the ACE inhibitors as well as for key intermediates. In this chapter we will describe four technically feasible routes for the synthesis of **1** that have been developed by us in the last few years

Asymmetric Catalysis on Industrial Scale: Challenges, Approaches and Solutions
Edited by H.U. Blaser, E. Schmidt
Copyright © 2004 Wiley-VCH Verlag GmbH & Co. KGaA, Weinheim
ISBN: 3-527-30631-5

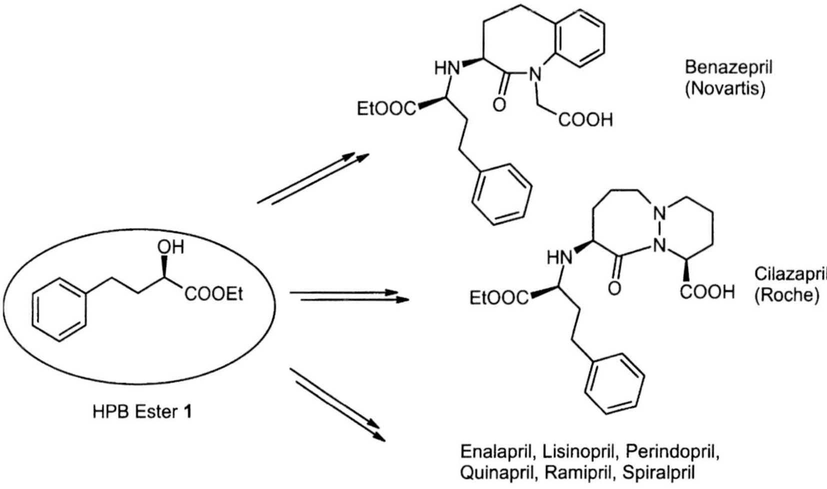

Fig. 1 Structures of ethyl (R)-2-hydroxy-4-phenylbutyrate and of selected ACE inhibitors.

[2] and will compare various chemical, ecological and economical aspects as a significant extension of the work described previously [3].

5.2
Synthetic Pathways to HPB Ester

In a recent literature survey we found 91 papers and patents reporting the synthesis of enantiomerically pure or racemic HPB ester, more than 20 of these appearing in the last two years. This shows the enormous interest in this valuable synthon and Fig. 2 depicts the three best pathways to **1**.

a) Synthesis of the racemic ester or acid, usually obtained by racemic reduction of the corresponding ketone, followed by resolution. For the HPB acid, various methods have been reported such as kinetic resolution using (immobilized) enzymes, e.g., lipases [4], crystallization of the corresponding acid via salt formation with a chiral amine [5], optionally followed by racemization of the undesired enantiomer [6], or dynamic kinetic resolution of a cyclic carbonate catalyzed by modified cinchona alkaloids [7].
b) Catalytic enantioselective reduction of a prochiral ketone by chemical [2a–c, 3b, 8], microbial [2d, 3b, 9] or enzymatic reduction [2e, 3b, 10].
c) Chiral pool synthesis, starting for example from malic acid derivatives [11].

From the descriptions in the literature it is hard to judge which of the many methods described are feasible on a technical scale, and procedures for large-scale preparations are described only very rarely.

This chapter will focus on the comparison of four enantioselective reduction routes **A–D** developed over the last two decades by Ciba-Geigy and/or Solvias in

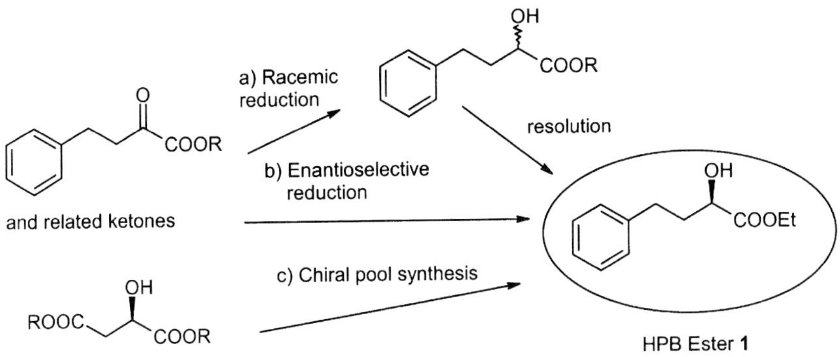

Fig. 2 Routes to HPB ester **1**.

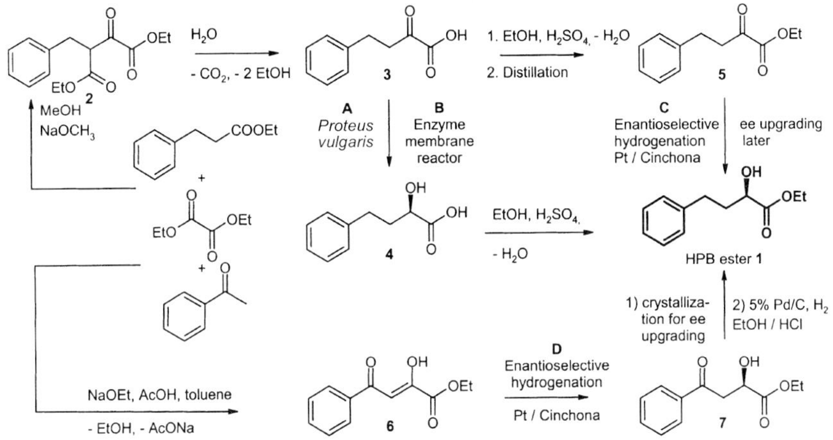

Fig. 3 Synthetic routes **A–D** to HPB ester **1**.

collaboration with Ciba Specialty Chemicals (see Fig. 3). All four processes have been carried out on the multi-kg to ton scale and in all cases there are enough data available for a critical assessment from an economical as well as from an ecological point of view. The comparison of the overall syntheses starts with the relatively simple building blocks ethyl hydrocinnamate (routes **A–C**) and acetophenone (route **D**). It should be noted that this choice is not unproblematic, since the price of acetophenone is significantly lower compared with ethyl hydrocinnamate. Special attention will be paid to the enantioselective reduction step, by far the most complicated and sophisticated in all routes.

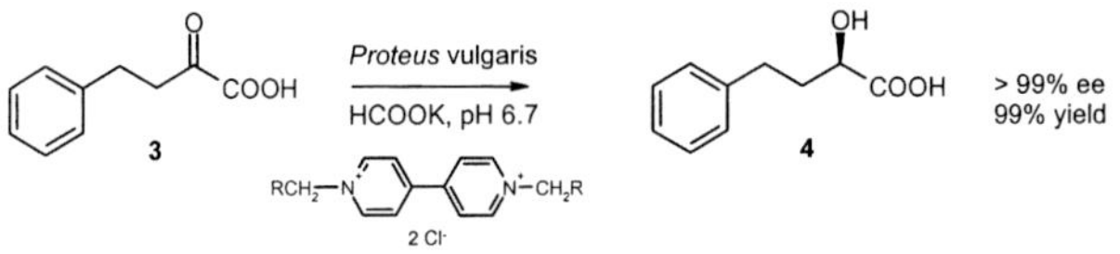

Fig. 4 Reduction step in route **A**.

5.2.1
Route A: Synthesis and Enantioselective Reduction of Keto Acid 3 with Immobilized *Proteus vulgaris* Followed by Esterification

The synthesis of the keto acid intermediate **3** is the same for routes **A–C** and it can be obtained in a one-pot reaction in essentially quantitative yield [12]. The enantioselective reduction of **3** to **4** was carried out using immobilized *Proteus vulgaris* [2d] (Fig. 4). For this, the microorganism was isolated from the culture medium by centrifugation. By adding polyphosphate, solid silicic acid and chitosan to the cell suspension, beads with a diameter of 2–3 mm consisting of the entrapped cells in a gel-matrix were produced. After washing and a reductive pretreatment, a volume of 4 L of beads was transferred to a 6 L glass column. An aqueous solution of **3** (112 mM in a phosphate buffer at pH 6.7 with 300 mM $KHCO_3$ and 1 mM benzyl viologen) was passed continuously over the beads. The addition rate was adjusted so that conversion was always >99.5%. After 25 days, when the reaction was stopped, the microorganism still had 40% of its original activity. The isolation of the product from the aqueous solution was achieved by extraction of the acidified solution with EtOAc and subsequent precipitation with cyclohexane. The isolated yield of crystalline **4** was 99.7%, and the ee was >99.8%. In the last step, the esterification of **4** was carried out. For this reaction, the same parameters were assumed as for the esterification of **3** in route **C**.

5.2.2
Route B: Enantioselective Reduction of 3 with D-LDH in a Membrane Reactor

Route **B** differs from **A** only in the reduction method (Fig. 5). D-Lactate dehydrogenase (D-LDH) from *Staphylococcus epidermidis* was chosen to carry out the NADH-dependent reduction of **3** [2e]. Formate dehydrogenase (FDH) was used for NADH regeneration and the reaction was carried out continuously in a 220 mL membrane reactor. The substrates and the coenzyme NAD were pumped through a sterile filter into the reactor, and the product left the reactor at the same rate through an ultrafiltration membrane. Enzymes were supplemented periodically (dependent on deactivation rates) via a sterile filter to the membrane reactor. Owing to their size enzymes were retained behind the ultrafiltration membrane. The ee was >99.8% at high conversion, and the chemical yield was 79%. The deactivation of the enzyme was minor (1.5% per day). It is important to note that this reduction is not optimized, and that there is significant room for improvement, especially in the amount of buffer, water, and solvent (for extraction) used.

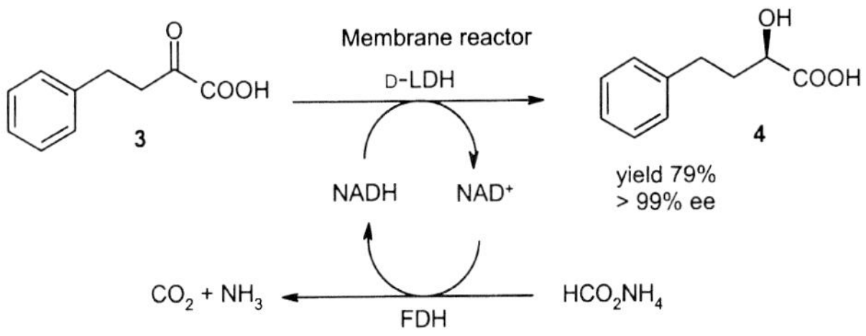

Fig. 5 Reduction step in route **B**.

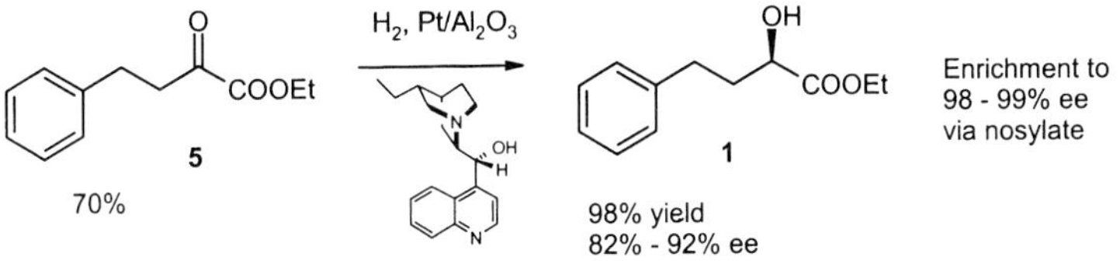

Fig. 6 Hydrogenation step in route **C**.

5.2.3
Route C: Synthesis and Enantioselective Hydrogenation of Keto Ester 5

The esterification of the α-keto acid **3** to **5** proceeded in high yield, but the distillation, which was necessary to obtain a material with good hydrogenation properties, yielded only about 70% of **5** [12] (Fig. 6). The enantioselective hydrogenation of α-keto ester **5** with cinchona-modified catalysts gave **1** in high chemical yield and an ee of 82–93% depending on the solvent and modifier used [2b,c]. In contrast with all other routes, the resulting HPB ester had an insufficient ee and upgrading was not possible because **1** is not crystalline at room temperature. As reported in [2c], the enrichment was carried out later in the sequence of the active ingredient. This means that route **C** is not suitable for the preparation of **1** in >99% ee and for this reason can only be used if there is a crystalline intermediate later on. For the comparison below, this upgrading was taken into consideration in the hydrogenation step where 82% ee and a yield of 74% instead of 98% were assumed.

5.2.4
Route D: Synthesis and Enantioselective Hydrogenation of Diketo Ester 6, Followed by Hydrogenolysis

In the most recent of all routes discussed, HPB ester was synthesized in only three steps (Fig. 3) [2a]. In this variant, a different ketone intermediate was synthesized in one step from the low cost acetophenone and diethyl oxalate in high yield. For the reduction, a similar catalytic system as described for route **C** was used. However, the maximum ee values and the catalyst activity were some-

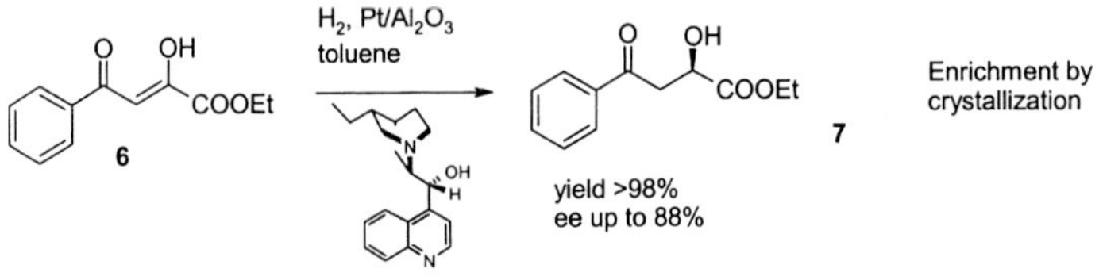

Fig. 7 Diketo ester hydrogenation of route **D**.

what lower for diketo ester **6**. Nevertheless, route **D** (Fig. 7) has several advantages compared with route **C**: high chemical yields of **6**, beginning with low price starting materials and crystallinity of the reduced intermediate **7**. This allowed ee upgrading before hydrogenolysis of the remaining keto group, which occurred without any racemization.

5.3
Comparison of Routes A–D with Respect to Mass Consumption, Environmental, Health and Safety Aspects

In this section we compare various parameters for the four different routes in general and for the reduction processes in particular. The main emphasis is placed on the quantitative comparison of the mass consumption, whereas environmental, health and safety aspects are only treated briefly and in a qualitative way.

5.3.1
Definitions

The following definitions are used in this discussion:

Input: Mass index S^{-1} [13], also MI [14] $S^{-1} = \Sigma$ raw materials (kg)/product (kg)
Output: Environmental factor E [15] $E = \Sigma$ waste (kg)/product (kg)

For clarity, the categories given in the following figures are briefly explained here, for more information refer to [13].

Sewage/water: water used and sewage produced in the synthesis, e.g., for extractions.

Solvent: organic solvents, in which the reactions take place.

Auxiliaries (isolation): chemicals used for the isolation, e.g., the organic solvent for extractions.

Catalysts: e.g., Pt-catalysts, modifiers, H_2SO_4 for esterification.

Auxiliaries (reaction): e.g., EDTA used for step 3 of route **B**.

Substrates: starting material, e.g., acetophenone, formates, a.s.o.

Byproducts: unwanted byproducts formed (unknown: not identified, but calculated).

Coupled products: e.g., CO_2 from KCOOH.

Loss of product: product not isolated.

Isomeric product: product isomer removed during upgrading.

5.3.2
Overall Material Masses Consumed and Produced for Routes A–D

Fig. 8 depicts mass indices S^{-1} and environmental factors E and gives an indication of the mass efficiency for the four routes **A–D**. The amounts of chemicals needed for the production of 1 kg HPB ester varied between approximately 40 kg and 105 kg. In all cases, the major components were water and the solvents needed for the reactions and/or extractions, a picture that is typical for fine chemical synthesis. Route **D** clearly had the lowest consumption of materials. The main drawback for the two biochemical routes **A** and **B** were the need for rather large amounts of water and solvents (for extraction), even though it has to be pointed out that these processes were not optimized. Comparable variations were observed for the substrate consumption, i.e., how much starting material was needed to produce 1 kg of HPB ester (see Fig. 8b). In this case, however, both the highest (**C**) and the lowest (**D**) consumptions were observed for the chemical routes. The

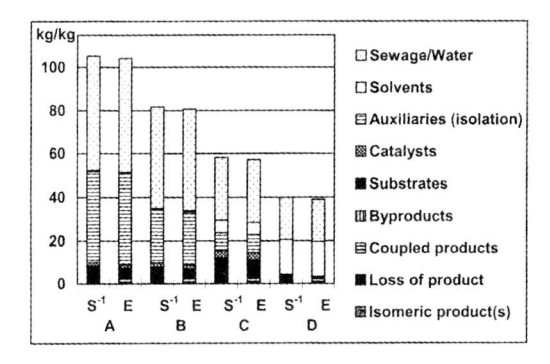

Fig. 8a Mass indices S^{-1} and environmental factors **E** of routes **A–D** using EATOS.

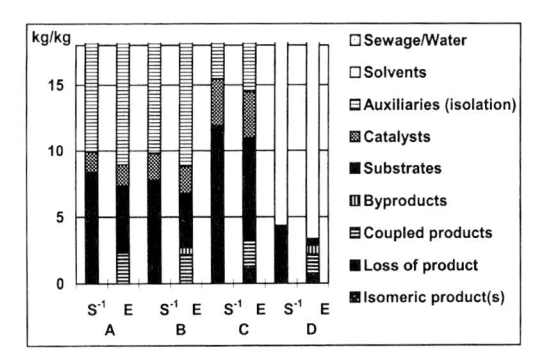

Fig. 8b Detailed view of Fig. 8a.

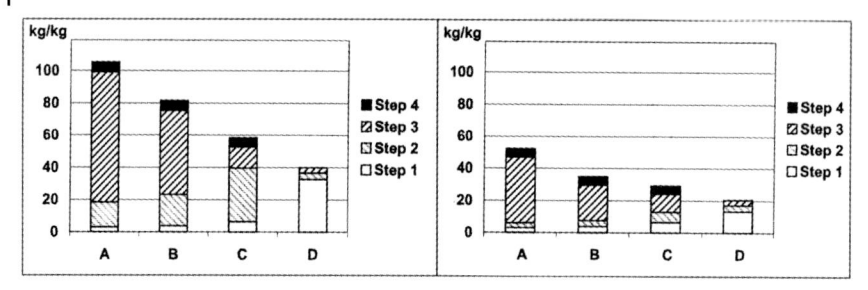

Fig. 9 Mass indices S^{-1} of the different reaction steps of routes **A–D** using EATOS: **a** water included, **b** without water.

nature of the output in all processes was similar to that of the input, again mainly water and solvents.

The same trends can be seen in Fig. 9a and 9b, which show the mass indices S^{-1} for the different reaction steps with and without taking water into account. In most cases, one step clearly dominated the mass consumption. For the biochemical routes, this was the reduction (step 3), and for the chemical routes, it was one step of the intermediate synthesis.

5.3.3
Mass Consumption of the Reduction Systems A–D

In order to make all four reduction methods comparable, the enrichment via crystallization for **C** and **D** was lumped into the hydrogenations steps. As can be inferred from Fig. 10a and b, the much higher overall mass consumption of the routes using biochemical reductions was mostly due to the reduction steps. Solvent demand in **A** and **B** was 5 to 13 times higher than in **C** and **D**. This is because of the relatively low concentration of the substrates in the aqueous solutions for the biocatalytic reduction and the need for extraction/precipitation of the reduction product. For the catalytic hydrogenations, the major input was the solvents for the hydrogenation (route **C**) and for recrystallization step (route **D**), respectively. The biochemical reductions needed significantly higher amounts of "substrates" since the formates have a much higher molecular weight than hydrogen. In addition, stoichiometrically used phosphate buffer in **A** and **B** contributes significantly to the "substrate volume". The demand for the key-substrate (keto ester and keto acid) differed significantly and reflects the chemical yields: it was lowest for **A**, and higher in **B** (1.3 times), in **C** (1.5 times) and in **D** (1.7 times). As for the input, the output in the reactions was mainly water and solvents. However, excess of reducing agents as well as coupled products such as CO_2 and phosphate salts contributed to the output in the biocatalytic reductions.

Fig. 10a Mass indices S^{-1} and environmental factors E of the enantioselective reduction steps **A–D** using EATOS.

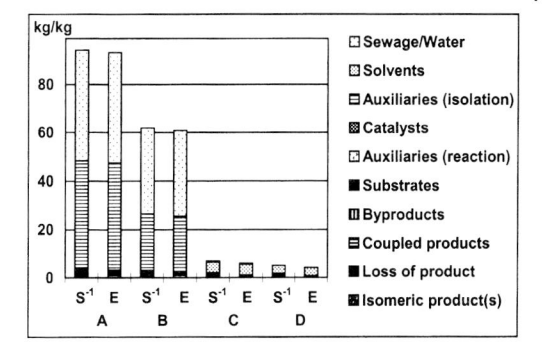

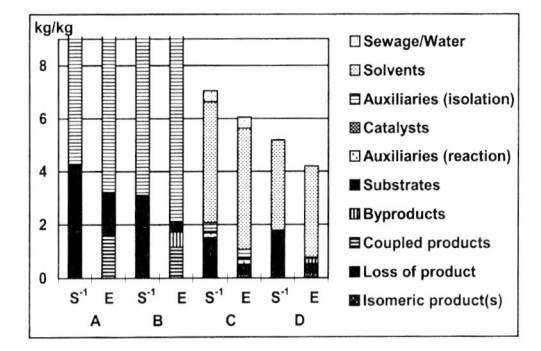

Fig. 10b Detailed view of Fig. 10a.

5.3.4
Problematic Chemicals: Environmental, Health and Safety Aspects (EHS)

As discussed in several papers and reviews [13, 15–17], it is apparent that not only the mass consumption but also substance-specific properties such as safety hazards, toxicology, ecotoxicology and eutrophication, etc., must be taken into account for a meaningful assessment. The nature of the chemicals used and the products formed determine the impact of a synthetic route on the environment. EATOS (Environmental Assessment Tool for Organic Synthesis) allows the estimating of this potential environmental impact. For this, the so-called unfriendliness quotient Q [18] varying from 0 (no effect) to 10 (largest effect) has been determined for several categories of effects according to the literature [13] in order to identify problematic substances. These calculations were carried out for all four routes but a detailed presentation is beyond the scope of this contribution and will be published elsewhere. The most important findings were, however, the following.

5.3.4.1 Safety Aspects
All routes required highly flammable or hazardous substances such as cyclohexane (**A**), methanol (**A–C**), toluene (**C, D**), diisopropyl ether (**D**), sodium ethoxide

and methoxide (all routes), sulfuric (A–C) and hydrochloric acid (D). In addition, routes C and D also used molecular hydrogen under pressure in combination with Pt- and Pd-catalysts, certainly a major safety issue. An additional concern was diisopropyl ether used in the crystallization in D, where explosive peroxides may be formed.

5.3.4.2 Health/Toxicology

All routes used toxic substances such as methanol, acetic acid (D), oxalic acid ethyl ester (A–D), cyclohexane (A), toluene, sulfuric acid and hydrochloric acid. Overall, no major differences were observed with the exception of viologen used in A: This substance is toxic to fish, meaning that the sewage of A has to be processed in a water treatment plant, which is not necessary in B.

5.3.4.3 Environment

The most dangerous substance with regard to the water hazard was toluene (C and D). All other substances only have a low water hazard potential. The potential for eutrophication was only given in A and B where phosphates were used. Persistent and accumulating substances were not used in any of the routes.

5.4
Overall Comparison of Routes A–D and General Conclusions

Tabs. 1 and 2 summarize the information described in Sections 5.2 and 5.3. For the four different reduction methods, the data are more detailed, also describing the nature of the chemicals used, the performance of the catalyst and major difficulties of the process (Tab. 1).

5.4.1
Conclusions for the Reduction Steps

- The yields and the ee values were significantly higher in the biocatalytic processes. Nevertheless, the chemical reductions needed substantially lower amounts of "substrates" because only hydrogen was used as a stoichiometric reducing agent and no buffer was necessary.
- Non-standard equipment was necessary for all enantioselective reductions; for the hydrogenations, a high pressure reactor, for the biocatalytic processes sophisticated continuous reaction systems with various feedback loops.
- All enantioselective reduction methods were rather demanding; either because a sensitive catalyst (in our case with a tedious reductive pretreatment [2a], even though this might not be necessary in the future [2f]) or a complex biological catalytic system had to be used.

Tab. 1 Comparison of reduction methods **A–D** shown in Figs. 4–7.

	A	B	C	D
Substrate	3	3	5	6
Enantioselective Catalyst	Immobilized *Proteus vulgaris*	Dehydrogenase enzyme	Pt-cinchona	Pt-cinchona
Reducing agent	HCOOH, viologen	HCOOH, NAD	H_2	H_2
Additives	(Poly)phosphates	Phosphates	–	–
Solvent	H_2O with buffer	H_2O with buffer	Toluene[a] or AcOH	Toluene
ee (%)	>99	>99.9	80[a]–92	76[a]–86
Isolated yield (%)	100	79	70[b]	66[b]
s/c (w/w)	50–100	25 000	200	40
s/c (mol/mol)	N/A	N/A	4000	700
ton	Living	High	4000	700
tof (h^{-1})	N/A	N/A	1000	300
Space-time yield [mol/(L · d)]	1.7–0.64	1.0	24	12
Eutrophication potential	Yes	Yes	No	No
Production scale	>20 kg	>1 kg	>500 kg	>100 kg
Problems	Complicated work-up	Complicated work-up	Very sensitive to substrate purity	Very sensitive to substrate purity

a) Used for the calculation.
b) Enrichment by crystallization lumped into the hydrogenation step.

- Space-time yield was high and work-up easy for chemical catalysts, whereas the two biocatalytic processes exhibited relatively low space-time yields and needed an elaborate work-up.
- Hydrogenation **C** had a higher catalyst productivity and a higher space-time yield compared with **D**.
- All reduction methods have a significant potential for improvement. This is especially true for **B** (lower buffer and solvent consumption seems feasible) and **C** (use of AcOH instead of toluene significantly increases the ee [2 b]).

5.4.2
Conclusions for the Overall Syntheses

- The variation in mass consumption for the four routes was significant and generally due to the reduction steps. However, these differences might be smaller when taking into account that most of the solvents can be recycled and that the biochemical routes would profit most in this respect.

Tab. 2 Overall comparison of the four routes **A–D** according to Fig. 3.

	A	B	C	D
Number of steps	4	4	4 [a]	3 [a]
Difficult steps [b]	1	1	2	2
Chemical yield in steps 1–4 (%)	100/100/99/100	100/100/79/100	100/100/70/66	90/58/96
Overall chemical yield (%)	99	79	46	50
Overall selectivity (%) [c]	12 (18 [d])	13 (20 [d])	8.4 (17 [d])	24
Overall atom economy (%)	30	34	51	52
Total kg material needed per kg **1** [e]	105	82	58	40
kg H_2O needed per kg **1** [e]	53	47	29	19
kg organic solvent needed per kg **1** [e]	42	25	14	16
kg $(EtOOC)_2$ needed per kg **1**	1.4	1.8	3.0	1.6
Safety issues (besides solvents)	–	–	Catalyst, H_2	Catalyst, H_2
Price of starting materials	Medium	Medium	Medium	Low
Overall economy	OK	OK	OK	Good

[a] ee upgrading included in the hydrogenation step.
[b] Reductions/hydrogenations.
[c] Mass of product/Σ mass of substrates.
[d] Mass of product/Σ (mass of substrates – excess EtOH).
[e] Recycling of solvents not taken into account.

- All routes going through intermediate **3** need a relatively expensive starting material. Alternative pathways to **3** or a related intermediate could significantly improve both mass consumption and economy.
- Among all EHS aspects, the safety concerns were the most significant. The use of flammable substances, especially hydrogen in combination with noble metal catalysts in **C** and **D** and possible peroxide formations in **D**, has to be addressed. Toxicity was a minor issue in all routes, except perhaps for sulfuric and hydrochloric acid, which have a very low workplace threshold value. In contrast, the eutrophication potential could be a major issue for the biochemical routes **A** and **B**.
- The shortest chemical route (**D**) had the lowest mass consumption. This was somewhat surprising, because the chemical yields and the ee in the enantioselective reduction step were lower compared with the other pathways. However, the shorter route, the high overall selectivity and a lower solvent demand obviously compensated for these shortcomings. It is most likely that this route is also the most economic one, even though this was not investigated in detail. The low mass consumption and the low price for acetophenone compared with ethyl hydrocinnamate strongly pointed into that direction. The main problems were associated with two hydrogenation steps which require sophisticated high pressure equipment, a rather expensive and not very active catalyst, use corrosive HCl, and have very high purity requirements for the substrates.

- The other chemical route **C** had the second lowest mass consumption. The main reason for the higher mass consumption compared with **D** was the relatively low yield in the distillation of **5**, which was necessary to obtain the required quality of **5**. The hydrogenation in **C** has a significant potential for optimization but faces the same problems as discussed for **D**. This route is only feasible if an enrichment later in the synthesis is possible.
- Both biochemical routes (**A** and **B**) are attractive alternatives to the chemical routes, provided that space-time yields and the solvent/water/buffer consumption can be improved.

5.5 References

1. R. A. Sheldon, Chimica Oggi 1991, May, 35.
2. a) P. Herold, A. F. Indolese, M. Studer, H. P. Jalett, H. U. Blaser, Tetrahedron 2000, 56, 6497. b) H. U. Blaser, H. P. Jalett, Stud. Surf. Sci. Catal. 1993, 78, 139. c) G. H. Sedelmeier, H. U. Blaser, H. P. Jalett, 1986; EP 206993, assigned to Ciba-Geigy AG, CAN 107:77434. d) G. Sedelmeier, P. F. Fauquex, EP 371408 A2, assigned to Ciba-Geigy AG, CAN 114:77770. e) E. Schmidt, O. Ghisalba, D. Gygax, G. Sedelmeier, J. Biotech. 1992, 24, 315. f) Degussa AG, oral communication.
3. a) G. Jödicke, O. Zenklusen, A. Weidenhaupt, K. Hungerbühler, J. Cleaner Production 1999, 7, 159. b) E. Schmidt, H.U. Blaser, P.F. Fauquex, G. Sedelmeier, F. Spindler in Microbial Reagents in Organic Synthesis (S. Servi, Ed.), Kluwer 1992, p. 377. c) H. U. Blaser, M. Studer, Chirality 1999, 11, 459.
4. A. Liese, U. Kragl, H. Kierkels, B. Schulze, Enzyme and Microbial Technology 2002, 673m and references cited.
5. H. Nohira, S. Yoshida, EP 329156, assigned to Kuraray Co., Ltd, 1989, CAN 112:55241.
6. N. Kumagai, H. I. Ohira, T. Onishi, K. Yamamoto, EP 421472, assigned to Kuraray Co., Ltd, 1991, CAN 115:71140.
7. L. Tang, L. Deng, J. Am. Chem. Soc. 2002, 124, 2870.
8. a) R. Schmid, E. A. Broger, Proceedings of the ChiralEurope '94 Symposium; Spring Innovation: Stockport, UK, 1994; p. 79. b) T. Saito, T. Yokozawa, T. Ishizaki, T. Moroi, N. Sayo, T. Miura, H. Kumobayashi, Adv. Synth. Catal. 2001, 343, 264. c) T. Benincori, O. Piccolo, S. Rizzo, F. Sannicolo, J. Org. Chem. 2002, 65, 8340.
9. a) D. H. Dao, M. Okamura, T. Akasaka, Y. Kawai, K. Hida, A. Ohno, Tetrahedron: Asymmetry 1998, 9, 2725, and references cited.
10. H. Kruszewska, I. Makuch, G. Grynkiewicz, J. Cybulski, Pol. J. Chem. 1996, 70, 1301.
11. W.-Q. Lin, Z. He, Y. Jing, X. Cui, H. Liu, A.-Q. Mi, Tetrahedron: Asymmetry 2001, 12, 1583.
12. K. Itoh, M. Kori, Y. Inada, K. Nishikawa, Y. Kawamatsu, H. Sugihara, Chem. Pharm. Bull. 1986, 34, 1128.
13. M. Eissen, J. O. Metzger, Chem. Eur. J. 2002, 8, 3580.
14. D. J. C. Constable, A. D. Curzons, V. L. Cunningham, Green Chem. 2002, 4, 521–527.
15. M. Eissen, J. O. Metzger, E. Schmidt, U. Schneidewind, Angew. Chem. 2002, 114, 402; Angew. Chem. Int. Ed. 2002, 41, 414.
16. G. Koller, U. Fischer, K. Hungerbühler, Ind. Eng. Chem. Res. 2000, 39, 960–972.
17. C.-P. Mak, H. Mühle, R. Achini, Chimia 1997, 51, 184.
18. R. A. Sheldon, Chemtech 1994, 3, 38.

6
Biocatalytic Approaches for the Large-Scale Production of Asymmetric Synthons

Nicholas M. Shaw, Karen T. Robins, and Andreas Kiener

Abstract

Lonza is a major custom manufacturer of intermediates for the life science industries, and uses biocatalysis in many of its processes. Some of the products for which a biotransformation is used are achiral, however, an important characteristic of enzymes is their chirality, and this characteristic is used by Lonza to produce a number of chiral synthons. Biotransformations for the synthesis of asymmetric compounds can be divided into two types of reactions: those where an achiral precursor is converted into a chiral product (true asymmetric synthesis) and those involving the resolution of a racemic mixture. Both types of reaction are used at Lonza, and several examples of each type of reaction are described for large-scale syntheses of chiral molecules that use a biotransformation in one or more of the steps.

6.1
Introduction

Biotransformations for the synthesis of asymmetric compounds can be divided into two types of reactions: those where an achiral precursor is converted into a chiral product (true asymmetric synthesis) and those involving the resolution of a racemic mixture. Both types of reaction are used at Lonza, which is a leading producer of intermediates for the life science industry. Lonza also uses biocatalysis for the synthesis of achiral molecules, for example, an immobilized whole-cell biocatalyst is used for the nitrile hydratase-catalyzed synthesis of thousands of tons per year of nicotinamide from 3-cyanopyridine.

The following examples illustrate Lonza's activities in large-scale biocatalysis for asymmetric synthesis. There are more examples of resolutions than true asymmetric syntheses, and this reflects the general current situation in the field of large-scale biotransformations. However, considerable research work is underway, both in universities and industry, to broaden the types of biocatalytic reactions that can be applied on a commercial scale.

Asymmetric Catalysis on Industrial Scale: Challenges, Approaches and Solutions
Edited by H. U. Blaser, E. Schmidt
Copyright © 2004 Wiley-VCH Verlag GmbH & Co. KGaA, Weinheim
ISBN: 3-527-30631-5

6.2
Asymmetric Biocatalysis

6.2.1
L-Carnitine

L-Carnitine is synthesized in the human liver from lysine and methionine, and functions in the transport of activated fatty acids across the inner mitochondrial membrane. It has pharmaceutical, food, and feed applications. Lonza produces L-carnitine in a whole-cell biocatalytic process that utilizes a naturally occurring pathway of four enzymes (Fig. 1). The pathway is analogous to the β-oxidation of fatty acids and occurs in a new genus of microorganism that is related to *Agrobacterium* and *Rhizobium* [1]. The introduction of the chiral center is by crotonobetainyl-CoA-hydrolase, which belongs to the lyase class of enzymes. The natural degradation of L-carnitine by carnitine dehydrogenase is prevented in the production strain by the inactivation of the L-carnitine dehydrogenase by frame shift mutagenesis [2]. Initially the production strain was inhibited by relatively low concentrations of L-carnitine, which made this mutant strain unsuitable for a large-scale industrial process. It was necessary to isolate other mutants that were resistant to higher concentrations of L-carnitine. Pharmaceutical quality L-carnitine is now produced at the multi-ton scale with a conversion of substrate to product of 99.5%, an ee value of 100% and <0.1% butyrobetaine. Less than 0.1% impurities and no D-carnitine are found in the final product.

The first generation process developed for the production of L-carnitine was a single-stage continuous process with cell-recycling that utilized the maintenance phase production that is characteristic of this strain. This process had a very high volumetric productivity of 130 g/L/d in the steady state [3]. Owing to the kinetics of this process the product solution contained 92% L-carnitine and 8% unconverted butyrobetaine. L-Carnitine and butyrobetaine have very similar physiochemical properties, which makes their separation from the product solution diffi-

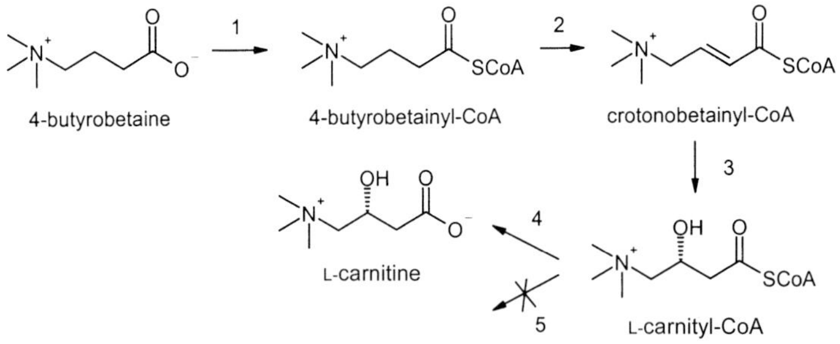

Fig. 1 The pathway for L-carnitine production in *Agrobacterium/Rhizobium* HK4. (1) 4-butyrobetainyl-CoA synthetase; (2) 4-butyrobetainyl-CoA dehydrogenase; (3) crotonobetainyl-CoA hydrolase; (4) thioesterase; (5) carnitine dehydrogenase.

Fig. 2 Comparison of the amount of waste produced by the chemically and biosynthetically produced L-carnitine.

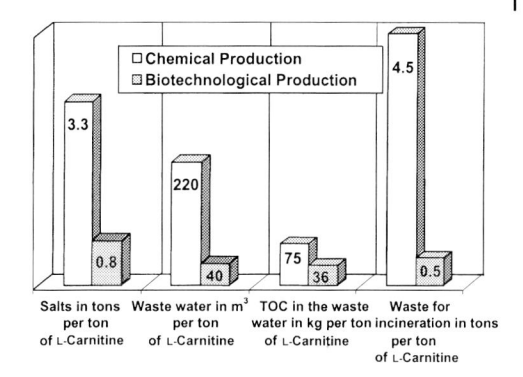

Chemical Production / Biotechnological Production

- Salts in tons per ton of L-Carnitine: 3.3 / 0.8
- Waste water in m³ per ton of L-Carnitine: 220 / 40
- TOC in the waste water in kg per ton of L-Carnitine: 75 / 36
- Waste for incineration in tons per ton of L-Carnitine: 4.5 / 0.5

cult. Two additional crystallization steps were required to produce >99% L-carnitine, which is required for the market. The goal of the second-generation process was to reduce the residual butyrobetaine concentration in the product solution so that the down-stream processing costs could be reduced. Comprehensive analysis based on the kinetic data obtained from the single stage continuous process with cell recycling was carried out so that the volumetric productivity, the bioconversion yield, and the potential savings of alternative processes could be calculated. These calculations revealed that a multi-stage continuous process, would have a much reduced volumetric productivity compared with the first-generation process, but double the productivity of a fed-batch process. In both cases the bioconversion of butyrobetaine is >99%. Both processes lead to a simplified down-stream processing and reduced costs by 40%. However, the higher volumetric productivity of the multi-stage continuous process is cancelled by the requirement of a technically complex plant. Despite the considerably lower productivity of the fed-batch process the simplified down stream processing, the simplified plant requirements, and the favorable economics makes the fed-batch process the process of choice [3, 4].

The environmental considerations for the biotechnological process are also important. A comparison of the waste stream (Fig. 2) from the bio-process with that from the Lonza chemical synthesis [5] showed that the amount of waste water, total organic carbon (TOC), salts, and waste for incineration were all considerably lower for the bio-process [6].

6.2.2
Asymmetric Reduction: (R)-Ethyl-4,4,4-trifluorohydroxybutanoate

(R)-ethyl-4,4,4-trifluoro-3-hydroxybutanoate is a building block for pharmaceuticals such as Befloxatone, an anti-depressant monoamine oxidase-A inhibitor from Synthelabo [7]. The process utilizes whole cells of *Escherichia coli* that contain two plasmids [8, 9]. One carries an aldehyde reductase gene from the yeast *Sporobolomyces salmonicolor*, which catalyzes the reduction of ethyl-4,4,4-trifluoroacetoacetate, and the second carries a glucose dehydrogenase gene from *Bacillus megater-*

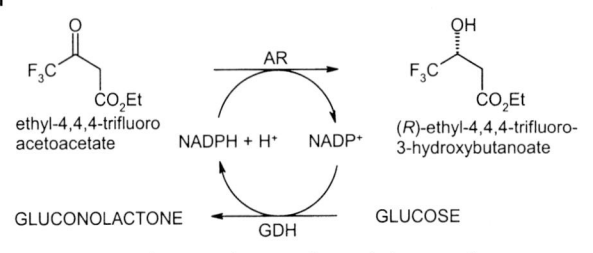

Fig. 3 Stereoselective reduction of (R)-ethyl-4,4,4-trifluoro-3-hydroxybutanoate by aldehyde reductase.

ium to generate NADPH from NADP$^+$ (Fig. 3). The strain was originally constructed to catalyze the stereoselective reduction of ethyl 4-chloro-3-oxobutanoate to ethyl 4-chloro-3-hydroxybutanoate, a precursor for L-carnitine synthesis [10]. Productivities of up to 300 g/L, and ee values of up to 92% were reported for the ethyl 4-chloro-3-hydroxybutanoate process.

The Lonza process is carried out in a water/butyl acetate two-phase system to avoid inhibition of the reductase by the substrate and product. An advantage of the two-phase system is that the cells are permeabilized, allowing the transfer of NADP$^+$ and NADPH through the cell wall. Cells of *Escherichia coli* JM109 containing the two plasmids are grown at 22 °C to avoid inclusion body formation. The cells are then washed and stored frozen before use in the biotransformation. The product has an ee value of >99%, and the yield is about 50%. Lonza is the world's leading manufacturer of ethyl-4,4,4-trifluoroacetoacetate, so there is optimal backward integration of the process.

NADP$^+$ is both essential and a major cost factor in the biotransformation, so securing a supply at a reasonable price was essential. The concentration of NADP$^+$ required for maximum activity is about 0.5 g/L. Experiments with a number of other esters showed that the isopropyl ester of 4,4,4-trifluoroacetoacetate is at least as good a substrate as the ethyl ester.

6.3
Resolution of Racemic Mixtures

6.3.1
(R)- and (S)-3,3,3-Trifluoro-2-hydroxy-2-methylpropionic Acid

(R)- and (S)-3,3,3-trifluoro-2-hydroxy-2-methylpropionic acid (Fig. 4) are intermediates for the synthesis of a number of potential pharmaceuticals, which include ATP-sensitive potassium channel openers for the treatment of incontinence [11], and inhibitors of pyruvate dehydrogenase kinase for the treatment of diabetes [12].

The aim was to develop efficient syntheses for these intermediates that yielded pure products with high ee values, and that were suitable for large-scale production. Several possible synthetic routes were evaluated. Some were purely chemical

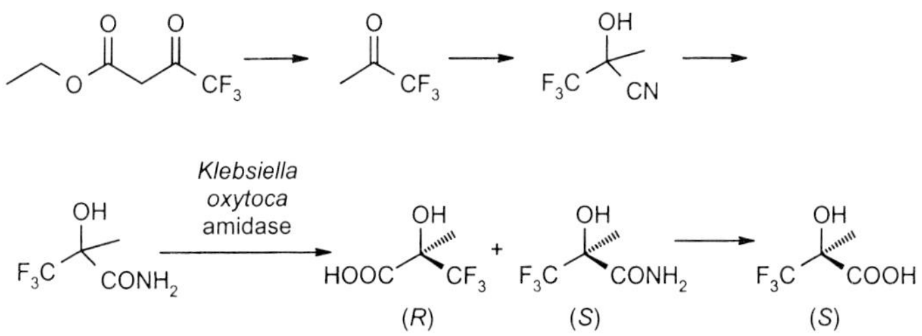

Fig. 4 Process scheme for the production of (R)- or (S)-3,3,3-trifluoro-2-hydroxy-2-methylpropionic acid.

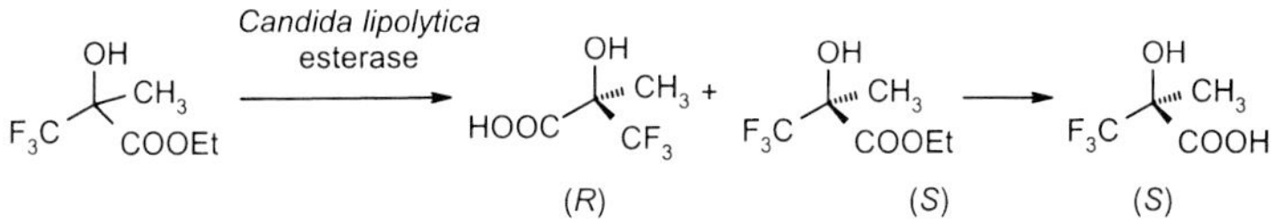

Fig. 5 The resolution of (R,S)-ethyl-3,3,3-trifluoro-2-hydroxy-2-methylpropionate with the esterase from *Candida lipolytica*.

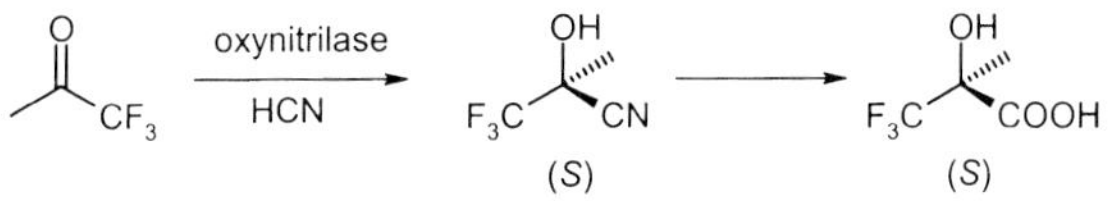

Fig. 6 A possible route to 3,3,3-trifluoro-2-hydroxy-2-methylpropionic acid using an oxynitrilase. The route to the (S)-cyanohydrin and (S)-acid is shown.

and others included biocatalytic steps. Of the routes tested, the biocatalytic routes were the most promising. For example, the resolution of (R,S)-ethyl-3,3,3-trifluoro-2-hydroxy-2-methylpropionate with the esterase from *Candida lipolytica* resulted in the (S)-ester with an ee value of 99% [13] (Fig. 5).

However, this route was not developed further because of the amount and resulting cost of the enzyme required to complete the reaction in a reasonable time. Other possible routes with one or more biocatalytic steps included those involving an enantioselective oxynitrilase reaction (Fig. 6). According to the choice of enzyme, it could be possible to form either the (R)- or the (S)-enantiomer. Fig. 7 depicts various routes starting from the racemic cyanohydrin. Nitrilases convert nitriles into the corresponding acids and are sometimes stereospecific. Nitrile hydratases convert nitriles into amides, and are also sometimes stereospecific. Amidases convert amides into the corresponding acids and are often stereospecific. Screening for enantioselective oxynitrilases [14] and for enantiospecific nitrilases [15] was started, but discontinued when the amidase route (below) was found to be successful.

For the successful route an enantiospecific amidase from *Klebsiella oxytoca* was isolated, characterized, and cloned and used to resolve (R,S)-3,3,3-trifluoro-2-hy-

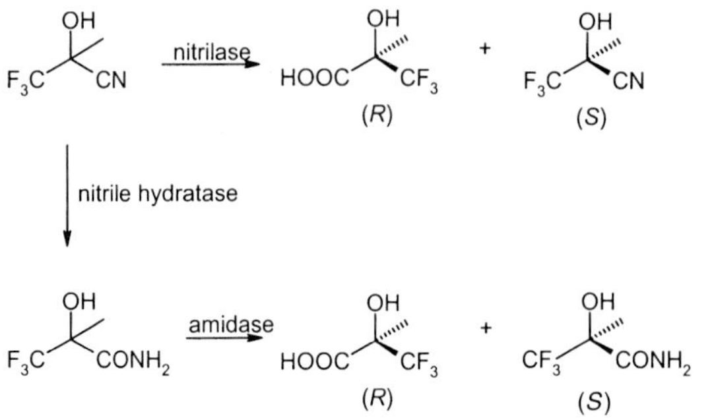

Fig. 7 3,3,3-Trifluoro-2-hydroxy-2-methylpropionic acid: possible routes starting from the racemic cyanohydrin. The routes to the (R)-acid are shown. The corresponding routes to the (S)-acid could also be possible.

droxy-2-methylpropionamide, giving (R)-3,3,3-trifluoro-2-hydroxy-2-methylpropionic acid and (S)-3,3,3-trifluoro-2-hydroxy-2-methylpropionamide. The (S)-amide could then be hydrolyzed chemically to (S)-3,3,3-trifluoro-2-hydroxy-2-methylpropionic acid. The process can therefore be adapted to produce both R- and S-enantiomers of 3,3,3-trifluoro-2-hydroxy-2-methylpropionic acid, or (S)-3,3,3-trifluoro-2-hydroxy-2-methylpropionamide. The biocatalytic step is part of a combined chemical and biocatalytic route (Fig. 4) that starts from the Lonza product ethyl-4,4,4-trifluoroacetoacetate, so again backward integration of the process is optimal. The products typically have a purity of greater than 98% and ee values of essentially 100% after isolation. The process has been used to produce 100 g amounts of the (S)-acid, and has been successfully scaled-up to produce 100 kg amounts of the (R)-acid, with the biotransformation carried out at the 1500 L scale [16, 17].

Cloning of the amidase gene into *Escherichia coli* was carried out for a number of reasons:

(i) To improve safety; *Klebsiella oxytoca* is a risk class 2 microorganism. Transfer of the amidase gene to a GRAS host such as an *Escherichia coli* K12 derivative facilitated handling and official registration procedures for the production process.
(ii) The productivity of the biotransformation was improved by cloning the amidase gene under the control of a strong promoter to improve its expression.
(iii) To avoid the slime-capsule problem encountered when using *Klebsiella oxytoca* as the production strain.
(iv) To have the possibility of using other microorganisms with, for example, higher substrate or product tolerances, as hosts for the cloned gene.

The amidase from *Klebsiella oxytoca* is robust, stable, and does not require cofactors. It can be heated to 70 °C for 10 min without loss of activity, and therefore, heat treatment of the biomass used in the biotransformation was used to stabilize the enzyme, presumably by inactivating proteases.

For the whole process attention was paid to safety, efficiency, and economics. The biotransformation was carried out in aqueous solution (which is often the case with biotransformations). Large volumes of aqueous solution containing relatively low product concentrations are difficult to handle in chemical plants (compared with normal processes in organic solutions), so particular attention was paid to the delivery of a biotransformation product solution suitable for downstream processing. It was subjected to ultra-filtration through a 70 kDa membrane to remove proteins that could cause foaming during extraction, and contamination of the product. Concentration by thin-film evaporation then removed as much water as possible before transfer to the chemical plant for product isolation.

6.3.2
(S)-2,2-Dimethylcyclopropane Carboxamide

(S)-2,2-dimethylcyclopropane carboxamide is an intermediate for the production of the Merck dehydropeptidase inhibitor Cilastatin, which is administered with penem and carbapenem antibiotics to prevent their degradation in the kidney [18]. Chemical and biotechnological processes were developed in parallel, but the bioprocess was simpler and resulted in a higher quality product.

The first generation process started with the chemical synthesis of the (R,S)-amide. There were several possible synthetic routes (Fig. 8) via the (R,S)-nitrile or (R,S)-acid [19, 20]. A microbial screening program resulted in the isolation of several bacterial strains containing amidases that could specifically hydrolyze the (R)-amide. One of these strains, *Comomonas acidivorans* A:18 was particularly effective [21]. After the hydrolysis of the unwanted isomer the product (S)-2,2-dimethylcyclopropane carboxamide was isolated from the bio-solution using a combination of salting-out and solvent extraction. This process had some intrinsic problems:

(i) The solubility of the racemic amide was only 0.5%, whereas that of the (S)-amide was 2.5%.
(ii) The (R,S)-amide crystals were not miscible with aqueous solutions. It was possible to improve their miscibility by the addition of detergent or by heating the solution to 90 °C to dissolve the crystals and then quickly cooling the resulting solution. This resulted in the formation of small miscible crystals.
(iii) The biomass yields were low and the (R,S)-amide was required as an inducer.
(iv) The amidase was susceptible to substrate and product inhibitions.
(v) The down-stream processing used organic solvents which caused safety and environmental considerations and resulted in a low yield of the isolated (S)-amide.
(vi) The (R)-acid was wasted.

The second-generation process introduced the hydrolysis of the racemic nitrile to the racemic amide (Fig. 9) by nitrile hydratase containing whole cells. The nitrile hydratase step led to several improvements:

(i) The yield was increased by 20% compared with the chemical synthesis.
(ii) The number of unit operations for the substrate synthesis was reduced.
(iii) Small miscible crystals of the (R,S)-amide were produced.
(iv) The kinetic behavior and specificity of the nitrile hydratase and amidase allowed the two biotransformation steps to be integrated.

Fig. 8 Possible synthetic routes for the production of (R,S)-2,2-dimethylcyclopropane carboxamide.

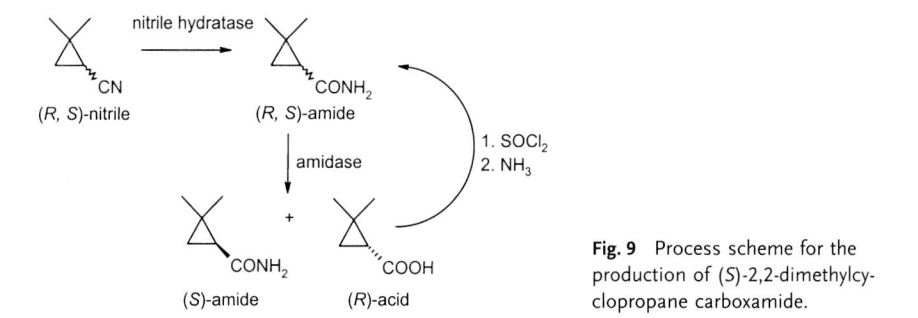

Fig. 9 Process scheme for the production of (S)-2,2-dimethylcyclopropane carboxamide.

This resulted in a two-step, one pot biotransformation. In this process the gene for the amidase was also cloned into *Escherichia coli*, resulting in a number of additional advantages:

(i) Only a minimal growth medium was required.
(ii) No amide was required in the growth medium to induce the amidase.
(iii) The biomass yield was five times higher.
(iv) The activity of the amidase was four times higher in the recombinant strain compared with the wild type strain.

(v) The production cost of the amidase was reduced by 90%.
(vi) Unexpectedly, the substrate and product inhibition of the amidase was removed.

In the first step the (R,S)-nitrile is rapidly and quantitatively hydrolyzed to the (R,S)-amide. The amidase containing biomass is then added so that the (R)-amide is specifically hydrolyzed to the (R)-acid, and the product, the (S)-amide, remains. A completely new isolation process was developed: ultra-filtration, electrodialysis, ion-exchange chromatography, reverse osmosis, crystallization, centrifugation, and drying. The product has been produced at a 15 m^3 scale and has an ee value of >98%. The isolated yield calculated from the nitrile was >35%. The (R)-acid can be recycled by reacting it with thionyl chloride and ammonia to produce the (R,S)-amide. This results in minimal waste and higher yields.

6.3.3
CBZ-D-proline [(R)-N-CBZ-proline]

A further example of a cyclic amino acid that is produced by Lonza for use as a pharmaceutical intermediate is CBZ-D-proline [22, 23]), which is used for the synthesis of Eletriptan, a drug from Pfizer for the treatment of migraine [24]. L-Proline is chemically racemized and derivatized to give racemic N-CBZ-proline, which is then stereospecifically hydrolyzed by a proline acylase in a strain of Arthrobacter sp. (Fig. 10). This strain was newly isolated from soil samples using enrichment methods. A fed-batch process was developed for the production of (R)-N-CBZ-proline. The cells are first grown on glucose and L-proline, and then induced with racemis N-CBZ-proline. After the induction of the acylase the (R,S)-N-CBZ-proline is fed into the fermenter at a constant rate. The (S)-N-CBZ-proline is converted into (S)-proline (L-proline), benzylalcohol, and carbon dioxide. The (R)-N-CBZ-proline is obtained with a product concentration of up to 70 g/L, and with an ee value of >99.5%. The process has been scaled-up to produce 100 kg amounts. The chemistry for the preparation of the racemic N-CBZ-proline is carried out in water to avoid solvent changes, and the final isolation procedure is a simple extraction that yields (R)-N-CBZ-proline and an aqueous solution of (S)-proline that can be recycled as starting material.

An alternative synthesis starting from L-arginine, and involving the biotransformation of (S)-5-[(amino-iminomethyl) amino]-2-chloropentanoic acid (L-Cl-arginine) to (S)-5-amino-2-chloropentanoic acid (L-Cl-ornithine), which then sponta-

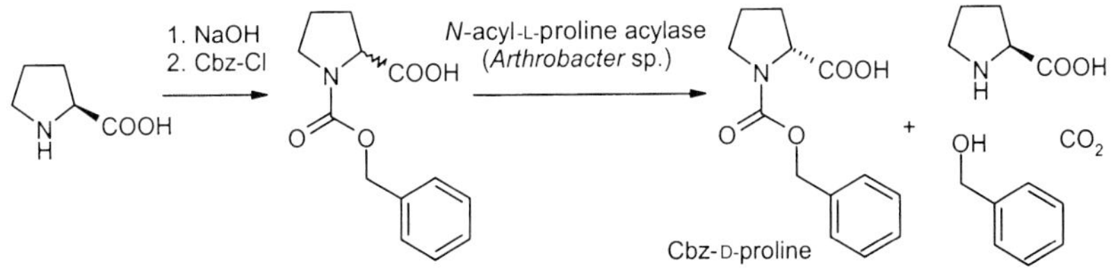

Fig. 10 Process scheme for the production of CBZ-D-proline.

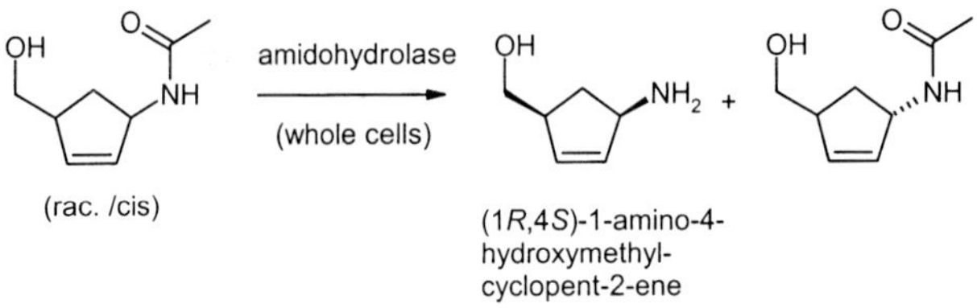

Fig. 11 Process scheme for the production of (1R,4S)-1-amino-4-hydroxymethyl-cyclopent-2-ene.

neously converted into D-proline with inversion of configuration, was not developed further due to low yields [25].

6.3.4
(1R,4S)-1-Amino-4-hydroxymethyl-cyclopent-2-ene

(1R,4S)-1-amino-4-hydroxymethyl-cyclopent-2-ene is an intermediate for the Glaxo anti-HIV drug Abacavir [26]. The racemic (cis) N-acetyl amino alcohol was used as the substrate for the selection and screening of microorganisms that could release the acetyl group by hydrolysis and then grow with acetate as the carbon source. In this way microorganisms were isolated that contained a stereospecific amidohydrolase for the hydrolysis of the racemic N-acetyl amino alcohol (Fig. 11). The reaction yields the amino alcohol product [(1R,4S)-1-amino-4-hydroxymethyl-cyclopent-2-ene] and the non-hydrolyzed (1S,4R)-N-acetyl amino alcohol, which cannot be recycled [27]. For scale-up to ton amounts the microorganism of choice was *Rhodococcus erythropolis* CB101 (DSM10686). Cells were grown on acetate as the carbon source, and then the amidohydrolase was induced with the reaction substrate. Incubation of the whole cells containing the induced amidohydrolase with substrate then gave product with a yield of about 43% and an ee value of 90%. After crystallization an ee value of about 98% was achieved.

An alternative process with commercially available penicillin G acylases was also investigated. The enzyme cleaved the substrate, N-phenylacetyl-1-amino-4-hydroxymethyl-cyclopent-2-ene, with high activity, but yields of up to only 40% were reached, with ee values of 80%. Consequently this process was not investigated further.

6.4
Summary

The examples given above show that biocatalysis is used on an industrial scale (100 kg to ton amounts) for the synthesis of chiral molecules. Both true asymmetric synthesis and the resolution of racemates are employed to produce compounds that are difficult or inefficient to synthesize by purely chemical means. All of the processes described use both chemical steps and one or more biocatalytic steps. A synergy between chemistry and biotechnology is therefore essential, and this is one of Lonza's strengths.

6.5 References

1 T.P. Zimmermann, K.T. Robins, J. Werlen, F.W.J.M.M. Hoeks, in Chirality in Industry II (A.N. Collins, G.N. Sheldrake, J. Crosby, Eds.), John Wiley and Sons Ltd, Chichester, 1997.
2 H. Kulla, P. Lehky, EP 0 158 194 B1.
3 F.W.J.M.M. Hoeks, H. Kulla, H.P. Meyer, J. Biotechnol. 1992, 22, 117–127.
4 H. Kulla, Chimia 1991, 45, 81–85.
5 R. Voeffray, J.-C. Perlberger, L. Tenud, J. Gosteli, Helv. Chim. Acta 1987, 70, 2058–2064.
6 H.-P. Meyer, A. Kiener, R. Imwinkelried, N. Shaw, Chimia 1997, 51, 287–289.
7 V. Rovei, D. Caille, O. Curet, D. Ego, F.-X. Jarreau,. J. Neural. Transm. 1994, 41, 339–347.
8 M. Kataoka, L.P.S. Rohani, M. Wada, K. Kita, H. Yanase, I. Urabe, S. Shimizu, Biosci. Biotechnol. Biochem. 1998, 62, 167–169.
9 M. Petersen, O. Birch, S. Shimizu, A. Kiener, M.-L. Hischier, S. Thoni, WO 99/42590.
10 S. Shimizu, M. Kataoka, A. Morishita, M. Katoh, T. Morikawa, T. Miyoshi, H. Yamada, Biotechnol. Lett. 1990, 12, 593–596.
11 C.J. Ohnmacht, K. Russell, J.R. Empfield, C.A. Frank, K.H. Gibson, D.R. Mayhugh, F.M. McLaren, H.S. Shapiro, F.J. Brown, D.A. Trainor, C. Ceccarelli, M.M. Lin, B.B. Masek, J.M. Forst, R.J. Harris, J.M. Hulsizer, J.J. Lewis, S.M. Silverman, R.W. Smith, P.J. Warwick, S.T. Kau, A.L. Chun, T.L. Grant, B.B. Howe, J.H. Li, S. Trivedi, T.J. Halterman, C. Yochim, M.C. Dyroff, M. Kirkland, K.L Neilson, J. Med. Chem. 1996, 39, 4592–4601.
12 T.D. Aicher, R.C. Anderson, G.R. Bebernitz, G.M. Coppola, C.F. Jewell, D.C. Knorr, C. Liu, D.M. Sperbeck, L.J. Brand, R.J. Strohschein, J. Gao, C.C. Vinluan, S.S. Shetty, C. Dragland, E.L. Kaplan, D. DelGrande, A. Islam, X. Liu, R.J. Lozito, W.M. Maniara, R.E. Walter, W.R. Mann, J. Med. Chem. 1999, 42, 2741–2746.
13 W. Brieden, A. Naughton, K. Robins, N. Shaw, A. Tinschert, T. Zimmermann, DE 19725802 A1.
14 F. Effenberger, B. Hörsch, F. Weingart, T. Ziegler, S. Kühner, Tetrahedron Lett. 1991, pp. 2605–2608.
15 M. Wieser, T. Nagasawa, in Stereoselective Biocatalysis (R. Patel, Ed.), Marcel Dekker, New York, 2000.
16 W. Brieden, A. Naughton, K. Robins, N. Shaw, A. Tinschert, T. Zimmermann, WO 98/01568.
17 N.M. Shaw, A. Naughton, K. Robins, A. Tinschert, E. Schmid, M.-L. Hischier, V. Venetz, J. Werlen, T. Zimmermann, W. Brieden, P. de Riedmatten, J.-P. Roduit, B. Zimmermann, R. Neumüller, Org. Proc. Res. Dev. 2002, 6, 497–504.
18 J. Birnbaum, F.M. Kahan, H. Kropp, J.S. Macdonald, Am. J. Med. 1985, 78, 3–21.
19 P. Hanselmann, EP 0491323 B1.
20 T. Meul, U. Kaempfen, EP 0491330 B1.
21 T. Zimmermann, K. Robins, O.M. Birch, E. Boehlen, EP 0524604 B1.
22 M. Petersen, M. Sauter, Chimia 1999, 53, 608–612.
23 M. Sauter, D. Venetz, F. Henzen, D. Schmidhalter, G. Pfaffen, O. Werbitsky, WO 97/33987.
24 J. Ngo, X. Rabasseda, J. Castaner, Drugs Future 1997, 22, 221–224.
25 C. Bernegger-Egli, K.-S. Etter, F. Studer, F. Brux, O.M. Birch, J. Mol. Cat. B: Enz. 1999, 6, 359–367.
26 R.H. Foster, D. Faulds, Drugs 1998, 55, 729–736.
27 C. Bernegger-Egli, O.M. Birch, P. Bossard, W. Brieden, F. Brux, K. Burgdorf, L. Duc, K.-S. Etter, Y. Guggisberg, WO 97/45529.

7
7-Aminocephalosporanic Acid –
Chemical Versus Enzymatic Production Process
Thomas Bayer

Abstract

7-Aminocephalosporanic acid (7-ACA) is a key intermediate for the synthesis of semisynthetic cephalosporin antibiotics. It is produced from the fermentation product cephalosporin C. Today, 7-ACA is mainly produced via a chemical method. In this process, highly purified cephalosporin C is required as a raw material. The carboxy and amino functions are protected with trimethylsilyl groups. The amide is activated by chlorination with phosphorous pentachloride. The reaction steps are carried out at –40 to –60 °C in chlorinated hydrocarbons as the solvent. After hydrolysis the 7-ACA is isolated. Because of the hazardous chemicals used, the aqueous phase has to be incinerated.

The enzymatic process uses water as the solvent and two immobilized enzymes as catalysts at room temperature. In a first step cephalosporin C is deaminated to a-ketoadipyl-7-ACA using a carrier-fixed d-amino acid oxidase in the presence of oxygen. Under reaction conditions the a-keto intermediate is oxidatively decarboxylated to glutaryl-7-ACA. In a second step the glutaryl-7-ACA is then hydrolyzed to 7-ACA by a carrier-fixed glutaryl amidase.

The main advantages of the enzymatic process are that: i) no hazardous chemicals are used, ii) emissions and waste for incineration are reduced and iii) reactions are carried out at room temperature thus reducing energy consumption.

7.1
Introduction

Cephalosporins belong, like penicillins, to the β-lactams, medicinal antibiotics produced by yeast, fungi and bacteria, but in general, they can treat a broader range of infections. Cephalosporins are the drugs of choice for postoperative care (preventing infections) and for fighting infections caused by a broad spectrum of gram positive and gram negative bacteria that had grown resistant to penicillins. Cephalosporins interfere with bacterial cell wall synthesis and result in the formation of defective cell walls and osmotically unstable spheroplasts. In the antibiotics market this widely used class of antibiotics has a share of about 30% by value. In 2000 cephalosporins and combinations of them totaled US$6.9 billion in sales, making it the seventh-largest therapy class worldwide [1].

The original filamentous fungus *Cephalosporium acremonium* (now *Acremonium chrysogenum*) was discovered in a sewer outlet in Sardinia by G. Brotzu in 1945. The weakly antibiotic compound cephalosporin C (Ceph C), which showed some activity against penicillin-resistant cultures, was isolated by Newton and Abraham [2]. By cleavage of the Ceph C side-chain 7-aminocephalosporanic acid (7-ACA) is obtained, which like its congener 6-aminopenicillanic acid (6-APA) is used as a starting material for most of the synthetic cephalosporin drugs available today. Chemical modifications of the cephem ring structure of 7-ACA (Fig. 1) are possible at the 7-amino group, the 3-acetyl group or the 2-carboxy group leading to a broad variety of drug substances. Some oral cephalosporin antibiotics are obtained from the intermediate 7-aminodeacetoxycephalosporanic acid (7-ADCA), which is derived chemically or enzymatically from penicillin G or V.

The demand for 7-ACA rose by 9% in 1999, due to increased consumption for the production of several semisynthetic cephalosporins for which patents are expiring, and increasing demand from China. The market volume was estimated to reach about 2000 tons per year by the year 2000 representing a market value of about US$400 million [3, 4]. Like several other pharmaceutical companies the former Hoechst AG produced 7-ACA as a starting material for the production of their cephalosporin antibiotics Claforan® (cefotaxime-Na), Modivid® (cefodizime-di-Na), Cefrom® (cefpirome-sulfate) and Cobactan® (cefquinome-sulfate) with a sales volume up to about US$400 million per year.

At that time 7-ACA was produced by chemical synthesis with a capacity of about 80 metric tons per year. Beginning at the end of the 1980s, the development of a two-step biocatalytic process with the enzymes, D-amino acid oxidase and glutaryl-7-ACA acylase, was started. Because of low titers of the glutaryl acylase in the natural strain the genes were cloned, sequenced and expressed in *E. coli*. Until 1993 public and legal restraints existed in Germany for the use of genetically modified organisms (GMOs). Since then, clear regulations have been in place for handling GMOs in enclosed systems, largely based on harmonized EC guidelines.

In 1994 Hoechst decided to invest about US$16 million into a biocatalytic production plant in Frankfurt, Germany, and to close the chemical 7-ACA plant. The capacity was increased to 180 metric tons per year, to ensure the availability for captive use

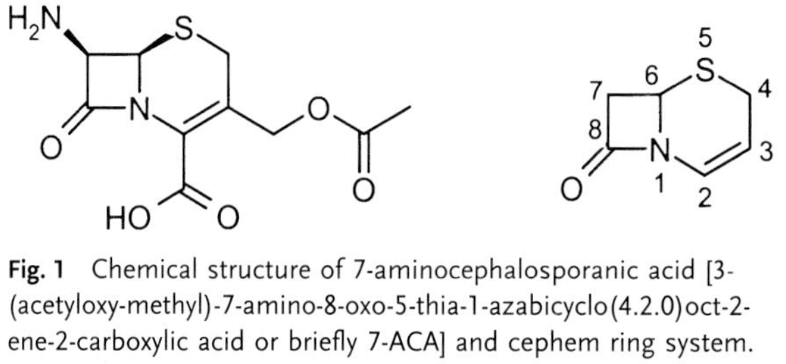

Fig. 1 Chemical structure of 7-aminocephalosporanic acid [3-(acetyloxy-methyl)-7-amino-8-oxo-5-thia-1-azabicyclo(4.2.0)oct-2-ene-2-carboxylic acid or briefly 7-ACA] and cephem ring system.

and for supply on to the world market. The new two-step enzymatic 7-ACA plant started in 1996. With the formation of Aventis Pharma, a joint venture of Hoechst and Rhone-Poulenc, it was decided that the business with fermentative produced antibiotics and precursors was no longer a core business. In November 1998 the Austrian company Biochemie, a member of the Novartis Group, bought the fermentation and downstream processing plants, including the 7-ACA plant.

Owing to the increasing demand, Biochemie invested about US$ 50 million in a new biocatalytic 7-ACA production plant in Frankfurt, including an enzyme production plant, with a capacity exceeding 400 metric tons per year. The new plant went on stream at the beginning of 2001. Today Biochemie supplies about 25% of global demand for the intermediate 7-ACA [5].

7.2
Synthesis of 7-ACA

Various methods are known to produce 7-ACA from cephalosporin C (Ceph C) by removing the α-aminoadipyl side-chain. They can be classified into three types, the chemical process, a two-step enzymatic process and an enzymatic process in which the side-chain is directly removed from Ceph C. Today two processes are running commercially on an industrial scale, the classical chemical process and the modern two-step biocatalytic process (Fig. 2). Until now the favorable direct process is less effective, because of low conversion.

7-ACA synthesis starts with the fermentation product Ceph C, a weakly antibiotic compound, generated by cultivation of the filamentous fungus *Acremonium*

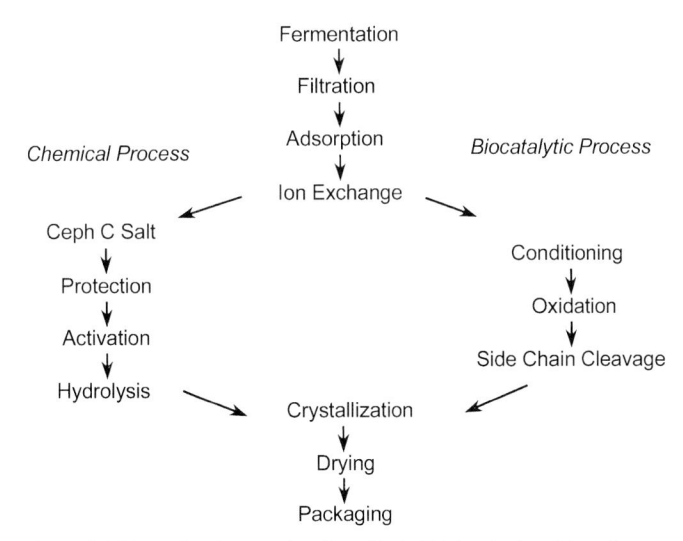

Fig. 2 7-ACA production starting from Ceph C (chemical and two-step biocatalytic process).

chrysogenum (formerly *Cephalosporium acremonium*). After removal of the biomass by filtration, laborious adsorption and ion exchange steps are necessary to purify and concentrate the strongly hydrophilic Ceph C [6]. The substance can be isolated as free acid or precipitated as a salt (e.g., sodium, potassium or zinc). The quality of Ceph C is important, in particular the content of desacetyl and desacetoxy cephalosporin C compounds, precursors of the biosynthesis of Ceph C, have to be limited, because they are transformed into the corresponding 7-ACA derivatives and thus influence the quality of the isolated product. The desired product becomes available by cleavage of the *a*-aminoadipyl side-chain and is isolated by crystallization at the isoelectric point. Chemical and biocatalytic processes will be described and compared.

7.2.1
The Chemical 7-ACA Route

The chemical synthesis of 7-ACA (Fig. 3) is a classical organic route, which uses equimolar amounts of chemicals. A preferred process is the so-called iminohalide/iminoether deacylation route, involving formation of the iminochloride by reaction with PCl_5, conversion into the iminoether by treatment with alcohol and hydrolysis with water [7]. Highly purified cephalosporin C, isolated, for example as the zinc salt, is required as a dried solid as the starting material.

In a water-free process in dichloromethane as the solvent, trimethylchlorosilane (3 mol per mol Ceph C) is used to protect the carboxy and amino functions with trimethylsilyl groups. *N,N*-dimethylaniline is used to bind the released HCl. The solution is cooled down to $-40\,°C$ to $-60\,°C$ and the amide is activated by chlorinating with phosphorous pentachloride; humidity has to be avoided. After hydrolysis and phase separation 7-ACA is isolated by crystallization from the aqueous phase with high quality and yield.

Glass-lined equipment is necessary with respect to the corrosiveness of the chemicals used. Approximately 7.5 tons of chemicals (without solvents) are used to produce 1 ton of 7-ACA. Because of the low biodegradability and bacterial toxicity of the chemicals used, biological waste water treatment is prohibited and the aqueous phase has to be incinerated. The organic phase (chlorinated hydrocarbons) has to be distilled and recycled. Distillation residues are incinerated too. In total about 31 tons of residues per ton of 7-ACA have to be incinerated.

With increasing environmental standards and a new law that forced Hoechst in Frankfurt to pay additional taxes on wastes to be incinerated, the chemical process came under more and more economic pressure. The challenge was to develop a sustainable process or to shut down the 7-ACA production plant in Frankfurt. Hoechst decided to search for alternative processes, in particular for biocatalytic ones.

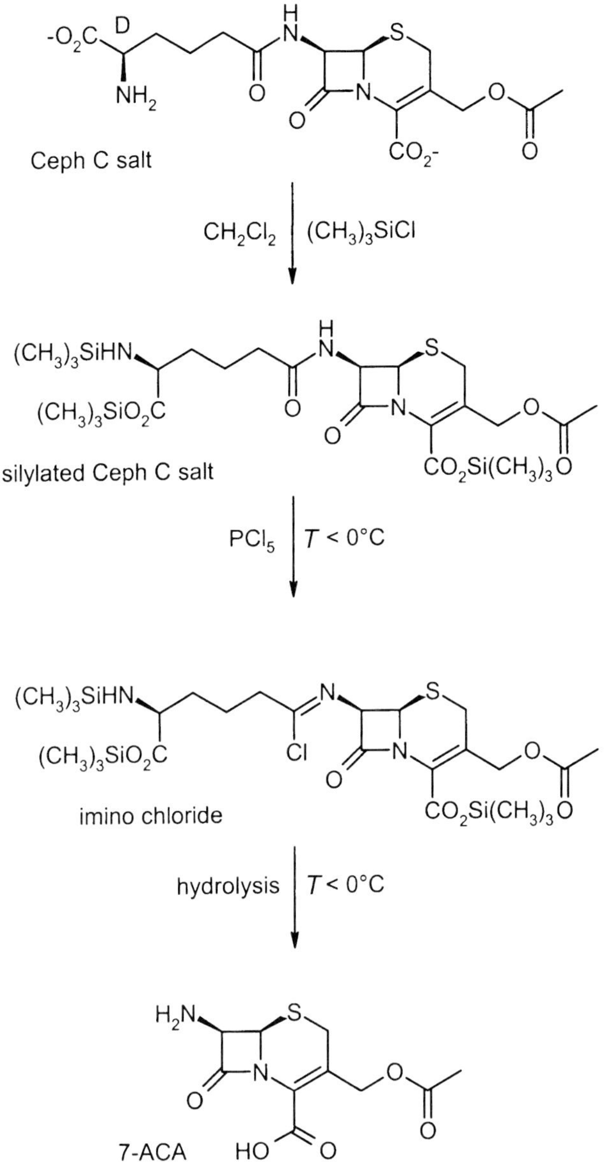

Fig. 3 Chemical synthesis of 7-ACA.

7.2.2
Single Step Biocatalytic 7-ACA Synthesis

For the penicillins a one-step biocatalytic route for the production of 6-aminopenicillin acid (6-APA) from the fermentation product penicillin (Pen G or Pen V) was well known. Beginning at the end of 1970s the chemical process (almost similar to the

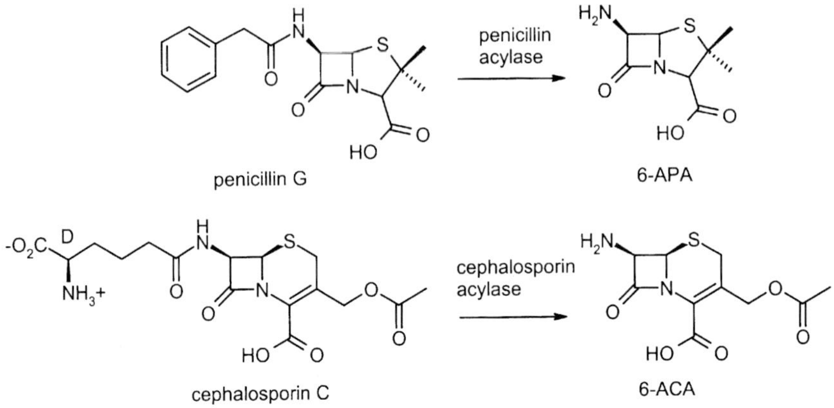

Fig. 4 Enzymatic side chain cleavage of β-lactams.

above described chemical 7-ACA process) was substituted world-wide by a one-step enzymatic process using penicillin acylase (Fig. 4). Using the experience of its own developments for a biocatalytic 6-APA route, in the 1980s Hoechst began research and development of an alternative, enzymatic process for the synthesis of 7-ACA.

The direct cleavage of Ceph C to 7-ACA and D-α-aminoadipic acid by a cephalosporin acylase would be the preferred route. A large number of enzymes were isolated from different microorganisms, e.g., *Pseudomonas diminuta*, *Bacillus megaterium*, but substrate specificity for Ceph C is low. The conversion obtained was on a low level, e.g., 8% in 4 hours [8–10]. Therefore, today, the direct one-step cleavage is not feasible on an industrial scale and a two-step process had to be developed.

7.2.3
Two-step Biocatalytic 7-ACA Synthesis

In a first step the chiral center at the α-aminoadipyl side-chain has to be eliminated. Besides commercially available D-amino acid oxidase (DAO) from pig kidney for laboratory use, the enzyme is also synthesized by bacteria, yeasts and fungi. This enzyme can be used for the resolution of D,L-amino acids in various solutions, but of particular interest is its capacity for oxidative deamination of Ceph C. The first research was carried out by Glaxo Laboratories Ltd. [11] where DAO isolated from different fungal cells was used. Under the reaction conditions used, in the presence of oxygen as a co-substrate, the enzyme deaminates cephalosporin C to give α-ketoadipyl-7-ACA, ammonia and hydrogen peroxide (Fig. 5).

The keto intermediate decarboxylates spontaneously to 7-β-(4-carboxybutanamido)-cephalosporanic acid (glutaryl-7-ACA) and carbon dioxide by chemical reaction with the hydrogen peroxide formed during the enzymatic reaction. The occurrence of the byproduct 7-ACA-sulfoxide is possible. In general a reaction mixture is produced, caused by the catalase also produced by the yeast. However, the α-ketoacid intermediate reacts poorly with the deacylation enzyme in the second step.

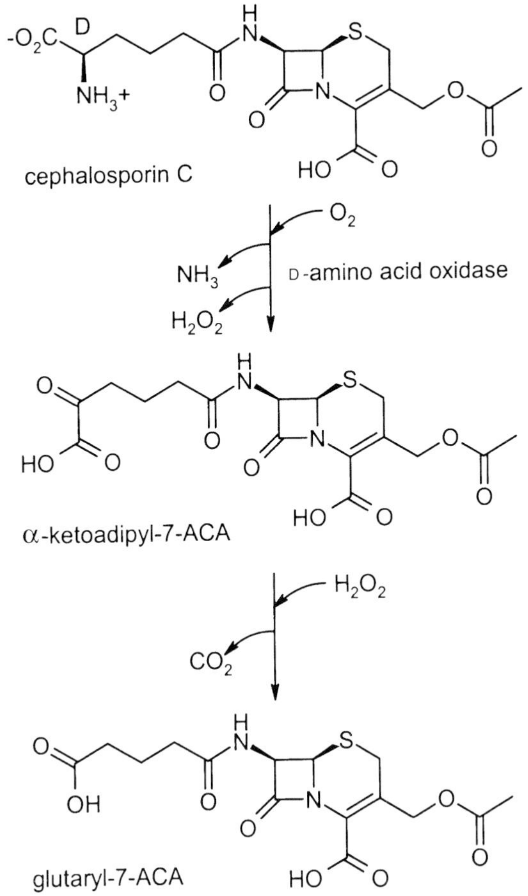

Fig. 5 Biocatalytic oxidation of Ceph C with D-amino acid oxidase (DAO).

Therefore it is necessary to reduce the amount of catalase in the purified DAO. Hoechst decided to develop a fermentation and downstream process by using the yeast *Trigonopsis variabilis*, which is distinctive in being a good DAO producer. To reduce the formation of the undesired desacetyl byproduct and accumulation of the intermediate α-ketoadipyl-7-ACA, the presence of esterases and catalase, which occur naturally in the yeast, had to be avoided. Selective reduction of these enzymes was possible, e.g., by heating the cells up to 60–85 °C for several seconds [12]. After several years of development of fermentation and downstream processing from lab to production scale this goal was achieved and the enzyme titer could be increased by a factor of about 100. After purification and isolation, DAO of high quality and in sufficient amount at an acceptable cost was available. For economic reasons, the reuse of the enzyme is recommended. Hence DAO is covalently immobilized on a polymeric carrier.

In a second step the glutaryl-7-ACA is hydrolyzed to 7-ACA and glutaric acid (Fig. 6) by means of a glutaryl-7-ACA acylase (GA). The enzyme can be isolated

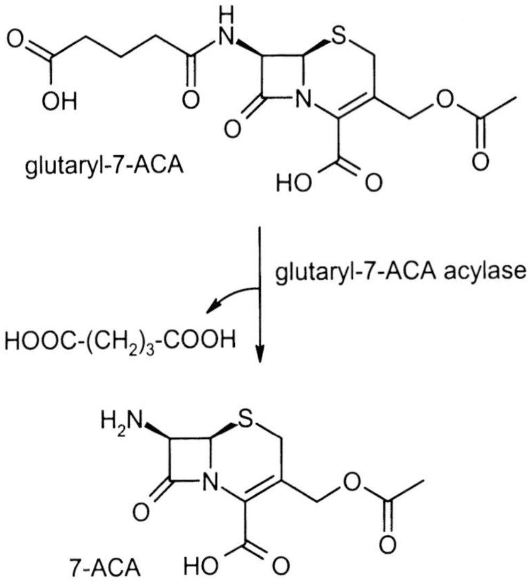

Fig. 6 Deacylation of glutaryl-7-ACA to 7-ACA with glutaryl-7-ACA acylase (GA).

from different organisms, e.g., *Acinetobacter* spp., *Pseudomonas* spp. Most of the glutaryl-7-ACA acylases consist of two non-identical subunits, the α-subunit and the β-subunit, which are generated from a common precursor protein in a processing pathway similar to that of penicillin G acylase in *Escherichia coli* [13].

Unfortunately the enzyme titer in the fermentation broth, even after classical mutation and several years of process development, was too low for the biocatalytic process. With respect to the economics, the gene encoding for glutaryl-7-ACA acylase has been cloned, sequenced and expressed in *E. coli* to produce sufficient amounts. The enzyme titer could be increased by a factor of more than 100. Even the purification of the enzyme became much easier and resulted in higher yields with less side-activities, e.g., esterases. Two chromatographic purification steps were substituted by crystallization of the enzyme. The enzyme crystals could be stored long term without deactivation. To allow for reuse, the glutaryl-7-ACA acylase was immobilized on a polymeric carrier.

This effort reduced the number of fermentations and the raw material consumption dramatically, simplified the downstream process and therefore reduced the enzyme costs considerably (Tab. 1). Even the size of the fermenters could be reduced to a pilot plant scale.

Tab. 1 Production of glutaryl-7-ACA acylase natural source versus GMO (natural source = 100%).

	Natural source (%)	GMO (%)
Culture broth	100	2.3
Raw materials	100	3.9
Enzyme cost	100	15.3

Tab. 2 Comparison of the raw material costs (chemical process = 100%).

Process	
Chemical	100%
Biocatalytic (GA from natural source)	114%
Biocatalytic (GA from GMO)	27%

All things considered, the application of recombinant DNA technology for the production of the glutaryl-7-ACA acylase made the biocatalytic process economically superior to the chemical route (Tab. 2) and allowed for the investment in a new biocatalytic 7-ACA production plant, after overcoming restraints against usage of GMOs in Germany.

7.2.4
Two-Step Biocatalytic Process

The biocatalytic process takes place in stirred tank reactors with sieves in the bottom and a working volume of several cubic meters (Fig. 7). The first biocatalytic step is a three-phase reaction with a solid biocatalyst, immobilized DAO on a spherical carrier, the liquid Ceph C solution and oxygen or air as the gaseous phase. The second step is a two-phase reaction with immobilized GA on a solid carrier and the glutaryl-7-ACA in solution.

The two-stage biocatalytic reaction can be performed in a single reactor [14], but the separation of the two reactions is preferred because of different reaction parameters (e.g., pH value, temperature, oxygen) and stability of the enzymes used. With water as the solvent and enzymes fixed on a carrier, the process runs in a repeated batch mode at room temperature (20–30 °C). Higher temperatures lead to increased reaction rates, but also to higher byproduct formation and reduced stability of the biocatalysts. A pH value between 7.0 and 8.5 is recommended with respect to thermodynamics, enzyme activities and stability and formation of byproducts. The use of cells is not recommended with respect to operational stability and possible product contamination. Therefore purified enzymes covalently immobilized on a polymeric carrier are chosen for the industrial process for both steps. The particle diameter of the spherical biocatalyst is about 100–300 µm, to allow for acceptable mass transfer and filtration times.

The reuse of the expensive biocatalysts is a prerequisite for the economy of the biocatalytic process. On a lab-scale the carrier-fixed enzymes can be used for more than 100 cycles (DAO) and 180 cycles (GA), before reaching half of the starting activity [15]. Prolonging the reaction time can compensate for the decreasing activity. As claimed by reference [15] for the lab-scale preparation of 1 kg 7-ACA about 1.2 kU D-amino acid oxidase and 1.5 kU glutaryl-7-ACA acylase are consumed, but operational stability is dependent on scale. In production vessels gradients, e.g., pH value and shear stress, are different and could influence the operational stability of the biocatalysts, therefore a higher biocatalyst consumption is usually realistic.

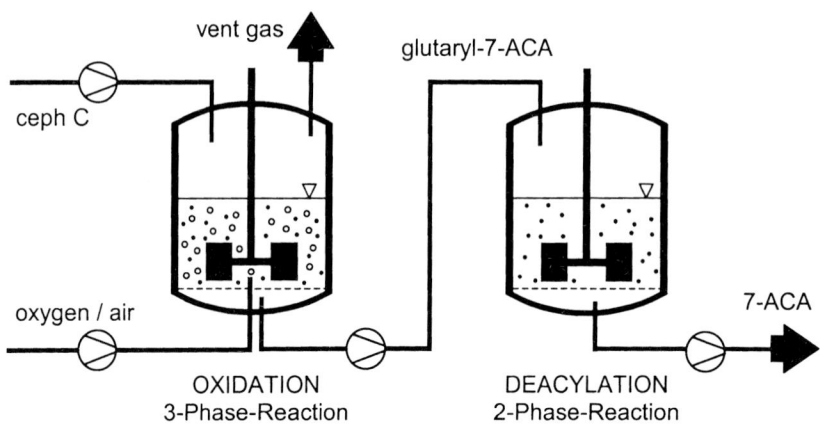

Fig. 7 The two-step enzymatic process (oxidation and deacylation reactor).

7.2.4.1 Oxidation

The purified cephalosporin C solution, an eluate from the ion exchange step of the downstream processing of the fermentation broth, is equilibrated with respect to the pH value (7.0–7.5) and concentration of Ceph C. Alternatively Ceph C salts, e.g., potassium or sodium, could be used. Heavy metal salts, e.g., zinc, deactivate the biocatalyst and are therefore not applicable. In general a Ceph C starting concentration of about 50–100 mM is used [15]. The reaction time is between 1 and 2 hours depending on the activity of the biocatalyst. Deactivation of the biocatalyst is compensated for by prolongation of the reaction time. During the reaction the addition of base is recommended to neutralize the produced acids glutaryl-7-ACA and carbon dioxide. At the end of the reaction the pH is rising, due to carbon dioxide evolving from the system. Process control could be easily achieved by measuring the pH or by measuring the dissolved oxygen content (Fig. 8). A costly reaction monitoring by on-line HPLC is not required.

For the oxygen supply, air, oxygen enriched air or pure oxygen can be used. In general a pressure range between 1 and 5 hPa is sufficient. For reactor design the oxygen transfer from the gaseous to the aqueous phase is important, therefore a stirrer with high oxygen dispersion capacity is recommended. On the other hand the mechanical stability of the enzyme carrier has to be taken into account. A high dissolved oxygen level is necessary to avoid diminishing of the reaction rate, at least when high Ceph C concentrations exist. When Ceph C becomes the rate-limiting component the dissolved oxygen level increases.

The glutaryl-7-ACA-containing solution is separated by filtration from the immobilized biocatalyst and directly transferred to the deacylation step in the second reactor. The oxidation reactor is ready to start for the next cycle. If production stops, storage of the DAO catalyst in a buffer solution is possible.

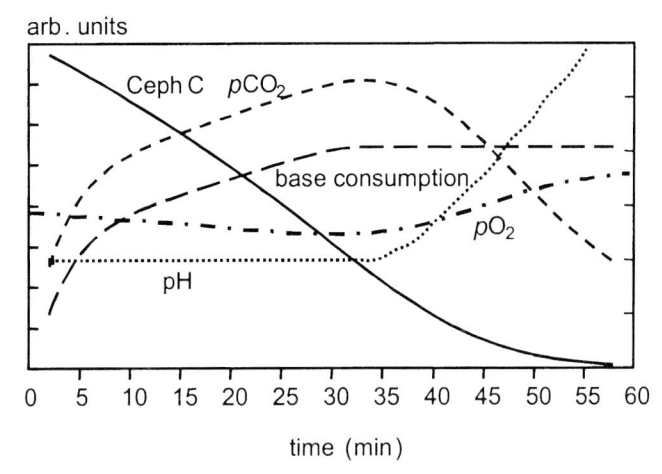

Fig. 8 Kinetics of the oxidation reaction.

7.2.4.2 Deacylation

In the second stirred tank reactor glutaryl-7-ACA is deacylated to 7-ACA and glutaric acid. The solution from the oxidation step is used directly. The reaction starts immediately, therefore a fast pH regulation is recommended. At higher pH values the enzyme activity is increased and the thermodynamic equilibrium changes from acylation to deacylation, while decomposition of the beta-lactam ring occurs at high pH values. A pH value between 7.5 and 8.5 is therefore recommended. For pH control equimolar amounts of base are necessary. Process control is achieved by following the titration curve, avoiding on-line HPLC. When conversion is completed the 7-ACA solution is separated by filtration, the immobilized biocatalyst can be reused for the next reaction cycle. Storage of the GA catalyst in a buffer solution is possible.

The 7-ACA produced is purified and isolated by crystallization at the isoelectric point. Of course, the quality of the 7-ACA produced by the biocatalytic synthesis has to be comparable to the chemical product with approximately 95% purity. Therefore the occurrence of byproducts, such as desacetyl-7-ACA, desacetoxy-7-ACA, 7-ACA-sulfoxide, glutaryl-7-ACA or α-ketoadipyl-7-ACA generated during the process or depending on the quality of the starting material, has to be watched very critically during the process development.

7.3
Comparison of Chemical and Two-Step Biocatalytic 7-ACA Process

Both processes are compared regarding their advantages and disadvantages. A great deal of effort was necessary to develop the biocatalytic process and fermentation and downstream processing of the enzymes in order to compete with the established chemical process, but the results are remarkable (Tab. 3). The chemical iminohalide/iminoether deacylation process is a classical organic synthesis route using equi-

Tab. 3 Chemical versus biocatalytic process.

Chemical process	Biocatalytic process
Isolation of Ceph C salt necessary	Ceph C in aqueous solution can be used
Equimolar amounts of chemicals used	Catalytic quantities of immobilized enzymes used
Toxic chemicals used (N,N-dimethylaniline)	No toxic chemicals
Hazardous chemicals used {$(CH_3)_3SiCl$, PCl_5}	No hazardous chemicals
Use of heavy metals (zinc)	No heavy metals
CH_2Cl_2 as solvent, avoidance of moisture (recycling and waste incineration)	Use of water as solvent
Expensive exhaust gas cleaning system (due to methylene chloride)	Easier exhaust gas cleaning (no methylene chloride as solvent)
Glass-lined equipment and/or highly corrosion-resistant alloys (due to corrosive chemicals)	Stainless steel equipment
High energy costs (due to low temperature reaction)	Low energy costs (reaction at room temperature)
Incineration of waste water (due to toxic chemicals)	Biological waste water treatment (no toxic chemicals)
Waste water with low COD	Waste water with high COD
Low process development costs	High process development costs

molar amounts of toxic and hazardous reagents such as N,N-dimethylaniline, trimethylchlorosilane and phosphorous pentachloride. It also uses dichloromethane as the solvent that has to be reworked by expensive distillation procedures. It requires inert gas conditions, moisture has to be avoided and expensive waste gas cleaning systems are necessary. The composition of the waste water prohibits or endangers conventional biological waste water treatment and therefore the waste water has to be incinerated. Some processes use Ceph C zinc salt as the raw material with the resulting problems of heavy metal disposal or reworking. The necessary very low reaction temperature gives rise to high energy costs. Owing to corrosive chemicals, e.g., HCl, glass-lined equipment and/or highly corrosion-resistant alloys are used.

The biocatalytic process uses immobilized enzymes in catalytic amounts. No isolation of cephalosporin C as a salt is necessary; the purified solution from the downstream processing can be used directly. No toxic or corrosive chemicals are required. It is an aqueous process at room temperature; stainless steel equipment is sufficient. There are no restraints for biological waste water treatment. Disposal of liquids by incineration and the amount of waste water are also reduced, while the COD (chemical oxygen demand) is increased, but highly biodegradable (Tab. 4). No hazardous chemicals or heavy metals are involved. Disposal or reworking of zinc salts is avoided. No chlorinated solvents are required, therefore recycling (distillation of CH_2Cl_2) is avoided and gas emissions are reduced.

Tab. 4 Comparison of the amount of waste per ton of 7-ACA.

	Chemical process	Biocatalytic process
Waste for incineration	31 tons	0.3 tons
Solvent distillation	100%	5%
Solvent emissions	7.5 kg	1.0 kg
Zinc disposal	1.8 tons	0.0 tons
Waste water volume	100%	90%
Waste water COD	0.1 tons	1.7 tons

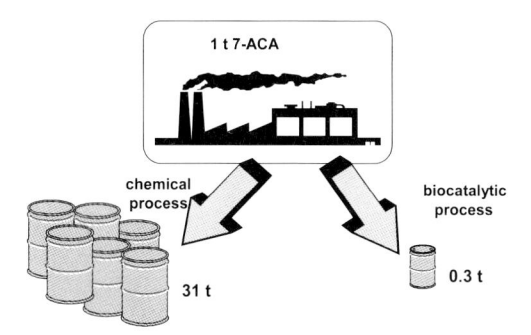

Fig. 9 Waste to be incinerated per ton of 7-ACA produced.

In summary, the biocatalytic process reduces raw material costs by a factor of four and the percentage of process costs for environmental protection (incineration, waste water treatment and waste gas cleaning) is cut down considerably from 21% to 1%. Compared with the chemical process the environmental protection costs were reduced by about 90% per ton of 7-ACA.

7.4 Conclusion

Today biocatalysis is not only used in laboratories, but is also broadly accepted for the production of several thousands of tons in the β-lactam antibiotics business. The biocatalytic synthesis produces 7-ACA with less environmental impact (Fig. 9) and, depending on the production site, with substantially lower costs compared with the classical chemical route.

This success was only possible by using recombinant DNA technology for the enzyme production. Of course, the development of a biocatalytic process is very complex. It is not only the biocatalytic steps that have to be developed. A lot of effort is necessary to gain access to the enzymes in the desired quality and at acceptable costs. The economics are mainly determined by the stability of the biocatalysts and product quality by the enzymatic activities in the cells of the producing organisms and an effective downstream process. As one can imagine the different parts of the development project dramatically influence the final result. However, the overall economy determines whether a classical chemical synthesis or a biocatalytic process will be realized in practice. Very often there is no chance for a new approach while an existing chemical plant is still running. Fortunately there was the chance to show that the two-step biocatalytic 7-ACA process not only fulfills the requirements of sustainable development by integrated environmental protection, but also the economic requirements.

Research is in progress, by means of recombinant DNA technology, to gain access to an industrial relevant cephalosporin C acylase [16]. This one-step synthesis would further improve the performance of the process with respect to economy, quality and ecology.

7.4
References

1 IMS Health, World Review 2001.
2 G. G. F. Newton, E. P. Abraham, Nature (London) 1955, 175, 548.
3 W. Riethorst, A. Reichert, Chimia 1999, 53, 600.
4 OECD Report 2001, The Application of Biotechnology to Industrial Sustainability, pp. 55–57.
5 Europa Chemie 1999, 19, 16.
6 W. Voser, US Patent 3725400, 1970.
7 G. Ascher, US Patent 4322526, 1980.
8 S. Isikawa, K. Yamamoto, K. Matsuyama, US Patent 4774179, 1985.
9 M. Morita, I. Aramori, M. Fukagawa, T. Isogai, H. Kojo, European Patent 0475652, 1990.
10 M. S. Crawford, D. B. Finkelstein, J. A. Rambosek, US Patent 5104800, 1989.
11 B. H. Arnold, R. A. Fildes, D. A. Gilbert, US Patent 3658649, 1969.
12 Th. Bayer, U. Holst, U. Wirth, US Patent 5403730, 1993.
13 B. S. Deshpande, S. S. Ambedkar, V. K. Sudhakaran, J. G. Shewale, World J. Microbiol. Biotechnol. 1994, 10, 129.
14 B. L. Wong, Y. Q. Shen, US Patent 5618687, 1994.
15 Roche Diagnostics, product brochure CC2 Twin enzyme process for the splitting of cephalosporin C (1999).
16 K. Fritz-Wolf, K.-P. Kollet, G. Lange, A. Liesum, K. Sauber, H. Schreuder, W. Aretz, W. Kabsch, Protein Science 2002, 11, 92.

8
Methods for the Enantioselective Biocatalytic Production of L-Amino Acids on an Industrial Scale

Harald Gröger and Karlheinz Drauz

8.1
Introduction

Among the most important chiral building blocks, proteinogenic and non-proteinogenic L-amino acids play an important role [1]. These compounds have found a wide range of applications, and are required in technical quantities by the life sciences industry. Typical methods to produce L-amino acids are fermentation, and "classic" chemical processes, in particular asymmetric catalytic syntheses and kinetic resolutions. In addition, several industrially feasible biocatalytic concepts have been developed for the enantioselective production of optically active amino acids [2, 3]. The range of commercially applied methods covers resolution processes using racemates as well as asymmetric biocatalytic conversions starting from prochiral compounds. The basic principles of the resolution type reactions, and asymmetric biocatalytic processes, respectively, are shown in Fig. 1. This section focuses on those methods that are based on an enantioselective formation of an L-amino acid (methods which are based on a modification of an existing L-amino acid, e.g., the manufacture of L-tryptophan using an L-tryptophan synthase, are not the focus of this particular section).

A "classic" example of a typical enzymatic resolution on an industrial scale is the acylase-mediated production of L-methionine. This method has also been applied for the production of L-phenylalanine and L-valine. In addition to acylases, amidases, hydantoinases, and β-lactam hydrolases represent versatile biocatalysts for the production of optically active L-amino acids. A schematic overview of the different type of enzymatic resolutions for the synthesis of L-amino acids is given in Fig. 2.

In addition, the use of a whole-cell biocatalyst consisting of a racemase, hydantoinase, and carbamoylase allows a dynamic biocatalytic resolution. Besides resolution processes, asymmetric (bio-)catalytic concepts have been applied successfully on an industrial scale. The different types of asymmetric (bio-)catalytic syntheses of L-amino acids, based on the use of prochiral starting materials, are shown in Fig. 3.

Using an L-amino acid dehydrogenase in the presence of a formate dehydrogenase for cofactor regeneration, a prochiral keto acid is converted with high yield and enantioselectivity. Furthermore, biocatalytic transaminations as well as Michael-additions are important reactions for the large-scale synthesis of L-amino acids.

Asymmetric Catalysis on Industrial Scale: Challenges, Approaches and Solutions
Edited by H. U. Blaser, E. Schmidt
Copyright © 2004 Wiley-VCH Verlag GmbH & Co. KGaA, Weinheim
ISBN: 3-527-30631-5

8 Methods for the Enantioselective Biocatalytic Production of L-Amino Acids on an Industrial Scale

Enzymatic resolution of racemates

Prochiral substrate → Racemate → Biotransformation → 50 % unwanted enantiomer / 50 % desired enantiomer

Racemization (subsequent or *in situ*) of the unwanted enantiomer back to racemate.

Asymmetric (bio-)catalysis

Prochiral substrate → Direct access via biotransformation → 100 % desired enantiomer

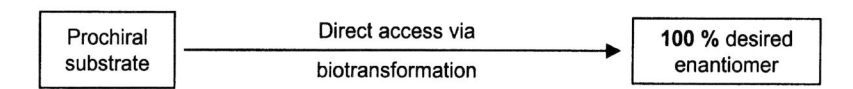

Fig. 1 Biocatalytic concepts for the production of L-amino acids.

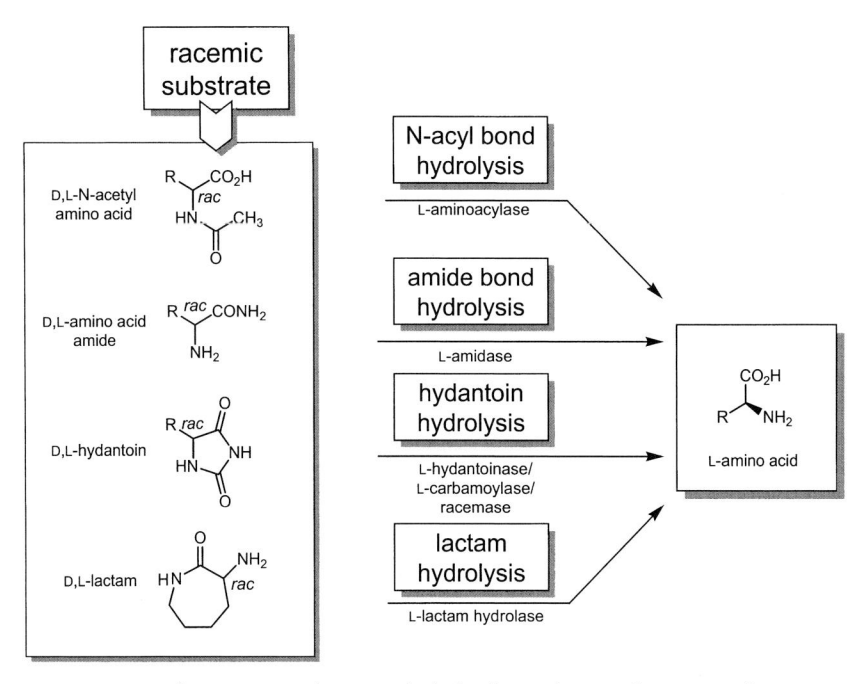

Racemic substrate:

- D,L-N-acetyl amino acid → N-acyl bond hydrolysis (L-aminoacylase) → L-amino acid
- D,L-amino acid amide → amide bond hydrolysis (L-amidase) → L-amino acid
- D,L-hydantoin → hydantoin hydrolysis (L-hydantoinase / L-carbamoylase / racemase) → L-amino acid
- D,L-lactam → lactam hydrolysis (L-lactam hydrolase) → L-amino acid

Fig. 2 Overview of enzymatic resolution methods for the production of L-amino acids.

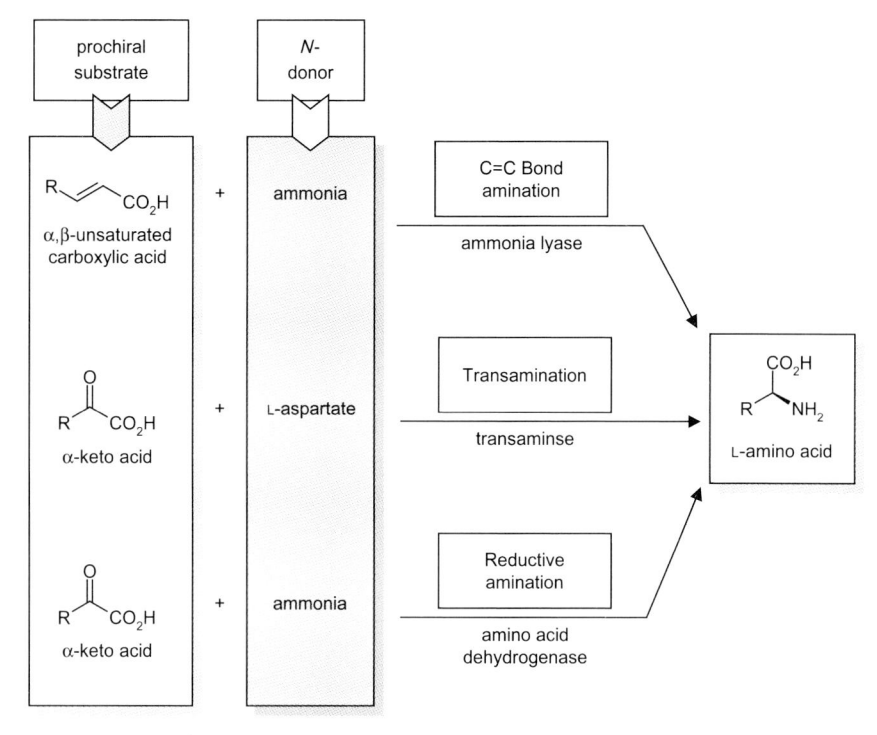

Fig. 3 Overview of asymmetric biocatalytic methods for the production of L-amino acids.

In the following discussions the individual biocatalytic methods applied in the production of L-amino acids are described in more detail.

8.2
L-Amino Acids via Enzymatic Resolutions

8.2.1
Processes Using L-Amino Acylases

The enantioselective hydrolysis of racemic N-acetylated α-amino acids D,L-**1** at Degussa represents a long established large-scale process for the production of L-amino acids, L-**2** [4]. This enzymatic resolution requires an L-aminoacylase as the biocatalyst. The starting materials for this process are readily available, since racemic N-acetyl amino acids D,L-**1** can be economically synthesized by acetylation of racemic α-amino acids with acetyl chloride or acetic anhydride under alkaline conditions via the so-called Schotten-Baumann reaction [5]. The enzymatic resolution reaction of N-acetyl D,L-amino acids, D,L-**1**, is achieved by a stereospecific L-aminoacylase which hydrolyzes only the L-enantiomer and produces a mixture of the corresponding L-amino acid, L-**2**, acetate, and N-acetyl D-amino acid, D-**1** (Fig. 4) [6].

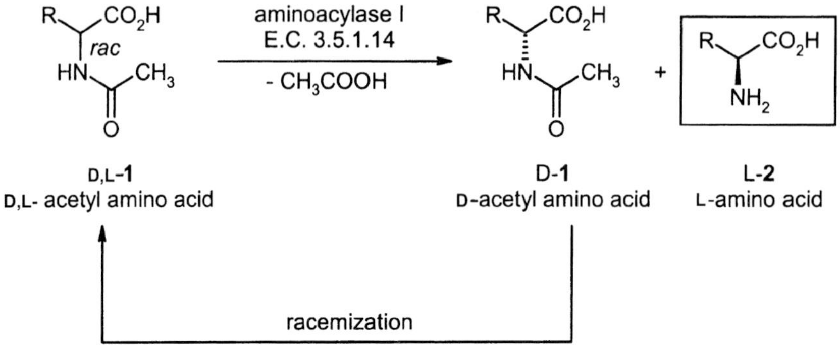

Fig. 4 Enzymatic resolution of racemic N-acetyl amino acids using an L-aminoacylase.

Subsequently, the L-amino acid, L-2, is separated and isolated by a crystallization step, and the remaining N-acetyl D-amino acid is recycled by, e.g., thermal racemization. As a preferred L-aminoacylase, the amino acylase I from *Aspergillus oryzae* [E.C.3.5.1.14] turned out to be particularly useful.

Several large-scale applications of aminoacylase-based processes have been reported. Many years ago Tanabe developed a process using an immobilized enzyme [7], whereas at Degussa a process using natural amino acylase I from *Aspergillus oryzae* in homogeneous solution is still being applied today. For this purpose, enzyme membrane reactor (EMR) technology has been designed, based mainly on the work of Kula and Wandrey [8].

The substrate specificity of the amino acylase from *Aspergillus oryzae* is very broad, and a wide range of proteinogenic and non-proteinogenic N-acetyl and N-chloroacetyl amino acids are transformed in the presence of the L-amino acylase. The enzyme membrane reactor (Fig. 5) is operated continuously as a loop reactor, and the enzyme is retained by an ultrafiltration hollow-fiber membrane (molecular weight cut off: 10 000 Dalton).

Since the early 1980s this process has been scaled-up to a production level of hundreds of tons per year. Fig. 6 shows the Degussa process for manufacturing L-methionine, L-6 [9a]. The biocatalyst is produced in bulk quantities and its operational stability is high; hence this continuous EMR-acylase process demonstrates high efficiency, especially on a large-scale [9].

In addition, the amino acylase process can be also applied in the production of other proteinogenic and non-proteinogenic L-amino acids such as L-valine and L-phenylalanine. It is worth noting that racemases have recently been developed by several companies which allow (in combination with the L-aminoacylases) an extension of the existing process towards a dynamic kinetic resolution reaction [10]. It should be mentioned that the same concept can be also applied for the synthesis of D-amino acids when using a D-aminoacylase as an enzyme.

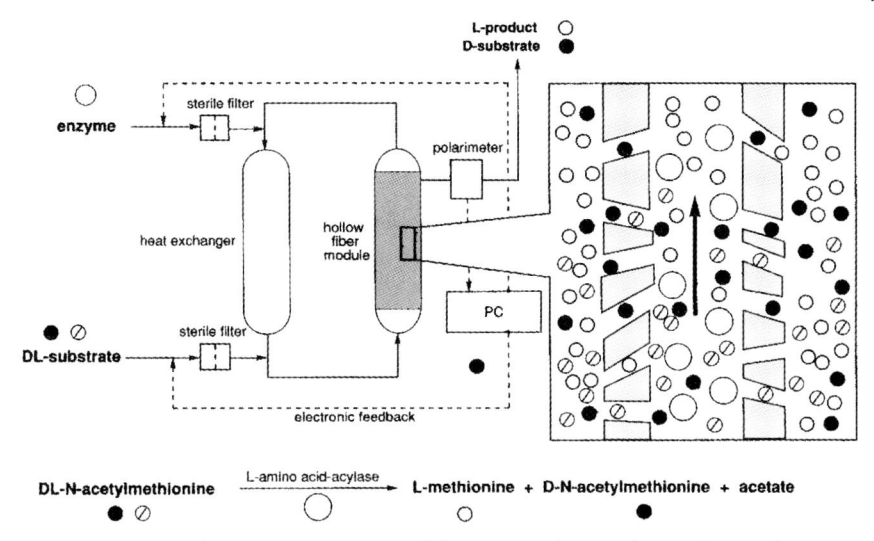

Fig. 5 Enzyme membrane reactor (EMR) used for amino acylase resolution of N-acetyl-D,L-amino acids.

8.2.2
Processes Using L-Amidases

DSM has developed a widely applicable industrial process for the production of enantiomerically pure D- and L-α-amino acids by enantioselective hydrolysis of racemic amino acid amides, D,L-**7** [11]. These precursor compounds, D,L-**7**, can easily be obtained by alkaline hydrolysis of α-amino nitriles. An efficient L-specific amidase (=aminopeptidase) has been found in *Pseudomonas putida* (ATCC 12633), and shows a broad substrate specificity [11, 12]. It should be noted that the enzyme activity is in a high range of >300 U/mg of protein. Using suspended whole-cells as a biocatalyst led to the formation of the optically active L-amino acids, L-**2**, with excellent enantioselectivities of >99% ee. The L-amino acids, L-**2**, can easily be separated from the remaining D-amide, D-**7**, subsequently. Typically the reaction is carried out at a pH of 8–10. This method allows an efficient industrial production of both L- and D-amino acids (Fig. 7). It is significant that additional amounts of magnesium ions enhance the enzymatic activity (up to 12 times) [13]. Methods for racemizing the undesired enantiomer are also integrated into the process [14].

From a broad variety of L-α-amino acids, L-phenylalanine as well as L-homophenylalanine were successfully produced using the amidase technology.

It should also be noted that at DSM this amidase process has been extended towards the synthesis of optically active α,α-disubstituted amino acids. For example, the antihypertensive drug L-methyl-dopa, L-**10**, has been produced successfully (Fig. 8) [14, 15].

As a biocatalyst that is capable of converting α,α-disubstituted amino acids efficiently, an aminopeptidase in the strain *Mycobacterium neoaurum* [ATCC 25795]

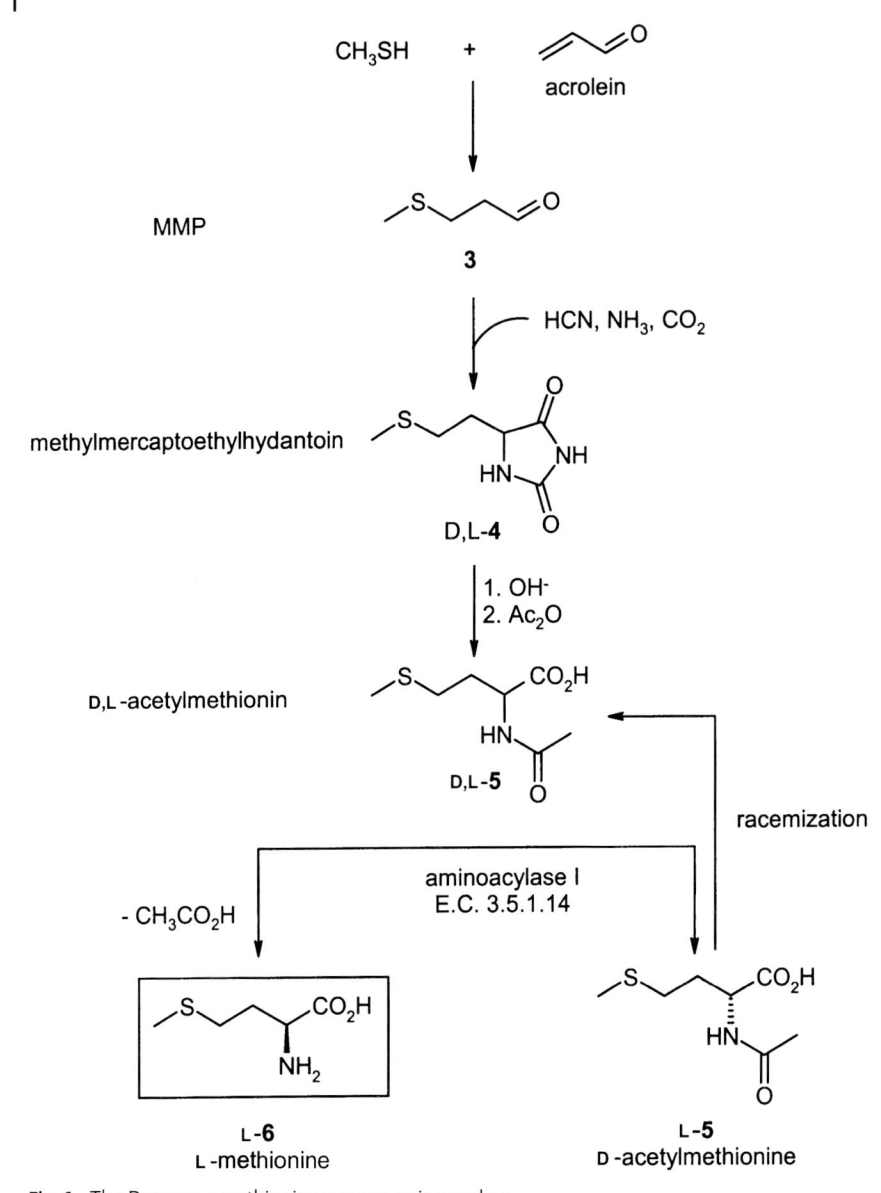

Fig. 6 The Degussa L-methionine process using acylase.

was used either as a crude enzyme preparation or as permeabilized whole cells. In general, high enantioselectivities of >98% ee are obtained.

In addition, at Ube an analogous process for the enantioselective synthesis of α-methyl phenylalanine has been developed based on an amidase from *Pseudomonas fluorescens* [IFO 3081] [14].

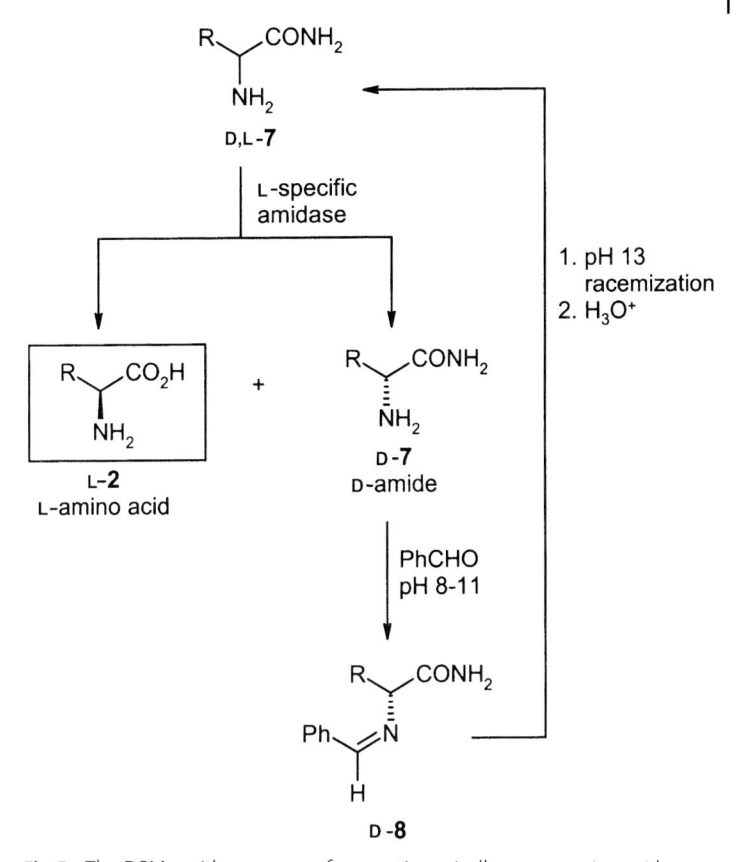

Fig. 7 The DSM amidase process for enantiomerically pure α-amino acids.

Fig. 8 Synthesis of L-α-methyl-dopa.

8.2.3
Processes Using L-Hydantoinases

Resolution processes based on the use of hydantoinases have been well-known for a long time, in particular for the production of D-amino acids [16]. By the 1970s

D-phenylglycine and *p*-hydroxy phenylglycine, which are used as side-chains for the β-lactam antibiotics ampicillin and amoxicillin, were produced using this method, and currently their annual production scale is in the region of >1000 tons. The corresponding L-enantioselective large-scale applications of the hydantoinase process have not been available for a long time due to low space-time yields and high biocatalyst costs.

Very recently, however, a remarkably improved whole-cell biocatalyst coexpressing an L-carbamoylase, a hydantoin racemase, and an L-hydantoinase has been developed which showed a 50-fold higher productivity and is suitable for large-scale applications [17]. A prerequisite of this efficient whole-cell biocatalyst, which is applied at Degussa, was the successful inversion of the enantiospecificity of a previously D-selective hydantoinase by means of directed evolution [18]. These improvements have already been confirmed at a m^3-scale using a batch reactor concept [17].

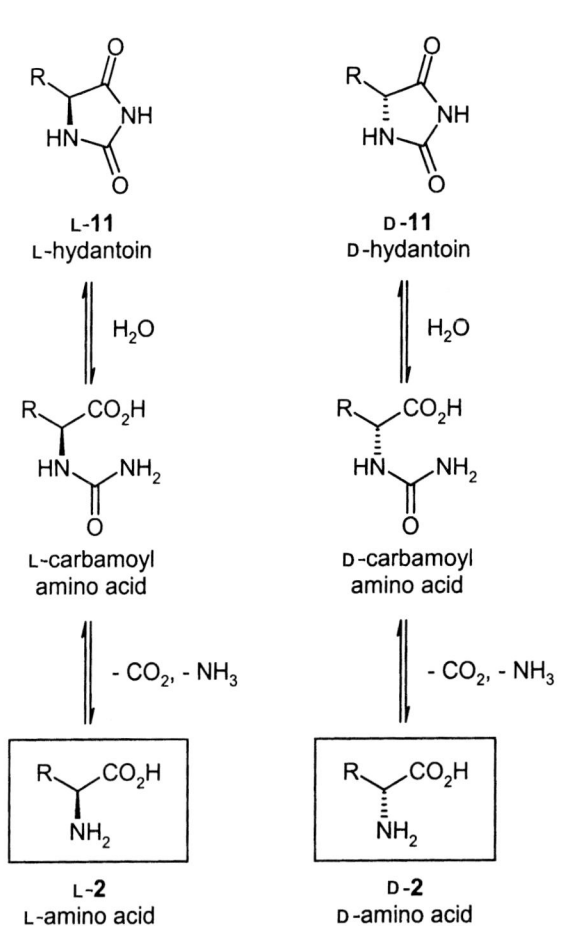

Fig. 9 Dynamic kinetic resolution of racemic hydantoins.

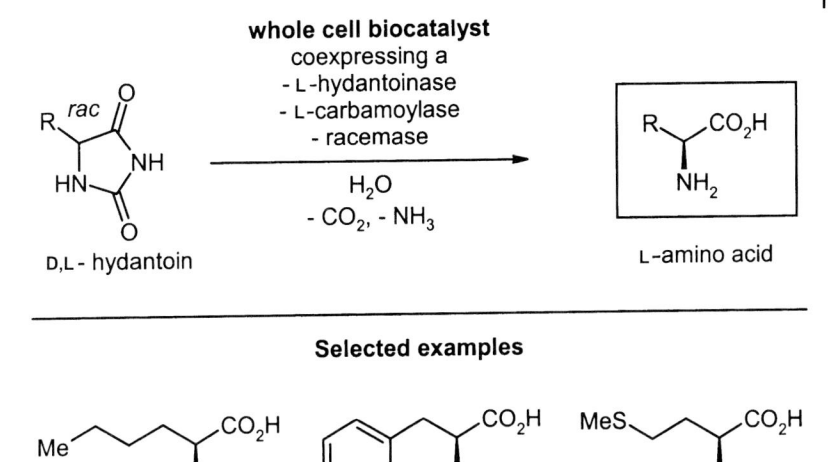

Fig. 10 Hydantoinase-based whole cell catalysts in L-amino acid synthesis.

The reaction concept with this new hydantoinase-based biocatalyst is economically highly attractive since it represents a dynamic kinetic resolution process converting a racemic hydantoin (theoretically) quantitatively into the enantiomerically pure L-enantiomer [19]. The L-hydantoinase and subsequently the L-carbamoylase hydrolyze the L-hydantoin, L-**11**, enantioselectively forming the desired L-amino acid, L-**2**. In addition, the presence of a racemase guarantees a sufficient racemization of the remaining D-hydantoin, D-**11**. Thus, a quantitative one-pot conversion of a racemic hydantoin into the desired optically active α-amino acid is achieved. The basic principles of this biocatalytic process in which three enzymes (hydantoinase, carbamoylase, and racemase) are integrated is shown schematically in Fig. 9.

It is worth noting that this whole-cell biocatalyst accepts a broad range of substrates, thus allowing access to numerous enantiomerically pure L-α-amino acids, L-**2**. Selected examples for successfully converted hydantoins are shown in Fig. 10. A further advantage of the hydantoinase-based technology is that hydantoins are readily accessible and economically attractive starting materials.

8.2.4
Processes via Lactam Hydrolysis

The enantioselective ring-opening of lactams can also be used for an efficient production of L-amino acids, as has been demonstrated by the production process for L-lysine, L-**16**. For some years Toray's enzymatic process for L-lysine, L-**16**, was

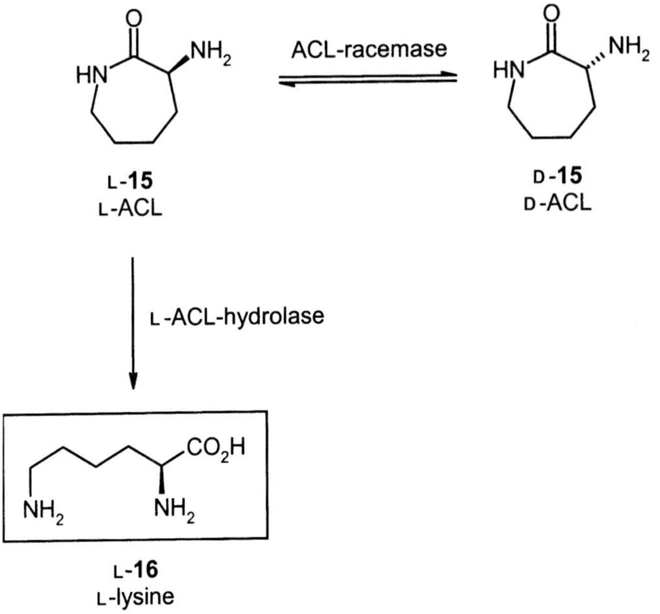

Fig. 11 The Toray enzymatic L-lysine process using α-aminocaprolactam as starting material.

competitive compared with fermentation. This chemoenzymatic L-lysine production was established with an annual capacity of 5000–10 000 tons. The key intermediate is *rac-α*-amino-ε-caprolactam (ACL), D,L-**15**, produced from cyclohexanone in a modified Beckmann rearrangement.

The enantiospecific hydrolysis forming L-lysine is based on two enzymes: L-ACL-hydrolase opens the ring of D,L-**15** to L-Lys and in the presence of the ACL-racemase the D-ACL, D-**15**, is racemized. When incubating D,L-**15** with cells of *Cryptococcus laurentii* having the L-ACL lactamase activity together with cells of *Achromobacter obae* with ACL-racemase activity, L-lysine, L-**16**, could be obtained in a yield of nearly 100% (Fig. 11) [20].

This process has now been abandoned because, on a multi-100 000 t/y scale, fermentation of L-lysine based on sugars has proven to be superior to the Toray process.

8.3
L-Amino Acids via Asymmetric Biocatalysis

8.3.1
Processes via Reductive Amination

The biotransformation of prochiral α-keto acids into the corresponding L-amino acids represents an interesting asymmetric reaction for the production of L-α-amino acids. In this reaction two enzymes, namely leucine dehydrogenase (LeuDH

8.3 L-Amino Acids via Asymmetric Biocatalysis

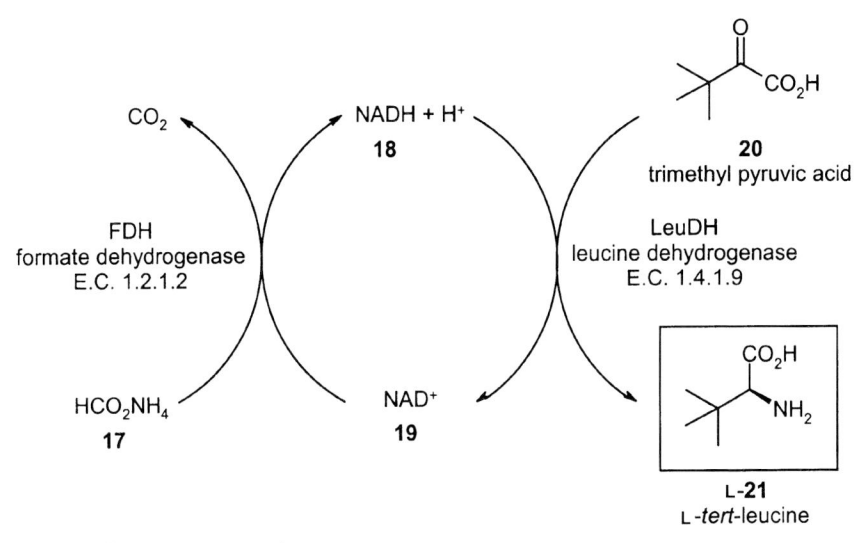

Fig. 12 L-Tle production by cofactor-dependent reductive amination.

[E.C.1.4.1.9]) and formate dehydrogenase (FDH [E.C.1.2.1.2]), and the cofactor NAD^+ act simultaneously on the transformation of the keto acid into the L-amino acid [21]. The reaction concept is shown in Fig. 12 for the formation of L-*tert*-leucine, L-**21**. The cofactor is used in catalytic amounts and is recycled *in situ* during the reaction. Furthermore, the cofactor-regeneration is a (practically) irreversible process since carbon dioxide is produced, which can easily be removed. The leucine dehydrogenase accepts a broad range of aliphatic keto acids with hydrophobic carbon chains. Notably, a further amino acid dehydrogenase, namely the phenylalanine dehydrogenase, is available which shows a complementary substrate range tolerating pyruvic acid derivatives containing an aromatic substituent.

In particular, the biocatalytic reductive amination turned out to be a very effective method for synthesizing acids with bulky side-chains. Such a process is carried out at Degussa [22]. For example, the bulky side-chain-containing amino acid L-*tert*-leucine (L-Tle) could be produced in high enantioselectivities of >99% ee starting from the corresponding *a*-keto acid (Fig. 12) [23]. In addition, the homologous L-neopentylglycine (L-Npg) can be produced in the same manner. The L-amino acids L-*tert*-leucine, L-**21**, and L-neopentylglycine, respectively, are of increasing importance because of their extended use as building blocks in pharmaceutical drugs and as a chiral auxiliary in asymmetric synthesis [21].

The use of an LeuDH as an amino acid dehydrogenase showed a high L-enantiospecificity [24]. In this connection, an L-leucine dehydrogenase from *Bacillus sphaericus* has been applied very efficiently. The FDH from *Candida boidinii* is the preferred formate dehydrogenase for this process. The stability of this enzyme, which is available in technical quantities, has been remarkably improved by protein engineering and directed evolution [25]. In particular the replacement of cysteine moieties (which have been found in positions 23 and 262) by other amino acids, e.g., serine and alanine, turned out to be beneficial for an improved stability (since an oxidative thiol-coupled inactivation process can be avoided).

8.3.2
Processes via Transamination

The synthesis of chiral α-amino acids starting from α-keto acids by means of a transamination has been reported by NSC Technologies [26, 27]. In this process, which can be used for the preparation of L- as well as D-amino acids, an amino group is transferred from an inexpensive amino donor, e.g., L-glutamic acid, L-22, or L-aspartic acid, in the presence of a transaminase (=aminotransferase). This reaction requires a cofactor, most commonly pyridoxal phosphate, which is bound to the transaminase. The substrate specificity is broad, allowing the conversion of numerous keto acid substrates under formation of the L-amino acid products with high enantioselectivities [28].

A typical example for an efficient transamination process is the production of L-alanine, L-25, which is carried out in a continuous manner starting from pyruvate, 24, and L-glutamate, L-22, with a high space-time yield of 4.8 kg/(L · d) (Fig. 13) [28]. In addition, several non-proteinogenic α-amino acids, e.g., L-phosphinothricine, L-homophenylalanine, and L-tert-leucine have been also produced via transamination.

A drawback of many transaminations has been the limitation of the yields (around 50%) due to thermodynamic reasons [28, 29]. This problem of incomplete reactions has been solved by coupling the transamination reaction to another reac-

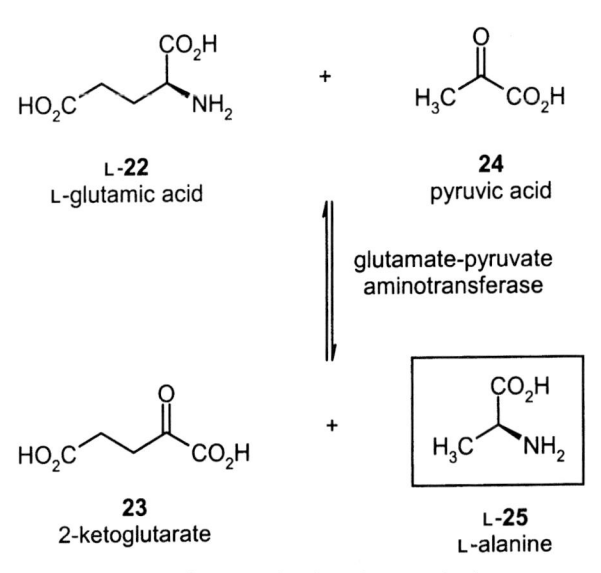

Fig. 13 Aminotransferase-catalyzed production of L-alanine.

tion that consumes, in an irreversible step, the α-keto acid formed [28a]. A selected successful example for such a second "coupled" reaction is the decarboxylation of oxaloacetate, which is the α-keto acid byproduct when using L-aspartic acid as an amino donor [29a, b].

In a recent optimization study, an alternative coupling system has been developed [29c]. The chemical yield in the synthesis of two non-proteinogenic α-amino acids was remarkably improved by applying a coupled transamination process using additionally an ornithine ω-aminotransferase to couple L-ornithine ω-transamination to L-glutamate α-transamination [29c].

8.3.3
Processes via Addition of Ammonia to α,β-Unsaturated Acids

A biocatalytic enantioselective addition of ammonia to a C=C bond of an α,β-unsaturated compound, namely fumaric acid, makes the manufacture of L-aspartic acid, L-27, possible [30]. This L-amino acid represents an important intermediate for the production of the artificial sweetener aspartame. The biocatalytic production process, which is applied on an industrial scale by, e.g., Kyowa Hakko Kogyo and Tanabe Seiyaku, is based on the use of an aspartate ammonia lyase [E.C.4.3.1.1] [31]. As a biocatalyst, an immobilized L-aspartate ammonia lyase from *Escherichia coli* [32, 33] as well as *Brevibacterium flavum* whole-cell catalysts [32a, 34] have been applied successfully.

In the presence of the immobilized enzyme, the reaction can be carried out at a substrate concentration of up to 2 M, as has been demonstrated by Kyowa Hakko. In addition, an excellent conversion of >99% ee was obtained, accompanied by impressive enantioselectivities of >99.9% ee [32]. Immobilized whole cells from *Escherichia coli* have been used by Tanabe Seiyaku for this reaction [34]. It is worth noting that this reaction represents not only the first application of immobilized whole cells in industrial α-amino acid production, but also proved to be economically superior to fermentation methods (Fig. 14). A high yield of >95% has been obtained for L-aspartic acid, L-27, and the annual capacity is several hundred tons.

The analogous biocatalytic process using whole-cells from *Brevibacterium flavum* also led to L-aspartic acid, L-27, in high yields of >99%, and excellent enantioselectivities of >99% ee [33]. This process that is carried out under basic conditions at pH 9–10 has been developed by Mitsubishi Petrochemical.

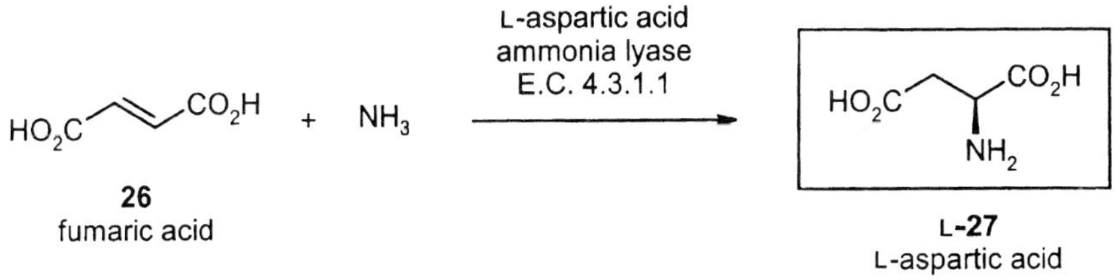

Fig. 14 L-Aspartic acid production using an L-aspartic acid ammonia lyase.

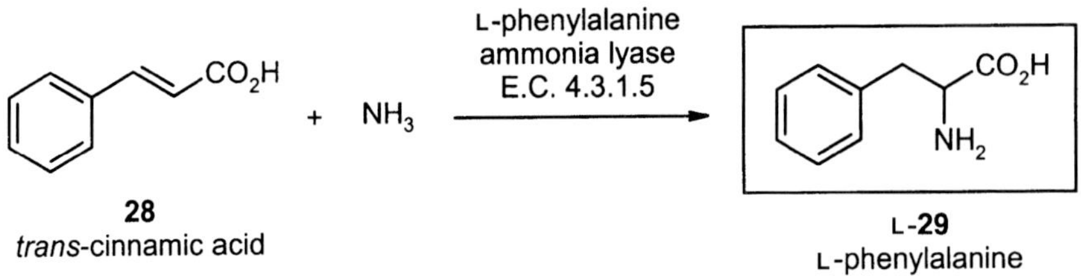

Fig. 15 L-Phenylalanine production using an L-phenylalanine ammonia lyase.

The substrate tolerance of this process, however, is narrow. Although some structural modifications of the substrate are acceptable, e.g., halogen substituents, the yields are lower in these cases (e.g., 60% for the 3-chloro-, and 61% for the 3-methyl-substituted L-aspartate compared with 90% for L-aspartate) [35]. It should be pointed out that in addition to ammonia as a nitrogen nucleophile, amines, e.g., small alkyl amines and protected hydroxylamine, are also tolerated [36].

The biocatalytic synthesis can be extended towards the production of L-alanine when using additionally an L-aspartate decarboxylase for a subsequent decarboxylation step. Such a process has been reported by Tanabe Seiyaku, and gave the desired L-alanine in 86% yield and with 99% ee [37]. A key feature of the decarboxylase process is the high substrate concentration of 2.5 M, and a space-time yield of 170 g/(L · d) of L-alanine. Based on this two-step approach, L-alanine has been produced in annual amounts of ca. 60 tons [32 b, 37].

Another comparable reaction is the asymmetric biocatalytic addition of ammonia to *trans*-cinnamic acid, **28** (Fig. 15) [38]. This reaction represents a technically feasible process for the production of L-phenylalanine as has been shown by the Genex Corporation [38 a, b]. The amino acid L-phenylalanine is required in technical quantities as an intermediate, e.g., for the manufacture of the artificial sweetener aspartame.

The biocatalytic addition of ammonia to *trans*-cinnamic acid, **28**, proceeds enantioselectively in the presence of whole-cells containing L-phenylalanine ammonia lyase. Of the suitable strains, wild-type strains of *Rhodococcus rubra* as well as *Rhodotorula rubra* were found to be very efficient. The biotransformation is carried out in basic media at pH 10.6. In addition, analogous reactions furnishing non-natural substituted derivatives of L-phenylalanine starting from substituted *trans*-cinnamic acid derivatives were reported by Mitsui, and Great Lakes, respectively [38 c].

A further industrially important lyase for the production of L-amino acids is the tyrosine phenol lyase [39]. This biocatalyst is used by Ajinomoto in the production of the pharmaceutically important L-3,4-dihydroxyphenylalanine (L-dopa), **32**, which is applied in the treatment of Parkinson's disease. The reaction concept is based on a one-pot three-component synthesis starting from catechol, **30**, pyruvic acid, **31**, and ammonia in the presence of suspended whole cells (strain: *Erwinia herbicola*) containing the tyrosine phenol lyase biocatalyst (Fig. 16). A key feature of this process is the high volumetric productivity of 110 g/L of the desired L-dopa product. Notably, this reaction runs with an annual capacity of 250 tons.

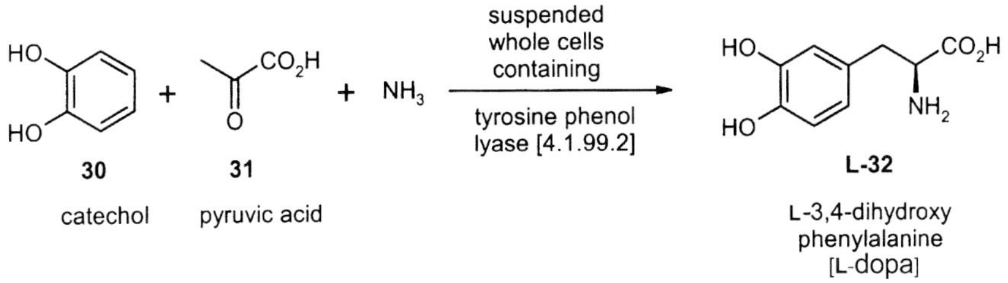

Fig. 16 Production of L-dopa using tyrosine phenol lyase.

8.4
Conclusion

In summary, a broad range of large-scale applicable biocatalytic methodologies have been developed for the production of L-amino acids in technical quantities. Among these industrially feasible routes, enzymatic resolutions play an important role. In particular, L-aminoacylases, L-amidases, L-hydantoinases in combination with L-carbamoylases, and β-lactam hydrolases are efficient and technically suitable biocatalysts. In addition, attractive manufacturing processes for L-amino acids by means of asymmetric (bio-)catalytic routes has been realized. Successful examples are reductive amination, transamination, and addition of ammonia to α,β-unsaturated carbonyl compounds, respectively.

In the future, certainly the improvement and further extension of such biocatalytic routes, comprising enzymatic resolutions as well as asymmetric (bio-)catalysis will be key issues for industrial research. The development of novel synthetic routes by means of metabolic pathway engineering will also be of increasing importance.

8.5
References

1 G. Lubec, G. A. Rosenthal (Eds.), Amino Acids: Chemistry, Biology, Medicine, ESCOM Science Publishers B.V., 1990.
2 For a comprehensive handbook with detailed reviews about biocatalytic approaches, see: K. Drauz, H. Waldmann (Eds.), Enzyme Catalysis in Organic Synthesis, Vol. 2, 2nd edn., Wiley-VCH, Weinheim, 2002.
3 J. D. Rozzell, F. Wagner (Eds.), Biocatalytic Production of Amino Acids and Derivatives, John Wiley & Sons, New York, 1992.
4 For a detailed review about aminoacylase-catalyzed resolutions, see: a) A. S. Bommarius in: Enzyme Catalysis in Organic Synthesis (K. Drauz, H. Waldmann, Eds.), Vol. 2, 2nd edn., Wiley-VCH, Weinheim, 2002, chapter 12.3; b) A. S. Bommarius, K. Drauz, H. Klenk, C. Wandrey, Ann. N. Y. Sci. 1992, 672, 126.
5 N. O. V. Sonntag, Chem. Rev. 1953, 52, 237.
6 For an excellent contribution about the scope of aminoacylase-catalyzed synthesis of amino acids, see: H. K. Chenault, J. Dahmer, G. M. Whitesides, J. Am. Chem. Soc. 1989, 111, 6354.
7 a) T. Tosa, T. Mori, N. Fuse, I. Chibata, Agric. Biol. Chem. 1969, 1047; b) I. Chibata, T. Tosa, T. Sato, T. Mori, Methods Enzymol. 1976, 746.

8 a) C. Wandrey, E. Flaschel, Adv. Biochem. Eng. 1979, 147; b) C. Wandrey, R. Wichmann, W. Leuchtenberger, M. R. Kula, US 4304858, 1981; W. Leuchtenberger, M. Karrenbauer, U. Plöcker, Enzyme Eng. 7 Ann. N.Y. Acad. Sci. 1984, 434.

9 a) A.S. Bommarius, M. Schwarm, K. Drauz, Chimica Oggi 1996, 14, 61; b) A. S. Bommarius, K. Drauz, H. Klenk, C. Wandrey, Enzyme Eng. 11 Ann. N. Y. Acad. Sci. 1992, 126; c) A. S. Bommarius in: Biotechnology (H.-J. Rehm, G. Reed, A. Pühler, P. Stadler, Eds.), Vol. 3, 1993, VCH, Weinheim, p. 427.

10 a) S. Tokuyama, K. Hatano, T. Takahashi, Biosci. Biotechnol. Biochem. 1994, 58, 24; b) S. Verseck, A. Bommarius, M.-R. Kula, Appl. Microbiol. Biotechnol. 2001, 55, 345.

11 For detailed reviews, see: a) J. Kamphuis, W. H. J. Boesten, Q. B. Broxtermann, H. F. M. Hermes, J. A. M. van Balken, E. M. Meijer, H. E. Schoemaker in: Advances in Biochemical Engineering/Biotechnology (A. Fiechter, Ed.), Vol. 42, Springer, Berlin, 1991, p. 133 ff; b) J. Kamphuis, W. H. J. Boesten, B. Kaptain, H. F. M. Hermes, T. Sonke, Q. B. Broxtermann, W. J. J. van den Tweel, H.E. Schoemaker in: Chirality in Industry (A. N. Collins, G. N. Sheldrake, J. Crosby, Eds.), John Wiley & Sons, Chichester, 1992, p. 187 ff.

12 F. P. J. T. Rutjes, H. E. Schoemaker, Tetrahedron Lett. 1997, 38, 677.

13 A. Liese, K. Seelbach, C. Wandrey, Industrial Biotransformations, Wiley-VCH, Weinheim, 2000, p. 236 ff.

14 B. Schulze, E. de Vroom in: Enzyme Catalysis in Organic Synthesis (K. Drauz, H. Waldmann, Eds.), Vol. 2, 2nd edn., Wiley-VCH, Weinheim, 2002, chapter 12.2.3.

15 J. Kamphuis, H. F. M. Hermes, J. A. M. van Balken, H. E. Schoemaker, W. H. J. Boesten, E. M. Meijer in: Amino Acids: Chemistry, Biology, Medicine (G. Lubec, G. A. Rosenthal, Eds.), ESCOM Science Publishers, Leiden, 1990, p. 119 ff.

16 a) For a review of d-hydantoinase applications, see: C. Syldatk, R. Müller, M. Siemann, F. Wagner in: Biocatalytic Production of Amino Acids and Derivatives (D. Rozzell, F. Wagner, Eds.), Hanser Publishers, München, 1992, p. 75 ff; b) F. Cecere, G. Galli, G. Della Penna, B. Rappuoli, 1976, DE 2422737.

17 O. May, S. Verseck, A. Bommarius, K. Drauz, Org. Proc. Res. Dev. 2002, 6, 452.

18 O. May, P. T. Nguyen, F. H. Arnold, Nat. Biotechnol. 2000, 18, 317.

19 M. Pietzsch, C. Syldatk in: Enzyme Catalysis in Organic Synthesis (K. Drauz, H. Waldmann, Eds), Vol. 2, 2nd edn., Wiley-VCH, Weinheim, 2002, chapter 12.4.

20 a) T. Fukumura, Agric. Biol. Chem. 1976, 40, 1687; T. Fukumura, Agric. Biol. Chem. 1976, 40, 1695; b) T. Fukumura, Agric. Biol. Chem. 1977, 41, 1327; J. Crosby, Tetrahedron 1991, 4789.

21 a) K. Drauz, W. Jahn, M. Schwarm, Chem. Eur. J. 1995, 538; b) A. S. Bommarius, M. Schwarm, K. Stingl, M. Kottenhahn, K. Huthmacher, K. Drauz, Tetrahedron: Asymmetry 1995, 6, 2851; c) reference [22].

22 A. S. Bommarius, K. Drauz, W. Hummel, M.-R. Kula, C. Wandrey, Biocatalysis 1994, 10, 37.

23 A. S. Bommarius, M. Schwarm, K. Drauz, J. Mol. Catal. B: Enzym. 1998, 5, 1.

24 G. Krix, A. S. Bommarius, K. Drauz, M. Kottenhahn, M. Schwarm, M.-R. Kula, J. Biotechnol. 1997, 53, 29.

25 H. Slusarczyk, S. Felber, M.-R. Kula, Eur. J. Biochem. 2000, 267, 1280.

26 a) D. J. Ager, I. G. Fotheringham, S. A. Laneman, D. P. Pantaleone, P. P. Taylor, Chim. Oggi 1997, 15, 11; b) P. P. Taylor, D. P. Pantaleone, R. F. Senkpeil, I. G. Fotheringham, TIBTECH 1998, 16, 412.

27 D. J. Ager, I. G. Fotheringham, T. Li, D. P. Pantaleone, R. F. Senkpeil, Enantiomer 2000, 5, 235.

28 a) J. D. Rozzell, A. S. Bommarius in: Enzyme Catalysis in Organic Synthesis (K. Drauz, H. Waldmann, Eds.), Vol. 2, 2nd edn., Wiley-VCH, Weinheim, 2002, chapter 12.7; b) S. P. Crump, J. D. Rozzell in: Biocatalytic Production of Amino Acids and Derivatives (J. D. Rozzell, F. Wagner, Eds.), Wiley, New York, 1992, p. 43 ff.

29 a) J.D. Rozzell, Methods Enzymol. 1987, 136, 479; b) S.P. Crump, J.S. Heier, J.D. Rozzell in: Biocatalysis (D.A. Abramowicz, Ed.), Van Nostrand Reinhold, New York, 1990, p. 155 f; c) T. Li, A. Kootstra, I.G. Fotheringham, Org. Proc. Res. Dev. 2002, 6, 533.

30 I. Shibata, T. Tosa, T. Sato, Methods Enzymol. 1976, 44, 739.

31 U.T. Bornscheuer in: Biotechnology (H.-J. Rehm, G. Reed, A. Pühler, P. Stadler, Eds.), Vol. 8b, VCH, 2000, p. 285.

32 a) A. Tanaka, T. Tosa, T. Kobayashi, Industrial Application of Immobilized Biocatalysts, Marcel Dekker, Inc., New York, 1993; b) A. Liese, K. Seelbach, C. Wandrey, Industrial Biotransformations, Wiley-VCH, Weinheim, 2000, p. 375 ff.

33 a) I. Chibata, T. Tosa, T. Sato, Appl. Microbiol. 1974, 27, 878–885; b) T. Tosa, T. Shibatani, Ann. N.Y. Acad. Sci. 1995, 750, 364–375; c) reference [32b].

34 M. Terasawa, H. Yukawa, Y. Takayama, Proc. Biochem. 1985, 20, 124–128.

35 K. Faber, Biotransformations in Organic Chemistry, 4th edn., Springer, Berlin, 2000, chapter 2.5.2.

36 M.S. Gulzar, M. Akhtar, D. Gani, J. Chem. Soc., Perkin Trans I, 1997, 649.

37 a) S. Takamatsu, J. Umemura, K. Yamamoto, T. Sato, T. Tosa, I. Chibata, Eur. J. Appl. Biotechnol. 1982, 15, 147–152.

38 a) J. Crosby, Tetrahedron 1991, 47, 4789-4846; b) reference [32b]; c) M. Wubbolts in: Enzyme Catalysis in Organic Synthesis (K. Drauz, H. Waldmann, Eds.), Vol. 2, 2nd edn., Wiley-VCH, Weinheim, 2002, chapter 12.6.

39 a) T. Tsuchida, Y. Nishimoto, T. Kotani, K. Iizumi (Ajinomoto), JP 5123177, 1993; b) see reference [32b], pp. 342–343.

II
New Catalysts for Existing Active Compounds

1
The Large-Scale Biocatalytic Synthesis of Enantiopure Cyanohydrins

Peter Poechlauer, Wolfgang Skranc, and Marcel Wubbolts

Abstract

This section will focus on the development of (R)- and (S)-hydroxynitrile lyases as biocatalysts and their large-scale application.

1.1
Introduction

The addition of hydrogen cyanide to carbonyl compounds such as aldehydes or ketones leads to 2-hydroxynitriles (cyanohydrins). This reaction as depicted in Fig. 1 is remarkable in several ways: it represents one of the easiest routes to carbon–carbon bond formation, and in many cases it creates a new stereocenter.

Forming and breaking this carbon–carbon bond requires base catalysis. In many cases the activation energy is low enough to allow a rapid establishment of equilibrium for cyanohydrin formation and decay: this is most remarkable for a carbon–carbon bond, the strength of which is frequently said to be at the heart of organic chemistry and biology. The stereoselective version of this reaction is an example of an asymmetric synthesis (in contrast to a racemate resolution), and in this respect is similar to an aldol addition. The formation of cyanohydrins is probably one element of a small set of reactions that led to the building blocks of early life. Consequently nature still uses cyanohydrin derivatives in many different ways. For instance, a cyanohydrin participates in the synthesis of L-asparagine from L-cysteine in plants [1].

1.2
Chiral Cyanohydrins as Building Blocks in the Synthesis of Fine Chemicals

Industry is very interested in chiral cyanohydrins because they are valuable starting materials for several classes of polyfunctional molecules that are useful in the synthesis of agrochemicals and pharmaceuticals. By use of a chiral catalyst the pro-chiral aldehydes or ketones, a large number of which are readily available, are

Asymmetric Catalysis on Industrial Scale: Challenges, Approaches and Solutions
Edited by H.U. Blaser, E. Schmidt
Copyright © 2004 Wiley-VCH Verlag GmbH & Co. KGaA, Weinheim
ISBN: 3-527-30631-5

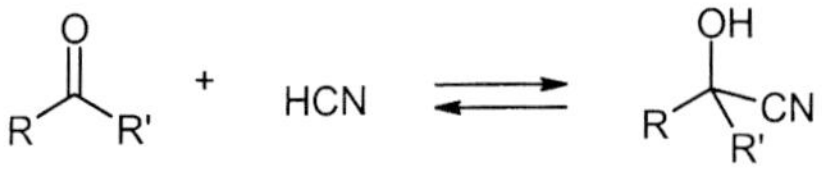

Fig. 1 The addition of hydrogen cyanide to carbonyl compounds leads to cyanohydrins.

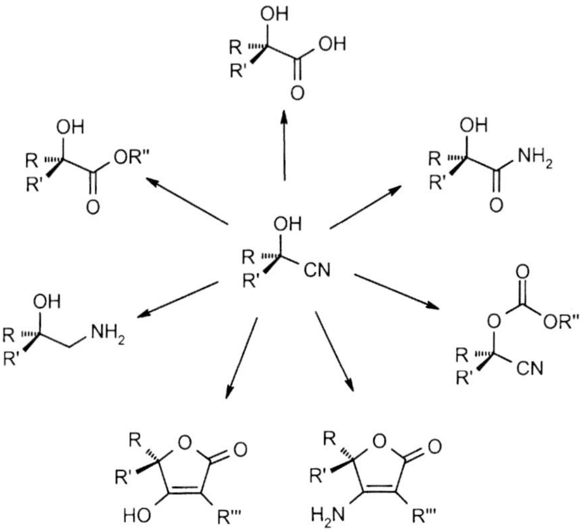

Fig. 2 Transformations of cyanohydrins leading to enantiomerically pure compounds.

transformed into the chiral cyanohydrins for which various further transformations are known. Fig. 2 shows a selection of transformations of cyanohydrins leading to classes of enantiomerically pure compounds.

1.3
Synthesis of Enantiopure Cyanohydrins

1.3.1
General Methods

Chiral cyanohydrins have been synthesized by a number of routes, which differ mainly in the form in which HCN enters the molecules and in the source of chirality.

1.3.1.1 Sources of HCN
Along with the addition of HCN as a liquid or as an alkali metal cyanide, followed by neutralization and liberation of HCN, a number of HCN carriers have been used. The most convenient of these carriers is acetone cyanohydrin, and re-

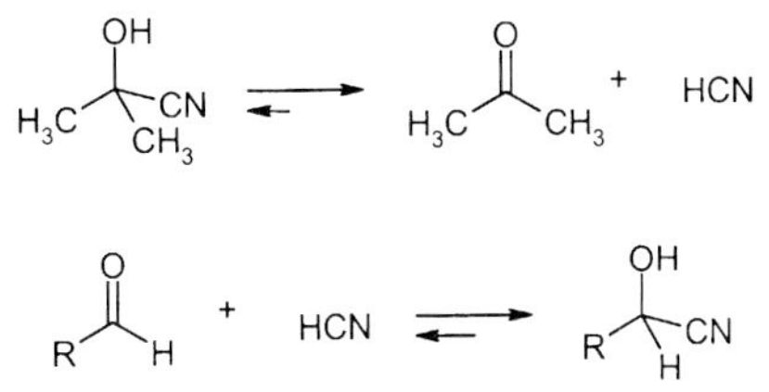

Fig. 3 Formation of cyanohydrins via transhydrocyanation.

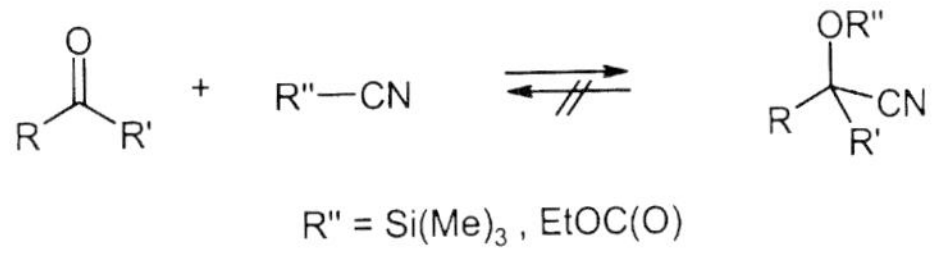

Fig. 4 Cyanohydrin formation and derivatization to complete the reaction.

actions employing this reagent could be called transhydrocyanations. These reactions proceed in two steps: cleavage of acetone cyanohydrin and reaction of the liberated HCN with the carbonyl substrate. This reaction is driven by the difference in equilibrium position of cyanohydrin formation between the different carbonyl compounds. As the equilibrium favors the formation of cyanohydrins from aldehydes, but not of ketones, this reaction works especially well for cyanohydrins of aldehydes. Fig. 3 shows the overall transformation.

There are two reasons to use reagents in cyanohydrin formation that act both as the HCN source and the derivatizing agent for the newly formed cyanohydrin:

i) The frequently unfavorable equilibrium position leads to low conversions of ketone cyanohydrins unless the cyanohydrin formed is removed from the equilibrium by an irreversible follow-up reaction.
ii) Many chemocatalysts do not work on HCN itself, but require derivatives such as silyl cyanides.

Fig. 4 shows methods of cyanohydrin formation and subsequent derivatization with various HCN carriers.

1.3.1.2 The Sources of Chirality of Cyanohydrins

The toolbox of catalyst systems for enantiopure cyanohydrin synthesis comprises racemate resolutions (yielding 50% of the desired product), dynamic kinetic racemate resolutions (enantioselective derivatization and racemization of the non-derivatized starting material, yielding 100% of the desired product) and enantioselective synthesis (yielding 100% of the desired product). Fig. 5 gives an overview of the different catalysts. Advantages and disadvantages of the four methods outlined in Fig. 5 are summarized in Tab. 1.

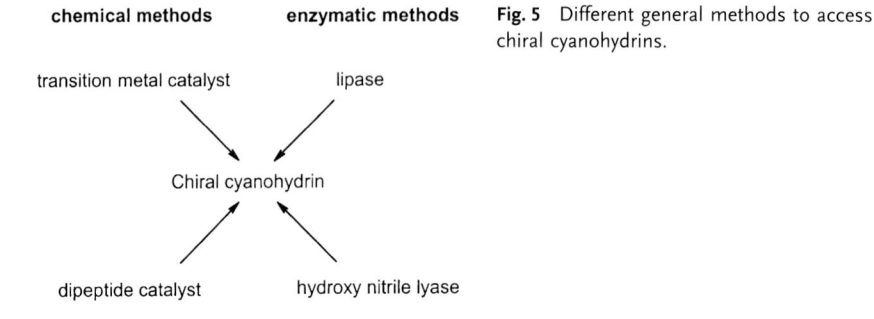

1 The Large-Scale Biocatalytic Synthesis of Enantiopure Cyanohydrins

Fig. 5 Different general methods to access chiral cyanohydrins.

Tab. 1 Advantages and disadvantages of different catalysts.

	HCN source	Advantage	Disadvantage
Chiral metal catalyst	Trimethylsilyl cyanide Alkyl cyanoformates NaCN + anhydride	All carbonyl functions accessible	Expensive cyanide source, expensive and sensitive catalyst, low temperatures required ($T<-20\,°C$)
Dipeptide catalyst	HCN	Cheap base chemicals	Difficult to recycle
Lipase	HCN	Large amount of enzyme available	Not all carbonyl functions accessible Frequently resolution, not asymmetric synthesis
HNL	HCN Alkyl cyanoformates	Large amount of enzyme available	Not all carbonyl functions accessible

1.3.2
Syntheses via Chiral Metal Catalysts

The synthesis of chiral cyanohydrins using tailor-made chiral metal catalysts has been investigated extensively, and Tab. 2 gives an overview of the most important methods [2–7].

1.3.3
Cyanohydrin Synthesis via Cyclic Dipeptides

The classical work of Inoue et al. [8] is an early example of a highly enantioselective chiral synthesis effected by a fairly simple chiral catalyst, cyclo-[(S)-Phe-(S)-His], which is shown in Fig. 6.

There has been a lot of discussion about the mechanism of the catalytic step. The high ee (90%) of mandelonitrile formation through this cyclic catalyst (acyclic

1.3 Synthesis of Enantiopure Cyanohydrins

Tab. 2 The most important synthetic methods for chiral cyanohydrins using tailor-made chiral metal catalysts.

Catalyst	Conditions	Benzaldehyde		2-Cl-Benzaldehyde		Phenylpropanal		3-Phenoxybenzaldehyde		Acetophenone	
		% ee	% yield	% ee	% yield	% ee	% yield	% ee	% yield	% ee	% yield
10% (R)- and (S)-BINOLAM×AlCl-derived catalyst [2]	3 eq. TMSCN Ph$_3$PO mol sieve 4 Å, toluene, 20°C to −40°C; 3.5–48 h	S: >98	99	S: >96	99	S: 88	99				
1% Salen-Ti-diimer type catalyst [3]	3–4 eq. KCN, t-BuOH, CH$_2$Cl$_2$, (CH$_3$CO)$_2$O 24°C to −78°C; 2–7 h	S: 90	93	S: 86	87	S: 84	80				
16 mol% Camphor-derived Ti catalyst 15 mol% Ti(OiPr)$_4$ [4]	TMSCN mol sieve 4 Å, CH$_2$Cl$_2$ −78°C	S: 93	87			S: 97	73	S: 95	78		
100% Taddol-Zr-catalyst [5]	2 eq. NC(OH)C(CH$_3$)$_2$, CH$_2$Cl$_2$ −40°C; 5–18 h	R: 63	45			R: 80	80				
10 mol% carbohydrate-based phosphine (oxide) 10 mol% Ti(Oi-Pr)$_4$ [6]	1.5 eq. TMSCl, THF −30°C to −50°C; 36–80 h									R: 92	85
10–30 mol% carbohydrate-based phosphine (oxide) 5–15 mol% Gd(Oi-Pr)$_3$ [7]	1.5 eq. TMSCl, THF −30°C to −60°C; 2–55 h									S: 92	92

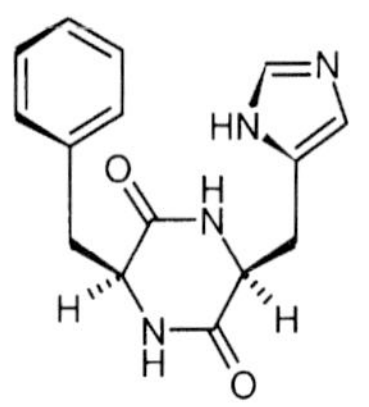

Fig. 6 Cyclo-[(S)-Phe-(S)-His], a chiral catalyst for the synthesis of non-racemic cyanohydrins.

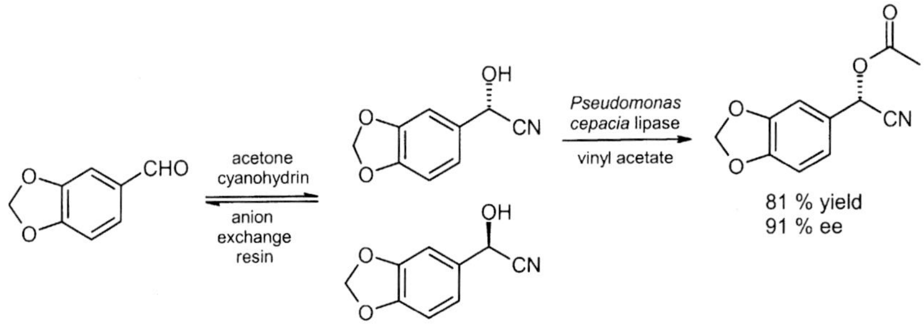

Fig. 7 Dynamic kinetic racemate resolution via lipase-catalyzed enantioselective acylation.

catalysts had yielded only 10% ee) [9] and the geometry of the catalyst led to the hypothesis that it would activate the substrate by positioning it between its imidazole and phenyl rings, which were thought to be in parallel and would cause preferential cyanide attack from one side of the carbonyl group. This hypothesis was later challenged. In its active form, the phenyl and the imidazole rings of the catalyst point away from each other. At the present time the enantioselectivity of the cyanide addition is ascribed to the cooperative effect of a number of catalyst molecules, which form a three-dimensional grid that macroscopically appears as a jelly, in which the reaction takes place. The catalyst has been industrially applied in the synthesis of (S)-3-phenoxybenzaldehyde cyanohydrin. After use, the catalyst had to be separated from the reaction mixture and treated in a specific way to regenerate it in its active form. Gregory has recently extensively reviewed this area [10].

1.3.4
The Cyanohydrin Synthesis via Lipases

Lipases have been used to effect the enantioselective esterification of cyanohydrins or the enantioselective hydrolysis of cyanohydrin esters. This works for aldehyde cyanohydrins. Selective (S)-cyanohydrin esterification is effected by the enzyme from *Pseudomonas sp.* [11]. There is also an example of selective (R)-cyanohydrin esterification by *Candida cylindracea* lipase [12]. Effenberger has shown the feasibility of this approach in principle to produce a number of enantiopure cyanohydrins derived from aldehydes. *In situ* derivatization with racemization as shown in Fig. 7 is possible, resulting in theoretically 100% yield of the desired enantiomer [13]. Ketone cyanohydrins, which are tertiary alcohols, do not easily undergo this reaction.

1.3.5
Synthesis of Chiral Cyanohydrins Using Hydroxy Nitrile Lyases (HNLs)

Hydroxynitrile lyases or oxynitrilases are enzymes that catalyze, in a reversible reaction, the stereospecific cleavage of cyanohydrins into aldehydes or ketones and hydrogen cyanide. The first descriptions of this particular reaction were published almost a century ago [14, 15]. It is generally accepted that the natural role of the hydroxynitrile lyases in plants is to serve as a defense mechanism against herbivores, whereby HCN is released from cyanogenic glucosides, such as amygdalin and vicianin [16]. This cyanogenesis defense mechanism is common to the higher plants, and enzymes from numerous plant sources have been explored for their synthetic applicability including the hydroxynitrile lyases from the rubber tree, almonds, cassava, and sorghum [17, 18]. About 3000 plant species and a small number of other organisms are known to possess these enzymes. One may be tempted to suggest, from the semi-ubiquitous presence of hydroxynitrile lyases in the plant kingdom, that these enzymes have evolved from a common ancestor and that they would have been conserved over the course of evolution. In contrast, several distinct classes of hydroxynitrile lyases have been described that seem to have evolved from hydrolytic enzymes [19] up to enzymes that stem from an oxidoreductase evolutionary branch (e.g., hydroxynitrile lyase from *Prunus amygdalus* [20]). Evidently, hydroxynitrile lyases represent a functional class of enzymes that originated from different ancestral genes by convergent evolution [21].

The reason why these enzymes have received considerable attention over the years is that they display a high degree of enantiotopic selectivity on the prochiral aldehyde and ketone substrates. The selectivity of these enzymes is in many instances masked by the rate of spontaneous racemization of the cyanohydrins, which are prone to racemization under non-acidic conditions. This balance of selectivity of the enzymes versus the competition with the spontaneous racemization reaction as a function of the pH was described as early as 1921 using the hydroxynitrile lyase enzyme from peach leaves [22]. These early experiments describe one of the challenges of applying hydroxynitrile lyases on an industrial scale.

So far HNLs from more than 11 cyanogenic plants have been characterized. Nature has evolved (R)- and (S)-selective HNLs, in contrast to other enzymes (for instance lipases) which act only on one stereoisomer. Interestingly, some enzymes, for example the one from *Hevea brasiliensis*, show pronounced stereoselectivity in the cyanohydrin formation, although their natural substrate is symmetric (acetone cyanohydrin) and their natural task is cyanohydrin cleavage.

The structures of HNLs from several plant species have been elucidated, giving insight into the respective mechanisms of cyanohydrin formation. The origin of these enzymes and thus the mechanisms of their action differ widely. The (S)-HNL from *Hevea brasiliensis* has a catalytic triad in its active center, which effects deprotonation of HCN and site-selective attack of CN^- to the substrate, which is fixed in the active site. In contrast, the HNL from *Prunus amygdalus* is a flavo-enzyme. However, the role of the flavin group in the catalytic cycle is unclear. The present hypothesis states that the enzyme seems to act via electrostatic change of

Tab. 3 Recombinant HNLs, their sources and selectivity [23–26].

Natural source		Recombinant source	Chirality of cyanohydrin
Hevea brasiliensis	Rubber tree leaves	*Pichia pastoris*	(S)
Manihot esculenta	Cassava tissue	*E. coli*	(S)
		Pichia pastoris	
Linum usitatissimum	Flax seedlings	*Pichia pastoris*	(R)
Prunus amygdalus	Almond	*Pichia pastoris*	(R)

the micro-pH environment in the active site, which leads to deprotonation of HCN and attack on the carbonyl group. Even if the pH of the bulk aqueous phase does not allow cyanohydrin formation, this reaction takes place in a stereo-controlled way in the active site.

Four different HNLs are now available on a large scale via fermentation. Their origins, host microorganism, and selectivity are depicted in Tab. 3.

Although nature provides plentiful amounts of some of these enzymes, for instance from almonds or from rubber trees, there are various reasons to produce the enzyme by fermentation once it is applied industrially:

i) Consistent quality of the enzyme and its protein matrix.
ii) Production and supply on demand and adjustable activity.
iii) Production in existing multi-purpose fermentation and down-stream processing equipment.
iv) Possibility of tuning enzyme characteristics, for example via genetic engineering.

1.4
Large-Scale Cyanohydrin Production

1.4.1
General Considerations

There are various process concepts to perform HNL-catalyzed cyanohydrin synthesis in the literature, these concepts comprise:

i) The use of immobilized HNL; among the first applications of an enzyme immobilized on a column was the (R)-HNL-catalyzed synthesis of mandelonitrile as described by Becker and co-workers [27].
ii) Enzyme membrane reactors [28].
iii) Biphasic systems.

Before we decided to enter into the development of an industrial process to produce chiral cyanohydrins and to embark on one of these process concepts for large-scale application, we considered the following factors:

1.3 Synthesis of Enantiopure Cyanohydrins

i) HCN carriers (for example acetone cyanohydrin, cyanoformate, and trimethylsilyl cyanide) may be useful for small-scale or high-price applications only.
ii) Large-scale application of HNL requires large-scale handling of HCN.
iii) Every scale-up involves the risk of introducing a new technology (in this case a new enzyme).
iv) Thorough reconfiguration of existing and building of new plants working with HCN has to get approval from the authorities, which may cost a lot of time.
v) On top of these technical risks there are economic risks of product failure.

Each of the above sources of risk is independent of the others, so the concomitant risks multiply. This quickly results in an overall economical project risk high enough to deter every new entrant.

Given the novelty of the technology (the enzyme), the other risks had to be minimized. This was achieved by:

i) Avoiding the use of new equipment.
ii) Avoiding expensive modifications of existing dedicated equipment (the enzyme reaction had to fit into an existing plant).
iii) Application to an existing product in a sizable and known market via replacement of an existing process by its technical merits and cost position.
iv) A significantly cheaper and more productive catalyst: this did not allow for the extra cost of enzyme immobilization or recycling.

Taking into account all of these factors, the rise of modern biotechnology and thus the possibility of large-scale production of the enzyme proved to be one of the key success factors. Schwab et al. had been able to produce recombinant (S)-HNL enzyme from *Hevea brasiliensis* in *Pichia pastoris* as host with an unprecedented productivity of protein [23]. Similarly, Glieder and Schwab cloned (R)-HNL into *Pichia pastoris* as host and laid the foundations for large-scale (R)-HNL applications in conventional reactor systems [26].

Instead of investing money into the realization of new plant concepts (vessels and work-up equipment) we took measures to tailor the process to existing equipment that had been dedicated to exclusive use with HCN – just a pot and a stirrer!

1.4.2
Scale-up of HNL-Catalyzed Cyanohydrin Formation

HNL, like most proteins, is exclusively soluble in water or buffer systems. Aldehydes and ketones, as substrates, are usually soluble in organic solvents. HCN, being a weak acid, is essentially undissociated at neutral to acidic pH values and distributes well between the organic and the aqueous phase with a slight preference for the organic phase. Again by virtue of its protein nature, HNL acts as a surfactant (remember "emulsin" is the ancient name for almond (R)-HNL!). Intense stirring of the reaction mixture is essential to effect a smooth reaction [29]. Indeed we witnessed that the reaction increases linearly with the stirrer energy input.

Tab. 4 The changes from laboratory to industrial production.

Important issues	Laboratory	Scale up	Industrial production
Source of enzyme	Natural source	Recombinant protein developed	High expression and cheap enzyme
Achieve high ee	Use bi-phasic system, emulsion, check pH, T, concentration	Use bi-phasic system and intense stirring to form an emulsion	Use bi-phasic system and intense stirring without shearing of enzyme ($P > 3$ kW/m^3)
Optimal conditions to maximize enzyme productivity	Use diisopropyl ether (DIPE) and aqueous buffer, constant temperature, use more than 2 eq. HCN	Use MTBE, H_3PO_4 and 2 eq. HCN, isothermic reaction	No temperature control (adiabatic reaction) and methanol free MTBE, less than 1.5 eq. HCN
Phase separation	Use filter aid	Filtration, centrifugation	Optimize stirring regime
Work-up	Gentle removal of DIPE	Wiped film evaporator with high vacuum	Wiped film evaporator with low vacuum and high temperature ($T > 60\,°C$)
Stabilization	1–2% acid	0.5% acid	Less than 0.2% acid
Quality of material		According to specification	

The following additional factors had to be taken into account:

i) Intense stirring could cause enzyme denaturation due to shear stress.
ii) Vigorous stirring leads to high reaction rates, but also to formation of a stable emulsion, which has to be broken into its components during work up, in the presence of HCN.
iii) The enzymatic reaction is always paralleled by spontaneous chemical HCN addition to the aldehyde or ketone, resulting in a lower ee. Lowering the pH prevents chemical HCN addition, but may also harm the enzyme.
iv) The enzyme itself is denatured by a high concentration of HCN, which puts a lower limit upon the amounts of aqueous and organic phases.

The scale-up of the HNL-catalyzed reaction thus reflects a multi-dimensional optimization problem with more than ten variables. Tab. 4 gives an overview of the necessary changes to the process during development from the laboratory scale to plant scale.

Transfer of the developed process into the plant led to the following process scheme, which is applicable to a variety of aldehyde and ketone substrates. An outline of the industrial process is depicted in Fig. 8.

The cyanohydrins are produced in an organic-aqueous phase mixture. Before starting the reaction via addition of HCN, the phases are thoroughly mixed in order to produce an emulsion for optimal contact between the phases. HCN is

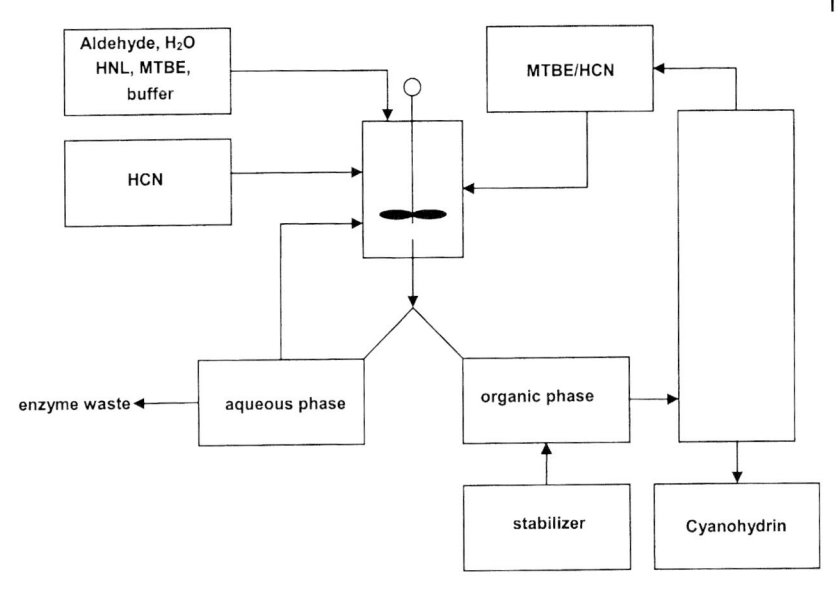

Fig. 8 General production figure for the cyanohydrins on a large-scale.

added within 1 to 3 hours, depending upon the aldehyde. The reaction temperature is normally between 0 °C and 27 °C (boiling point of HCN). After the reaction, the emulsion is broken and the phases are allowed to separate. The aqueous phase containing the enzyme is recycled. Eventually, only a little enzyme and water are added to it. The cyanohydrin in the organic phase is stabilized with acid and sent through a wiped film evaporator in order to remove MTBE, HCN, and water.

1.5
Examples of Large-Scale Production of Cyanohydrins

The combination of the knowledge of genetics, fermentation, biocatalysis, and use of hazardous materials such as HCN enabled DSM to develop and to implement processes using both (R)- and (S)-HNL for large-scale synthesis. Although both enzymes differ completely in their amino acid sequence, prosthetic group, and mechanism of action, we found that their application technology is strikingly similar (possibly more enzymes could be used in this manner):

i) Both enzymes work best in emulsions at high substrate concentration.
ii) Both enzymes are very resistant to shear and low pH values.
iii) Both enzymes can be recycled without immobilization.

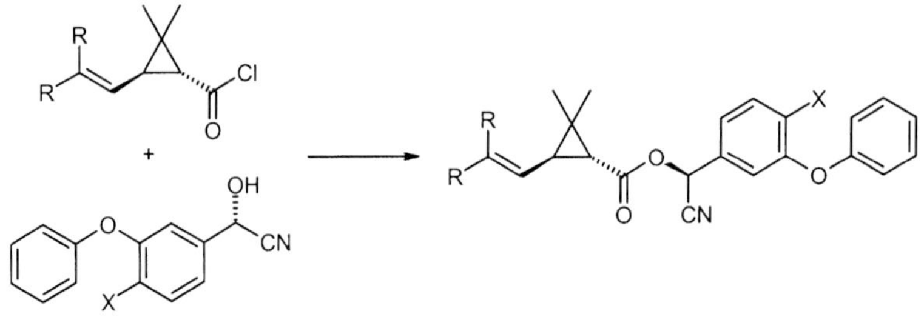

Fig. 9 Use of chiral cyanohydrins in the synthesis of pyrethroids.

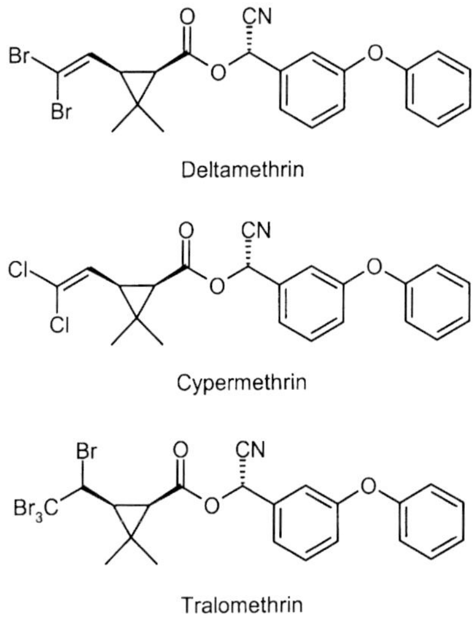

Deltamethrin

Cypermethrin

Tralomethrin

Fig. 10 Pyrethroids containing chiral cyanohydrin moieties.

1.5.1
Production of (S)-3-Phenoxybenzaldehyde Cyanohydrin (SCMB)

The largest commercial production of a chiral cyanohydrin is that of (S)-3-phenoxybenzaldehyde cyanohydrin, which is used for the production of synthetic pyrethroids, i.e., coupling of a chiral pyrethrum acid chloride derivative with the (S)-configured cyanohydrin yields insecticides, which represent a valuable market.

The production of the chiral cyanohydrin that is used for pyrethroid synthesis as depicted in Fig. 9 is performed using recombinant (S)-HNL from *Hevea brasiliensis*. Under the above-mentioned boundary conditions a space-time yield of 1000 g/L/d (ee=98.5%) is routinely reached. Fig. 10 shows a selection of products, which are currently on the market.

Of these products, Deltamethrin is produced from racemic CMB: the chiral pyrethrum acid effects dynamic resolution of the cyanohydrin.

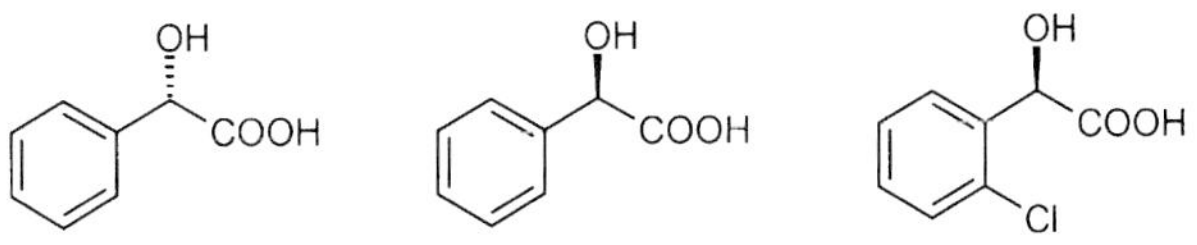

Fig. 11 A selection of chiral mandelic acids with applications in fine chemistry.

1.5.2
Production of (R)- and (S)-Mandelic Acid Derivatives

Chemical or enzymatic hydrolysis of the chiral cyanohydrin gives access to 2-hydroxycarboxylic acids. The most prominent examples are (S)- and (R)-mandelic acids, which are mainly used for racemate resolution. Other chiral acids, such as (R)-2-chloromandelic acid, are used as precursors for pharmaceuticals (see Fig. 11). Scale-up and production of (R)-2-chloromandelic acid was successfully performed with a space-time yield of 250 g/L/d (ee = 95%).

1.6
Summary

Both recombinant (R)- and (S)-HNL have been successfully used in the synthesis of chiral cyanohydrins at the plant-scale level. Their availability on a large-scale via fermentation and their striking similarities in reaction technology and chemical behavior have been crucial for the development of robust, cost-effective processes applicable to a wide variety of substrates. Exploitation of the possibilities of HNL technology has just begun. The large number of substrates and follow-up products with applications in fine chemistry reflects the attractiveness of this transformation.

1.7
References

1 R. Lieberei, D. Selmar, B. Biehl, Plant Syst. Evol. 1985, 150, 49–63.
2 J. Casas, C. Nájera, J. M. Sansano, J. M. Saá, Org. Lett. 2002, 4, 2589–2592
3 M. North, Y. Belokon, WO 02/ 10095 A2.
4 C.-W. Chang, C.-T. Yang, C.-D. Hwang, B.-J. Uang, J. Chem. Soc., Chem. Commun. 2002, pp. 54–55.
5 T. Ooi, T. Miura, K. Takaya, H. Ichikawa, K. Maruoka, Tetrahedron 2001, 57, 867–873.
6 Y. Hamashima, M. Kanai, M. Shibasaki, J. Am. Chem. Soc. 2000, 122, 7412–7413.
7 K. Yabu, S. Masumoto, S. Yamasaki, Y. Hamashima, M. Kanai, W. Du, D. P. Curran, M. Shibasaki, J. Am. Chem. Soc. 2001, 123, 9908–9909.
8 J. Oku, N. Ito, S. Inoue, Makromol. Chem. 1982, 183, 579–586.
9 J. Oku, N. Ito, S. Inoue, Makromol. Chem. 1979, 180, 1089–1091.
10 R. J. H. Gregory, Chem. Rev. 1999, 99, 3649–3682.
11 Y.-F. Wang, S.-T. Chen, K.-C. Liu, C. H. Wong, Tetrahedron Lett. 1989, 15, 1917–1920.

12 F. Effenberger, B. Gutterer, T. Ziegler, E. Eckhart, R. Aichholz, Liebigs Ann. Chem. 1991, 47–54.
13 M. Inagaki, J. Hiratake, T. Nishioka, J. Oda, J. Org. Chem. 1992, 57, 5643–5649.
14 L. Rosenthaler, Arch. Pharm. 1911, 248, 534.
15 V. K. Krieble, J. Am. Chem. Soc. 1913, 35, 1643–1647.
16 D. R. Haisman, D. J. Knight, M. J. Ellis, Phytochemistry 1967, 6, 1501–1505 and D. R. Haisman, D. J. Knight, Biochem. J. 1967, 103, 528–534.
17 D. V. Johnson, H. Griengl, Adv. Biochem. Eng./Biotechnol. 1999, 63, 31–55.
18 V. Gotor, Org. Proc. Res. Dev. 2002, 6, 420–426.
19 U. G. Wagner, M. Hasslacher, H. Griengl, H. Schwab, C. Kratky, Structure (London) 1996, 4, 811–822.
20 I. Dreveny, K. Gruber, A. Glieder, A. Thompson, C. Kratky, Structure (Cambridge, MA, US) 2001, 9, 803–815.
21 H. Lauble, B. Miehlich, S. Foerster, H. Wajant, F. Effenberger, Biochemistry 2002, 41, 12043–12050 and H. Lauble, B. Miehlich, S. Foerster, C. Kobler, H. Wajant, F. Effenberger, Protein Sci. 2002, 11, 65–71.
22 V. L. Krieble, W. A. Wieland, J. Am. Chem. Soc. 1921, 43, 164–175.
23 M. Hasslacher, M. Schall, M. Hayn, R. Bona, K. Rumbold, J. Luckl, H. Griengl, S. D. Kohlwein, H. Schwab, Protein Express. Purific. 1997, 11, 61–71.
24 S. Foerster, J. Roos, F. Effenberger, H. Wajant, A. Sprauer, Angew. Chem., Int. Ed. Engl. 1996, 35, 437–439.
25 K. Trummler, J. Roos, U. Schwaneberg, F. Effenberger, S. Forster, K. Pfizenmaier, H. Wajant, Plant Sci. (Shannon, Ireland) 1998, 139, 19–27.
26 H. Schwab, A. Glieder, C. Kratky, I. Dreveny, P. Poechlauer, W. Skranc, H. Mayrhofer, I. Wirth, R. Neuhofer, R. Bona, Prunus genes HNL1 and HNL5 and encoded hydroxynitrile lyase and their use in cyanohydrin synthesis. DSM Fine Chemicals Austria NFG GmbH & Co. KG, Austria 2001-130058(1223220), 49. 18-12-2001. EP.
27 W. Becker, H. Freund, E. Pfeil, Angew. Chem., Int. Ed. Engl. 1965, 4, 1079.
28 U. Niedermeyer, U. Kragl, M.-R. Kula, C. Wandrey, K. Makryaleas, K. Drauz: Enzymic preparation of optically active cyanohydrins. Kernforschungsanlage, Juelich and Degussa AG 89-101127 (326063), 9. 23-1-1989. EP.
29 P. Poechlauer, M. Schmidt, I. Wirth, R. Neuhofer, A. A. Zabelinskaja-Mackova, H. Griengl, C. van den Broek, R. W. E. G. Reintjens, H. Jelle Wories: Enzymatic process for the preparation of (S)-cyanohydrins. DSM Fine Chemicals, Austria G. 98-124003(927766), 14. 17-12-1998. EP.

2
Industrialization Studies of the Jacobsen Hydrolytic Kinetic Resolution of Epichlorohydrin

Larbi Aouni, Karl E. Hemberger, Serge Jasmin, Hocine Kabir, Jay F. Larrow, Isidore Le-Fur, Philippe Morel, and Thierry Schlama

Abstract

The hydrolytic kinetic resolution (HKR) of racemic terminal epoxides catalyzed by chiral (salen)-Co(III) complexes provides efficient access to epoxides and 1,2-diols, valuable chiral building blocks, in highly enantioenriched forms. While the original procedure has proved scalable for many substrates, several issues needed to be overcome for the process to be industrially practical for one of the most useful epoxides, epichlorohydrin. Combined with kinetic modelling of the HKR of epichlorohydrin, novel solutions were developed which resulted in linearly scalable processes that successfully addressed issues of catalyst activation, analysis and reactivity, control of exothermicity, product isolation, racemization, and side-product formation.

2.1
Background

First reported in 1997 [1], the Jacobsen Hydrolytic Kinetic Resolution (HKR) of racemic epoxides has become the standard method for the production of highly enantioenriched terminal epoxides and 1,2-diols (Fig. 1). The reaction shows a high degree of generality and functional group tolerance, as most epoxides evaluated (>50) are successfully resolved to ≥99% ee [2]. The HKR technology has found ready application in the commercial production of a number of chiral building blocks derived from inexpensive racemic epoxides such as propylene oxide, epichlorohydrin, and methyl glycidate [3].

The original procedures for catalyst preparation and use have proven sufficient for large-scale production of many building block targets, but improvements were necessary to enhance the industrial and economic practicality of batch-wise production utilizing this technology. Active catalyst preparation was a general problem because it remained coupled to the resolution reaction, due to the inability to physically isolate the catalyst by means other than concentration to dryness. More specifically, the HKR of epichlorohydrin was additionally handicapped by the presence of a racemization pathway with concomitant side-product formation (see Sec-

Asymmetric Catalysis on Industrial Scale: Challenges, Approaches and Solutions
Edited by H. U. Blaser, E. Schmidt
Copyright © 2004 Wiley-VCH Verlag GmbH & Co. KGaA, Weinheim
ISBN: 3-527-30631-5

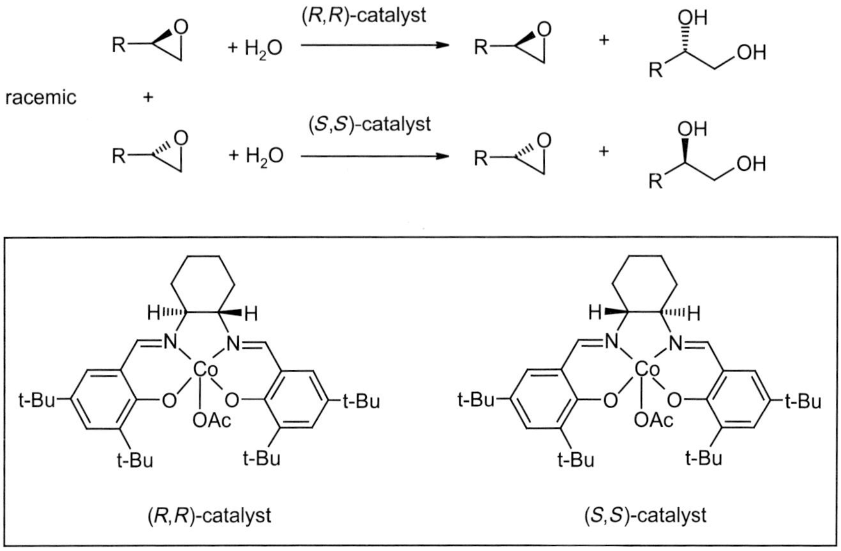

Fig. 1 General HKR of terminal epoxides catalyzed by chiral (salen)-Co(III) complexes.

tion 2.3.2) [4]. Our efforts to overcome these limitations [5], combined with a rigorous characterization of the HKR of epichlorohydrin, have resulted in the development of linearly scalable processes for the preparation and isolation of the active catalyst, the HKR of racemic epichlorohydrin, and the subsequent isolation of the resolved epoxide.

2.2
The HKR Catalyst

The HKR reaction is catalyzed by Co(III) complexes of the well-known Jacobsen salen ligand formed by the condensation of two equivalents of 3,5-di-*tert*-butylsalicylaldehyde with either enantiomer of *trans*-1,2-diaminocyclohexane [6]. These catalysts show remarkable levels of selectivity and reactivity, and they can be indefinitely recycled with most substrates (epichlorohydrin being a notable exception). The counterion can be varied to enhance the reactivity of the catalyst with slow-reacting substrates, but the initially reported acetate complex (X=OAc in Fig. 1) has proved to be the most broadly applicable catalyst [2].

Typically, the active (salen)-Co(III)-OAc complex has been accessed via air oxidation of the unreactive Co(II) complex in the presence of acetic acid according to Eq. (1). This process requires the use of a solvent in order to ensure proper mixing of the reagents, each of which is in a different physical state. Owing to the presence of air in the reaction, the solvent must meet certain flashpoint restrictions for use at scale, and dichloromethane is particularly well-suited because it has no flashpoint and the Co(II) complex has a relatively high solubility in it (ca.

40 g/L). Analytical methods were also needed to follow the activation reaction quantitatively. Following activation, the solvent is removed prior to, or it must be separated from the product after, the direct use of the catalyst in the HKR reaction. In either case, the catalyst activation and use stages have been historically coupled, and the productivity of the total process is compromised.

$$2\,(salen)Co(II) + 2\,HOAc + \frac{1}{2}O_2 \rightarrow 2\,(salen)Co(III)OAc + H_2O \quad (1)$$

Since the Co(II) complex is inactive in the HKR reaction, one might question its isolation in the first place. A method of preparation and isolation of the active complex starting from the ligand would be practically advantageous because the problem of solvent removal and the isolation of an additional intermediate would be eliminated. With these objectives in mind, we sought to develop a one-pot process for metal complexation and oxidation, whereby the active catalyst could then be isolated by conventional means. Gratifyingly, we have successfully developed a process for preparation of the active catalyst from the ligand with subsequent isolation by crystallization [5]. The facile isolation of the active catalyst results in a disconnection between the activation step and the HKR reaction that makes the technology more simple and flexible to apply. Moreover, an independent process to isolate a stable and storable form of the active catalyst allows for its production on a large scale, with the ready-to-use material available for any HKR campaign.

2.2.1
Preparation and Isolation of Active Catalyst

The active (salen)-Co(III)-OAc catalyst is prepared by reacting the salen ligand with cobalt(II) acetate and oxygen in a large excess of glacial acetic acid. Acetic acid serves a dual role as reactant and solvent, and its flashpoint is sufficiently high (40 °C) to be compatible with a continuous air sparge (Fig. 2). The initial reddish-yellow slurry gradually changes to a homogeneous dark brown solution characteristic of the (salen)-Co(III)-OAc complex. The progress of the isothermal reaction can be monitored qualitatively by thin layer chromatography (to follow consumption of the ligand), by quantitative monitoring of oxygen consumption, or by an electrochemical method that allows the direct quantitative measurement of Co(II) and Co(III) species (Section 2.2.2.1).

Upon completion of the reaction, the catalyst is released from solution in three steps. First, the reaction medium is heated under vacuum to partially remove acetic acid (75% of the initial amount used). At 100 mbar pressure, the pot temperature remains below 60 °C throughout the distillation, minimizing the potential for thermal degradation of the catalyst. The vacuum is then released to nitrogen and the reaction medium is cooled to room temperature. Next, methanol is back added to the reactor in an amount equivalent to the volume of acetic acid removed. Finally, water, used as a non-solvent, is added continuously drop wise to precipitate the catalyst. Seeding the mixture with catalyst crystals during water ad-

Fig. 2 Improved process for preparation and isolation of the active (salen)-Co(III)-OAc HKR catalyst.

Reagents/conditions:
1. excess HOAc, air, Co(OAc)$_2$-tetrahydrate
2. distill HOAc (75%)
3. back add MeOH
4. precipitate with H$_2$O
5. filter, wash with H$_2$O

(S,S)-salen ligand → (S,S)-(salen)Co(III)OAc

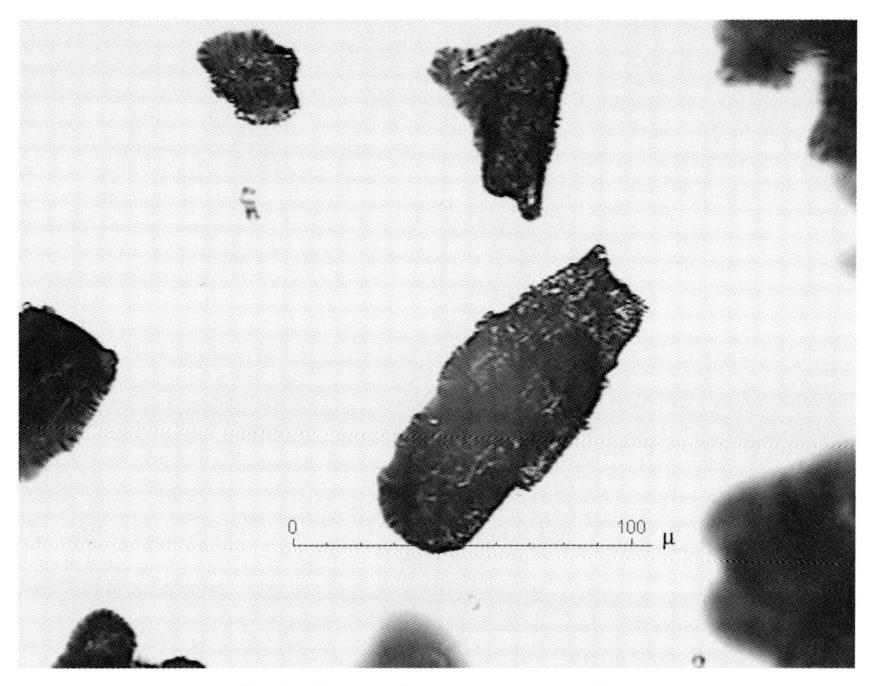

Fig. 3 Microscopic view of isolated (S,S)-(salen)-Co(III)-OAc crystals.

dition encourages crystallization, and the slurry is matured under moderated stirring after complete addition to enhance crystal growth.

Owing to the large crystal size, the subsequent filtration, washing, and cake drying steps are facilitated, and the catalyst stability is enhanced. The isolated yield of product is typically in the range of 95–97% of theory, an improvement over that for the Co(II) complex. Ironically, oxidation of the Co(II) complex to the Co(III)OAc catalyst is problematic during isolation of the Co(II) species and results in non-trivial losses. Fig. 3 shows an enlarged view of representative catalyst crystals prepared via this new method.

2.2.2
Activation Step Characterization

Despite the simplicity of the process described above, its application on an industrial scale requires careful control of each step of the process. This is especially true for the activation step during which the Co(II) to Co(III) oxidation requires accurate monitoring and the completion time must be precisely determined. The kinetic law that governs the oxidation reaction, whether chemical resistance or mass transfer limited, is also of fundamental importance for scale-up as it applies to the design of the vessel for proper mixing of the gas and liquid phases. Suitable analytical methods were required for the characterization of the activation step before scale-up could be undertaken.

2.2.2.1 Quantitative Analysis of Co(II) and Co(III) Species

Since the Co(II) to Co(III) complexes were redox active, an electrochemical method of analysis seemed viable for the quantification of the two species in the reaction. The specific electrochemical technique developed to monitor the activation reaction allowed the simultaneous quantitative measurement of (salen)Co(II) and (salen)Co(III) species in the medium. The principle of the method is based on the electro-oxidation of both species on a platinum-rotating electrode linearly polarized with respect to a standard electrode [7]. The electrochemical reactions operative with this cyclic voltammetry technique involve the single electron oxidation of each species and occur at the revolving surface of the electrode. With this salen ligand system, the Co(II) to Co(III) transformation was determined as being fully reversible, while the Co(III) to Co(IV) reaction was irreversible.

Applying a variable potential-independent value between the electrodes induces the occurrence of the electrochemical reactions. Each cobalt species responds at a specific potential with a resulting diffusion-limited current proportional to its concentration in the medium. A representative voltammogram is reproduced in Fig. 4. The use of standard solutions of Co(II) and Co(III) complexes for calibration allowed for the quantitative determination of the concentration of each component. This methodology was found to be applicable for in-process reaction monitoring purposes, as well as for establishing Co(II) impurity levels in the isolated product (typically not detected).

2.2.2.2 Kinetic Limitation of the Co(II) to Co(III) Oxidation Reaction

As described below, the metal insertion into the salen ligand with acetic acid as solvent was determined to be fast and not the rate-limiting step of the global reaction. The activation step was then studied to determine if a pure chemical resistance limited the reaction rate or if oxygen transfer from the gas phase into the liquid mixture controlled the reaction. The kinetic rate of oxidation should be independent of the rate of oxygen introduction into the organic liquid phase under chemical control, but directly related to the rate of oxygen introduction under mass transfer limitations.

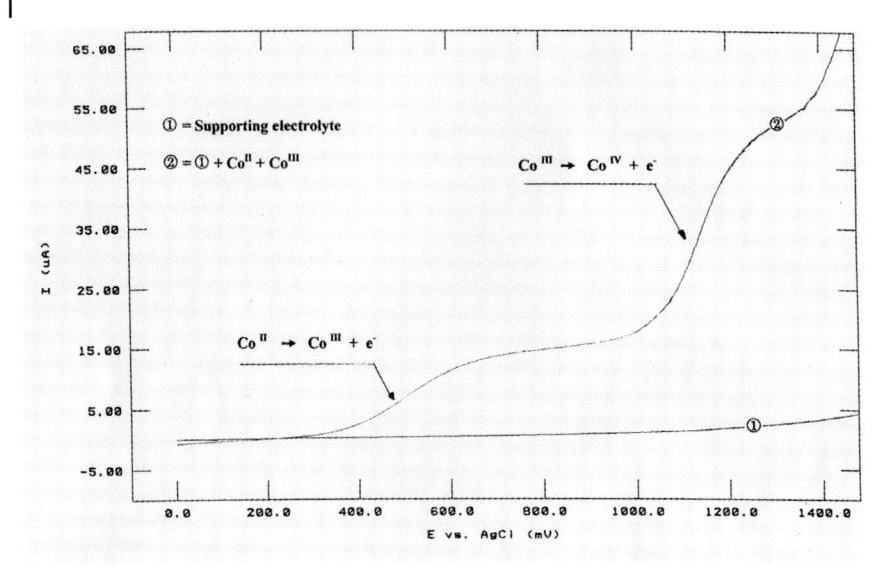

Fig. 4 Representative voltammogram showing Co(II) and Co(III) oxidation waves.

A linear driving force model is generally well adapted to characterize mass transfer from gas to liquid phase when the gas solubility in the liquid phase is low, which is the case for oxygen in most common organic solvents. This model is expressed mathematically in Eq. (2), where n_{O_2} is the number of moles of oxygen transferred from the gas phase to the liquid phase of volume V. For a given gas/solvent system, the kinetic mass transfer coefficient k_L is dependent on both the temperature and the turbulence level of agitation. The specific surface area a between the gas and liquid phases is directly related to the degree of mixing (stirrer type), the vessel geometry characteristics, and the mode of oxygen introduction into the medium. The thermodynamic Henry's constant H defines the relationship between gas phase concentration and the gas solubility in the liquid phase. It is dependent on the reaction temperature and the physical properties of the solvent. The variables $[O_2]_{Liq}$ and P_{O_2} are the oxygen concentration in the liquid phase and the partial pressure of oxygen in the gas phase, respectively.

$$\frac{1}{V}\frac{dn_{O_2}}{dt} = k_L a ([O_2]_{Liq} - HP_{O_2}) \qquad (2)$$

Initially, a simple Mariotte volumetric method was used to investigate the mass transfer behavior in the catalyst activation. The experimental Mariotte apparatus is represented schematically in Fig. 5 and consisted of a stirred vessel (300 mL) containing acetic acid (200 mL used as reactant and solvent) and the (salen)Co(II) complex to be oxidized. The system was hermetically in contact with pure oxygen at atmospheric pressure. Oxygen consumption was volumetrically determined over time for activation reactions using three different stirring speeds. The use of

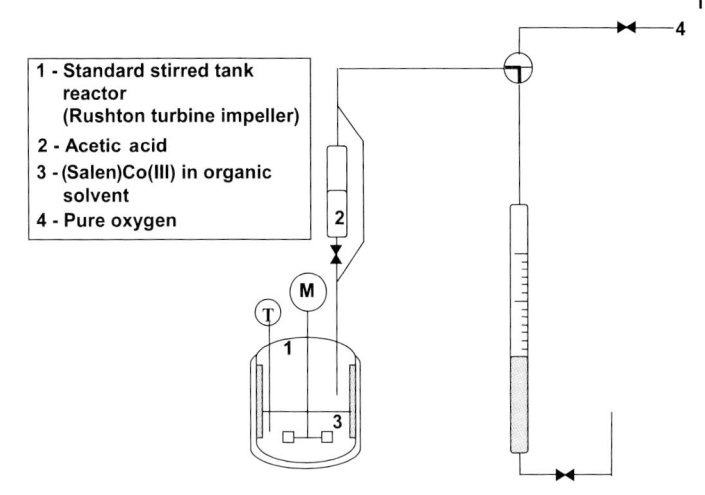

Fig. 5 Experimental Mariotte set-up for gas-liquid mass transfer measurements.

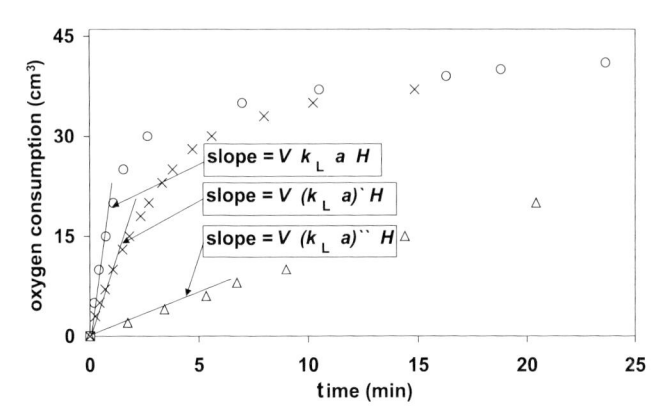

1 - Standard stirred tank reactor (Rushton turbine impeller)
2 - Acetic acid
3 - (Salen)Co(III) in organic solvent
4 - Pure oxygen

Fig. 6 Oxygen consumption monitoring versus time using Mariotte set-up at 3 different stirrer speeds: ○ 825 rpm, × 600 rpm, △ 300 rpm.

pure oxygen obviated the need for gas phase analysis and avoided any complicated apparatus to keep the gas phase homogeneous.

If a fast activation reaction is assumed, then $[O_2]_{Liq}$ is expected to be very low until completion. Mass transfer control of the reaction was indicated by the linear oxygen consumption behavior in the initial period, and by the overall kinetic dependency on stirring rate (Fig. 6).

The global mass transfer coefficients ($k_L a$) were then derived from the slopes of the linear initial rates of oxygen consumption with oxygen solubility in acetic acid (HP_{O_2}) equal to 0.17 cm^3/cm^3 at 20 °C (Tab. 1).

In another experiment, rate limitation by mass transfer was confirmed by the observed change in slope when switching from a low agitation speed to the highest speed experimentally tested (Fig. 7).

Tab. 1 Mass transfer coefficient.

Agitation speed (rpm)	$k_L a$ (s^{-1})
825	0.9×10^{-2}
600	4.9×10^{-3}
300	5.8×10^{-4}

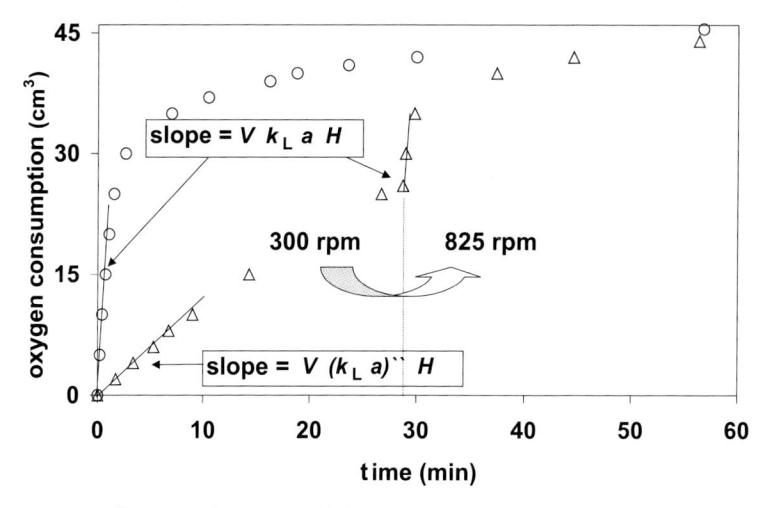

Fig. 7 Confirmation of mass transfer limitation on reaction rate using Mariotte set-up: ○ 825 rpm, × 300 rpm then 825 rpm.

This simple method of analysis was inappropriate for kinetic monitoring of an activation reaction using compressed air as the oxidant because a second method of measuring the oxygen content of the gas phase would be required. Under these realistic conditions, the direct analysis of the Co(II) and Co(III) species by the electrochemical method was used to follow the reaction kinetics. Using standard experimental conditions, the (salen)Co(II) complex was dissolved in glacial acetic acid (10 equiv. relative to catalyst) in a reaction vessel equipped with a Rushton turbine, thermometer, and an air-sparging dip tube. The Rushton turbine is designed to overcome mass transfer limitations between the gas and liquid phases by producing efficient phase mixing and high gas dispersion. The reaction mixture was stirred at room temperature while bubbling with compressed air at a rate of 3×10^{-6} m^3/s beneath the stirrer flow. The composition of the reaction was monitored every 15 min by electrochemical assay. The kinetic profile of Co(II) species consumption and Co(III) species production is depicted in Fig. 8. The data show good correlation between reactant disappearance and product formation, while the total Co species, as determined by inductively coupled plasma (ICP) spectrometric analysis, remains logically constant throughout the reaction.

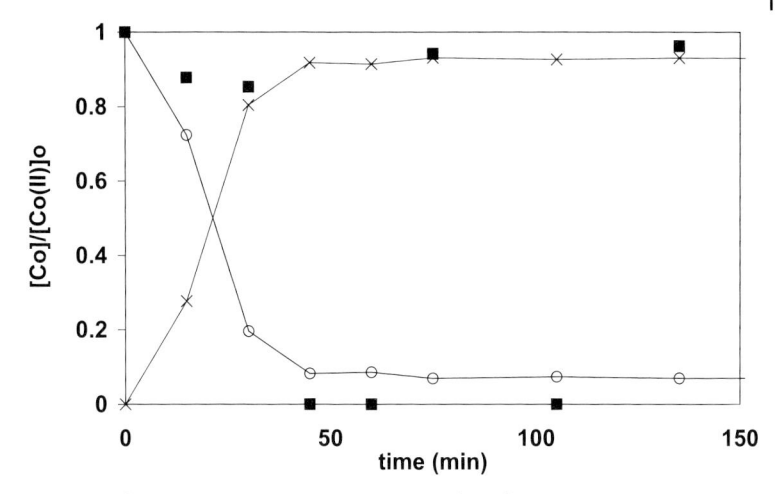

Fig. 8 Catalyst air activation reaction monitoring by voltammetry: × Co(III) species concentration, ○ Co(II) species concentration, ■ total Co concentration (ICP).

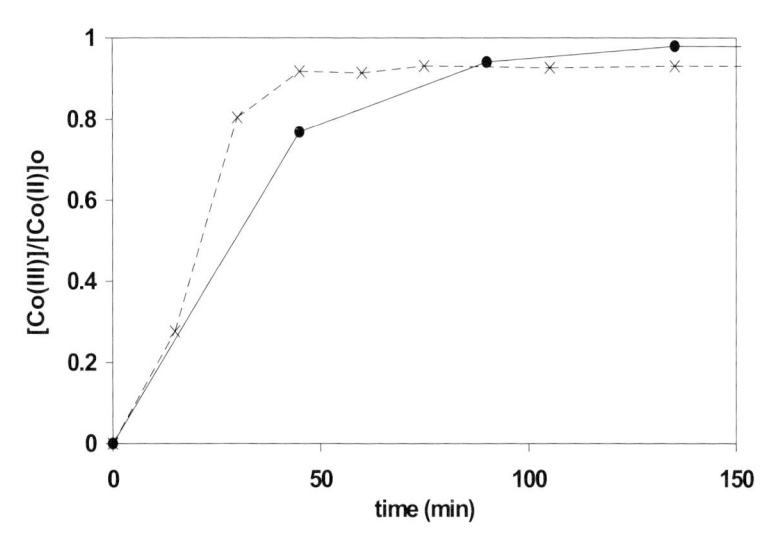

Fig. 9 Comparison of air activation reactions using voltammetry: × activation of pure (salen)Co(II) complex, ● simultaneous metal insertion and air activation.

As shown in Fig. 9, a similar experiment starting with the salen ligand and combining metal insertion and catalyst activation led to the same completion time as the previous experiment. This result demonstrates that the metal insertion step does not limit the overall rate of reaction. The similarities in reaction profile using either pure oxygen (non-scalable) or air at different stirrer speeds confirm the important role of the degree of mixing and gas–liquid dispersion in the activation reaction. Since the overall rate of the complexation and activation sequence is mass transfer limited, successful scale-up requires the identification and preservation of the critical mass transfer parameters.

2.2.3
Scale-up Considerations for the (Salen)-Co(III)-OAc Process

Agitation is a critical process parameter in mass transfer-limited processes, particularly ones such as this for which the reaction rates are purely gas transfer dependent. Moreover, control of reactant transfer from the gas phase to the liquid phase is vital to obtain consistent process performance. Thus, scale-up from the lab-scale to industrial production must focus on the critical parameters that preserve the desired mass transfer properties.

Gas–liquid-agitated vessel technology finds the most widespread application in the chemical industry due to its simplicity and its efficient mixing characteristics. The scale-up (or scale-down from an existing industrial vessel) generally requires maintaining the same geometric ratio between lab or pilot scale and the industrial vessels and finding operating conditions that closely reproduce the mixing regime that preserves the global transfer coefficient $k_L a$ (see Section 2.2.2.1). This global transfer coefficient is commonly related to the specific stirrer power input P_g/V and the superficial gas velocity in the vessel u_G. Several correlations are given in the literature, but the most common one is given in Eq. (3) [8]. Hickman further refined this correlation for agitated vessels of 0.6 and 2 m diameter using a Rushton turbine in the standard configuration (Eq. 4) [9]. Although the specific power input and the gas hold-up are readily conserved during scale-up, the gas injection operation must be carefully investigated to ensure the optimum dispersion mode.

$$k_L a = K \left(\frac{P_g}{V}\right)^A (u_B)^B \tag{3}$$

$$k_L a = 0.046 \left(\frac{P_g}{V}\right)^{0.54} (u_G)^{0.68} \tag{4}$$

The gas dispersion is characterized by the bubble/droplet formation at the outlet of the gas distributor. In a bubble regime, each droplet is gradually ejected from the distributor hole and can be considered individually. Conversely, a jet regime results from a bubble train being continuously propelled into the liquid. These two regimes can be distinguished by the Weber number (We), which relates the inertial force of the gas to the surface tension of the liquid. In a gas dispersion system using a gas-sparging dip tube or torus, the Weber number can be derived according to Eq. (5) in which d_{Hole} is the diameter (m) of the dip tube opening, ρ_{gas} is the gas density (kg/m³), v_G is the gas velocity (m/s) at the hole outlet, and σ_L is the surface tension of the liquid (kg/s²). For $We \gg 4$ the dispersion lies in a jet regime, while $We \ll 4$ indicates a bubble regime.

$$We = \frac{d_{Hole} \rho_{gas} V_G^2}{\sigma_L} \tag{5}$$

Tab. 2 Catalyst activation scale-up parameters.

Parameter	0.2 L Scale	50 L Scale	Unit
N (agitation speed)	800	150	rpm
u_G	1.4×10^{-3}	3×10^{-3}	m/s
P_g/V	0.33	0.32	kW/m^3
$k_L a$ (experimental)	0.9×10^{-2}	–	s^{-1}
$k_L a$ (estimated)	1.2×10^{-2}	2×10^{-2}	s^{-1}
v_G	22	20	m/s
We	9	94	(–)
T	20	20	°C

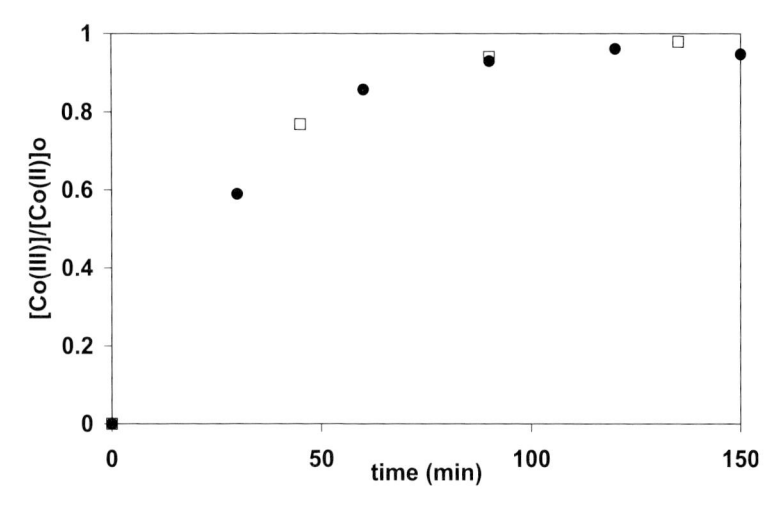

Fig. 10 Comparison of air activation reactions using voltammetry: ● pilot-scale, ☐ lab-scale.

When physical mass transfer governs the kinetic rate, gas dispersion is highly enhanced in a jet regime as the bubble train speed provides better gas–liquid intermixing outside the immediate area surrounding the impeller. Scale-up parameters for the catalyst activation for transfer from a 0.2 L to a 50 L scale utilizing an air-sparging dip tube as the gas distributor are given in Tab. 2. Based upon these parameters, the kinetic profiles of the activation reaction at both lab and pilot scale are comparable (Fig. 10).

2.3
Kinetic Modelling and Simulation of the HKR Reaction

2.3.1
Objectives and Approach

In order to facilitate scale-up and optimization, the HKR of epichlorohydrin using (salen)-Co(III)-OAc was investigated to provide a detailed understanding of the epoxide ring opening chemistry and to develop a global model that kinetically characterizes the HKR reaction. The objective was not to describe the kinetic mechanism at a molecular scale [10], but rather to develop a kinetic description of the global reaction system including the main epoxide hydrolysis reaction and lesser side-product forming reactions. A global model was deduced to provide a simplified macroscopic description of the reaction. The model was then refined through an iterative trial-and-error loop until the experimentally observed specifications were reached (Fig. 11).

The global reaction model was developed in four stages. First, all products of the reaction, including minor side products, were identified and quantitatively characterized. Potential reaction routes that logically lead to the observed products were experimentally evaluated and validated to provide material balance. Next, a simplified global mechanism was developed to fit the experimental observations. Finally, the proposed mechanism was validated with a suitable mathematical model to enable the prediction of the overall reaction kinetics and the estimation of reaction parameters based upon vessel configuration. The resulting model has allowed the complete characterization of the reaction system within the range of reaction conditions typical for this application. The work presented in this chapter focuses on the HKR of racemic epichlorohydrin with (S,S)-(salen)-Co(III)-OAc catalyst for the production of (R)-epichlorohydrin. Naturally, the results presented for

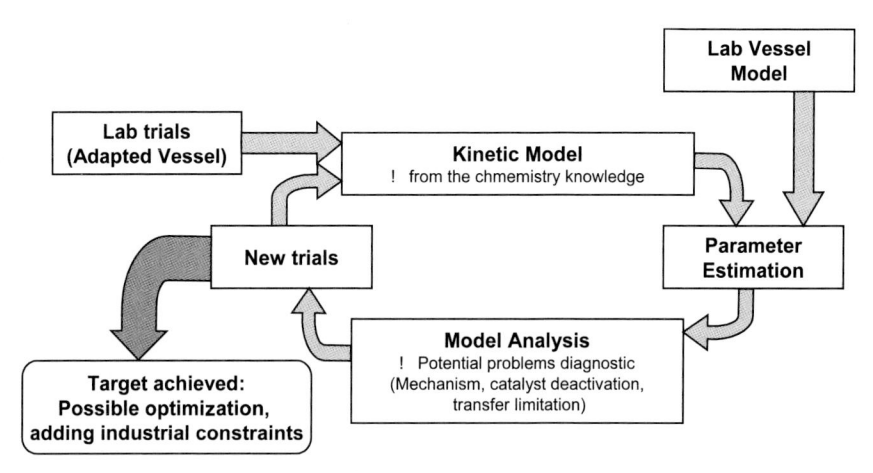

Fig. 11 Trial and error loop procedure for process optimization.

this system should be symmetrically identical for the production of the (S)-enantiomer using the (R,R)-catalyst.

2.3.2
Development of a Global Reaction Scheme

The HKR of epichlorohydrin using pre-activated catalyst basically involves the dissolution of the catalyst in racemic epichlorohydrin followed by water addition to the resulting homogeneous solution. The solubility of water in this substrate is sufficiently high (5.9 mol-%, 1.2% w/w) to begin the reaction in a homogeneous state. Formation of 3-chloro-1,2-propanediol (CPD) then facilitates water dissolution in the medium so that the mixture remains homogeneous throughout the reaction (provided the water is not added all at once). Reaction homogeneity is crucial, as kinetic measurements on a homogeneous system are independent of reactor size, type, and geometry. Thus, the measured kinetic parameters can be applied to any vessel design or set of operating conditions.

During the HKR reaction, five main components were measured, including water, epichlorohydrin, and CPD. Glycidol and 1,3-dichloro-2-propanol (DCP) are the primary impurities formed during the reaction, and constitute a total of approximately 2% of the mass in the mixture. Moreover, both side-products are produced at virtually the same level, suggesting that their formation is mechanistically related. Mass balance of the total chloride content throughout the reaction was relatively constant (Tab. 3), which is indicative of a chloride exchange mechanism between epichlorohydrin, CPD, and DCP. Other impurities, including 1- and 2-glyceryl acetate, were found in much smaller quantities and were considered secondary byproducts that were immaterial to the mechanism of interest.

Chloride exchange reactions between species were demonstrated by combining racemic epichlorohydrin, racemic CPD, and (S,S)-(salen)-Co(III)-OAc in the ab-

Tab. 3 HKR total chloride mass balance.

Time (min)	Cl in epichlorohydrin (g)	Cl in DCP (g)	Cl in CPD (g)	Total
0	15.73	0	0	15.73
54	14.06	0.02	0.23	14.31
99	11.35	0.03	3.69	15.07
130	9.31	0.03	5.94	15.26
176	9.08	0.04	5.57	14.69
221	8.00	0.05	5.78	13.83
260	8.55	0.06	6.28	14.89
305	7.26	0.07	6.81	14.15
360	8.12	0.09	6.74	14.96
409	7.46	0.11	6.38	13.95
1295	6.92	0.48	6.52	13.93
1549	7.01	0.6	6.35	13.95

Tab. 4 Equilibration of epichlorohydrin and CPD under anhydrous HKR conditions.

Time (day)	Epichlorohydrin ee (%)	Epichlorohydrin (mol-%)	Chloropropanediol (mol-%)	Glycidol (mol-%)	Dichloropropanol (mol-%)
0	0.5	38.58	32.62	0	0
3	7.0	23.36	17.51	16.78	12.76
5	5.5	21.26	15.75	17.45	14.9
6	6.5	19.81	15.65	17.68	16.39
7	6.3	20.76	13.99	16.59	16.38

sence of water at room temperature. The reaction was monitored over time and revealed the degradation of epichlorohydrin and CPD with simultaneous formation of glycidol and DCP (Tab. 4). The total material balance indicates a stoichiometric reaction that confirms the presence of chloride exchange reactions. A steady state was observed after three days, which implied that the components had reached their equilibrium concentrations.

These observations were mechanistically interpreted to result from ring closure of CPD to form glycidol and HCl. Subsequent reaction of epichlorohydrin with HCl to form DCP accounts for all of the major components in the HKR of epichlorohydrin. As indicated by the enantiomeric excess (ee) of epichlorohydrin in Tab. 4, these chloride exchange reactions proceed with low enantioselectivity. This interpretation was supported when considering the reverse reactions, whereby glycidol and DCP reacted in the presence of catalyst to form epichlorohydrin and CPD. The low ee of the epichlorohydrin produced confirm this as the racemization pathway [4].

The inclusion of the chloride exchange reactions with the primary epoxide hydrolysis reactions enables the development of a global reaction scheme for the HKR of racemic epichlorohydrin with glycidol and dichloropropanol as the primary impurities (Fig. 12). The irreversible hydrolysis of (S)-epichlorohydrin to form (S)-CPD (bold arrow) constitutes the major pathway of the system. Slow conversion of (S)-CPD to (S)-glycidol is driven toward glycidol formation by fast reaction of the HCl coproduct with epichlorohydrin to produce DCP. The ring closure of achiral DCP results in the non-selective formation of epichlorohydrin and HCl, providing the pathway for racemization. Because the HKR of epichlorohydrin is not completely selective, a minor mirror image set of reactions is observed for the enantiomeric compounds. Control experiments revealed that none of the forward or reverse reactions proceed to any measurable extent in the absence of active catalyst under ambient conditions.

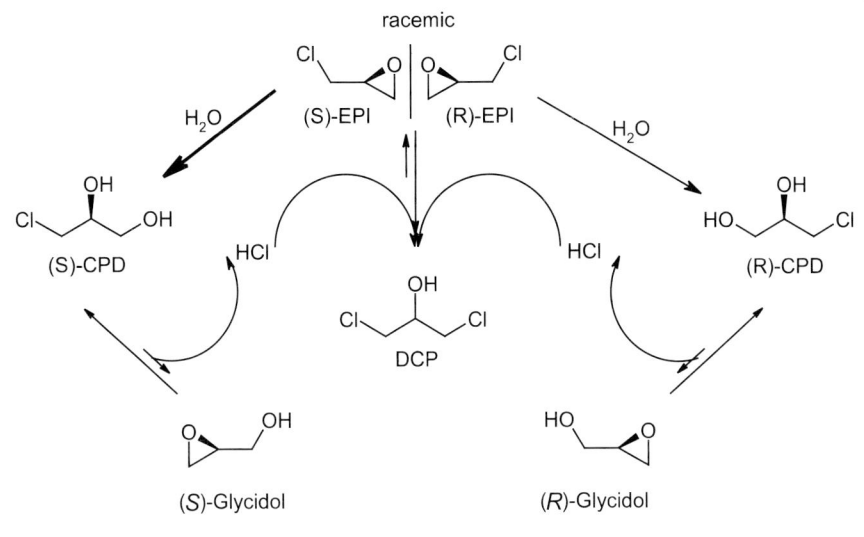

Fig. 12 Global reaction scheme for HKR of epichlorohydrin.

2.3.3
Kinetic Modelling

2.3.3.1 Kinetic Rate Equations and Assumptions

To facilitate the development of the kinetic parameters for the HKR of epichlorohydrin, the global mechanism scheme (Fig. 12) has been simplified by only considering the three main reactions: the hydrolysis of (S)-epichlorohydrin (r_S) and (R)-epichlorohydrin (r_R), and side-product formation (r_d) (Eqs. 6–8). Assuming that HCl transfer is fast, the formation of impurities can be considered as a single reaction. The result is the apparent reaction between epichlorohydrin and CPD to form glycidol and DCP.

$$\text{Cl-EPI} + H_2O \xrightarrow[k_S]{(S,S)\text{-catalyst}} \text{(S)-CPD} \tag{6}$$

$$\text{Cl-EPI} + H_2O \xrightarrow[k_R]{(S,S)\text{-catalyst}} \text{(R)-CPD} \tag{7}$$

$$\text{EPI} + \text{CPD} \underset{k_{-d}}{\overset{k_d}{\rightleftharpoons}} \text{Glycidol} + \text{DCP} \tag{8}$$

As mentioned in Section 2.3.2, glycidol and DCP are produced only in small amounts during the HKR reaction. Upon completion of the HKR, the equilibrium equations for DCP and glycidol formation are far from equilibrium and a non-equilibrated condition is assumed. Based upon these assumptions, the macroscopic kinetic law governing each reaction is expressed as a typical polynomial equation (Eqs. 9–11).

$$r_S = k_S[\text{Epi}_S]^{n_{S\text{Epi}}}[\text{H}_2\text{O}]^{n_{S\text{H}_2\text{O}}}[\text{Cat.}]^{n_S} \tag{9}$$

$$r_R = k_R[\text{Epi}_R]^{n_{R\text{Epi}}}[\text{H}_2\text{O}]^{n_{R\text{H}_2\text{O}}}[\text{Cat.}]^{n_R} \tag{10}$$

$$r_d = k_d[\text{Epi}_d]^{n_{d\text{Epi}}}[\text{CPD}]^{n_{\text{CPD}}}[\text{Cat.}]^{n_d} \tag{11}$$

For homogeneous systems, the kinetic parameters k_i and n_i are commonly determined by lab-scale batch experiments in which all reagents are combined at the start of the reaction. Following the concentration of all components (reactants and products) over the course of the reaction then allows for the estimation of the kinetic parameters. Since water has limited solubility in the reaction mixture at the start, conventional kinetic batch experiments could result in erroneous calculation of k_i and n_i if the limits for homogeneity are crossed. To ensure reaction homogeneity and reliable kinetic measurements, the gradual and continuous addition of water was selected as a suitable method for experimentation (semi-batch mode). The kinetic parameters were then recovered using an appropriate mathematical model with parameter estimation module.

2.3.3.2 Experimentation and Kinetic Model for a Fed-Batch Vessel

The fed-batch reactor model is commonly built using classical thermal and mass balance differential equations [11]. Under isothermal conditions, the material balance for each measured component in the HKR of epichlorohydrin with continuous water addition is expressed by one of the following equations (Eqs. 12–17). These equations can be solved using an appropriate solver package (Lsoda, Ddassl [12]) with a connected optimization module for parameter estimation.

$$(S)\text{-Epichlorohydrin:} \quad V\frac{\partial[\text{Epi}_S]}{\partial t} + [\text{Epi}_S]\frac{\partial V}{\partial t} + (r_S + r_d)V = 0 \tag{12}$$

$$(R)\text{-Epichlorohydrin:} \quad V\frac{\partial[\text{Epi}_R]}{\partial t} + [\text{Epi}_R]\frac{\partial V}{\partial t} + (r_R + r_d)V = 0 \tag{13}$$

$$\text{Chloropropane Diol (CPD):} \quad V\frac{\partial[\text{CPD}]}{\partial t} + [\text{CPD}]\frac{\partial V}{\partial t} - (r_S + r_R - r_d + r_d)V = 0 \tag{14}$$

$$\text{Water:} \quad V\frac{\partial[\text{H}_2\text{O}]}{\partial t} + [\text{H}_2\text{O}]\frac{\partial V}{\partial t} + (r_S + r_R)V = Q[\text{H}_2\text{O}]_0 \tag{15}$$

Tab. 5 Lab-scale standard epichlorohydrin HKR conditions.

Component	Condition	Ratio
Racemic epichlorohydrin	40 g	1 equiv.
(S,S)-(salen)Co(III)OAc	35 mol/m^3	0.5 mol-%/Epi
Water	5.8 mL	0.75 equiv.
1,2-DCB (solvent)	25 g	0.4 equiv./Epi
Time of water addition	2 h	2.9 mL/h
Temperature	5 °C	

$$\text{Glycidol:} \quad V\frac{\partial [Gly]}{\partial t} + [Gly]\frac{\partial V}{\partial t} + r_d V = 0 \tag{16}$$

$$\text{Dichloropropanol (DCP):} \quad V\frac{\partial [DCP]}{\partial t} + [DCP]\frac{\partial V}{\partial t} - r_d V = 0 \tag{17}$$

The kinetic order (n_i) for the reactions governed by Eqs. (9)–(11) were estimated with this model through standard experimentation designed to provide the kinetic orders of each reaction. The experimental set-up consisted of a temperature-controlled jacketed stirred-tank reactor (100 mL) equipped with a thermometer and a dip-tube for sampling. A water flask was connected to the vessel via a volumetric pump that allowed water addition at a constant flow-rate. A four-pitched blade agitator was used to provide efficient mixing and to ensure good heat transfer. The standard operating conditions for these experiments are provided in Tab. 5.

Because the epoxide hydrolysis reaction is exothermic, isothermal conditions were critical to obtain good data in this parameter study. A solvent (1,2-dichlorobenzene) was used to dilute the reaction mixture in order to help maintain a constant reaction temperature. This proved especially useful when exploring higher catalyst concentrations that accelerate the rate of reaction. It will be demonstrated in Section 2.3.3.3 that the use of solvent did not alter the kinetic parameters of the HKR reaction, and thus the estimated kinetic parameters remained valid under solvent-free conditions. Water retained sufficient solubility in the mixed system to maintain homogeneity at the desired rate of addition. A slight excess of water was used to ensure complete conversion of (S)-epichlorohydrin and to compensate for any competitive hydrolysis of the (R)-enantiomer.

It was assumed that catalyst degradation was inoperative (and subsequently confirmed by the kinetic study), thus maintaining a constant concentration and constant activity throughout the reaction. With catalyst concentration experimentally fixed, the kinetic rate equations (9–11) simplify to Eq. (18)–(20), respectively, where $K_i = k_i [\text{catalyst}]^{n_i}$.

$$r_S = K_S [\text{Epi}_S]^{n_{SEpi}} [H_2O]^{n_{SH_2O}} \tag{18}$$

$$r_R = K_R [\text{Epi}_R]^{n_{REpi}} [H_2O]^{n_{RH_2O}} \tag{19}$$

$$r_d = K_d[\text{Epi}]^{n_{\text{Epi}}}[\text{CPD}]^{n_{\text{CPD}}} \tag{20}$$

Figs. 13 and 14 plot the experimental (points) and simulated (solid lines) concentration profile versus time for (S)-epichlorohydrin, (R)-epichlorohydrin, and water under the standard HKR operating conditions using 0.5 and 0.28 mol-% catalyst (relative to racemate), respectively. Also shown is the simulated and observed evolution of epichlorohydrin ee and conversion over the course of the reaction. Fig. 15 gives the corresponding profiles for glycidol and DCP at both catalyst concentrations.

To disconnect the kinetic rate of hydrolysis for each enantiomer of epichlorohydrin, identical experiments were performed with each pure enantiomer

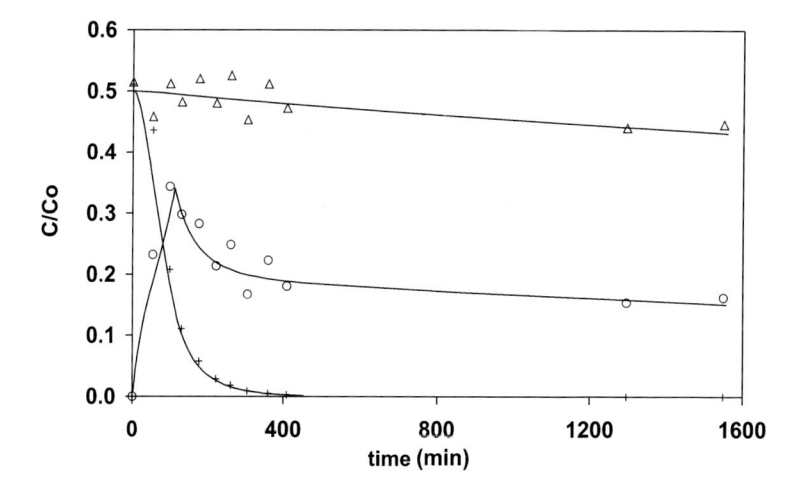

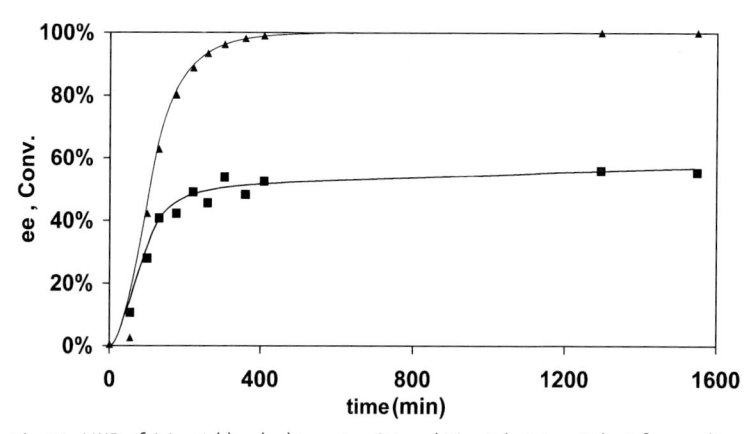

Fig. 13 HKR of (±)-epichlorohydrin using 0.5 mol-% catalyst (see Tab. 5 for conditions). Experimental: △ (R)-epichlorohydrin, + (S)-epichlorohydrin, ○ water, ▲ ee, ■ conversion. Numerical simulation: solid lines.

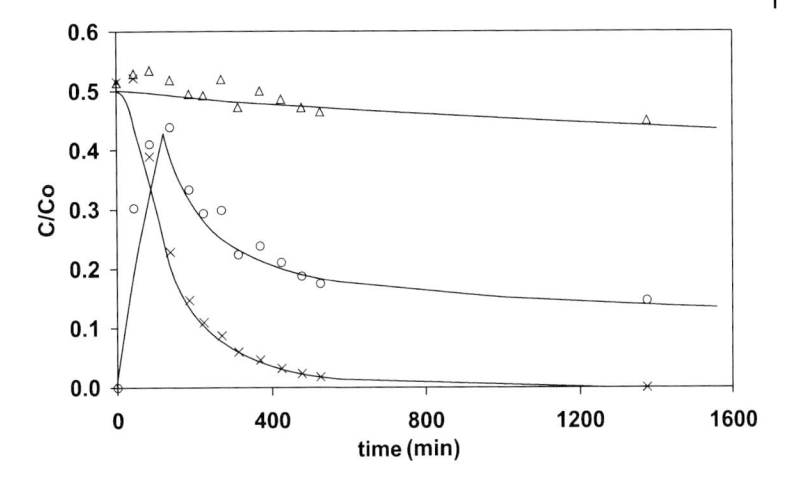

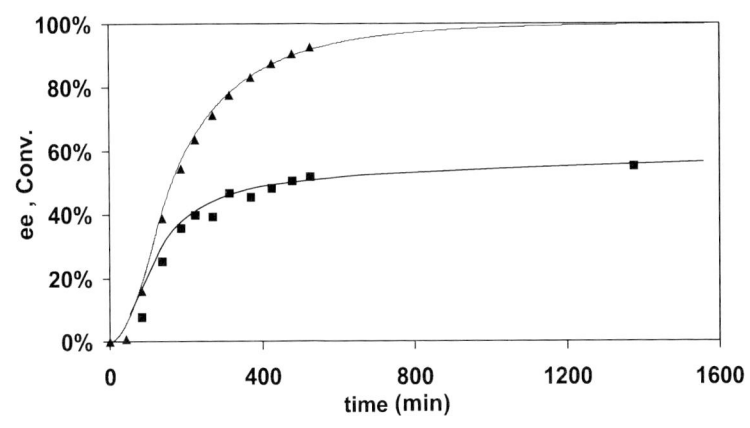

Fig. 14 HKR of (±)-epichlorohydrin using 0.28 mol-% catalyst (see Tab. 5 for conditions). Experimental: △ (R)-epichlorohydrin, × (S)-epichlorohydrin, ○ water, ▲ ee, ■ conversion. Numerical simulation: solid lines.

(>99.5% ee) using conditions shown in Tab. 6. Fig. 16 shows the experimental results and the simulated fit for (S)-epichlorohydrin and water, while the related results for (R)-epichlorohydrin are plotted in Fig. 17. It can be observed that water is always present in the system, thus proving that the observed rate of reaction is determined by the chemical resistance of the reaction rather than by the rate of water addition.

In all the experiments, the estimated parameters exhibited an overall second-order behavior for the three reaction rates. The model best fits the experimental data by assuming a first-order rate dependence on the concentration of each component in the rate equation (Eqs. 18–20). The global kinetic constants (K_S, K_R,

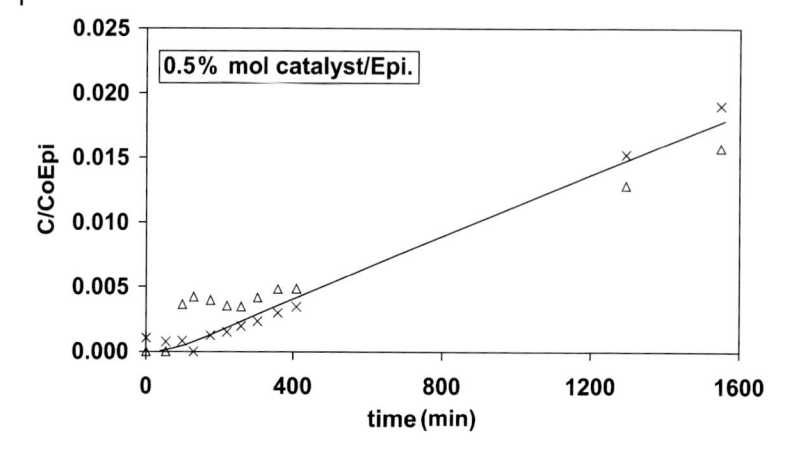

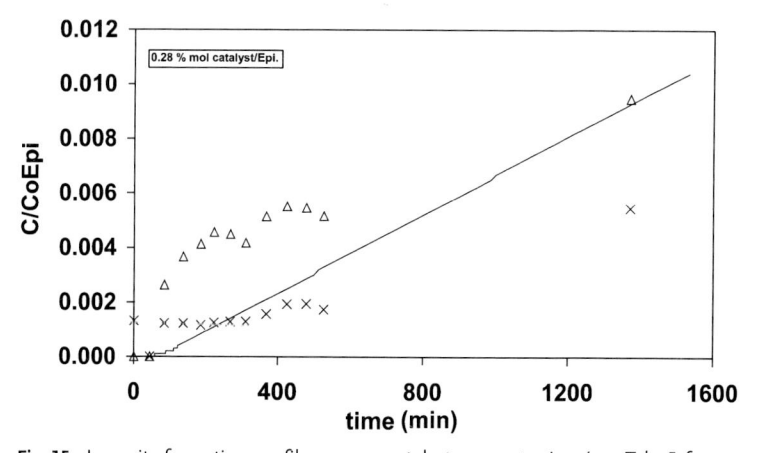

Fig. 15 Impurity formation profiles versus catalyst concentration (see Tab. 5 for conditions). Experimental: △ glycidol, × 1,3-dichloro-2-propanol (DCP). Numerical simulation: solid lines.

Tab. 6 Lab-scale chiral epichlorohydrin HKR.

Component	Condition	Ratio
Chiral epichlorohydrin	20 g	1 equiv.
(S,S)-(salen)Co(III)OAc	~50 mol/m^3	1 mol-%/Epi
Water	5.8 mL	1.5 equiv.
o-DCB (solvent)	25 g	0.8 equiv./Epi
Time of water addition	2 h	2.9 mL/h
Temperature	5 °C	

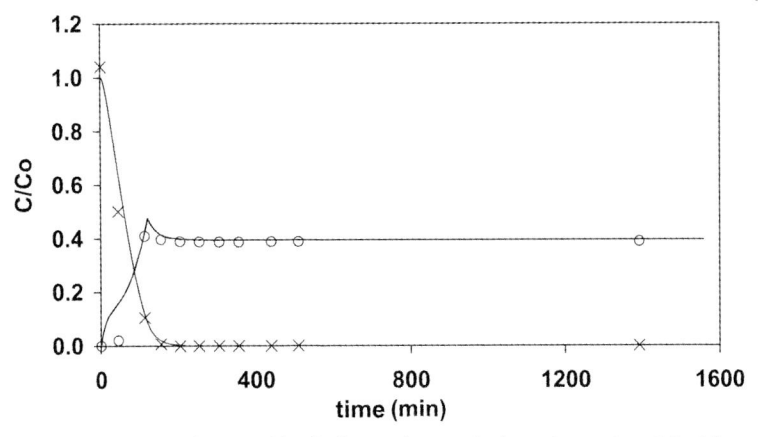

Fig. 16 Hydrolysis of (S)-epichlorohydrin under standard conditions (see Tab. 6 for conditions). Experimental: × (S)-epichlorohydrin, ○ water. Numerical simulation: solid lines.

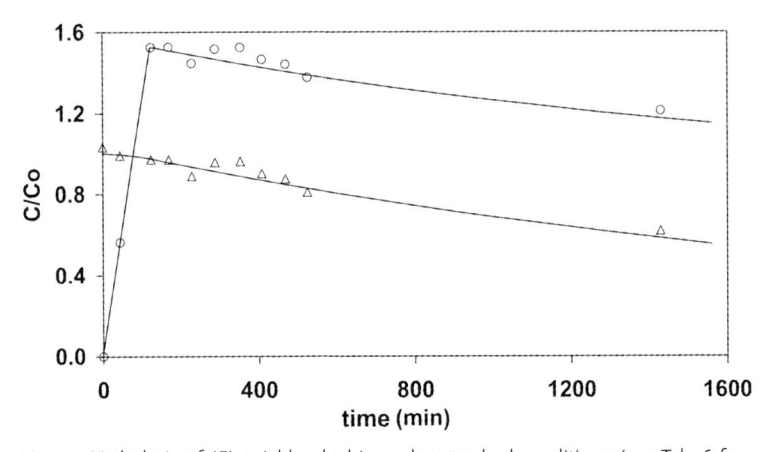

Fig. 17 Hydrolysis of (R)-epichlorohydrin under standard conditions (see Tab. 6 for conditions). Experimental: △ (R)-epichlorohydrin, ○ water. Numerical simulation: solid lines.

and K_d) for each set of conditions were then obtained from initial rates of the simulated runs (Tab. 7). The kinetic constants (k_i) and the catalyst kinetic orders ($n_{i(Cat)}$) were then recovered from the estimated global kinetic constants (K_i) as the y-intercept and the slope, respectively, of the double natural logarithm plot of K_i versus [catalyst] (Fig. 18).

With (S,S)-(salen)-Co(III)-OAc, the hydrolysis of (S)-epichlorohydrin exhibited a second order dependence on catalyst concentration, which is consistent with previous results obtained for HKR reactions [1, 10]. Significantly, hydrolysis of the slow reacting (R)-enantiomer was estimated to have a zero order dependence on catalyst concentration. Zero order dependence was also observed for the impurity

Tab. 7 Epichlorohydrin HKR global kinetic constants.

Epichlorohydrin starting material	HKR temp. (°C)	[Catalyst] (mol/m^3)	K_S (m^3/mol/s)	K_R (m^3/mol/s)	K_d (m^3/mol/s)
Racemic	5	35	1.33×10^{-7}	8.33×10^{-10}	1.11×10^{-10}
Racemic	5	22	5.08×10^{-8}	9.25×10^{-10}	9.60×10^{-11}
Pure (S)	5	53	3.87×10^{-7}	–	–
Pure (R)	5	50	–	8.80×10^{-10}	–
Racemic	25	37	2.59×10^{-7}	5.92×10^{-9}	1.16×10^{-9}

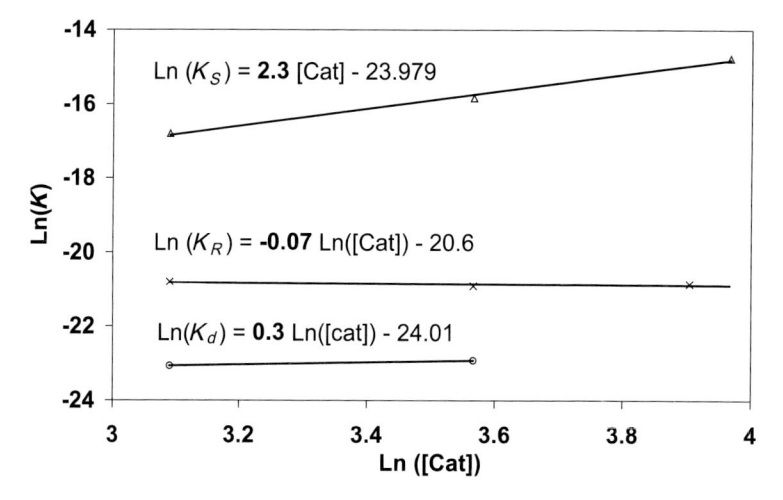

$$\mathrm{Ln}(K_S) = 2.3\,[\mathrm{Cat}] - 23.979$$

$$\mathrm{Ln}(K_R) = -0.07\,\mathrm{Ln}([\mathrm{Cat}]) - 20.6$$

$$\mathrm{Ln}(K_d) = 0.3\,\mathrm{Ln}([\mathrm{cat}]) - 24.01$$

Fig. 18 Kinetic parameter estimation: △ (R)-epichlorohydrin hydrolysis, × (S)-epichlorohydrin hydrolysis, ○ impurity formation.

Tab. 8 Epichlorohydrin HKR kinetic parameters.

Reaction	Catalyst kinetic order (n_i)	k_i (5°C)	k_i (25°C)	k_i^0	E_i (J/mol)
(S)-Epi. Hydrolysis	2	1.1×10^{-10}	1.9×10^{-10}	9.0×10^{-7}	−20910
(R)-Epi. Hydrolysis	0	8.8×10^{-10}	5.9×10^{-9}	2.1×10^{3}	−65875
Impurity formation	0	1.0×10^{-10}	1.2×10^{-9}	3.2×10^{4}	−77160

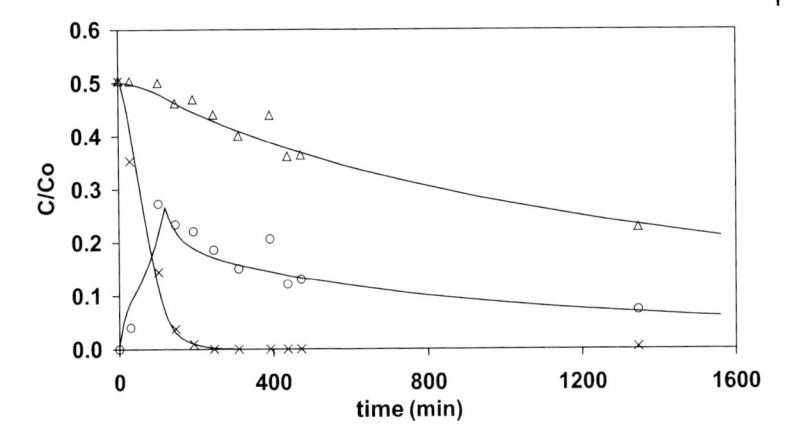

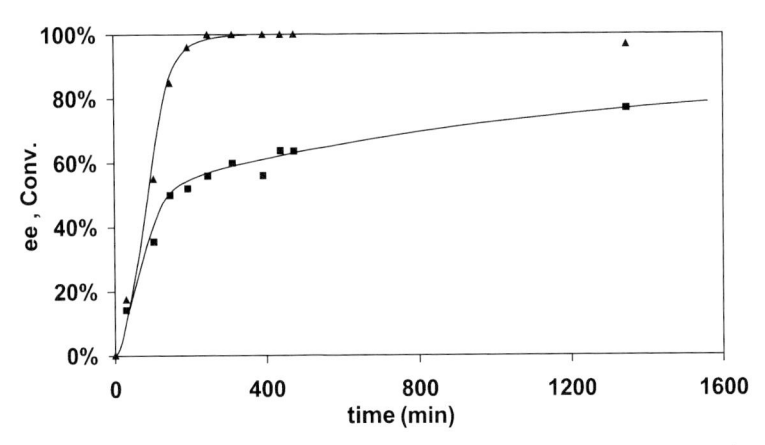

Fig. 19 HKR of (±)-epichlorohydrin using 0.5 mol-% catalyst at 25 °C. Experimental: △ (R)-epichlorohydrin, × (S)-epichlorohydrin, ○ water, ▲ ee, ■ conversion. Numerical simulation: solid lines.

forming reaction. As stated previously, these conclusions are only valid within the explored range of experimental conditions.

Complete characterization of the kinetic parameters for the HKR of epichlorohydrin was then obtained by evaluation of the reaction dependence on temperature. A standard experiment at 25 °C (Fig. 19) was numerically fitted, which allowed the expression of the kinetic constants in term of an Arrhenius law relationship (Eq. 21). From this relationship, the activation energy (E_i) and pre-exponential frequency parameter (k_i^0) were derived for each component of the reaction (Tab. 8). Of significant practical importance is the impact an increase in reaction temperature has by decreasing the selectivity of the HKR and increasing the level of impurity production. For this experiment, a maximum yield of only 44% (to reach ee > 99%) was possible compared with 48% when the reaction was performed at

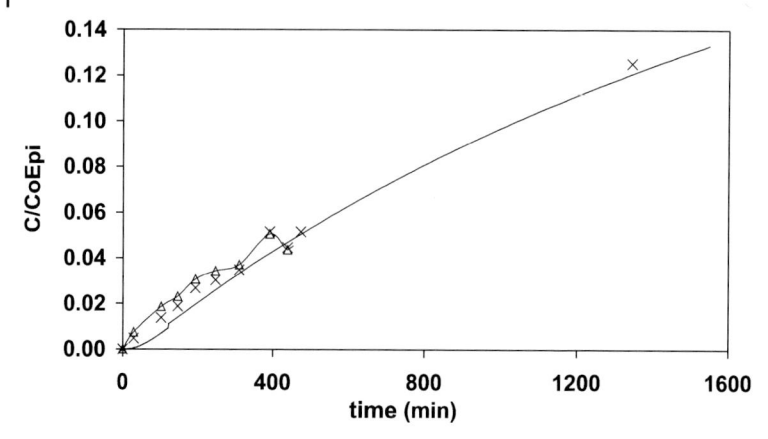

Fig. 20 Impurity formation during HKR of (±)-epichlorohydrin using 0.5 mol-% catalyst at 25 °C. Experimental: △ glycidol, × DCP. Numerical simulation: solid lines.

5 °C (Fig. 13). This loss in yield is in part due to a 10-fold increase in glycidol and DCP formation (Fig. 20). At 25 °C, a decrease in ee due to the racemization pathway is observable at extended reaction times, which also contributes to the lower possible yield (Fig. 19).

$$k_i = k_i^0 \mathrm{Exp}\left(\frac{E_i}{RT}\right) \tag{21}$$

2.3.3.3 Parameter Deviation Study

At this point, the kinetic model was fully characterized. The reliability of the model was then assessed by a parameter deviation study. As shown in Fig. 21, the observed data from a solvent-free experiment under standard conditions were accurately predicted using the simulated kinetic model. The increase in the overall rate of the reaction results from the effective increase in catalyst concentration by solvent removal, rather than from a change in the kinetic parameters. This indicates that the use of 1,2-dichlorobenzene during the kinetic investigation had no impact on the estimated kinetic values.

Figs. 22 and 23 show the experimental and simulated kinetic results obtained when varying the number of equivalents of water and the water addition rate, respectively. It is clear that water addition mode has a real impact on the time necessary to reach a given ee specification. The amount of water and the mode of addition will be discussed in greater detail in the next section in the context of process optimization and temperature control.

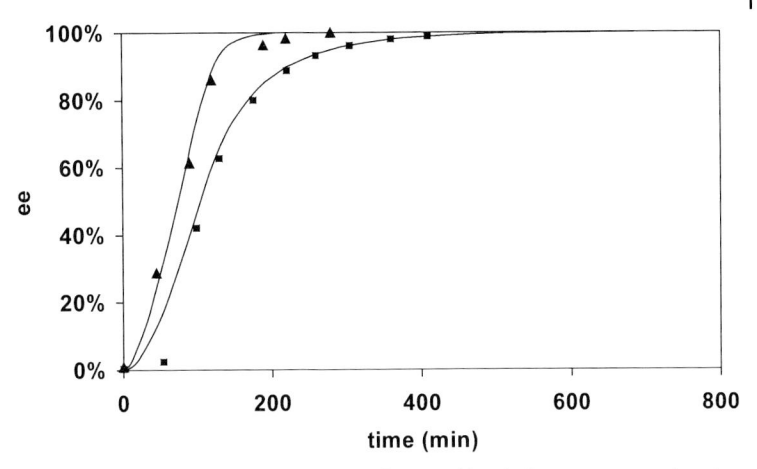

Fig. 21 Comparison of solvent-free HKR of (±)-epichlorohydrin. Experimental evolution of ee: ▲ solvent-free conditions (0.5 mol-% catalyst=50 mol/m^3), ■ Tab. 5 conditions (0.5 mol-% catalyst=37 mol/m^3). Numerical simulation: solid lines.

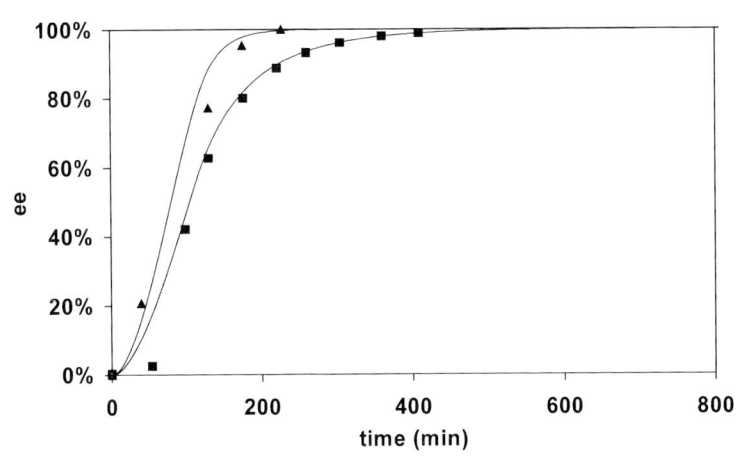

Fig. 22 Comparison of amount of water used in HKR of (±)-epichlorohydrin. Experimental evolution of ee: ■ Tab. 5 conditions (0.75 equiv. H$_2$O added over 2 h), ▲ 1.0 equiv. H$_2$O (added over 2 h, solvent-free). Numerical simulation: solid lines.

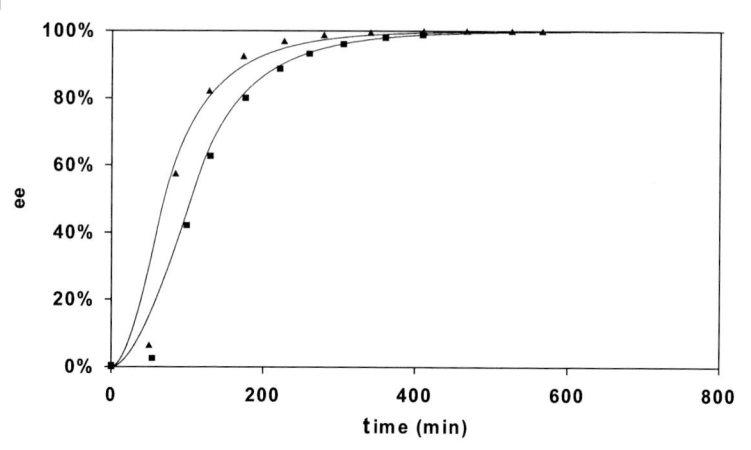

[Figure: Plot of ee (%) vs time (min), ranging from 0 to 800 min, with two curves rising to ~100%.]

Fig. 23 Comparison of rate of water addition in HKR of (±)-epichlorohydrin. Experimental evolution of ee: ■ Tab. 5 conditions (0.75 equiv. H_2O added over 2 h), ▲ 0.75 equiv. H_2O (added over 1 h, solvent-free). Numerical simulation: solid lines.

2.4
HKR Process Optimization and Scale-up

2.4.1
Process Description

With a complete characterization of the global kinetic behavior of the HKR of epichlorohydrin in hand, the entire HKR process, including the epoxide separation and isolation steps, was optimized to provide a realistic industrial scale process. This process can be divided into four main stages. The first stage entails the actual HKR reaction, in which water is continuously added to a solution of (salen)-Co(III)-OAc catalyst in racemic epichlorohydrin. Water is used in excess of the theoretically required amount to enhance the reaction rate near completion (see Section 2.4.2). Next, the reaction mixture is transferred into a second vessel and stabilized by chemical reduction of the cobalt species. In the third stage, water is removed from the mixture through an epichlorohydrin/water heteroazeotrope. The vapors condense in two layers, and the upper aqueous layer is removed while the organic phase is recycled. Finally, the highly enantiomerically enriched epichlorohydrin is isolated by flash vacuum distillation.

2.4.2
Process Optimization

As with any normal kinetic resolution, the maximum theoretical yield of chirally pure compound is 50% using the HKR technology. As the HKR of epichlorohydrin is not completely selective, process optimization consists of maximizing the isolated yield while minimizing the cycle time to obtain material of the desired ee and purity specification. Economic and safety criteria induce additional non-technical constraints to the full process design. These two fundamental aspects will not be discussed in detail in this chapter, although they have been an integral factor in the ultimate process design.

2.4.2.1 HKR Stage Optimization

The primary parameters that influence the performance of the HKR reaction include the catalyst loading, the amount of water, the reaction temperature, and the rate and mode of water addition. The yield, the time for completion and the amounts of side-products are directly connected to these parameters, as well as to the downstream processing conditions.

2.4.2.1.1 Shortcut Optimization Method

At this point, the global kinetic model could be used to estimate numerically the optimum parameter set that achieves a given objective. However, a further simplification of the fed-batch model can be made that leads explicitly to the general behavior trends of the HKR reaction under specific operating conditions. This shortcut method is developed using the global kinetic model (Eqs. 12–17) assuming isothermal conditions and neglecting the impurity formation reaction. The model is then solely based on the hydrolysis reaction of both enantiomers of epichlorohydrin and declined in three material balance relations (Eqs. 22–24).

$$(S)\text{-Epi:} \quad V\frac{\partial[\text{Epi}_S]}{\partial t} + [\text{Epi}_S]\frac{\partial V}{\partial t} + k_S[\text{Epi}_S][H_2O][\text{Cat.}]^2 V = 0 \quad (22)$$

$$(R)\text{-Epi:} \quad V\frac{\partial[\text{Epi}_R]}{\partial t} + [\text{Epi}_R]\frac{\partial V}{\partial t} + k_R[\text{Epi}_R][H_2O]V = 0 \quad (23)$$

$$\text{Water:} \quad V\frac{\partial[H_2O]}{\partial t} + [H_2O]\frac{\partial V}{\partial t} + (k_S[\text{Epi}_S][\text{Cat.}]^2 + k_R[\text{Epi}_R])[H_2O]V = Q[H_2O]_0 \quad (24)$$

The ratio of Eq. (22) to Eq. (23) provides a direct differential equation between epichlorohydrin enantiomer concentrations and the enantioselectivity factor, which is a function of temperature and catalyst concentration as variables (Eq. 25). This equation is readily solved to give the mathematical relation between enantiomeric excess and yield (R) as a function of a (Eq. 26). The maximum obtainable yield for a given ee specification (and vice versa) is provided straightforwardly in Eq. (26). Moreover, this equation shows that the yield–ee pair is independent of the amount of water used (provided the amount is sufficient to reach the desired ee) and of the water addition rate. Thus, the water addition mode will only impact the reaction time necessary to reach the desired yield and ee target.

$$\frac{\partial(V[\text{Epi}_S])}{\partial(V[\text{Epi}_R])} = a\frac{[\text{Epi}_S]}{[\text{Epi}_R]} \quad \text{with} \quad a = \frac{k_S^0}{k_R^0}\text{Exp}\left(\frac{E_S - E_R}{RT}\right)[\text{Cat.}]^2 \tag{25}$$

$$R = \frac{1}{2}\left[\frac{1-\text{ee}}{1+\text{ee}}\right]^{\frac{1}{a-1}} \tag{26}$$

Utilizing this shortcut method, catalyst concentration and reaction temperature are the critical parameters that impact the selectivity through the enantioselectivity factor. The effect of these two parameters on the yield and ee performance of the HKR is graphically illustrated in Fig. 24. The data clearly indicate that operating at the highest catalyst concentration and the lowest reaction temperature maximizes the difference in the rates of reaction for the two enantiomers, and thereby

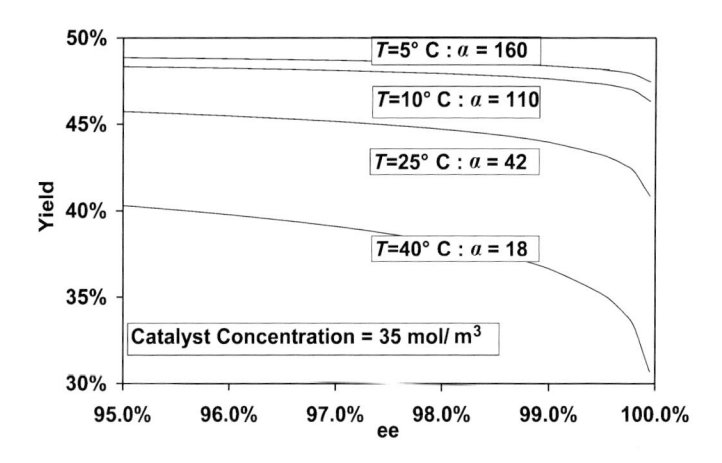

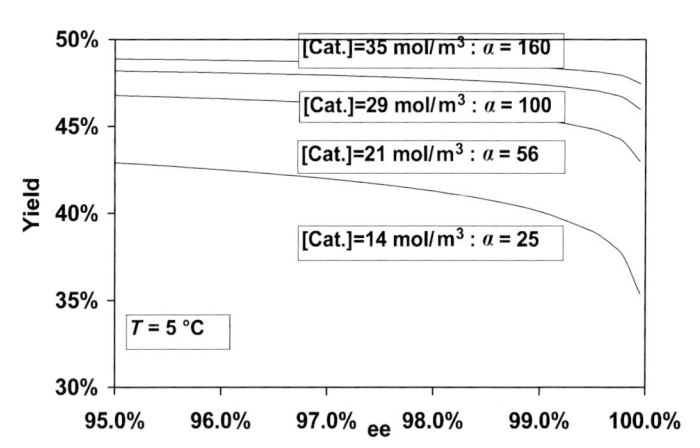

Fig. 24 Numerical simulation plots of yield (R) versus ee as a function of the selectivity factor (a) determined by reaction temperature and catalyst concentration.

Tab. 9 Water addition duration effect on epichlorohydrin HKR yield.

Run	Water addition (h)	ee (%)	HKR at 5 °C (a=340)		HKR at 15 °C (a=173)	
			Imp. (mol-%)	Yield (%)	Imp. (mol-%)	Yield (%)
Short-cut	–	99	–	49.2	–	48.5
1	5	99	0.8	48.5	2.1	46.5
2	10	99	1.4	48.1	4.3	45.1
3	15	99	1.9	47.7	6.5	44.0
4	20	99	2.4	47.2	8.6	42.1

Tab. 10 Optimal operating conditions for the HKR of epichlorohydrin.

T (°C)	Water (equiv./Epi)	Catalyst (mol-%/Epi)
5–0	0.65–0.75	0.4–0.5

increases the potential yield for a given ee. However, technical and economic considerations temper this conclusion. The production cost of epichlorohydrin via HKR is closely connected to the amount of catalyst used, and the reaction temperature has to be kept over a certain limit to enable a reasonable reaction rate and to avoid freezing of the water.

Nevertheless, this straightforward method allowed us to fix the starting optimum operating parameter set (T, catalyst loading) and indicated the boundary limits that could be expected in an ideal situation (isothermal conditions, no side-product formation). Certainly, this ideal performance is negatively affected by side-product formation, which is directly connected to the addition time of water (and hence, the total reaction time). Tab. 9 shows the simulated impact of water addition duration on yield and impurity formation using the global kinetic and reactor model. These results are compared with those obtained with the ideal shortcut model for epichlorohydrin HKR at 5 and 15 °C. The results indicate that adding water at a high rate (short addition time) leads to a higher yield of epichlorohydrin with lower amounts of impurities. The amount of water used must be sufficient to consume all of the fast-reacting enantiomer. For maximum yield, 0.55 equiv. of water relative to racemic epichlorohydrin are sufficient to drive the reaction to completion. In practice, however, slightly more water is used to increase the reaction rate when approaching the completion point. Tab. 10 summarizes the optimum operating parameter range considering the technical and industrial constraints.

2.4.2.1.2 Process Temperature Control

The heat of reaction for the hydrolysis of epichlorohydrin was determined to be –17.6 kcal/mol by DSC measurement. With a value of this magnitude, the uncontrolled temperature rise in the system could be significant (if not catastrophic), especially since epichlorohydrin is typically used as the bulk liquid reactant with no co-solvent. At an industrial scale, it is often difficult to maintain a low temperature of a medium in which an exothermic reaction occurs. This is particularly true for fed-batch reactors where heat exchange occurs mainly through the low surface area of the jacket. Since the volumetric surface area (m^2/m^3 batch) generally diminishes on scale-up, the thermal aspect must be carefully investigated to ensure control of the medium temperature. For the HKR, temperature control is achieved by management of the water addition profile and the coolant temperature profile (external loop heat exchangers can also be used for high heat transfer requirements). The complete thermal balance for the reaction and vessel system is provided in Eq. (27). The global heat transfer coefficient, U, is an intrinsic property of the vessel, while S_w is the surface area of the vessel wall covered by the reaction mixture of volume V. Values for U are typically in the range of 150–250 kcal/K/m²/h for glass-lined vessels and 400–500 kcal/K/m²/h for stainless steel reactors. The use of this equation completes the global model for the reaction, making the linear scale-up of the process possible.

$$V\rho_m C_p^m \frac{\partial T}{\partial t} + QC_p^{H_2O}(T - T_{H_2O}) + V\sum r_i \Delta H_i + US_w(T - T_w) = 0 \qquad (27)$$

Based upon the complete model, the simulated behavior of the HKR reaction in a 50 L glass-lined vessel is shown for the following temperature control scenario (Fig. 25). Water flow-rate is adjusted to obtain 5 °C in the medium (T) with a constant temperature of –5 °C in the jacket (T_w) during the initial period of the HKR. The jacket temperature is then gradually increased to compensate for the reduced heat output when near reaction completion.

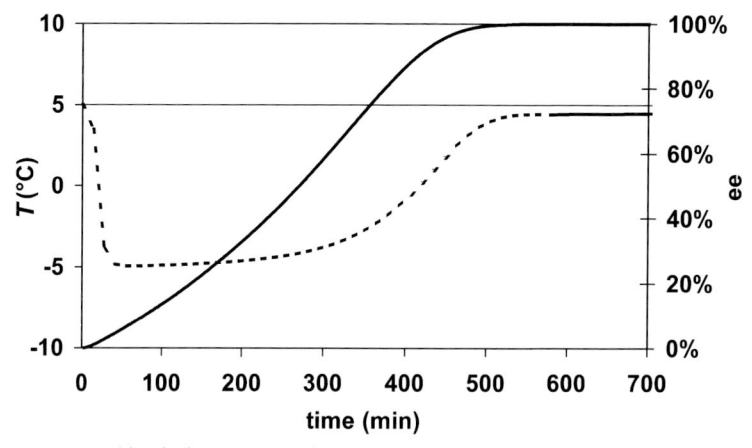

Fig. 25 Epichlorohydrin HKR simulation with constant temperature control. Thin solid line, medium temperature; dashed line, jacket temperature; thick solid line, epichlorohydrin ee.

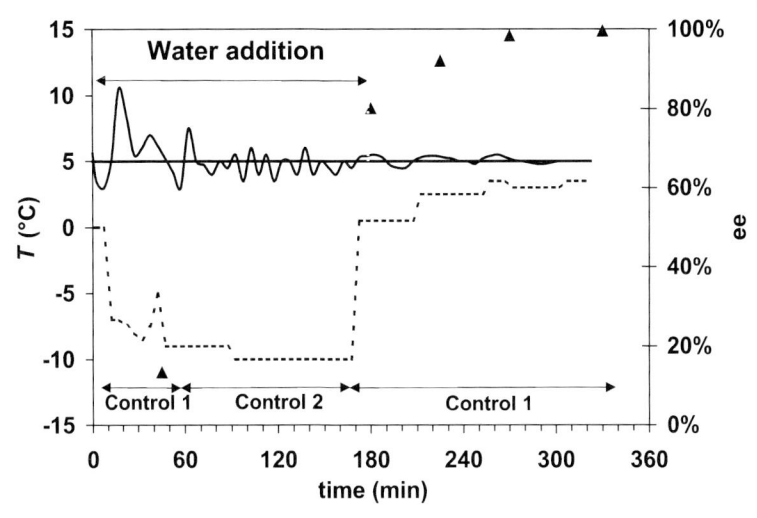

Fig. 26 Temperature control for a 2 L scale HKR experiment. ▲ ee. Solid line, medium temperature; dashed lined, jacket temperature. Control 1: Coolant temperature control. Control 2: Coolant flow-rate control.

The difficulty in maintaining temperature control using a jacketed, glass-lined vessel was demonstrated at the 2 L scale. The HKR of epichlorohydrin (0.5 mol-% cat. relative to racemate, 5 °C) was conducted using a constant water flow-rate over 3 hours (0.75 equiv.). Temperature control was attempted by tuning the coolant flow-rate and the coolant temperature. The medium was maintained more or less at the desired temperature, but the thermal fluctuations reveal the difficulty in managing this control, especially during the initial period of reaction (Fig. 26). The experimental batch composition upon reaching the desired specification (ee > 99%) is compared with the results for the corresponding simulated isothermal run.

At an industrial scale, heat removal is performed preferably under conditions of constant coolant temperature and flow-rate. Under these conditions, the rate of water addition is the only means of maintaining the reaction temperature within the desired range. To compensate for the diminished heat output as the reaction proceeds, the water addition rate can be increased accordingly. Ultimately, the coolant temperature must also be gradually increased when close to completion in order to maintain the desired working temperature. This mode of temperature control was demonstrated on a 50 L scale, and the temperature and rate of water addition profiles for this experiment are plotted in Fig. 27. Reaction completion is attained in 9 hours (ee = 99%) with a potential yield of epichlorohydrin of 47% (94% of theory). These results are consistent with those predicted by the kinetic model (see Tab. 9).

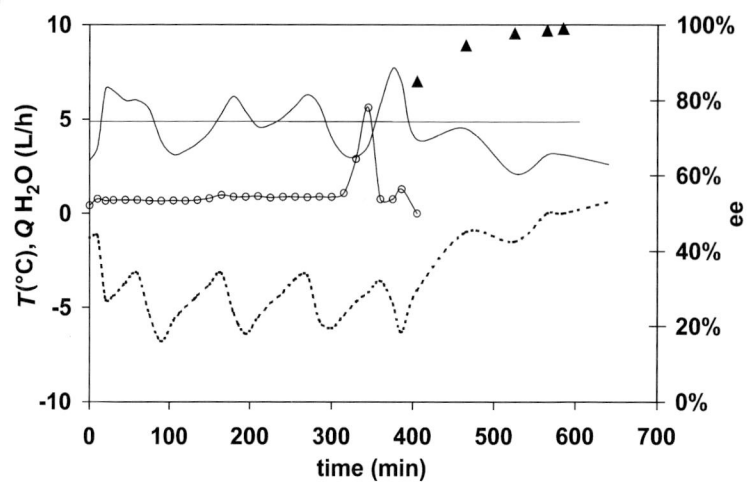

Fig. 27 Temperature control for a 50 L pilot-scale HKR experiment. ▲ ee, ⊖ water flow-rate, Solid line, medium temperature; dashed lined. jacket temperature.

2.4.2.2 Isolation of Resolved Epichlorohydrin

2.4.2.2.1 Reaction Medium Stabilization

Upon completion of the HKR reaction, the resolved epichlorohydrin is recovered from the reaction mixture by a distillation sequence. Since the catalyst is still active, the temperature increase during distillation results in an increase in the amounts of impurities and in the loss of the desired enantiomer through direct hydrolysis and conversion into the fast reacting enantiomer via the racemization pathway (water being in excess for HKR). Upon consumption of the excess water, racemization of the resolved product ensues. Originally, these issues were addressed with engineering solutions by separating the excess water and resolved epoxide from the catalyst and reaction mixture through short-path distillation using a wiped film evaporator (WFE). Subsequent azeotropic drying and final distillation steps recovered the epichlorohydrin. Although the short residence times obtained with the WFE allowed for the isolation of product with the desired quality, the product loss associated with the process led to a significant 15–20% decrease in the global isolated yield.

Industrially, these losses were unacceptable, so alternative measures were sought. Since the (salen)-Co(II) complex was known to be inactive in the HKR reaction, potential reductants were evaluated for their ability to stabilize the reaction medium by reduction of the (salen)-Co(III) species. Ascorbic acid was found to be the optimum reagent for this purpose [5]. Addition of ascorbic acid (2 mol equiv. relative to catalyst) to the HKR mixture upon reaching completion results in the precipitation of the red Co(II) complex within 2 hours. The resulting mixture proved to be stable at high temperature (<90 °C) over extended periods of time. This stabilization thus allows the direct distillation of the product from the reaction mixture without the need for catalyst separation or specialized equipment.

2.4.2.2.2 Azeotropic Drying and Epichlorohydrin Isolation

After ascorbic acid treatment, the excess water is removed from the reaction mixture by azeotropic distillation. The epichlorohydrin-water azeotrope is 24% (w/w) water at atmospheric pressure. For safety reasons, the maximum allowable pot temperature was 90 °C [13]. This necessitates conducting the drying step under vacuum, which reduces the productivity of the azeotrope. Depending on the amount of water to be removed (and vessel heat transfer capabilities), the azeotropic drying may comprise a significant portion of the global process cycle time. The amount of excess water used in the HKR is balanced between increasing the HKR reaction time (and impurity formation) with a small excess and extending the azeotropic drying time with a large excess.

Upon reaching the desired water content, the resolved epichlorohydrin is recovered by vacuum distillation. Fractionation is avoided due to the large difference in volatility between epichlorohydrin and CPD, DCP, and glycidol. Simple flash distillation is sufficient to isolate epichlorohydrin of high chemical and enantiomeric purity. Since the mixture consists mainly of epichlorohydrin and CPD, the pot temperature will gradually increase during distillation from the initial boiling point of the mixture at the operating pressure to near the boiling point of CPD by the end of epichlorohydrin removal. Again, the pot temperature is maintained below a certain limit (90 °C) for safety reasons. Because of this, the maximum yield obtainable from the distillation is defined by the working pressure as graphically represented in Fig. 28. Significantly, conducting the distillation at moderate vacuum will result in a low isolated yield of resolved epichlorohydrin, even if the HKR reaction has been optimized. At 35 mbar, 90% of the epichlorohydrin present can be recovered, leading to an overall yield of 42% for the global process. At a 50 L pilot scale, the actual yield was 41% with an additional 5% of material remaining

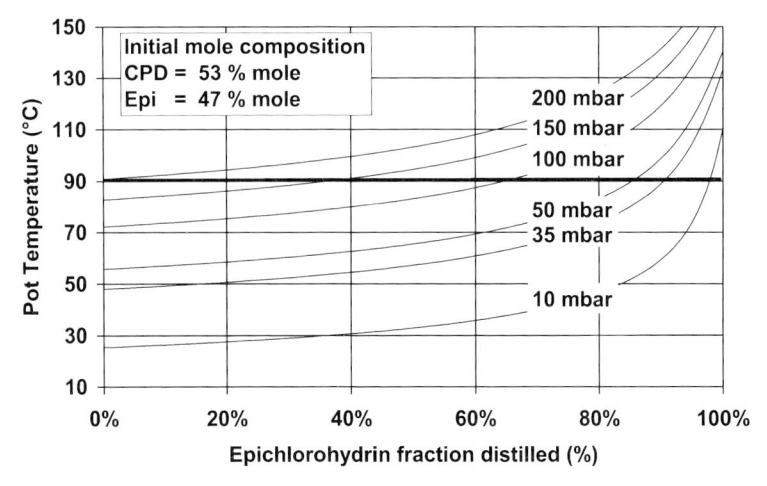

Fig. 28 Yield of (R)-epichlorohydrin by distillation as a function of pressure with 90 °C pot temperature constraint.

2.5
Conclusions

The practicality of the hydrolytic kinetic resolution technology for the commercial production of (R)- and (S)-epichlorohydrin was limited by the general inefficiencies associated with preparing the active HKR catalyst and by the presence of a specific catalyst-mediated racemization and the decomposition pathways. A new process for the preparation and isolation of the active HKR catalyst was developed, resulting in the successful decoupling of the HKR from the process to generate the catalyst. Furthermore, a novel method of reaction stabilization by catalyst reduction has been successfully integrated into the HKR process, thereby eliminating the racemization/decomposition pathways [5]. These technical improvements served to significantly increase the economic and practical efficiency of the HKR of epichlorohydrin. A thorough industrialization study of the overall process (catalyst preparation, HKR, and isolation) was conducted in order to ensure the smooth scale-up from lab to industrial scale. For this study, a global kinetic and thermodynamic model was generated for the HKR reaction, which allowed for the scale-independent evaluation and definition of all critical operating parameters. From this model, reaction temperature, catalyst concentration, the amount of water used, and its rate of addition were identified as the key parameters for HKR performance and temperature control. The model and overall process were then validated through a 50 L pilot trial that produced (S)-epichlorohydrin in 41% isolated yield and in > 99.5% chemical and enantiomeric purity (Fig. 29). Based on these results, the straightforward scale-up of this process could be reasonably anticipated. This study will also benefit the industrialization of other building blocks accessed by the HKR technology.

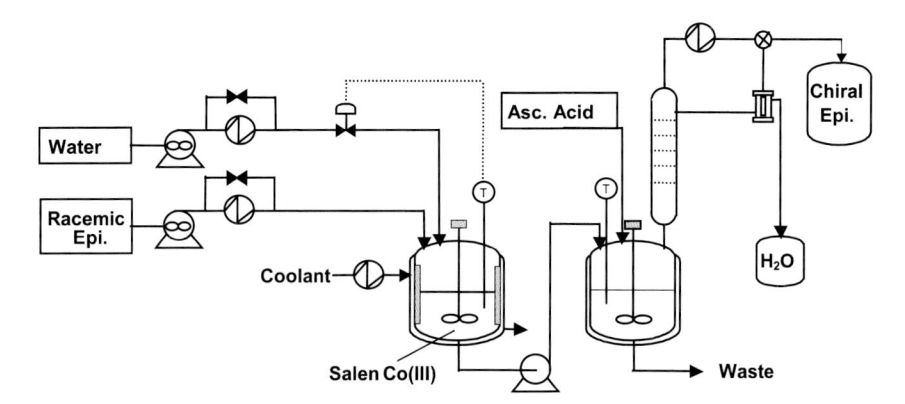

Fig. 29 Process flow diagram of HKR of epichlorohydrin using isolated catalyst.

2.6
References

1. M. Tokunaga, J. F. Larrow, F. Kakiuchi, E. N. Jacobsen, Science 1997, 277, 936.
2. S. E. Schaus, B. D. Brandes, J. F. Larrow, M. Tokunaga, K. B. Hansen, A. E. Gould, M. E. Furrow, E. N. Jacobsen, J. Am. Chem. Soc. 2002, 124, 1307.
3. For a list of chiral building blocks available via the HKR chemistry, see: *http://www.rhodia-pharmasolutions.com*.
4. M. E. Furrow, S. E. Schaus, E. N. Jacobsen, J. Org. Chem. 1998, 63, 6776.
5. K. E. Hemberger, S. Jasmin, H. Kabir, J. F. Larrow, P. Morel, Tetrahedron: Asymmetry, submitted. Patent applications covering catalyst synthesis and deactivation are pending. See: US Patent Application 20030088114 A1, published 05/08/03.
6. For the preparation of the ligand, see: a) J. F. Larrow, E. N. Jacobsen, Y. Gao, Y. Hong, X. Nie, C. M. Zepp, J. Org. Chem. 1994, 59, 1939; b) J. F. Larrow, E. N. Jacobsen, Org. Syn. 1997, 75, 1. For a review of asymmetric reactions catalyzed by (salen)metal complexes, see: J. F. Larrow, E. N. Jacobsen, Top. Organomet. Chem., in press.
7. a) M. J. Carter, D. P. Rillema, F. Bosolo, J. Am. Chem. Soc. 1974, 96, 392; b) R. F. Hammerschmidt, R. F. Broman, J. Electroanal. Chem. 1979, 99, 103; c) A. Kapturkiewitch, B. Behr, Inorg. Chim. Acta 1983, 69, 247; d) A. Kapturkiewitch, B. Behr, J. Electroanal. Chem. 1984, 163, 189; e) M. Opallo, A. Kapturkiewitch, B. Behr, J. Electroanal. Chem. 1985, 182, 427; f) A. Nishinaga, K. Tajima, B. Speiser, E. Eichorn, A. Rieker, H. Ohya-Nishiguchi, K. Ishizu, Chem. Lett. 1991, 1403.
8. a) K. van't Riet, Ind. Eng. Chem. Process Des. Dev. 1979, 18, 357; b) A. Bakker, J. M. Smith, K. J. Myers, Chem. Eng. 1994, 101, 98; c) M. Nocentini, D. Fajner, G. Pasquali, F. Magelli, Ind. Eng. Chem. Res. 1998, 37, 19.
9. G. B. Tatterson, Fluid Mixing and Gas Dispersion in Agitated Tanks, McGraw Hill, USA, 1991, pp. 417–497.
10. a) E. N. Jacobsen, Acc. Chem. Res. 2000, 33, 421; b) K. B. Hansen, J. L. Leighton, E. N. Jacobsen, J. Am. Chem. Soc. 1996, 118, 10924; c) M. H. Wu, E. N. Jacobsen, in Comprehensive Asymmetric Catalysis, E. N. Jacobsen, A. Pfaltz, H. Yamamoto, (Eds.) 1999, Springer, Berlin.
11. a) J. Villermaux, Génie de la Réaction Chimique. Conception et Fonctionnement des Réacteurs, Lavoisier, Paris, 1993; b) O. Levenspiel, The Chemical Reactor Omnibook, OSU Book Stores, Inc, Oregon, 1993.
12. L. R. Petzold, SAND82-8637, Unlimited release, printed September 1982. The software utilized in this study was React Op software by CISP, see: *http://www.CISP.spb.ru*.
13. While the 90 °C maximum temperature was acceptable at 50 l scale, subsequent experiments have shown the onset of decomposition reactions to occur between 70 and 80 °C in the absence of water. Therefore, a lower maximum pot temperature during the distillation is appropriate. This can be compensated for by lowering the pressure during the distillation (see Fig. 28).

3
Scale-up Studies in Asymmetric Transfer Hydrogenation
JOHN BLACKER and JULIETTE MARTIN

Abstract

Catalytic asymmetric transfer hydrogenation is an efficient method for producing optically active alcohols and amines. Analysis of launched and development pharmaceuticals show that these types of chiral centers occur most frequently, so this technology is proving particularly valuable. The chapter will describe the background, development, and application of asymmetric transfer hydrogenation, with particular emphasis on Avecia's proprietary CATHy^{TM} catalysts.

3.1
Background

Transfer hydrogenation has been practiced for several decades with heterogeneous catalysts, but so far these are unsuitable for asymmetric synthesis. Research by various groups in the 1980s showed homogeneous catalysts, based on group VIII metals, were suitable for asymmetric transfer hydrogenation of aldehydes, ketones, and alkenes. The hydrogen donors they used were isopropanol and formic acid and turnover numbers were in some cases impressive. Steckhahn recognized the mechanistic similarity of these catalysts to the biological cofactor NADH and indeed used rhodium-based transfer hydrogenation catalysts to reduce NAD [1]. Mestroni, Gladiali and others synthesized some chiral bipyridine ligands and showed these could be used to give products in low optical purity [2]. Despite many ligand modifications the enantiomeric excesses remained low, and work in the area tailed off until Noyori and co-workers discovered that saturated, chiral, 1,2-aminoalcohols ligated with ruthenium eta-6 arene complexes, gave catalysts that would reduce ketones with very high optical purities [3]. Isopropanol (IPA) was used as the hydrogen donor, and a base was used to activate the catalyst. One of the features of this reaction is reversibility, and in many ways it is analogous to a Meerwein-Pondorf-Verley reduction, Oppenauer oxidation, but the mechanism seems to be different. In order to get high yields of the optically enriched alcohol it is preferable to use isopropanol both as the hydrogen donor and as the solvent (IPA system), but also to use a low concentration of the substrate ketone. From an industrial perspective the reaction was not productive and therefore of little commercial interest.

Noyori was subsequently able to show that triethylamine salts of formic acid (TEAF) could be used to reduce ketones to alcohols and imines to amines with high enantioselectivities [4]. The byproduct of this reaction is carbon dioxide gas and this prevents the possibility of the reverse reaction. Strangely, aminoalcohol ligands are poor in this reaction, whilst unsymmetrical 1,2-diamines have proven very effective. A particularly effective ligand is mono-N-tosyl-1,2-diphenylethylenediamine.

Avecia first devised and patented the use of diamine or aminoalcohol-ligated rhodium cyclopentadienyl complexes for asymmetric transfer hydrogenation [5], and for commercial reasons, these have been named CATHy^{TM} catalysts (Catalytic Asymmetric Transfer Hydrogenation). Several groups have also made the same catalysts, and likewise they found they were very effective in asymmetric transfer hydrogenation reactions [6]. Initial comparisons with the ruthenium arene catalysts showed the latter were more active in the isopropanol system but gave slightly lower enantiomeric excesses. It was subsequently found that the rhodium catalysts were also more active in the reverse transfer oxidation process. This happens because the reaction is not perfectly enantioselective and the optical purity of the product falls to its lowest energy, the racemate. Both Avecia and the chemists at Mitsubishi Chemical Corporation have found a way of increasing the concentration of the process and preventing the racemization by continuously removing acetone [5, 7]. This is discussed later on in the chapter. Bäckvall has shown that a racemization process of this type can be harnessed in a dynamic kinetic resolution [8]. A lipase is used to resolve a chiral alcohol by enantioselective esterification, and the residual antipode is racemized by the transfer hydrogenation catalyst.

The CATHy^{TM} catalysts, like their ruthenium analogues are also active in the reduction of ketones and imines using the formate system [5]. Wills has published on the reduction of ketones [9], and Baker on the reduction of imines [6]. The development of this process for large-scale use has proven more complex and this will be described elsewhere in this chapter.

3.2
The Catalytic System

3.2.1
Catalyst

The CATHy^{TM} catalysts consist of a transition metal center, an ionic group (e.g., chloride), a negatively charged cyclopentadienyl group (e.g., pentamethylcyclopentadienyl) and a chiral bidentate ligand (Fig. 1). The 18-electron pre-catalyst dihydrochloride salt is prepared by mixing one mole equivalent of a dimeric dihalometalcyclopentadiene, and two mole equivalents of the ligand. In the isopropanol system, the catalyst is activated by adding four mole equivalents of base: two to

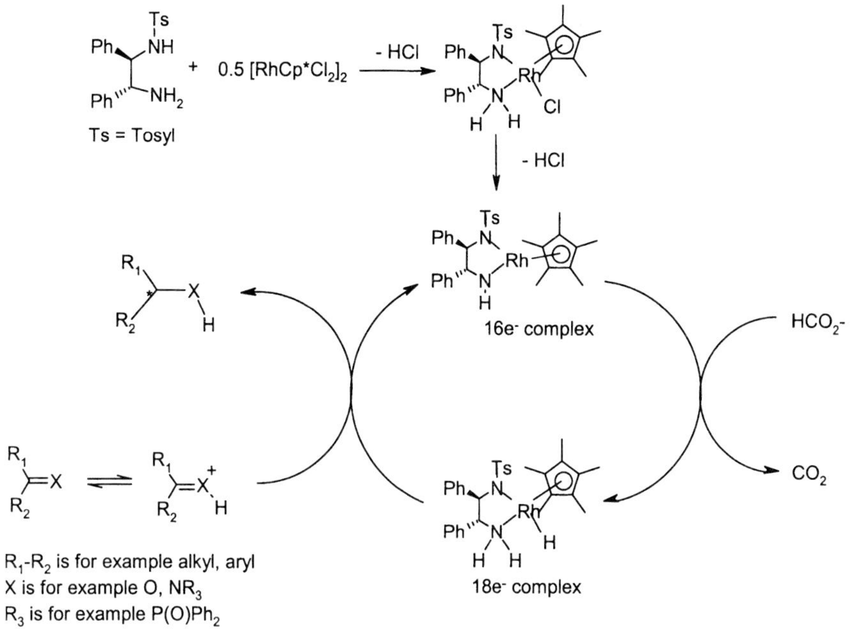

Fig. 1 Formation of a CATHy^{TM} catalyst and proposed catalytic cycle.

mop-up the hydrogen chloride, and two to dehydrochlorinate the rhodium. The base is commonly potassium hydroxide or more preferably sodium isopropoxide. In the TEAF system it is presumed that triethylamine serves the role of the base, despite the fact the medium is acidic.

The preferred metal is rhodium(III), however, some encouraging results have also been obtained with the isoelectronic iridium (III) catalysts. Ruthenium(II) catalysts have been prepared but display only moderate activity and enantioselectivity. The ligand is crucially important, not only in determining the optical induction of the substrate, but also of modulating the catalyst's activity. It is useful to discuss its role in more detail. The primary amine on the ligand is essential in allowing elimination of hydrogen chloride to generate the active 16-electron metal. After complexing with the metal, the amide thus formed is fairly basic, and will readily protonate in either the IPA or TEAF system. The nitrogen-metal bond changes from one that is covalent to one that is dative. At this point in the cycle, the metal is likely to become more oxidative and be reduced to the metal-hydride intermediate. The metal is now a chiral center and appears to be configurationally stable and optically active. Indeed when the substrate binds, the hydride is delivered selectively to one prochiral face. It seems likely that the ligand NH protons are involved in hydrogen bonding with the substrate. Substitution of the primary amine results in a dramatic loss of activity, for example $NH_2 > NHMe \gg NMe_2$. The work of Noyori on the analogous ruthenium arene catalysts also indicated that the ruthenium is optically active and that the substrate interacts by hydrogen bonding [10].

The sulfonamide proton on the ligand is fairly acidic and is likely to be that lost in formation of the pre-catalyst. In our experience, functional groups other than aryl sulfonamides give poor catalysts, for example methane sulfonamide and trifluoromethane sulfonamide. This is thought to be because of the pK_a provided by the aryl sulfonamide. Modification with electron withdrawing or donating substituents failed to provide a relationship that could be correlated with activity or optical induction, but might be related simply to the rate of ligation to the metal. Most influence on enantioselectivity was seen with aryl substituents that affected the steric environment or provided secondary coordination sites for the substrate.

It is easy to prepare a range of ligands with different sulfonamide groups, and screening, rather than mechanistic understanding, has provided a convenient way of determining the optimal catalyst for a given substrate (Fig. 2). For example, for the reduction of a dihydroisoquinoline the enantiomeric excess of the product was improved from 71% ee with *p*-toluenesulfonamide to 90% ee with 2,4-dichloro-3-methylphenylsulfonamide. Williams has synthesized phenylsulfonic sulfonamide ligands that provide the catalyst with water solubility [11]. At the end of a reaction the catalyst can be washed out, preventing contamination of the product [12]. In theory the catalyst could also be recycled though this has not been currently demonstrated. The water-soluble catalyst also enables reactions to be performed in aqueous solution, and a number of cases have been demonstrated where this is advantageous.

The diamine ligands are successful in the TEAF system, but are poor in the IPA system, whilst 1,2-aminoalcohol ligands work well in the IPA, but not the TEAF system. The reason for this is not well understood, but it illustrates the subtle nature of the electronic environment on the metal.

The backbone of the bidentate ligand is usually an ethylene bridge so that a 1,2-relationship between the heteroatoms provides a stable five-membered ring with the metal. As the ring is enlarged association with the metal is weakened and these ligands give lower optical inductions.

The other substituents that form the vicinal chiral centers are essential in inducing optical activity in the product. It is thought that their main role is in providing steric bulk that stabilizes a twisted conformation. The diphenylethylenediamine ligands are especially useful as they are relatively inexpensive and are easily made at the kilogram scale. *Trans*-N-arylsulfonyl-1,2-diphenylethylenediamines are particularly effective ligands and the tosyl analogue (TsDPEN) is generally an excellent ligand in the TEAF system. Other ligands such as N-arylsulfonyl-cyclohexane diamine have not given generally such high enantiomeric excesses. The most frequently used 1,2-aminoalcohols are those based on norephedrine and 1,2-aminoindanol, as again these are cheap and used without modification.

The role of the cyclopentadiene is less well understood, but its size seems to be important. Cyclopentadiene (Cp) may be too small and the tetraphenyl analogue too large, as these give only moderate optical inductions. Our work has shown that the extra stability and steric bulk imparted by the pentamethyl analogue (Cp*) make this consistently the best. Using a chiral cyclopentadiene such as neomenthylcyclopentadiene gave disappointing results, but tethered cyclopentadienyls are showing more promise [13].

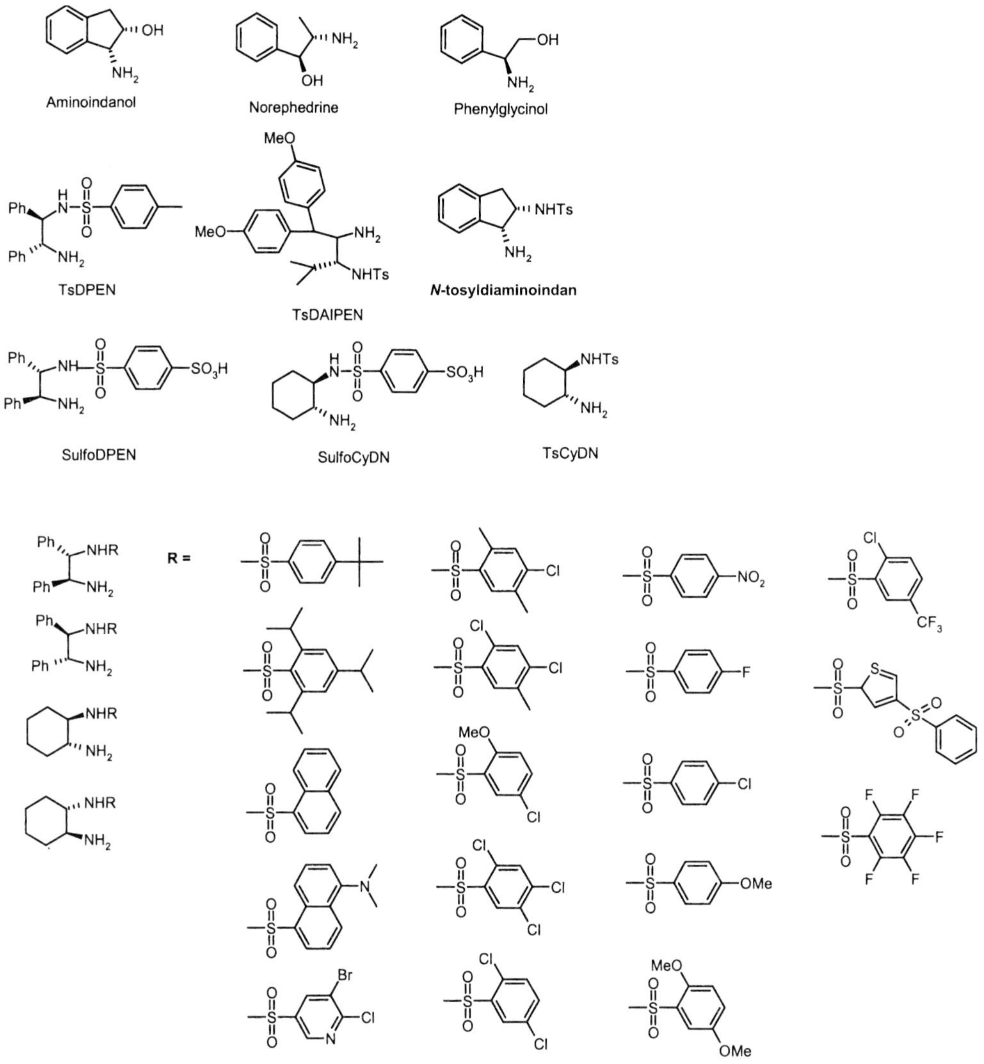

Fig. 2 Examples of ligands.

Different ligands and metal complexes can be combined in different ways to generate a library of catalysts. Since the reactions do not involve molecular hydrogen and are insensitive to air oxidation, it is a simple matter to set-up an experiment to screen catalysts against a substrate to find the most active and enantioselective for the reaction, and this is done conveniently using a robot.

Since the CATHy^{TM} catalysts involve precious metals, their activity is a key to providing an economic process. In our experience, for the average pharmaceutical intermediate, a substrate to catalyst ratio of <5000/1 is sufficient that the catalyst contribution is negligible, and these are regularly achieved. Consequently there has been little incentive to recover and recycle the catalyst. When making pharmaceuticals, one critical issue is to control and minimize metal impurities in the product, often to less than 10 ppm. Each product requires a different work-up and purification protocol and it is difficult to describe a general solution. Sometimes we find that washing removes the catalyst, other times the product is crystallized and the catalyst remains in the mother liquors, other times the product is distilled. Occasionally the catalyst is carried forward to the next stage and removed at this point. In our experience residual metal has not been problematic.

3.2.2
Hydrogen Donor

Alcohols will serve as hydrogen donors for the reduction of ketones. Isopropanol is frequently used, and during the process is oxidized into acetone. The reaction is reversible and the products are in equilibrium with the starting materials. To enhance formation of the product, isopropanol is used in large excess and conveniently becomes the solvent. Initially the reaction is controlled kinetically and the selectivity is high. As the concentration of the product and acetone increase, the rate of the reverse reaction also increases, and the ratio of enantiomers comes under thermodynamic control with the result that the optical purity of the product falls. To obtain high yields and to avoid the slow racemization of the product, originally it was initially necessary to work under dilute conditions. This limitation is overcome by removing continuously the acetone as it is formed by distillation (Fig. 3) [5, 7]. The catalyst is unstable above 40 °C so the distillation is best performed under reduced pressure at room temperature.

Avecia have scaled-up this process to the 200 L scale with acetophenone and 1-tetralone. A number of batches were run with consistently high enantiomeric excess and yield, with an acceptable concentration (see Example 1 later). Other alcohols such as methanol and ethanol will also work but are typically less effective as the aldehyde byproducts can interfere in the reaction. Isobutanol is an effective hydrogen donor and others such as glucose will also work but cannot be used in such high concentrations. Isopropanol can be mixed with an inert solvent, including water, but the rates of reaction fall linearly as expected.

A hydrogen donor for ketones and imines is formic acid and its salts. This reaction is always under kinetic control and no erosion of ee is observed. Carbon dioxide is generated as a byproduct and this gases out of the system preventing the re-

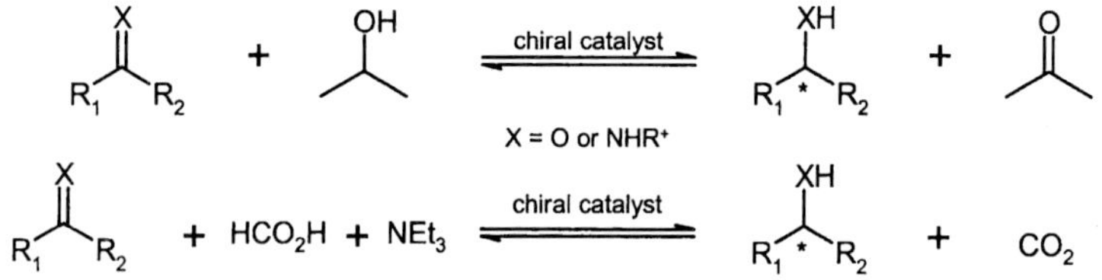

Fig. 3 IPA and TEAF systems.

Tab. 1 The effect of mixtures of triethylamine and formic acid on reaction rate.

Ratio of HCOOH/Et₃N (mole)	Time (h)	Conversion (%)
1:0	4	2
	15	26
3:1	4	50
	15	89
2.5:1	4	80
	15	100
2:1	4	50
	15	73
1.5:1	4	30
	15	73
1:1	4	50
	15	100
0.67:1	4	50
	15	95
0.5:1	4	50
	15	95
0.4:1	4	48
	15	100

verse reaction, that is in any case thermodynamically unfavorable. Early work on transfer hydrogenation recognized that mixtures of triethylamine and formic acid were particularly effective. The most commonly used mixture is a 5:2 molar ratio of formic acid to triethylamine (TEAF), which is the azeotrope of these two liquids. TEAF is a single phase at ambient temperature, it is soluble in most organic solvents and is frequently found to give the optimum reaction rate. Other ratios of formic acid and triethylamine form biphasic solutions. Tab. 1 shows an example where a series of formic acid, triethylamine ratios have been evaluated under identical conditions with a ketone.

The TEAF reagent is acidic, and the triethylamine may act to buffer the pH which changes as formic acid is consumed during the reaction. An excess of formic acid is often used, and as will be discussed later on, it is preferable to charge TEAF during the reaction to ensure a high yield of product. Other salts of formic acid have been used with good results, for example sodium and potassium formate in water with the water-soluble catalysts discussed above.

Rather surprisingly alcohols are poor at reducing imines, yet TEAF works well. During our studies we rationalized that the TEAF system was sufficiently acidic (pH approximately 4) to protonate the imine (pK_a approximately 6) and that it was an iminium that was reduced to an ammonium salt [14]. When an iminium was used in the IPA system, it was reduced albeit with a low rate and moderate enantioselectivity. Quaternary iminium salts were also reduced to tertiary amines. Hydrogen will not reduce ketones or imines using the CATHy™ catalysts, but hydrides such as sodium borohydride have been shown to work.

Tab. 2 Examples of the effect of solvent on enantioselectivity with two different ketones.

Solvent	ee (%)	Conversion (%)
Ketone 1		
Methanol	58	>90
Dimethylsulfoxide	72	>90
Dimethylformamide	52	>90
Acetonitrile	4	>90
Ethyl acetate	46	>90
Tetrahydrofuran	60	>90
Dichloromethane	16	>90
Toluene	0	>90
Ketone 2		
Neat	74	17
Methanol	78	14
Isopropanol	79	19
Dimethylformamide	80	100
Dimethylacetamide	79	98
1,4-Dioxane	82	52
Ethyl acetate	80	74
Tert-butyl acetate	77	50
Tetrahydrofuran	80	71
Glyme	74	59
Diglyme	74	18
Methoxyethanol	74	21
Tert-butylmethyl ether	73	30
Dichloromethane	80	51
Triethylamine	78	82
2-Pentanone	77	30
Toluene	63	13

Ketone 1: an electron-rich aryl alkyl ketone.
Ketone 2: an electron-deficient aryl alkyl ketone.
For the screening studies 1 mol-% catalyst was used and the reactions were analyzed after 1 hour.

The TEAF reagent can be used neat if the substrate is an oil, but it is usually more convenient to dissolve the substrate in a solvent. In many cases we have observed the solvent to have a large effect on the optical purity of the product. Tab. 2 illustrates this with a ketone and the Cp*-rhodium TsDPEN catalyst. Further optimization of this reaction improved the enantiomeric excess to 98% ee. A second example involved reduction of 4-fluoroacetophenone, in this case the enantioselectivity was largely unaffected but the rate of reduction changed markedly with solvent. Development of this process improved the optical purity to 98.5% ee.

A rationale for these results is that the catalyst interacts with the substrate through weak intermolecular association, and the strength and nature of this are sensitive to the solvent. Since it is difficult to predict the best solvent, the best approach is to screen a suitable range of them.

3.2.3
Substrate

Two useful aspects of transfer hydrogenation are the broad scope and high regioselectivity. The range of ketones that can be reduced includes diaryl, dialkyl, and arylalkyl ketones, alpha-, beta-unsaturated cyclic, heterocyclic, and alicyclic ketones. The substrates may bear many functional groups including halides, ethers, thioethers, alkenes, amines, alcohols, amides, acids, esters, and nitriles, yet selective reduction of the ketone or imine is still achieved. Strongly electron withdrawing substituents on aryl ketones tend to give lower enantiomeric excesses, whilst coordinating alpha groups to the ketone tend to give poor results because the product is a ligand and can interfere with the catalyst. Most substrates are reduced with excellent enantioselectivity and with good catalyst turnover frequencies.

The reduction of imines and iminium salts present a particular difficulty in that those that are *N*-substituted can exist in different geometrical isomers, which are reduced at different rates and with different selectivities. One way around this problem is to use cyclic imines that can exist only as *cis*-isomers. These are good substrates, however this is not a general solution. Primary amines are difficult to prepare directly as the imines are unstable in TEAF, so *N*-substitution is required. *N*-benzyl and *N*-alkyl imines are often reduced but with moderate ee values due to the geometrical isomer problem. Some *N*-acyl, *N*-sulfonyl, oximes, and hydrazones are reduced but the reaction appears to be substrate specific. Diphenylphosphinylimines are particularly good substrates [15]. The large steric size of the diphenylphosphinyl group may cause the imine to exist in predominantly one geometrical isomer, and this leads to high optical activities. An improved method for their preparation involves reacting diphenylphosphonylamide with the ketone and dehydrating with titaniumtetraisopropoxide. The imine may be isolated in good yield, or more conveniently the transfer hydrogenation can be carried out directly on this mixture. The diphenylphosphonylamine product is hydrolyzed with acidic ethanol to give the primary amine. In general, acyclic imines are reduced in lower yields and ee values than their cyclic counterparts and this is probably due to poor facial discrimination, however excellent results have been obtained with the phosphinylimines of 2-butanone, 2-hexanone, and 2-octanone substrates (Fig. 4).

A recent development is the transfer hydrogenation of heterocyclic systems such as pyrrole, pyridinium, and quinoline systems. Whilst the yields and enantioselectivities are modest at the moment, further development may improve this. For example, 1-methyl-isoquinoline has been reduced to the tetrahydro species and 1-picoline has been reduced to 1-methylpiperidine. Hydrogenation of alkenes has been reported [16], but in the CATHy^{TM}-catalyzed reaction alkene reduction has only been observed after extended reaction times. Whilst heterogeneous transfer hydrogenation of nitro groups is well known, homogeneous systems do not reduce these groups.

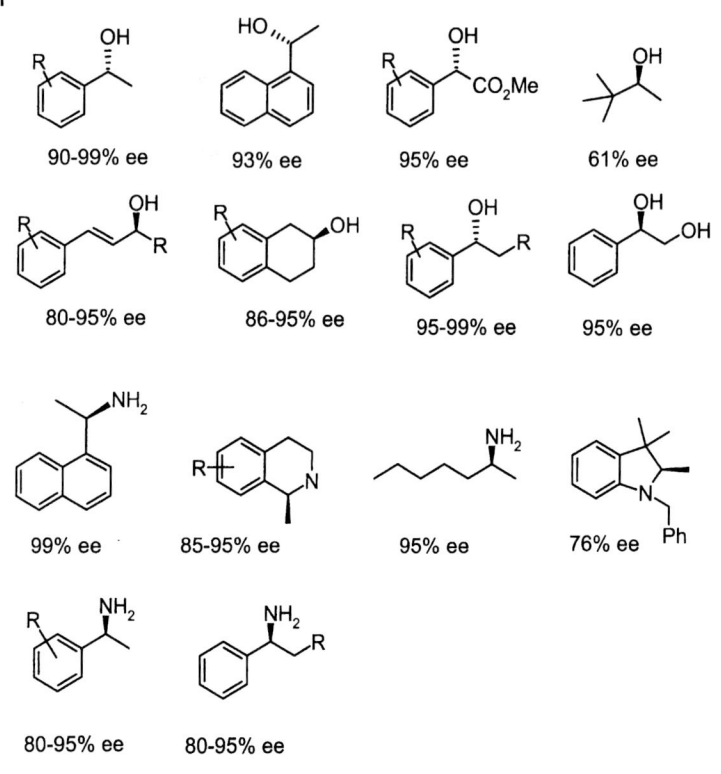

Fig. 4 Examples of alcohols and amines that are produced with high enantioselectivity. All acyclic amines have been prepared via the corresponding diphenylphosphinylimines.

3.3
Process

The transfer hydrogenation methods described above are sufficient to carry out laboratory studies, but it is unlikely that a direct scale-up of these processes would result in identical yields and selectivities. This is because the reaction mixtures are biphasic (liquid, gas): the gas being distilled is acetone from the IPA system and carbon dioxide from the TEAF system. The rate of gas disengagement is related to the superficial surface area, so as the process is scaled-up, or the height of the liquid increases, the ratio of surface area to volume decreases. To improve the de-gassing, parameters such as stirring, reactor design, and temperature are important and these will be discussed along with other factors that have been found to be important in scaling-up the process.

Tab. 3 An example of the effect of temperature on rate and enantiomeric excess.

Temperature (°C)	Enantiomeric excess (%)	Relative rate
20	95.7	1.00
5	97.8	0.60
0	98.4	0.37
–5	98.4	0.26

3.3.1
Temperature

The CATHy^{TM} catalysts are best used below 40 °C, above this temperature we have observed signs of decomposition. In the IPA system, preventing the back-reaction depends on how efficiently acetone is distilled. Normally this would be best done at around 80 °C, the boiling point of isopropanol, but an optimal performance of the catalyst requires ambient temperature or less, and reduced pressure. Whilst acetone can be fractionally distilled, it is simpler to distil the mixture with isopropanol and to maintain constant volume by continuously charging with fresh solvent. In the TEAF system the reaction is normally operated at ambient temperature. Operating at lower temperatures can improve the enantiomeric excess slightly but gives lower rates, for example with 4-fluoroacetophenone the results described in Tab. 3 were achieved.

3.3.2
Reactant Concentration

In industrial chemistry it is important to develop a productive process since the amount of material that can be processed in a given time is directly related to the product cost. Originally the concentrations used in the IPA system were low and reaction times fairly long. These have been considerably improved using the acetone vacuum distillation process. Depending upon the efficiency with which acetone is removed, concentrations up to 0.5 M and cycle times of less than 8 hours have been achieved. It is interesting to note that the batch process is clearly not the best means of operating this, rather a reactor that provides much larger surface areas, such as thin film evaporator, is better suited.

In the TEAF system there is no problem with a back-reaction and concentrations up to 10 M are possible. Neat TEAF has been used satisfactorily but is fairly viscous, so it is preferable to use a diluent. As mentioned above, the solvent can have a marked effect on the rate and enantioselectivity.

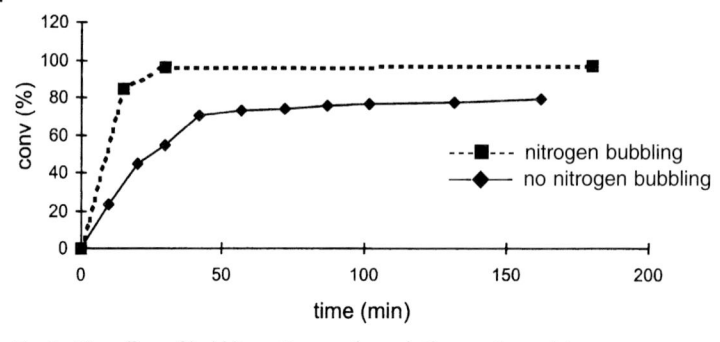

Fig. 5 The effect of bubbling nitrogen through the reaction mixture.

3.3.3
Reaction Control

The IPA system is unusual in that it is thermo-neutral, so all the components can be mixed safely at the start of the reaction and the reaction is initiated with small amounts of base.

The TEAF system is slightly exothermic, so in the laboratory all the components can be safely charged together. However, if one operates this same process at a larger scale, the reaction does not proceed with very high conversion despite a high initial rate. Analysis of the final reaction mixture shows that the reaction has sufficient TEAF and one possibility is that the catalyst has stopped working part way through the reaction. Addition of more catalyst only results in a small increase in conversion. We, and other workers, have found that bubbling nitrogen through the reaction improves the rate and enables complete conversion (Fig. 5) [7]. The origin of this effect is under investigation.

3.4
Case Studies

3.4.1
Example 1: (R)-1-Tetralol

Rhodium 1(R)-amino-2(S)-hydroxyindanyl pentamethylcyclopentadienyl chloride was selected from a screen of catalysts as giving the highest rate and enantiomeric excess for this reaction. A further screen showed isopropanol was the best hydrogen donor and solvent. In the laboratory a closed vessel was used and without actively removing the acetone, the maximum concentration that could be achieved with high conversion was 0.05 M, and during the reaction the enantiomeric excess fell markedly.

A key development target was to remove acetone efficiently. When acetone and isopropanol were distilled under vacuum at 10–50 mbar and 5–18 °C with efficient

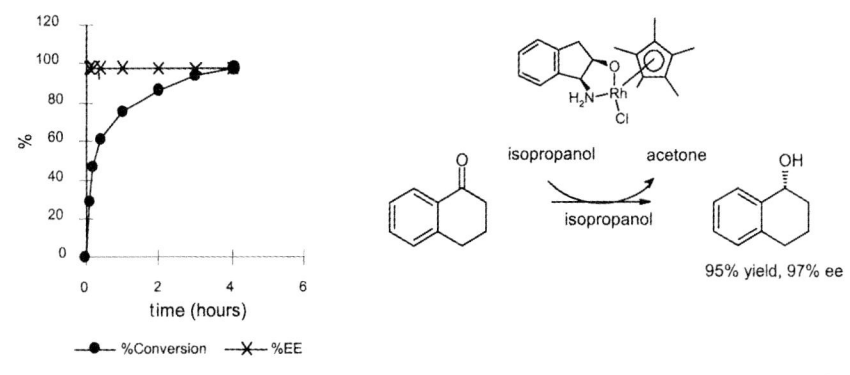

Fig. 6 Asymmetric transfer hydrogenation of 1-tetralone using the IPA system. Reaction profile and optical purity of (R)-1-tetralol.

agitation, a concentration of up to 0.5 M could be achieved with quantitative conversion. As the liquids were distilled, the reaction was concentrated, and to maintain constant volume fresh isopropanol was added. The process was successfully operated at the 200 L scale. In the developed reaction TOF values of 500–2500 per hour were achieved with reproducible ee results of 97% and 95% yield (Fig. 6).

3.4.2
Example 2: (S)-4-Fluorophenylethanol

A rhodium N-alkylsulfonyl-(1R,2R)-1,2-diphenylethylenediamine pentamethylcyclopentadienyl chloride was selected from a screen of catalysts as giving the highest rate and enantiomeric excess for this reaction. A further screen showed TEAF was the best hydrogen donor and THF the best solvent giving 96% ee.

A key development target was to increase the ee to >98%. It was found that if the reaction was operated at 0 °C then the product was produced in 98.5% ee. Despite the fact that the reaction is slower at 0 °C we found that by separately feeding TEAF and catalyst to a solution of the ketone good rates could be achieved. Bubbling nitrogen through the solution and agitating well enabled a TOF of 75 per hour (Fig. 7). A substrate concentration of 3.6 M was found to be most effective. This productive process was successfully operated at the 50 L scale, where the product was produced in 98.4% ee in 85% yield with very low levels of residual rhodium.

3.4.3
Example 3: (R)-N-Diphenylphosphinyl-1-methylnaphthylamine

Rhodium N-4-toluenesulfonyl-(1S,2S)-1,2-diphenylethylenediamine pentamethylcyclopentadienyl chloride was selected from a screen of catalysts as giving the highest rate and enantiomeric excess for this reaction. A further screen showed TEAF was the best hydrogen donor and acetonitrile the best solvent giving 99% ee.

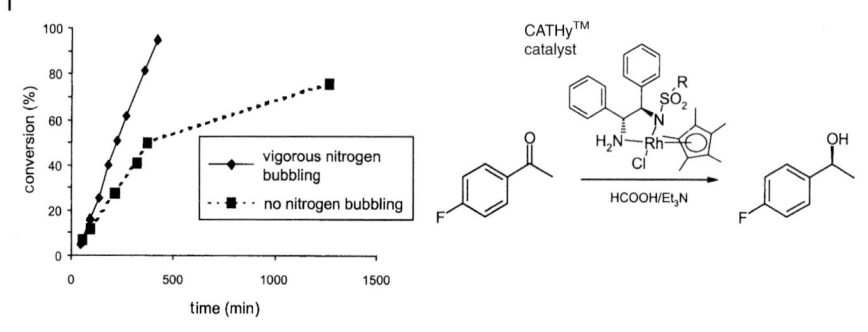

Fig. 7 Asymmetric transfer hydrogenation of 4-fluoroacetophenone using the TEAF system. The effect of bubbling nitrogen through the reaction mass.

Fig. 8 Asymmetric transfer hydrogenation of methylnaphthyl diphenylphosphinylimine using the TEAF system.

A key development target was to increase the activity of the catalyst. Initially an S:C ratio of 200:1 was used. We found that by separately feeding TEAF and catalyst to a solution of the imine good rates could be achieved. Bubbling nitrogen through the solution and agitating vigorously enabled a TOF of up to 1000 per hour. A substrate concentration of 0.6 M was found to be most effective. This process was successfully operated at larger scale, where the product was produced in 99% ee in 95% yield with very low levels of residual rhodium (Fig. 8).

3.5
Conclusions

When faced with the need to make a chiral pharmaceutical there are usually several technologies that can be envisaged to make the product. It must be recognized that each technology has strengths and weaknesses, moreover each project is different and requires its own solution. The complete fine chemical company should have expertise in a range of technologies that can be applied at different scales to cover diverse range of possibilities [17, 18].

There are numerous technologies that have been developed for manufacturing optically active alcohols and amines. Asymmetric hydrogenation is cost effective, well understood and has been used commercially at full-scale [19]. Likewise dehydrogenase enzymes are also economic, are used commercially at full-scale and display exquisite enantio- and regio-selectivity, however they require at least some solubility of the substrate in water [20]. Catalytic asymmetric transfer hydrogenation is a good technology that has been developed from a method of academic interest to one that is used for commercial manufacture. One of the main advantages is the avoidance of handling hydrogen, so that screening experiments can be done in test tubes, and at larger scale the processes can be sited in standard reactors rather than specialized and expensive autoclaves. On the downside, the catalyst turnover frequencies are generally lower than asymmetric hydrogenation or dehydrogenases, so require more catalyst or longer reaction times. The catalysts are, however, generally less expensive than those used in asymmetric hydrogenation, since the ligands are simpler to make than chiral diphosphines. The result is that for most pharmaceutical applications the catalyst does not contribute significantly to the raw material costs. There are a number of methods to prevent catalyst residues from contaminating the product, and typically levels of less than 10 ppm metal can be achieved. One way of improving the catalyst efficiency and product quality is to immobilize the catalyst on an insoluble support, and various techniques are currently being evaluated [18].

Transfer hydrogenation is an unusual case for scale-up, as the problem is how to transfer efficiently the byproducts from solution to the gas phase, compared with hydrogenation, which is the opposite. Unsurprisingly scale-up of the batch process is affected by physical aspects of the reaction such as agitation and surface area of the gas to liquid. One means of overcoming these limitations is continuous operation, and the fast reaction kinetics lends itself to this. With immobilization of the catalyst one can envisage a low cost and efficient continuous flow process.

Both the IPA and TEAF transfer hydrogenation processes have been successfully scaled-up on a number of occasions to make tens of kilograms of high quality products. Studies are currently underway to scale-up the process to the multi-ton scale.

Avecia undertakes screening and development programs and are ideally positioned to manufacture at the small- and large-scale if this is required [17]. CATHy^{TM} catalyst kits are available for research purposes from Strem Chemicals Inc.

3.6
Acknowledgements

The authors would like to thank the many Avecia employees that have been involved with development of the catalytic asymmetric transfer hydrogenation and especially Dr. Lynne Campbell, Dr. Ian Houson, Dr. David Moody. Professor Jonathan Williams of Bath University and Prof. Martin Wills of Warwick University.

3.7
References

1 R. Ruppert, E. Steckhahn, J. Chem. Soc., Chem. Commun. 1988, 17, 1150–1.
2 G. Zassinovich, G. Mestroni, S. Gladiali, Chem. Rev. 1992, 92, 1051.
3 S. Hashiguchi, A. Fujii, J. Takehara, T. Ikariya, R. Noyori, J. Am. Chem. Soc. 1995, 117, 7562; J. Takehara, S. Hashiguchi, A. Fujii, S. Inoue, T. Ikariya, R. Noyori, J. Chem. Soc., Chem. Commun. 1996, 233 p.
4 A. Fujii, S. Hashiguchi, N. Uematsu, T. Ikariya, R. Noyori, J. Am. Chem. Soc. 1996, 118, 2521.
5 J. Blacker, B. Mellor, WO9842643B1, Avecia Ltd, filed 26/03/97.
6 K. Mashima, T. Abe, K. Tani, Chem. Lett. 1998, 1199; J. Mao, D. Baker, Org. Lett. 1999, 1(6), 841; K. Murata, T. Ikariya, J. Org. Chem. 1999, 64, 2186.
7 M. Miyagi, J. Takehara, S. Collet, K. Okano, Org. Proc. Res. Dev. 2000, 4(5), 346.
8 S. Hashiguchi, A. Fujii, K.-J. Haack, K. Matsumura, T. Ikariya, R. Noyori, Angew. Chem. 1997, 36, 288; P. Dinh, J. Howarth, A. Hudnott, J. Williams, Tetrahedron Lett. 1996, 37, 7623; A. Larsson, B. Persson, J.-E. Backvall, Angew. Chem. 1997, 36, 1211.
9 M. Palmer, T. Walsgrove, M. Wills, J. Org. Chem. 1997, 62, 5226.
10 C. Casey, S. Singer, D. Powell, R. Hayashi, M. Kavana, J. Am. Chem. Soc. 2001, 123, 1090; D. Alonso, P. Brandt, S. Nordin, P. Andersson, J. Am. Chem. Soc. 1999, 121, 9580.
11 C. Bubert, J. Blacker, S. Brown, J. Crosby, S. Fitzjohn, J. Muxworthy, T. Thorpe, J. Williams, Tetrahedron Lett. 2001, 42, 4037.
12 T. Thorpe, J. Blacker, S. Brown, C. Bubert, J. Crosby, S. Fitzjohn, J. Muxworthy, J. Williams, Tetrahedron Lett. 2001, 42, 4041.
13 D. Dossett, I. Houson, M. Wills, unpublished results.
14 J. Blacker, L. Campbell, EP1117627, Avecia Ltd.
15 L. Campbell, J. Martin, EP1210305, Avecia Ltd.
16 H. Brunner, W. Leitner, Angew. Chem. 1988, 27(9), 1180; J. Brown, H. Brunner, W. Leitner, M. Rose, Tetrahedron Asymm. 1991, 2(5), 331.
17 http://www.avecia.com
18 J. Blacker, SP2, April 2002, 20; J. Blacker, SP2, September 2002, 28.
19 T. Ohkuma, R. Noyori, in Catalytic Asymmetric Synthesis (I. Ojima Ed.), Wiley VCH, 2000, pp. 1–110.
20 J. Blacker, R. Holt, in Chemistry in Industry (A. Collins, G. Sheldrake, J. Crosby Eds.), Wiley VCH, 1997, pp. 245–261.

4
Practical Applications of Biocatalysis for the Manufacture of Chiral Alcohols such as (R)-1,3-Butanediol by Stereospecific Oxidoreduction

Akinobu Matsuyama and Hiroaki Yamamoto

Abstract

A convenient biocatalytic process has been developed using a novel whole-cell biocatalyst for the preparation of (R)-1,3-butanediol (BDO) by stereo-specific oxidoreduction on an industrial scale. (R)-1,3-BDO is an important chiral synthon for the synthesis of various optically active compounds, such as azetidinone derivatives, which are used to prepare penem and carbapenem antibiotics for industrial usage.

Two approaches were studied to obtain (R)-1,3-BDO. The first was based on an enzyme-catalyzed asymmetric reduction of 4-hydroxy-2-butanone, and the second was based on enantioselective oxidation of the undesirable (S)-1,3-BDO in the racemate. As a result of screening for yeasts, fungi, and bacteria, the enzymatic resolution of racemic 1,3-BDO by *Candida parapsilosis* IFO 1396, which showed differential rates of oxidation for two enantiomers, was found to be the most practical process to produce (R)-1,3-BDO with high enantiomeric excess and yield.

An attempt was made to develop the bioprocess using a recombinant microorganism in order to produce (R)-1,3-BDO efficiently. A (S)-1,3-BDO dehydrogenase purified from a cell-free extract of *C. parapsilosis* was characterized. This enzyme was found to be a novel secondary alcohol dehydrogenase (CpSADH). A gene encoding CpSADH was cloned and expressed in *Escherichia coli*. The CpSADH activity (43.2 U/mL-medium) of a recombinant *E. coli* strain was 78-fold higher than that of *C. parapsilosis*. Under the optimized conditions using a racemate concentration of 15% the yield of (R)-1,3-BDO reached 72.6 g/L (48.4%) with an optical purity of 95% ee. Interestingly, the recombinant *E. coli* strain catalyzed the reduction of ethyl 4-chloroacetoacetate to ethyl (R)-4-chloro-3-hydroxybutanoate with high enantiomeric excess.

4.1
Introduction

Chiral 1,3-butanediol (1,3-BDO) is an important material for synthesizing various optically active compounds, such as azetidinone derivatives, which are intermediate materials for antibiotics [1, 2], pheromones [3], fragrances [4, 5], and insecti-

Asymmetric Catalysis on Industrial Scale: Challenges, Approaches and Solutions
Edited by H.U. Blaser, E. Schmidt
Copyright © 2004 Wiley-VCH Verlag GmbH & Co. KGaA, Weinheim
ISBN: 3-527-30631-5

cides [6] (Fig. 1). (R)-1,3-BDO, in particular, is a starting material for azetidinone derivatives. These are important intermediates in the synthesis of penem and carbapenem antibiotics, which have industrial uses (Fig. 2).

Biocatalysts are advantageous because when they are used environmentally hazardous reagents, solvents, and chemical catalysts can be avoided. Therefore, new biocatalytic processes have been developed to manufacture optically active compounds

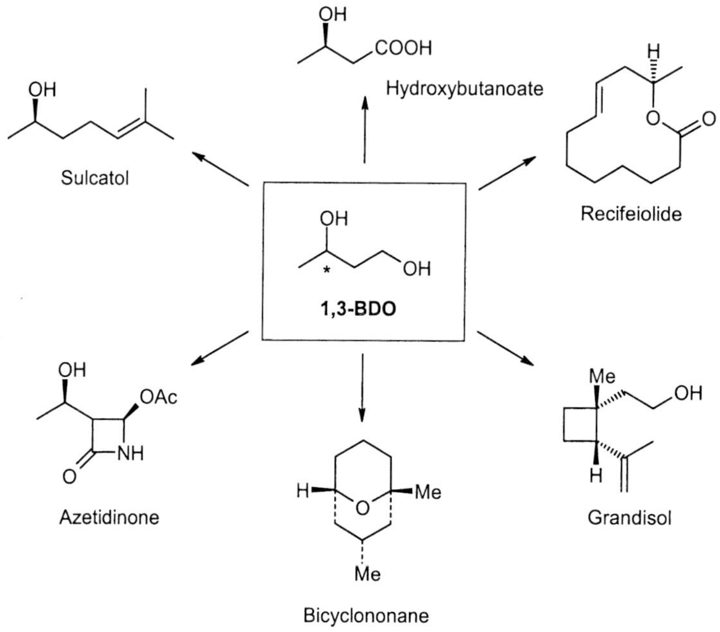

Fig. 1 Uses of chiral 1,3-BDO.

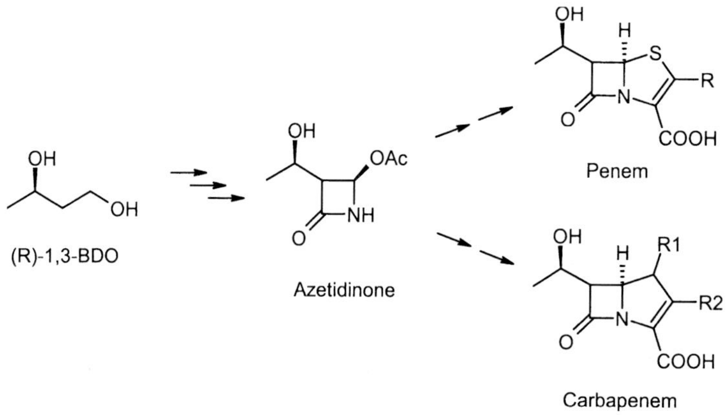

Fig. 2 Industrial uses of (R)-1,3-BDO.

4.2
Screening of Microorganisms Producing Optically Active 1,3-BDO from 4-Hydroxy-2-butanone (4H2B) by Asymmetric Reduction

with improved efficiency and environmental compatibility. Although microbial processes for optically active 1,3-BDO have been reported with the use of baker's yeast, the productivity was unsatisfactory and only yielded (R)-1,3-BDO [7, 8].

A program was begun to establish an economical and convenient microbial process from 4-hydroxy-2-butanone (4H2B) by asymmetric reduction. All microorganisms used were stock cultures preserved at the Daicel laboratories. The methods and results of screening are shown in Tab. 1. As a result of this screening, many strains of yeasts and bacteria produced optically active 1,3-BDO from 4H2B; however, lactic acid bacteria were unproductive. High productivity and high optical purity were found in 13 yeast strains. Eleven strains produced (R)-1,3-BDO, and

Tab. 1 Screening of microorganisms producing optically active 1,3-BDO from 4H2B by asymmetric reduction.

Microorganisms (yeasts, bacteria, and fungi)
↓
Cultivation (100 mL/500 mL flask at 30 °C for 24–48 h with shaking)
↓
Centrifugation and washing with saline
↓
Cells
↓
100 mM KPB (pH 7.5) 5 mL with 12% sucrose
↓
Preincubation at 30 °C for 10 min
↓ ←──────────────── 4H2B 50 mg
Incubation (in test tube at 30 °C for 48 h with shaking)
↓
Centrifugation
↓
Supernatant (2 mL) ──────────→ Saturation with NaCl
↓   ↓
Glc Analysis   Extraction with EtOAc (2 mL)
                                 ↓
Determination of formed 1,3-BDO and   Evaporation *in vacuo*
residual 4H2B   ↓
                                       Acetylation with AcCl
                                       ↓
                                       HPLC Analysis

                                       Determination of optical purity

Tab. 1 (continued)

Strain	Absolute configuration	% ee	Formed 1,3-BDO yield (%)
Candida utilis IFO 1086	R	81	88
Candida utilis IAM 4246	R	85	82
Candida utilis IAM 4277	R	95	82
Candida arborea IAM 4147	R	99	37
Kluyveromyces lactic IFO 1903	R	92	88
Kluyveromyces lactis IFO 1267	R	93	99
Hansenula fabianii IFO 1254	R	67	16
Hansenula polymorpha ATCC 26012	R	85	87
Issatchenkia scutulata IFO 10070	R	99	48
Issatchenkia scutulata IFO 10069	R	93	50
Puchia heedii IFO 10020	R	81	24
Candida parapsilosis IFO 1396	S	98	60
Geotrichum candidum IFO 4601_	S	88	78

$$\text{4-hydroxy-2-butanone (4H2B)} \xrightarrow{\text{microorganism}} \text{optically active 1,3-butanediol (BDO)}$$

two strains produced (S)-1,3-BDO with high optical purity. In particular, *Candida arborea* IAM 4147 and *Issatchenkia scutulata* IFO 10070 converted 4H2B into (R)-1,3-BDO with 99% ee, but in low yield. *Kluyveromyces lactis* IFO 1267 was selected as the best strain for producing (R)-1,3-BDO by asymmetric reduction because it also gave a high yield. *K. lactis* IFO 1267 produced (R)-1,3-BDO with an ee of 93% [9] (Tab. 1). Among the (S)-1,3-BDO-producing strains, *C. parapsilosis* IFO 1396 gave the best optical purity (98% ee).

The amounts of 1,3-BDO and 4H2B were measured by gas chromatography under the following conditions: column, PoraPak PS (Waters Corporation, Milford, MA, USA); column temperature, 165 °C; carrier gas, N_2; detection, flame ionization detector. The optical purity of 1,3-BDO was measured as 1,3-BDO diacetyl by chiral HPLC with a Chiralcel OB packed column (4.6×250 mm, Daicel Chemical Industries, Ltd., Tokyo, Japan) at 40 °C, eluted with n-hexane: 2-propanol (19:1) at a flow-rate of 1 mL/min, and detected at 220 nm.

4.3
Screening of Microorganisms Producing Optically Active 1,3-BDO from the Racemate

Enantioselective oxidation of the hydroxyl group at the 1-position of 1,3-BDO to optically active 3-hydroxybutyric acid has been described [10], but there has been no report of the enantioselective oxidation of 1,3-BDO to 4H2B. Thus, we screened microorganisms to produce optically active 1,3-BDO from racemic 1,3-BDO. The methods and results of screening are shown in Tab. 2. In the screening of more than 1000 strains, many strains of yeasts, fungi, and bacteria produced optically active 1,3-BDO from the racemate. High productivity and high optical purity were found in 16 strains. The strains that produced (S)-1,3-BDO from 4H2B by asymmetric reduction could produce (R)-1,3-BDO from the racemate by enantioselective oxidation: C. parapsilosis IFO 1396 and Geotrichum candidum IFO 4601. Similarly, the strain that produced (R)-1,3-BDO from 4H2B by asymmetric reduction could produce (S)-1,3-BDO from the racemate by enantioselective oxidation: K. lactis IFO 1267. C. parapsilosis IFO 1396 produced (R)-1,3-BDO with an ee of 97%, and K. lactis IFO 1267 produced (S)-1,3-BDO with an ee of 99% from the racemate [11] (Tab. 2).

Tab. 2 Screening of microorganisms producing optically active 1,3-BDO from the racemate.

Microorganisms (yeasts, bacteria, and fungi)
↓
Cultivation (10 mL/test tube at 30 °C for 24–48 h with shaking)
↓
Centrifugation and washing with saline
↓
Cells
↓
100 mM KPB (pH 7.0) 2.5 mL
↓ ←———————————————— Racemic 1,3-BDO 25 mg
Incubation (in test tube at 30 °C for 48 h with shaking)
↓
Centrifugation
↓
Supernatant (2 mL) ————————→ Saturation with NaCl
↓   ↓
Glc Analysis   Extraction with EtOAc (2 mL)
                                                ↓
Determination of residual 1,3-BDO and formed 4H2B   Evaporation *in vacuo*
                                                ↓
                                                Acetylation with AcCl
                                                ↓
                                                HPLC Analysis

                                                Determination of optical purity

Tab. 2 (continued)

Strain	Absolute configuration	% ee	Residual 1,3-BDO yield (%)
Pseudomonas putida IFO 3738	S	75	50
Candida utilis IFO 0639	S	91	43
Candida inconspicua IFO 0621	S	97	50
Hansenula subpelliculosa IFO 0808	S	71	54
Kluyveromyces lactis IFO 1267	S	99	45
Pichia pountiae IFO 10025	S	96	50
Fusarium solani IFO 5232	S	99	8
Ambrosiozyma philentoma IFO 1847	S	83	34
Talaromyces flavus IFO 7231	S	85	11
Candida parapsilosis IFO 1396	R	95	50
Candida intermedia IFO 0761	R	76	42
Candida maltosa IFO 1978	R	70	43
Aciculoconidium aculeatum IFO 10124	R	72	68
Geotrichum candidum IFO 4601	R	95	43
Lodderomyces elongisporus IFO 1676	R	72	49
Trichosporon cutaneum IFO 0743	R	69	48

racemic 1,3-BDO → (microorganism) → optically active 1,3-BDO

4.4
Preparation of (R)- and (S)-1,3-BDO by the Same Strain or from the Same Material

Figs. 3 and 4 show the routes for the production by K. lactis IFO 1267 and C. parapsilosis IFO 1396 from 4H2B and from racemic 1,3-BDO. Each strain was cultivated in 100 mL of a YM (2.0% glucose, 0.5% polypepton, 0.5% yeast extract, and 0.3% malt extract; pH 6) medium in a 500 mL Sakaguchi flask with reciprocal shaking at 30 °C for 48 hours. The cells in 100 mL of broth were harvested by centrifugation and resuspended in 25 mL of a 100 mM potassium phosphate buffer (pH 6.5): Cell concentration OD at 660 nm = 35 (C. parapsilosis IFO 1396), and 25 (K. lactis IFO 1267). Next, 0.3 g of 4H2B or racemic BDO was added, and the reaction mixture in a 500 mL Sakaguchi flask was incubated with reciprocal shaking at 30 °C for 43 hours.

K. lactis IFO 1267 produced (R)-1,3-BDO with an ee of 93% from 4H2B and (S)-1,3-BDO with an ee of 99% from the racemate. C. parapsilosis IFO 1396, on the other hand, produced (S)-1,3-BDO with an ee of 94% from 4H2B and (R)-1,3-BDO with an ee of 97% from the racemate. The methods of preparation of (R)- and (S)-1,3-BDO are illustrated by the same strain or from the same material (Fig. 5).

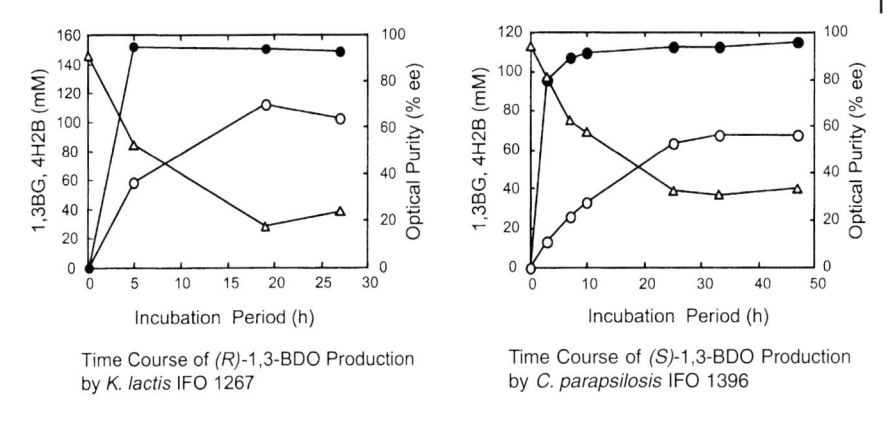

Time Course of (R)-1,3-BDO Production
by K. lactis IFO 1267

Time Course of (S)-1,3-BDO Production
by C. parapsilosis IFO 1396

○BDO; △4H2B; ●Optical purity

Fig. 3 Optically active 1,3-BDO production from 4H2B.

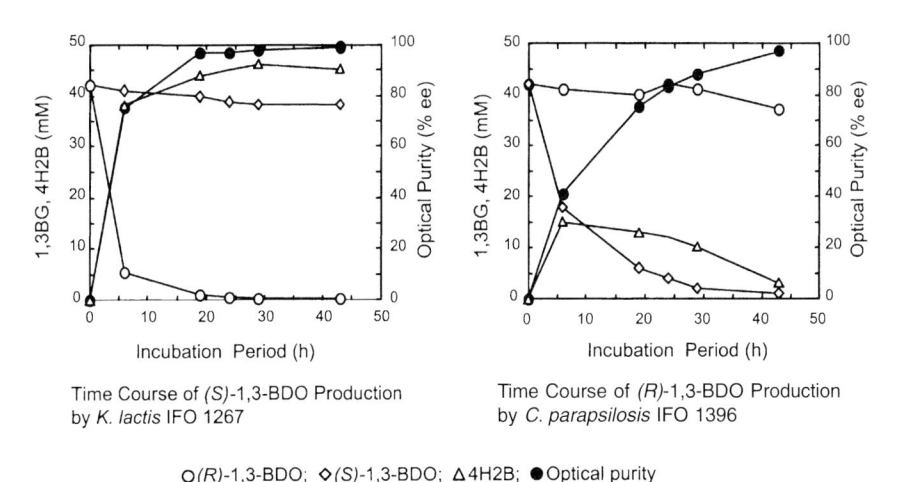

Time Course of (S)-1,3-BDO Production
by K. lactis IFO 1267

Time Course of (R)-1,3-BDO Production
by C. parapsilosis IFO 1396

○(R)-1,3-BDO; ◊(S)-1,3-BDO; △4H2B; ●Optical purity

Fig. 4 Optically active 1,3-BDO production from the racemate.

4.5
Large-Scale Preparation of (R)-1,3-BDO from the Racemate Using Candida parapsilosis IFO 1396

Large-scale preparation of (R)-1,3-BDO from the racemate by C. parapsilosis IFO 1396 [12] was carried out using a large fermentor with a working volume of 2000 L. The cells were harvested from 3000 L of a YMBG (2.0% glucose, 0.5% polypepton, 0.5% yeast extract, 0.3% malt extract, and 1.0% racemic 1,3-BDO; pH 6.0) culture (2 batches with 1500 L each) by centrifugation. Production of (R)-1,3-BDO was achieved in a reaction mixture containing 258 kg of the cells, 465 kg of water, 7.5 kg of calcium carbonate, and 20 kg racemic 1,3-BDO at 30 °C. Agitation

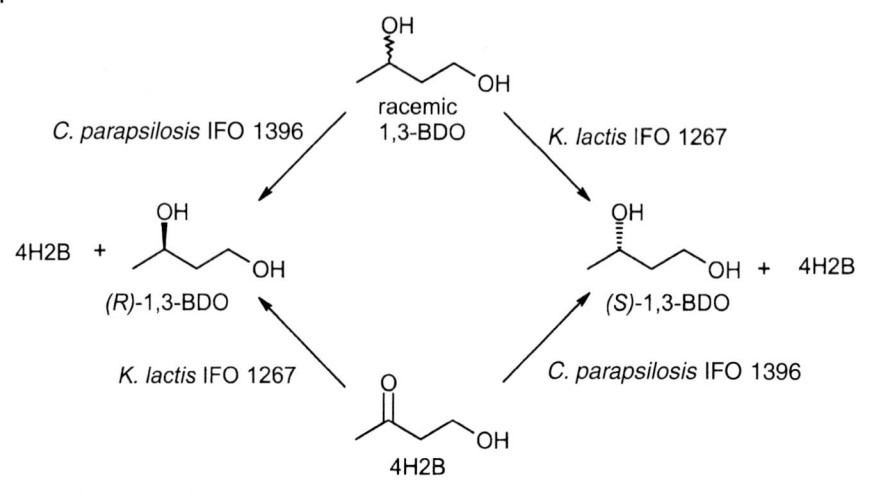

Fig. 5 Preparation of chiral 1,3-BDO with both configurations.

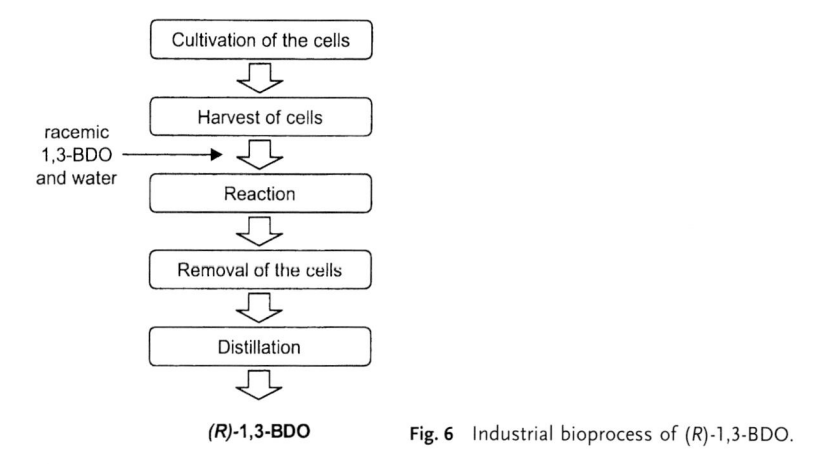

Fig. 6 Industrial bioprocess of (R)-1,3-BDO.

and aeration were 66 rpm and 29 m^3/h, respectively. After reaction for 81 hours (residual 1,3-BDO 41%), the reaction mixture was centrifuged to remove the cells and concentrated to 18.7 kg of slurry at 50 °C *in vacuo*. Next, 22.3 kg of methanol were added to the slurry. After filtration, the solvent was removed by evaporation, and 3092 g of (R)-1,3-BDO (an overall yield of 15.5%) were distilled *in vacuo*. The chemical purity was 98.8%, and the optical purity was 94% ee. The specific rotation was $[a]_D$ −29 (c 1, EtOH) (reference $[a]_D$ −29 [9]). The ^{1}H-NMR spectra were also identical with an authentic (R)-1,3-BDO NMR spectrum (270 MHz, in CDCl$_3$)δ_H 1.22 (3H,d,J=6.2 Hz), 1.67–1.69 (2H,m), 3.79 (3H,m), 4.00–4.07 (1H,m). The simple, industrial "Green bioprocess" of (R)-1,3-BDO is shown schematically in Fig. 6.

Although the activity of this wild strain was useful, it was necessary to construct a recombinant biocatalyst in order to improve the production efficiency since the use of a recombinant biocatalyst is more advantageous than the use of a wild strain for achieving high productivity. A recombinant whole-cell biocatalyst was a very convenient, high-performance, and stable source of enzymes, making it very efficient and much less expensive than a purified enzyme. To construct a more efficient whole-cell biocatalyst, a secondary alcohol dehydrogenase gene was cloned in an *Escherichia coli* host microorganism. *E. coli* is a suitable host for genetic engineering.

4.6
Purification and Characterization of (S)-1,3-Butanediol Dehydrogenase from *Candida parapsilosis* IFO 1396

A NAD^+-dependent (S)-1,3-BDO dehydrogenase (designated as CpSADH), which appears to participate in the production of (R)-1,3-BDO from the racemate, was purified from *C. parapsilosis* IFO 1396 [13]. The results of the enzyme purification are summarized in Tab. 3. The specific activity of the final preparation was about 5400-fold higher than that of the crude extract. The molecular weight of the enzyme was estimated to be approximately 40,000 by SDS-polyacrylamide gel electrophoresis and approximately 140,000 by gel filtration. The enzyme showed the maximal activity at 50 °C and pH 9 for the oxidation of (S)-1,3-BDO.

The substrate specificity of the enzyme is shown in Tabs. 4 and 5. The enzyme catalyzed the oxidation of a number of aliphatic alcohols enantioselectively and the reduction of a number of aliphatic aldehydes and ketones. The enzyme was shown to oxidize the hydroxyl group on the 3-position but not the 1-position of (S)-1,3-BDO, as the enzyme oxidized (S)-1,3-BDO to 4H2B.

From these results, the enzyme was found to be a NAD^+-dependent (S)-specific secondary alcohol dehydrogenase. So far, only a few (S)-specific secondary alcohol

Tab. 3 Purification of NAD^+-dependent (S)-1,3-butanediol dehydrogenase from *Candida parapsilopsis*.

Step	Volume (mL)	Total protein (mg)	Total activity (U)	Specific activity (U/mg)	Yield (%)
Crude extract	4800	157000	37140	0.045	100
Protamine sulfate	5200	94600	6258	0.066	87.6
0–70% Ammonium sulfate	550	78700	5460	0.069	76.5
Q-Sepharose FF	550	8870	1731	0.195	24.2
Phenyl-Toyopearl	22	191	969	5.07	13.6
Red-Sepharose	2.4	22.1	1095	49.6	15.3
Superdex 200	5.3	3.7	559	151	7.8
Mono Q	1.1	1.7	420	244	5.9

Tab. 4 Substrate specificity of (S)-1,3-BDO dehydrogenase for oxidative reaction.

Substrate	(mM)	Cofactor	Relative activity (%)
(S)-1,3-Butanediol	50	NAD$^+$	100.0
(R)-1,3-Butanediol	50	NADP$^+$	3.4
(S)-2-Butanol	50	NAD$^+$	562
(S)-2-Butanol	50	NAD$^+$	18.8
2-Propanol	100	NAD$^+$	337
2-Pentanol	100	NAD$^+$	191
3-Pentanol	100	NAD$^+$	58.3
2,4-Pentanediol	100	NAD$^+$	239
(2R,4R)-2,4-Pentanediol	50	NAD$^+$	0.7
4-Methyl-2-Pentanol	20	NAD$^+$	230
2-Hexanol	50	NAD$^+$	156
(S)-2-Octanol	5	NAD$^+$	381
(R)-2-Octanol	5	NAD$^+$	0.0
Cyclohexanol	20	NAD$^+$	297
(S)-1-Phenylethanol	50	NAD$^+$	502
(R)-1-Phenylethanol	50	NAD$^+$	6.4
Methyl (S)-3-hydroxybutanoate	50	NAD$^+$	267
Methyl (R)-3-hydroxybutanoate	50	NAD$^+$	9.9
Ethyl (S)-3-hydroxybutanoate	50	NAD$^+$	737
Ethyl (R)-3-hydroxybutanoate	50	NAD$^+$	3.2
Methanol	100	NAD$^+$	1.0
Ethanol	100	NAD$^+$	5.4
Allyl alcohol	100	NAD$^+$	13.4
1-Propanol	100	NAD$^+$	8.5
1-Butanol	100	NAD$^+$	13.1
1-Pentanol	100	NAD$^+$	6.8
2-Phenylethanol	100	NAD$^+$	0

The oxidative reaction was carried out as described in Materials and Methods for (S)-1,3-butanediol dehydrogenase, except that substrates and cofactors indicated in the table were used. Activity for (S)-1,3-butanediol was taken as 100%.

dehydrogenases have been reported from *C. parapsilosis* DSM 70125 [14], *Rhodococcus erythropolis* DSM 743 [15], and *R. erythropolis* ATCC 4277 [16]. The properties of these ADHs are summarized in Tab. 6.

4.7
Cloning and Expression of a Gene Coding for a Secondary Alcohol Dehydrogenase from *Candida parapsilosis* IFO 1396 in *Eschericha coli*

A gene encoding CpSADH was cloned from *Candida parapsilosis*. The CpSADH-gene consisted of 1,009 nucleotides coding for a protein with M_r 35 964 [17]. The overall amino acid sequence identity with alcohol dehydrogenases from *Saccharomyces cerevisiae* (ScADH1, ScADH2, ScADH3, and ScADH5), alcohol dehydroge-

Tab. 5 Substrate specificity of (S)-1,3-BDO dehydrogenase for reductive reaction.

Substrate	(mM)	Cofactor	Relative activity (%)
4-Hydroxy-2-butanone	100	NADH	100
	100	NADPH	0
Acetone	100	NADH	299
2-Butanone	100	NADH	243
Cyclopropyl methyl ketone	20	NADH	13.3
5-Chloro-2-pentanone	20	NADH	84.3
2-Acetylbutyrolactone	20	NADH	19.9
Ethyl acetoacetate	100	NADH	465
Ethyl 4-chloroacetoacetate	100	NADH	341
Acetophenone	20	NADH	296
Propionaldehyde	100	NADH	185

The reductive reaction was carried out as described in Materials and Methods for 4-hydroxy-2-butanone reductase, except that substrates and cofactors indicated in the table were used. Activity for 4-hydroxy-2-butanone was taken as 100%.

Tab. 6 Properties of (S)-specific secondary alcohol dehydrogenases.

Name	CpSADH	CPCR	RECR	Long-chain ADH
Origin	Candida parapsilosis IFO 1396	Candida parapsilosis DSM 70125	Rhodococcus erythropolis DSM 743	Rhodococcus erythropolis ATCC 4277
M_r				
Native	140 000	136 000	161 000	110 000
Subunit	40 000	67 000	40 000	48 000
Optimum for oxidation				
Temperature	50	52–56	ND	74
pH	9.0	7.8–8.6	9.5	9
Optimum for reduction				
Temperature	ND	36–40	40	ND
pH	6.0	6.5–7.2	5.5	ND
Cofactor	NAD^+/NADH	NAD^+/NADH	NAD^+/NADH	NAD^+/NADH
Specific activity (U/mg)				
(S)-1,3-Butanediol	244	ND	ND	ND
2-Propanol	822	2'926	5.38	ND
(S)-2-Octanol	929	ND	ND	266
Acetophenone	627	631	13.5	ND
Ethyl 4-chloroacetoacetate	726	160	69.9	ND

M_r, relative molecular mass; ND, not determined.

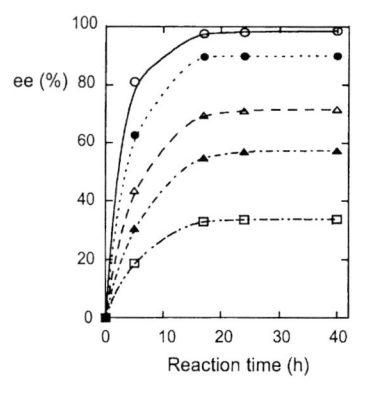

Fig. 7 Effects of the concentrations of 1,3-BDO on enantioselective oxidation.

A reaction mixture (25 ml), containing 7 -15% 1,3-BDO and cultured cells obtained from 25 ml of the medium, was incubated at pH 6.8. Other reaction conditions and analytical methods were the same as those indicated in Fig. 3. 1,3-BDO concentrations: 7% (○), 10% (●), 12% (△), 15% (▲), 20% (□)

nase I from *Zymomonas mobilis* (ZmADH1), and the NADP-dependent secondary alcohol dehydrogenase from *Thermoanaerobium brockii* (TbADH) were estimated to have 31.3%, 30.5%, 28.0%, 29.9%, 31.3%, and 23.9% identity, respectively.

A high-expression plasmid for CpSADH, pSE-CPA1, was constructed based on the pSE420 vector containing the trc promoter (Invitrogen Corporation, Carlsbad, CA, USA) [18]. *E. coli* cells harboring pSE-CPA1 were grown in 100 mL of a 2×YT medium (Bacto-Tryptone, 20 g/L; Bacto-Yeast extract, 10 g/L; NaCl, 10 g/L; pH 7.2) containing ampicillin (50 mg/L) in a 500 mL baffled shake flask at 30 °C on a rotary shaker (140 rpm) to OD600 (the optical density at 600 nm) of 3–4; after addition of 2% lactose as an inducer, the culture medium was further shaken for 11 hour. *E. coli* W3110 cells harboring pSE-CPA1 showed a CpSADH activity of 43.2 U/mL-medium, which was 78-fold higher than that of *C. parapsilosis*.

The reaction conditions for the synthesis of (R)-1,3-BDO were optimized using recombinant *E. coli* cells [19]. The optimal pH of a reaction mixture was found to be pH 6.8. This result suggested that the rate-limiting step of enantioselective oxidation was a regeneration step of the coenzyme, NAD^+, and that the optimal pH and/or the stablest pH for the regeneration of the coenzyme was around pH 6.8, since the optimal pH of CpSADH for the oxidation of (S)-1,3-BDO was 9.0 and CpSADH was stable within the range of 8.0–11.0.

The effects of the concentrations of 1,3-BDO in the reaction mixture were studied. At the reaction containing 7% 1,3-BDO, the optical purity of (R)-1,3-BDO reached 97.6% ee after incubation of 17 hours, as shown in Fig. 7.

However, in the reactions containing higher concentrations of 1,3-BDO, the optical purities of (R)-1,3-BDO did not increase from a 17-hour to a 40-hour incubation. To elucidate the reason for the cessation of enantioselective oxidation in the reaction mixtures containing 10% 1,3-BDO, the residual CpSADH activity of the

Tab. 7 Effects of the incubation in the reaction mixture containing 10% 1,3-BDO on the residual activities of CpSADH and the 1,3-BDO-producing activity.

Treatment	CpSADH	Specific 1,3-BDO-producing activity (% ee/OD_{600}/h)
None	100	1.9
After reaction	64.5	0
After storage	81.4	1.76

cell-free extract and the (R)-1,3-BDO-producing activity of E. coli cells were measured. After a 17-hour incubation, E. coli cells retained a CpSADH activity of 64.5% but had no 1,3-BDO-producing activity, as shown in Tab. 7. These findings suggested that the cessation of enantioselective oxidation resulted from the loss of NAD^+ and/or the NAD^+-regeneration activity of the cell. To overcome the loss of NAD^+ and/or NAD^+-regeneration activity, several additives were tested, such as acetone, sodium hydrogen sulfite, and several components of the medium. Among these additives, the addition of a YT medium (Bacto-Tryptone, 10 g; Bacto-Yeast extract, 5 g; NaCl, 5 g in a reaction mixture), which was expected to increase and maintain an amount of NAD^+ in the cells and/or the NAD^+-regeneration activity of cells by the growth of cells, brought about an increase in the turbidity of the reaction mixture at 600 nm after a 17-hour incubation and an increase of the optical purity of (R)-1,3-BDO from 89.8% to 96.1% after a 41-hour incubation. The addition of Bacto-Yeast extract alone could replace the effect of the addition of the YT medium.

Furthermore, in the reaction containing 15% 1,3-BDO, cells obtained from the 25 mL culture medium needed to be added to the reaction mixture after the 17 hour incubation to reach the optical purity of 95% ee. Under these optimized conditions, the yield of (R)-1,3-BDO reached 72.6 g/L, with a molar recovery yield of 48.4% from a racemate concentration of 15% and an optical purity of 95% ee.

The CpSADH was assayed spectrophotometrically. (R)-1,3-BDO-producing activity was measured as follows. A reaction mixture (25 mL) containing 5% 1,3-BDO and collected cells (45 OD_{600}) was incubated in a 500 mL Sakaguchi flask at 30 °C for 17 hours with reciprocal shaking (245 strokes per minute). The specific (R)-1,3-BDO-producing activity was expressed as the percent change of optical purity per OD_{600} at the start of the reaction and per hour of the reaction time (% ee/OD_{600}/h). The results for no treatment, after reaction, and after storage at 4 °C for 17 hours are shown in Tab. 7.

4.8
Preparation of Ethyl (R)-4-Chloro-3-hydroxybutanoate (ECHB) by Recombinant E. coli Cells Expressing CpSADH

(R)-ECHB is a chiral compound useful for the synthesis of biologically and pharmacologically important materials (Fig. 8): (R)-carnitine [20], (R)-4-amino-3-hydroxybutyric acid [21], and (R)-4-hydroxy-2-pyrrolidone [22]. The synthesis of (R)-ECHB using recombinant E. coli cells expressing CpSADH was studied, as CpSADH had both a strong reduction activity for ethyl 4-chloroacetoacetate (ECAA) and a strong oxidation activity for 2-propanol, as shown in Tabs. 4 and 5. 2-Propanol was used as a co-substrate to regenerate NADH by CpSADH itself. Under the optimized conditions [3.8% ECAA, 2 molar excesses of 2-propanol over ECAA in a 200 mM potassium phosphate buffer (pH 6.5); 20 °C], the yield of (R)-ECHB reached 36.6 g/L (more than 99% ee, 95.2% conversion yield) without the

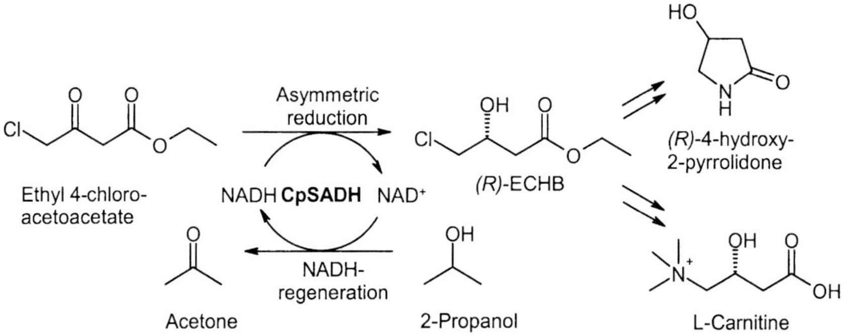

Fig. 8 Preparation of (R)-ECHB.

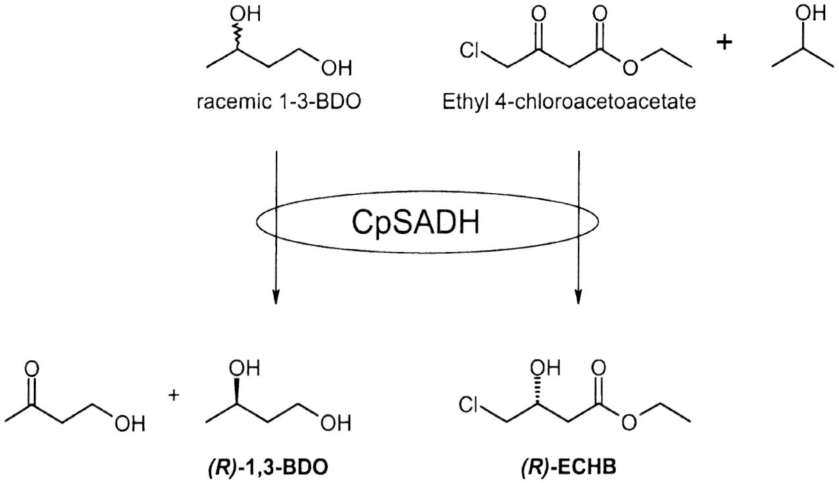

Fig. 9 Convenient system to synthesize (R)-1,3-BDO and (R)-ECHB using CpSADH.

addition of NADH to the reaction mixture [17]. The microbial reduction for the synthesis of (R)-ECHB could be scaled-up to a 30 L reactor scale.

4.9
Conclusion and Outlook

The modern tools of molecular genetics are currently applied to make enzymes available in recombinant hosts in larger quantities and at lower costs. The gene encoding the oxidoreductase with high stereoselectivity was isolated, and overproduction of CpSADH in *E. coli* was successfully carried out, resulting in a much improved biocatalyst that can be used for the large-scale preparation of chiral alcohols, as described above. The enantioselective oxidation and asymmetric reduction systems using the recombinant *E. coli* expressing CpSADH have been demonstrated to be efficient and convenient systems to synthesize many chiral alcohols, for instance, (R)-1,3-BDO and (R)-ECHB, on an industrial scale (Fig. 9). It is expected that the practical application of this recombinant whole-cell biocatalyst on more advanced intermediates in the synthesis of chiral pharmaceuticals will be expanded.

4.10
References

1 T. Nakayama, H. Iwata, R. Tanaka, S. Imajo, M. Ishiguro, J. Chem. Soc., Chem. Commun. 1991, 9, 662.
2 R. Tanaka, H. Iwata, M. Ishiguro, J. Antibiol. 1990, 43, 1608.
3 K. Mori, M. Miyake, Tetrahedron 1987, 43, 2229.
4 G. Ohloff, W. Giersch, R. Decorzant, G. Buechi, Helv. Chim. Acta 1980, 63, 1589.
5 J.T.B. Ferreira, B. Ferreira, F. Simonelli, Tetrahedron 1990, 46, 6311.
6 V.M.F. Choi, J.D. Elliott, W.S. Johnson, Tetrahedron Lett. 1984, 25, 591.
7 P.A. Levene, A. Walti, J. Biol. Chem. 1931, 94, 361.
8 C. Neuberg, A. Walti, Biochem. Z. 1918, 92, 96.
9 A. Matsuyama, Y. Kobayashi, H. Ohnishi, Biosci. Biotech. Biochem. 1993, 57, 348.
10 T. Yata, K. Makino, Japan Kokai Tokkyo Koho 8974999, 1989.
11 A. Matsuyama, Y. Kobayashi, H. Ohnishi, Biosci. Biotech. Biochem. 1993, 57, 685.
12 A. Matsuyama, Y. Kobayashi, Biosci. Biotech. Biochem. 1994, 58, 1148.
13 H. Yamamoto, A. Matsuyama, Y. Kobayashi, N. Kawada, Biosci. Biotech. Biochem. 1995, 59, 1769.
14 J. Peters, T. Minuch, M.-R. Kula, Enzyme Microb. Technol. 1993, 15, 950-958.
15 T. Zelinski, J. Peters, M.-R. Kula, J. Biotechnol. 1994, 33, 283–292.
16 B. Ludwig, A. Akundi, K. Kendall, Appl. Environ. Microbiol. 1995, 61, 3729-3733.
17 H. Yamamoto, N. Kawada, A. Matsuyama, Y. Kobayashi, Biosci. Biotechnol. Biochem. 1999, 63, 1051-1055.
18 H. Yamamoto, A. Matsuyama, Y. Kobayashi, Biosci. Biotechnol. Biochem. 2002, 66, 481-83.
19 H. Yamamoto, A. Matsuyama, Y. Kobayashi, Biosci. Biotechnol. Biochem. 2002, 66, 925-927.
20 B. Zhou, A.S. Gopala, F. van Middlesworth, W.R. Shieh, C.J. Sih, J. Am. Chem. Soc. 1983, 105, 5925–5926.
21 M.E. Jung, T.J. Shaw, J. Am. Chem. Soc. 1980, 102, 6304-6311.
22 E. Santaniello, R. Casati, F. Milani, J. Chem. Res., Synop. 1984, pp. 132–133.

5
Production of Chiral C3 and C4 Units via Microbial Resolution of 2,3-Dichloro-1-propanol, 3-Chloro-1,2-propanediol and Related Halohydrins

Naoya Kasai and Toshio Suzuki

Abstract

The study and development of microbial methods for the industrial-scale production of C3 and C4 chiral synthetic units such as 2,3-dichloro-1-propanol (DCP), epichlorohydrin (EP), 3-chloro-1,2-propanediol (CPD), glycidol (GLD), 4-chloro-3-hydroxybutyrate (CHB), 3-hydroxy-γ-butyrolactone (HL) is described. The following points are emphasized: overall strategy; screening, isolation, and cultivation of bacteria; control of fermentation reactions; and transfer from lab- to production-scale.

5.1
Introduction

We have been studying and developing the industrial-scale production of C3 and C4 chiral synthetic units such as 2,3-dichloro-1-propanol (DCP), epichlorohydrin (EP), 3-chloro-1,2-propanediol (CPD), glycidol (GLD), 4-chloro-3-hydroxy-butyrate (CHB), 3-hydroxy-γ-butyrolactone (HL), etc. These compounds were not available even on a laboratory scale, although they have various possible applications for chiral pharmaceuticals, agrochemicals, optically active liquid crystals or polymers, etc. (Fig. 1) [1]. Our concept of a study and the development for the industrial production is shown in Fig. 2.

These chiral synthetic units have the following characteristics.

From the point of view of compounds:

1) The units are small compounds.
2) The units are liquids that do not form crystals at normal temperatures.
3) The units have a chlorine atom and/or an OH group; most of them are halohydrins.
4) Racemization occurs readily under some conditions in the production process.
5) The starting racemates are very cheap (C3: 0.5–2 US $/kg; C4: < 10 US $/kg).

From the point of view of the production method:

1) It belongs to the field of bio-methods using microorganism or enzymes having dehalogenation activity.

Asymmetric Catalysis on Industrial Scale: Challenges, Approaches and Solutions
Edited by H. U. Blaser, E. Schmidt
Copyright © 2004 Wiley-VCH Verlag GmbH & Co. KGaA, Weinheim
ISBN: 3-527-30631-5

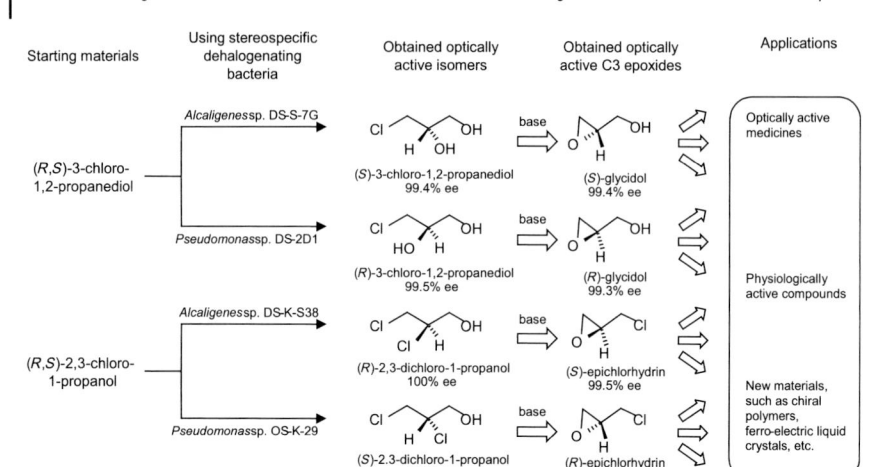

Fig. 1 Chiral C3 synthetic units; microbial resolution of (R)- and (S)-DCP, (R)- and (S)-CPD and the corresponding epoxides.

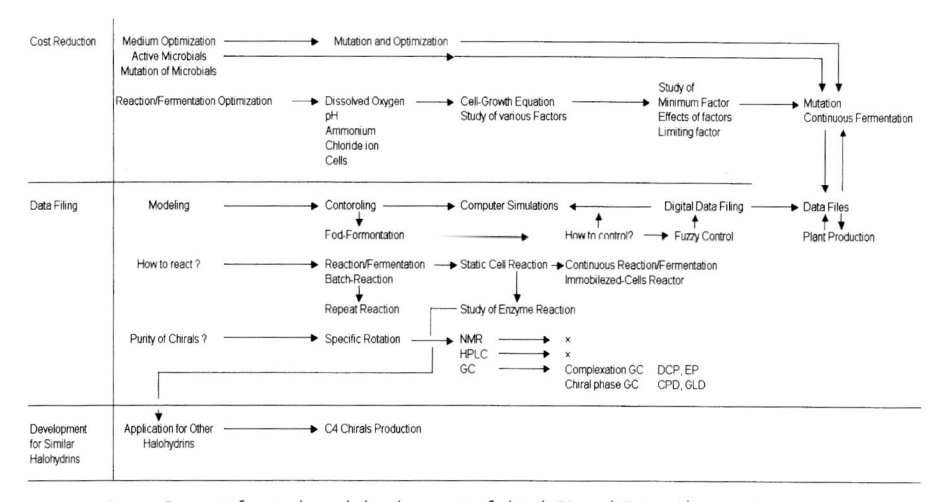

Fig. 2 Concept for study and development of chiral C3 and C4 synthetic units.

2) For the C3 chiral units production, one optical isomer is removed via assimilation by the microorganism.
3) The microorganisms, considered to be the catalysts, are alive and multiply.
4) The reaction is safe.

The most important features of the production can be summarized as follows. The actual operation of the production is easy to carry out, although crystallization can not be used for isolation or purification. Therefore, the products are isolated or purified by adsorption, extraction or distillation at low temperature. Surprisingly, the economics are not a problem with C3 chirals, even though half of the raw material is lost, because the racemates are produced in huge amounts and are very inexpensive. There is no need for further derivatization. The products have excellent optical purity, >99.5% ee in one step (if these chiral units were solid, the resolution by fractional crystallization would be the most economic way). The costs for production and the catalyst (i.e., the multiplying microorganism itself) are low.

Production of some C3 and C4 chirals such as EP, CPD, GLD, propanediol, 4-chloro-3-hydroxybutyrate is also being carried out or has been reported by other means such as chiral metal catalysts [2], reduction using microorganisms [3] or enzymatic resolution of suitable derivatives [4]. A comparison of these methods can be found in [1].

5.2
C3 Chiral Synthetic Units

In this section, we describe the problems for the industrial production of our C3 chiral units and our solutions to these problems. Chiral DCP and CPD are produced from racemic DCP and CPD by microbial resolution, chiral EP from chiral DCP, and chiral GLD from chiral CPD by the microbial resolution via synthetic conversion, respectively.

5.2.1
(R)- and (S)-2,3-dichloro-1-propanol (DCP)

5.2.1.1 (R)-2,3-Dichloro-1-propanol (DCP) Assimilating Bacterium for (S)-2,3-Dichloro-1-propanol (DCP) and (R)-epichlorohydrin (EP) Production [5, 6]

Chiral EP is a valuable C3 synthetic unit, Baldwin et al. reported the first synthesis from D-mannitol [7], however, the method was not used for production purposes. We investigated the microbial resolution of DCP, a precursor for the industrial production of EP. Racemic DCP was used as the single source of carbon in a synthetic medium containing 0.1% ammonium sulfate and ammonium nitrate, 0.1 M phosphate buffer, and traces of inorganic salt; the screening for the desired microorganisms, and enrichment cultivation with soils were carried out for 2 weeks. The desired bacterium was isolated from soil, identified and named *Pseudomonas* sp. OS-K-29. The chiral resolution of DCP is based on the stereoselective assimilation by the microorganism of (R)- or (S)-DCP in (R,S-)-DCP. Consequently, the medium is very simple and inexpensive. The resulting (S)-DCP was worked-up by using a charcoal column and distillation. The conversion to (R)-EP was achieved by alkaline epoxidation. The problems in the conversion are described under "Racemization" (see Section 5.5).

5.2.1.2 (S)-Epichlorohydrin (EP) Production and Isolation of a DCP Assimilating Bacterium [8]

Before the (S)-DCP assimilating bacterium had been isolated, (S)-EP was converted synthetically from (R)-EP. However, the synthesized (S)-EP has to be considered to be a derivative of (R)-EP, so there was a discrepancy between value and cost. Also, the catalyst for the reduction of the mesylated halohydrin by hydrogen was not easy to remove on an industrial scale. Therefore, a screening for an (S)-DCP assimilating bacterium was started. The existence of an (S)-DCP assimilating microorganism was considered likely because of the observation that the (R)-DCP assimilation bacterium changed in time into an (S)-DCP assimilating bacterium (see Section 5.2.2.1).

In the first screenings all microorganisms isolated were (R)-DCP and (R,S-)-DCP assimilating bacteria. However, we finally succeeded in isolating a true (S)-DCP assimilating bacterium, identified it, and named it *Alcaligenes* sp. DS-K-S38. The production of the actual (R)-DCP was easily carried out by the same method as for the (S)-DCP described above. The optical purity of the resulting (R)-DCP was 100% ee and the (S)-EP produced via alkaline treatment had an optical purity of 99.5% ee. This was the first time that (R)-DCP could be produced as well as (S)-DCP and (S)- and (R)-EP production from (R)- and (S)-DCP was established. Some years later, another (S)-DCP assimilating bacterium, which is a true reverse mutant of the (R)-DCP assimilating bacterium, was also isolated.

5.2.2
Large-Scale Trials Using Immobilized Cells [8]

The immobilized bacterium was used for producing (S)-DCP. Investigations as well as production trials were performed on a 100 L and 5000 L scale.

5.2.2.1 Production on a 100 L Scale

The (R)-DCP assimilating bacterium (the catalyst) was grown in 20 L fermentor (1/5 of the volume of the reactor) on a rich medium composed of 1% polypeptone, yeast extract, and glucose. The harvested and washed cells were collected by centrifugation, and the cells were immobilized in beads using 2% calcium alginate. The beads were packed in a stainless steel palanquin, and the resolution reaction was carried out under aeration (see Fig. 3). The immobilized bacterium performed well and it was possible to reuse it. So, repeated examination of a continuous semi-batch reaction was performed. In addition, the medium composition was simplified further. Though the bacterium continued growing and the excess growing cells were removed from the beads and the free cells were lost in the reaction medium, the reaction could be continued. The resolution of DCP was continued until 50% of the DCP had been consumed. The reaction solution containing (S)-DCP was obtained from the bottom of the reactor and the operation was repeated by adding fresh synthetic medium with (R,S)-DCP through the top of

Fig. 3 Cell-immobilized reactor and continuous batch-reaction.

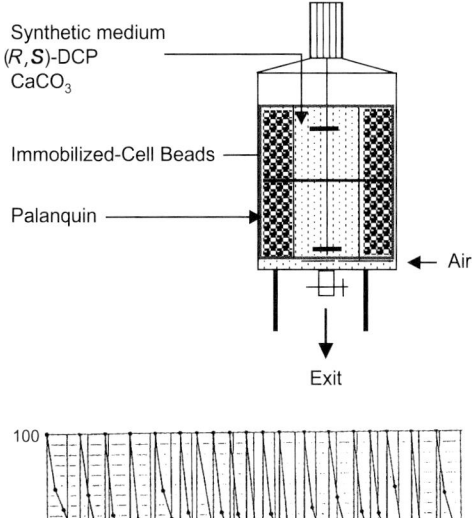

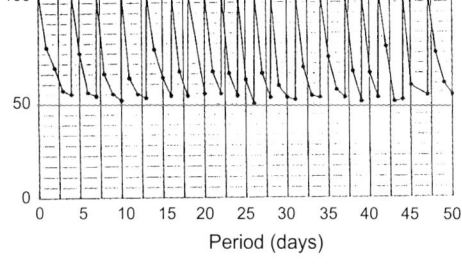

the reactor. The average operating time for one batch was 63–83 hours, and 19 batches could be completed in 50 days. The (S)-DCP production was found to be very stable. The first use of the cells required some extra time after the activation, but the assimilating activity was not decreased and the reaction continued in a stable manner.

However, we observed some contamination by other bacteria during production. Furthermore, we found that the (R)-DCP assimilating bacteria lost stereoselectivity and started assimilating the remaining (S)-DCP over longer reaction periods, and both isomers of DCP were finally lost. Obviously, the (R)-DCP assimilating bacteria gradually changed in nature. This change in stereoselectivity directed us to the opposite isomer production as described above.

5.2.2.2 Production on a 5000 L Scale

Production was also examined on a 5000 L scale in the same way. However, damage due to stirring was found, which was fundamentally the same as on the 100 L scale. Although there was basically no problem concerning the scale-up of the fixed bed reactor and the actual production, the reaction was contaminated by bacteria which in fact degraded the calcium alginate beads during the production [9]. Fixation by acrylamide could be achieved; however, scale-up was difficult, and an optical purity decline was sometimes observed. It was considered that acrylamide interferes with the bacterium. Therefore, the practical scale-up of the immobilized bed reactor was stopped. The batch reaction is easier to handle compared with a continuous reactor.

5.2.3
Production of (R)- and (S)-3-Chloro-1,2-propanediol (CPD) and (R)- and (S)-Glycidol (GLD) [10]

5.2.3.1 Screening of (R)- and (S)-3-Chloro-1,2-propanediol (CPD) Assimilating Bacteria [11, 12]

Chiral GLD is a C3 epoxide compound as well as a chiral EP equivalent, and it is a useful chiral synthetic unit. The old synthesis from D-mannitol by Sowden and Fischer [13], and recently the Sharpless asymmetric epoxidation of allyl alcohol, is well-known for the production of chiral GLD [14]. However an optical purity of 91% ee is low, even though it is an excellent industrial method [15]. Some microbial or enzymatic preparations had also been reported but production of high optical purity for both isomers was not known. We considered that microbial resolution of CPD, as well as of DCP, was possible, and had started a screening program. The first screening was carried out by enriched cultivation using various soils and a synthetic medium containing (R,S-)-CPD as the sole carbon source. The second screening was carried out using (R)- and (S)-CPD made from highly optical pure (R)- and (S)-EP. The isolated microorganisms were evaluated for their stereoselectivity and activity. This screening could be carried out efficiently because each CPD optical isomer was available. Many CPD assimilating microorganisms were isolated from soils polluted with EP or CPD. Typical soils were obtained by tank-yards of F.P. Finally, three (R)- and (S)-CPD stereoselective assimilating bacteria were isolated. All of the former belong to the genus of *Pseudomonas*, the latter to the genus of *Alcaligenes*. Using these strains and a synthetic medium containing (R,S-)-CPD, cultivation similar to the DCP procedure was carried out and (R)- and (S)-CPD with high % ee values were obtained [10]. Since there are fewer chlorine atoms in CPD than in DCP, the compounds are less toxic, and the production was consequently easier than that of DCP.

A competing production method was reported and carried out by the Kaneka group [3]. They used microorganisms which metabolize CPD via 3-chloro-2-keto-1-propanol using microorganisms, i.e., the starting raw material was (R,S)-CPD, and metabolism occurs by stereoselective oxidation and reduction, whereas our strains have assimilating activity for dehalogenation. Assimilating activity implies that only one cell, the first starter, can grow and multiply. Therefore the cost of catalyst is very low.

5.2.4
Scale-up of the Production of C3 Chiral Synthetic Units

5.2.4.1 Fed-Batch Fermentation and Control

We consider that the production of the C3 chiral unit is relatively easy, because a simple synthetic medium can be used. However, we had to reduce the reaction time in order to decrease the production cost, and establish a stable and reliable reaction for the industrial production. Generally, the fermentation process is like a black box making reliable production difficult, but on the other hand, failures should not occur in industrial production. This means that the fermentation reaction must be well known and a control system established. As for the C3 chiral units production, this is a reaction in which either optical isomer is converted by a microorganism. In as far as it multiplies, the microorganism itself acts like a catalyst. As for the status of the reaction, the cell multiplication, and the phase of the fermentation are not constant but vary over time. The cells as catalysts are formed and multiplied by consuming oxygen, DCP or CPD, ammonium sulfate and a trace of inorganic substance.

The multiplication pattern changes depending on the fermentation conditions and is not fixed for the transformation reaction. The purity of the racemates, the status of the seed culture or the cultivation itself are variable, and simple feeding is not sufficient. Therefore, problems may occur and the final optical purity at >99% cannot be achieved. Some successful and failed examples are shown in Fig. 4. Though the C3 chiral production is now run on a 30 kL scale, relevant basic data for this production such as KL_a, which defines the oxygen supply by aeration, the consumption of dissolved oxygen, respiration calories, and calculation of agitation power, and so on, were collected using a test reactor of 5 kL scale. Data

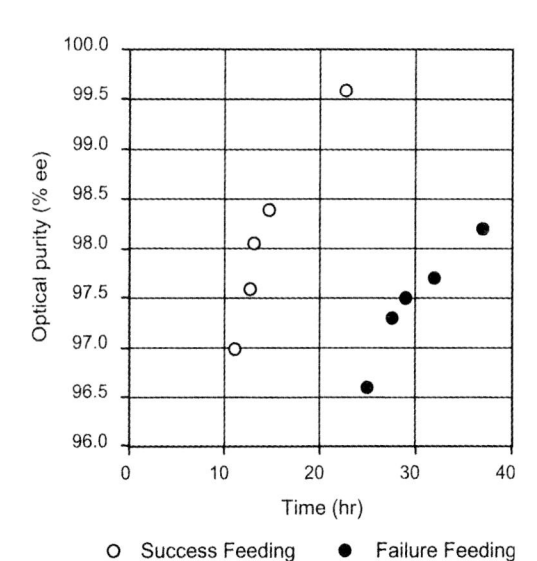

Fig. 4 Success and failure for fed-cultivation; time dependence of % ee of CPD.

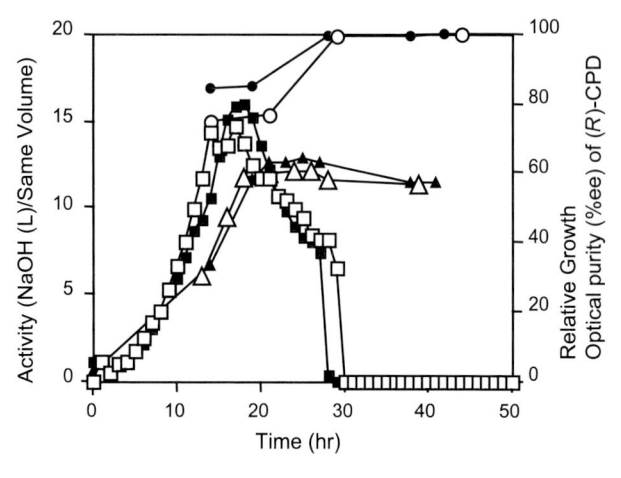

□ Activity (LAB) ■ Activity (INDUSTRIAL)
○ CPD %ee/LAB ● CPD %ee/INDUSTRIAL
△ Growth (OD at 660 nm)/LAB ▲ Growth (OD at 660 nm)/INDUSTRIAL

Fig. 5 Comparison of the time course of the microbial CPD resolution in the laboratory (2 L) and on an industrial scale (30 kL).

for cultivation control were collected using a 1–2 L reactor equipped with several monitoring sensors and a personal computer. We applied the resulting data and the system directly to the 30 kL reactor, and obtained very good reproduction. Fig. 5 shows the comparison of (R)-CPD production by fed-batch fermentation on a laboratory scale of 2 L and a plant scale of 30 kL.

5.2.4.2 Control Logic System

In all cases, hydrochloric acid is formed due to stereoselective dehalogenation of CPD or DCP by the bacteria. Therefore, we have to feed racemic raw materials using a pH-stat system; we detect the decline in pH by the hydrochloric acid formed and add $NaOH_{aq}$ to neutralize the reaction mixture, and then more racemate is added for further reaction (Fig. 6). When the reaction is finished, the % ee of the product should be >99.5–99.8%. The total added racemate should be as high as possible, and the time should be as short as possible. The production efficiency and the final optical purity are essential factors in the plant. The problem is detecting when the final feeding occurs and how much of the racemate is in the production (the fed-cultivation).

To paraphrase this problem, it originates in the non-linear form of the microorganism cultivation. Theoretically, it is necessary for the racemate quantity fed into the reactor to be slightly lower than the total resolution activity of the microorganism at a given time. However, the total resolution activity is not fixed but is changed by various factors. Therefore, we investigated a number of ways to detect the state of the cultivation.

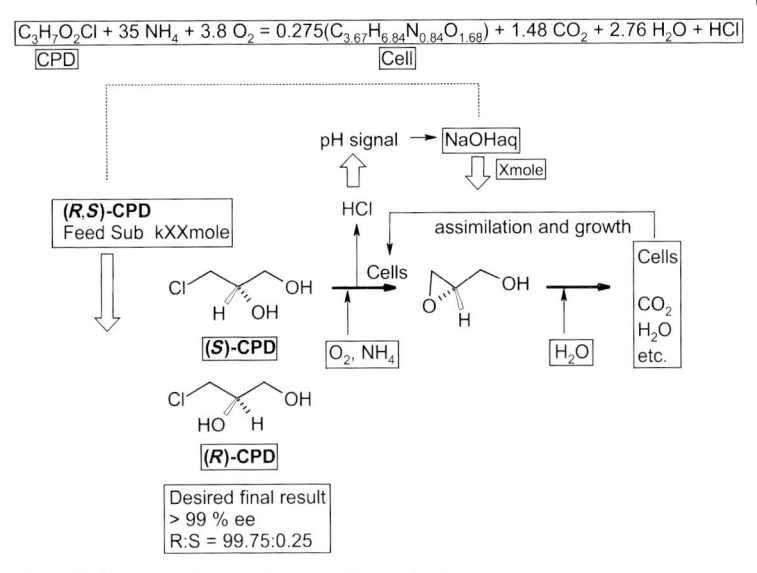

$$C_3H_7O_2Cl + 35\,NH_4 + 3.8\,O_2 = 0.275(C_{3.67}H_{6.84}N_{0.84}O_{1.68}) + 1.48\,CO_2 + 2.76\,H_2O + HCl$$
CPD — Cell

Fig. 6 Fed-batch cultivation in production of (R)-CPD

Finally, we understood that digitization of the response and fuzzy control were very effective. These days fuzzy control and learning control are becoming useful methods. This is a way of learning and scientific control using the knowledge of an expert. In the fermentation field, the methodology has been studied and many successful examples have been reported. Measured parameters were the pH, the consumption of DO (dissolved oxygen), the optical purity, the amount of fed racemate, added NaOH, bacteria and carbon dioxide formed at various reaction times. Monitoring was done digitally every 15 seconds, all data were analyzed and calculated, and we attempted to feed the reactor depending on the state of a control signal. Our analysis showed the following results:

1) CPD is dehalogenated and assimilated aerobically.
2) The oxygen consumption by assimilation is significant throughout the entire fermentation period.
3) The hydrogen chloride formation by assimilation is significant throughout the fermentation period.

Considering that CPD is the sole carbon source and that the fermentation is aerobic, these signals and data were the expected results. The data were collected in digital form which was very important for a reproducible reaction. Computer simulations and microorganism mutation were automated using this technology. A specific explanation is as follows. The amount of alkali necessary for the neutralization of the hydrochloric acid formed was obtained by weight or volume; this quantity was used to control the proportions necessary to obtain an optical purity of 60–70% ee in the reaction. A change in the dissolved oxygen and the amount

PROGRAM OF C3 CHIRAL PRODUCTION

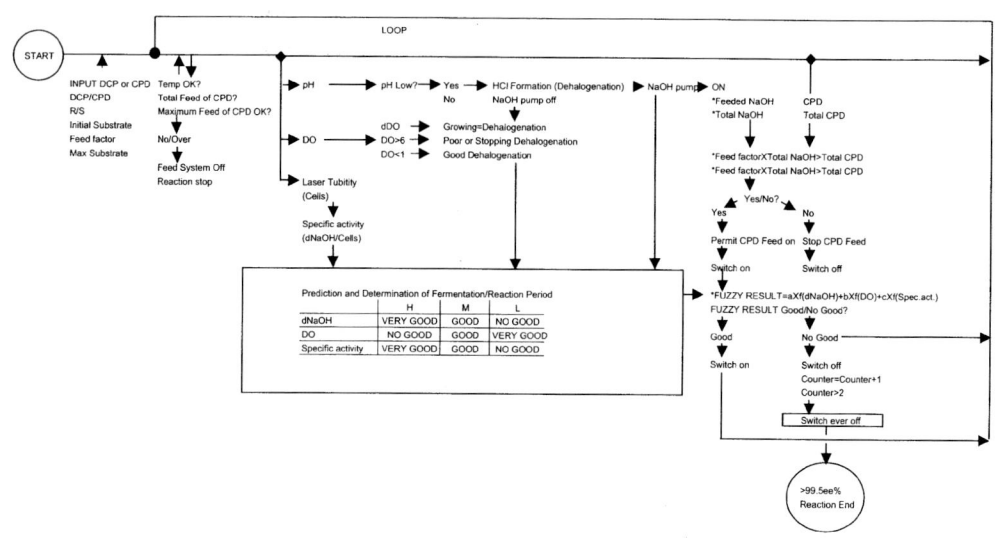

Fig. 7 Concept scheme for fed-cultivation for C3 chiral production.

of alkali were used to determine the status and to judge the period of the reaction. This concept is summarized and illustrated in Fig. 7.

The amount of alkali consumption and dissolved oxygen were divided into three steps, High, Medium and Low, and then the status was calculated as a membership function. The residual substrate concentration was anticipated from the amount of alkali. In this system, the data were evaluated for H, M, and L and obtained as a point rating. The amount of the racemate feed is proportional to the amount of alkali; the status of activity in the reaction was evaluated by readings of consumed alkali, of dissolved oxygen, laser turbidity, pH, etc., and a judgement was made at 15 min intervals as to whether the reaction was to be continued. A typical cultivation time course and the evaluation example for the method of control is shown in Tab. 1 and Figs. 8 and 9.

As shown in the figures, the state of the reaction can be judged well and reproducibly by the evaluation score FZ based on the dissolved oxygen (DO) and the specific activity (SA), even though a complicated reaction actually occurred in each batch. The time-dependent FZ is defined as $FZ(t) = \text{DO-factor} \times DO(t) + \text{SA-factor} \times SA(t)$. The evaluation score was calibrated by examining the data from some failures in the 2 L scale production, for example, the final % ee and the added amounts of racemate in detail. It was automated and showed no failure in a real plant. On a laboratory scale, a continuous feed reaction (over 1 year) was possible. This continuous reaction system was very effective for a continuous semi-batch system and for obtaining more active natural mutants of the production microorganism. It was very useful for isolating powerful natural mutants and for computer simulations in comparison with the plant production (Fig. 10).

Tab. 1 Success and failure for CPD fed-cultivation on 5 kL scale.

Test run No.	Relative fed CPD	% ee of finished reaction	Theoretical feed from % ee	Fuzzy program predicted feed
Success 1	100	99.9	100	99.9
Success 2	100	99.1	99.1	98.2
Success 3	100	99.9	100	99.9
Failure 1	100	92.9	93	92.6
Failure 2	100	91.2	91	90.8
Failure 3	100	77.4	77	71.7

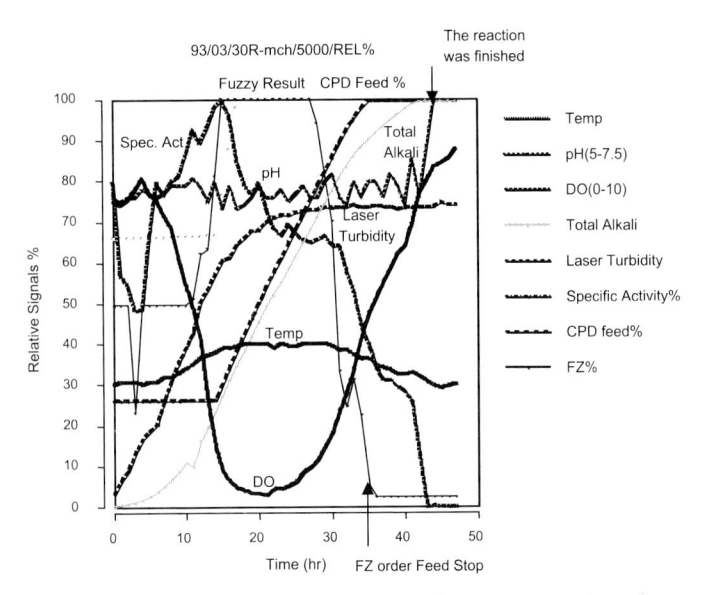

Fig. 8 Typical monitoring of a fed-cultivation of (R)-CPD on a 5 kL scale.

The same results were obtained from the 1–2 L scale reaction; therefore, different lots of the racemate, water, ratio of the feed, and a new active starter were evaluated. The prediction system was also applied to the C4 chiral production. Even though the C4 chiral production using a static cell is not the same as the C3 chiral production reaction, good reliability was actually obtained. Fermentation is a type of chemical reaction where scale-up is difficult, and reproduction is sometimes not achieved. We thought that this might be difficult before the control system was established, but we do not think so now, as the signals are digitized. The status is evaluated as one number, no matter whether the reaction is a fermentation or a synthesis.

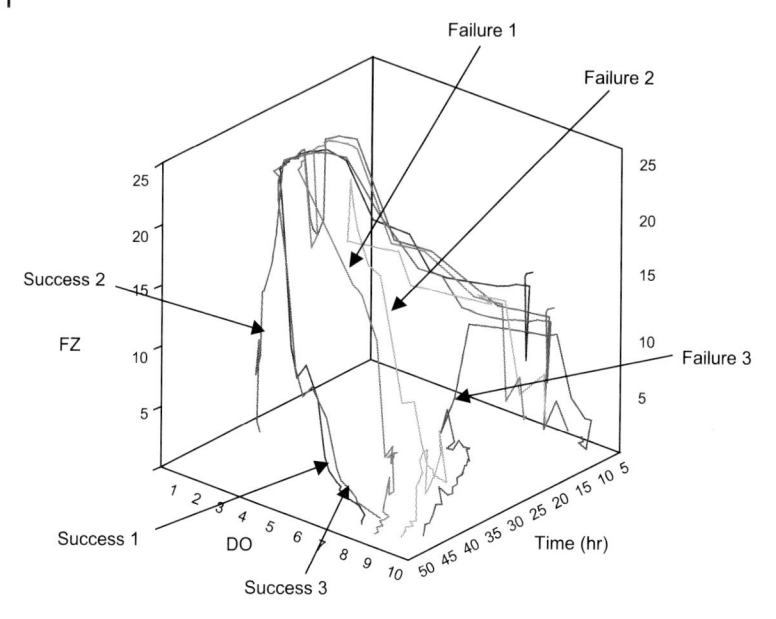

Fig. 9 Time course of fuzzy evaluation and DO in (R)-CPD fed-cultivation (5 kL).

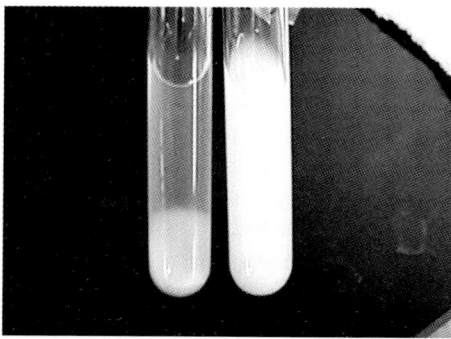

Fig. 10 Top: Automatic data collection with a PC on a laboratory scale of C3 chiral synthetic production. Bottom: a mutant of (S)-CPD producing bacteria. This system can also be used to obtain a powerful or convenient mutant. Left: the bacterium was precipitable and a powerful mutant. The cells of the catalysts are quickly removed.

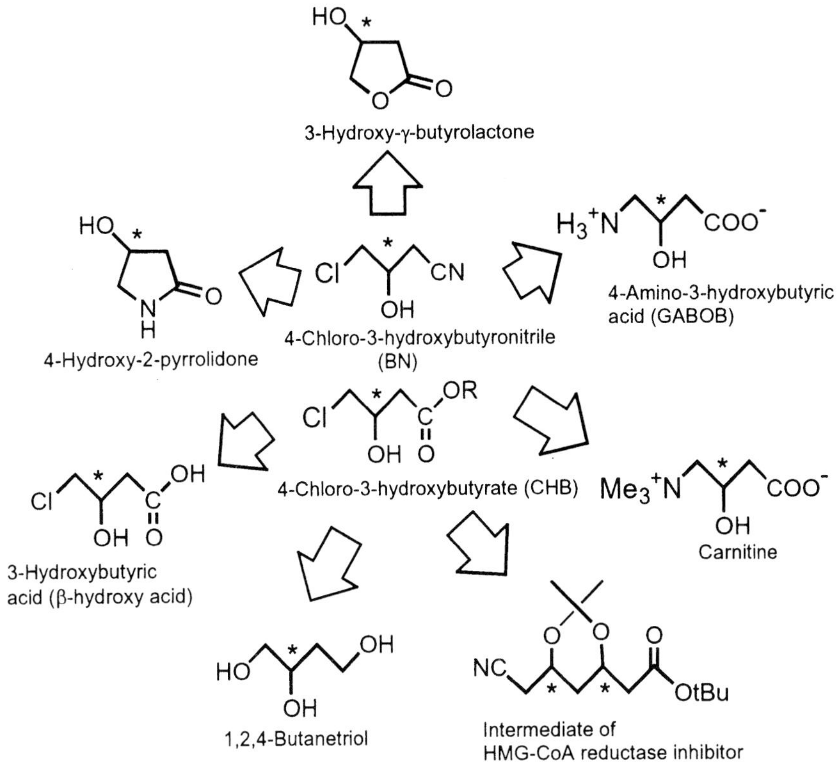

Fig. 11 Potential applications of C4 chiral chlorohydrins.

5.3
Production of C4 Chiral Synthetic Units

C4 chiral synthetic units are also important for the syntheses of pharmaceuticals and their intermediates. For example, optically active 4-chloro-3-hydroxybutyrate (CHB) and 4-chloro-3-hydroxybutyronitrile (BN) are key compounds as C4 chiral building blocks for the syntheses of L-carnitine [16], L-GABOB [17], β-hydroxybutyric acid, 3-hydroxy-γ-butyrolactone, and 4-hydroxy-2-pyrrolidone. Recently, CHB* has been reported as being used for synthesizing an intermediate for HMG-CoA reductase inhibitor for hyperlipidemia (Fig. 11) [18].

C3 chiral production was based on the microbial resolution using the asymmetric assimilation by various bacteria, belonging to the *Pseudomonas* sp. and *Alcaligenes* sp. In contrast, C4 chiral production is based on the stereoselectively enzymatic resolution using resting cells. Initially we thought that because the desired C4 chirals have the same 4-chloro-3-hydroxy moiety as the C3 halohydrins, it should be possible to use the same C3 chiral producing bacteria for producing C4 chirals as well. However, this turned out not to be possible, the main reason being that in the screening we found that the active strains stereoselectively converted one enantiomer and that this activity was stronger than the required assimilation process. In this section, the production of BN using resting cells with the secondary alcohol esterase, and the dual production of CHB and HL with the stereoselective carboxylic acid esterase are described.

5.3.1
Production of 4-Chloro-3-acetoxybutyronitrile (BNOAc) by Ester-Degrading Enzymes

Preparation of optically active BN and its derivatives using our haloalcohol dehalogenases and the halohydrin halide-hydrogen lyase of Nakamura et al. was already known [19]. Our results are summarized in Tab. 2. Thus, the (R)-DCP assimilating bacterium, *Pseudomonas* sp. OS-K-29, was found to have the stereoselective dechlorinating activity for the chloroalcohol and we succeeded in producing (S)-BN and (S)-CHB. Moreover, we searched for more active and powerful microorganisms to establish cost-effective methods of C4 chirals production. Tab. 2 shows the results of screening samples from different soils. Several types of bacteria capable of stereoselective ester conversion were discovered. *Pseudomonas* sp. DS-K-19, which left (S)-BNOAc, was unique in its properties and we would like to describe it here [20]. This ester-degrading enzyme was induced and enhanced by some ester compounds (Tab. 3). Especially, BNOAc was the best inducer for the ester-de-

Tab. 2 Stereoselectivity for several C4 chlorohydrins using isolated bacteria.

Strain and substrate	Residual ratio (%)	Optical purity (% ee)	Products
Pseudomonas sp. OS-K-29			
4-Chloro-3-hydroxybutyronitrile	40	94.5 (S)	(R)-3,4-Dihydroxybutyronitrile
Methyl 4-chloro-3-hydroxybutyrate	56	56.0 (S)	(R)-Methyl 3,4-dihydroxybutyrate
Ethyl 4-chloro-3-hydroxybutyrate	35	98.5 (S)	(R)-Ethyl 3,4-dihydroxybutyrate
Pseudomonas sp. DS-K-19			
4-Chloro-3-acetoxybutyronitrile	45	99.2 (S)	(R)-4-Chloro-3-hydroxybutyronitrile
Pseudomonas sp. DS-mk3			
4-Chloro-3-acetoxybutyronitrile	27	99.0 (R)	(S)-4-Chloro-3-hydroxybutyronitrile
Pseudomonas sp. DS-K-717			
4-Chloro-3-acetoxybutyronitrile	43	99.4 (R)	(S)-4-Chloro-3-hydroxybutyronitrile
Bacillus sp. DS-ID-819			
Methyl 4-chloro-3-hydroxybutyrate	43	91.9 (S)	(R)-3-Hydroxy-γ-butyrolactone
Ethyl 4-chloro-3-hydroxybutyrate	37	99.0 (S)	(R)-3-Hydroxy-γ-butyrolactone
Pseudomonas sp. DS-K-NR818			
Methyl 4-chloro-3-hydroxybutyrate	40	98.5 (R)	(S)-3-Hydroxy-γ-butyrolactone
Ethyl 4-chloro-3-hydroxybutyrate	42	98.4 (R)	(S)-3-Hydroxy-γ-butyrolactone
Pseudomonas sp. DS-S-75			
Methyl 4-chloro-3-hydroxybutyrate	48	99.5 (R)	(S)-3-Hydroxy-γ-butyrolactone
Ethyl 4-chloro-3-hydroxybutyrate	43	99.8 (R)	(S)-3-Hydroxy-γ-butyrolactone

Tab. 3 Effect of medium compositions on ester-hydrolyzing activity of Pseudomonas sp. DS-K-19.

Poly-peptone	Yeast extract	Glycerol	Glucose	Additives	Relative activity
1	1	1	–		100
2	–	1	–		55
–	2	1	–		67
1	–	–	–		36
–	1	–	–		18
1	1	–	1		87
1	1	1	–	BNOAc (0.2%(v/v))	2060
1	1	1	–	Butyl acetate (0.2%(v/v))	50
1	1	1	–	Monoacetin (0.2%(v/v))	69
1	1	1	–	Olive oil (1.0%(v/v))	0
1	1	1	–	Soybean oil (1.0% (v/v))	16
1	1	1	–	Tri-n-butylin (1.0%(v/v))	43
1	1	1	–	Triacetin (1.0%(v/v)))	0
1	1	1	–	Lecithin (1.0%(v/v))	0
1	1	1	–	Triolein (1.0%(v/v))	46
1	1	1	–	Sodium acetate (1.0% (v/v))	0
1	1	1	–	Sodium citrate (1.0% (v/v))	75

grading enzyme, but had harmful effects on the growth of this strain. As a consequence, a 0.2% concentration was chosen for enzyme induction and cell growth. Under these conditions, the activity was increased 20 times compared with the reaction without inducer. This enhanced activity by addition of BNOAc was also observed for the strain Pseudomonas sp. DS-K-717, exhibiting the opposite stereoselectivity.

We then carried out the optical resolution of (R,S-)-BNOAc (10%) on a 500 mL scale to produce (S)-BNOAc (Fig. 12), and (S)-BONAc was obtained with >98% ee in a yield of 32% in 24 hours. This resolution reaction was reproduced up to a 1 kL scale, so that this technology should be suitable for industrial-scale production. Moreover, in order to make this method more efficient, if (R)-BN with 45% ee was reacted to (R)-BNOAc after extraction of (S)-BNOAc, and treated by the reverse type of strain, Pseudomonas sp. DS-K-717, the productivity would be substantially increased.

5.3.2
Production of (R)-4-Chloro-3-hydroxy-butyrate (CHB) and (S)-3-Hydroxy-γ-butyrolactone (HL) by Enterobacter sp.

(S)-CHB is considered to be a very important compound, which can easily be converted into an intermediate for the HMG-CoA reductase inhibitor [18]. Furthermore, (R)-CHB is used for the synthesis of L-carnitine and (R)-4-hydroxy-2-pyrrolidone. Therefore, several methods have been developed. Specifically, ethyl (S)-CHB

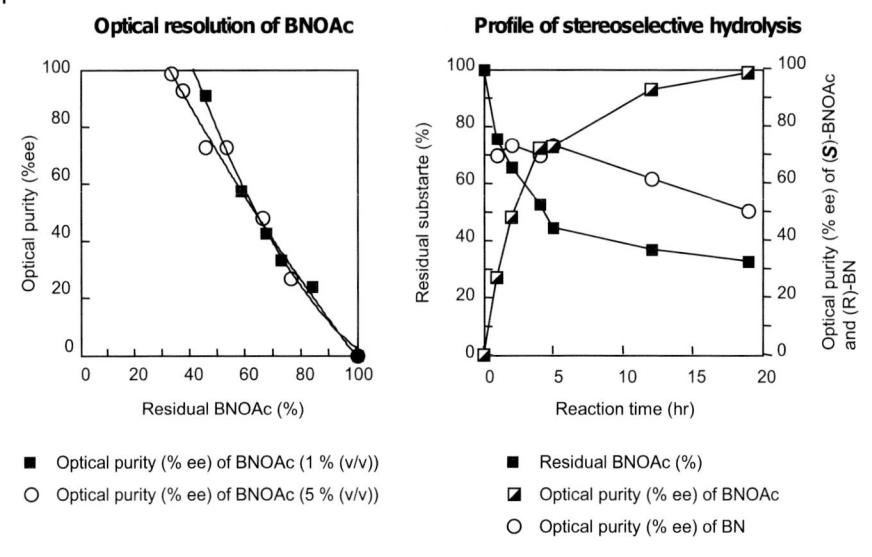

Fig. 12 Production of optically active BNOAc on a 500 mL scale.

was produced by means of reduction methods using chemical catalysts and enzymes such as BINAP and reductases from yeasts, respectively. We carried out screening experiments for bacteria capable of stereoselectively dehalogenating each enantiomer (Tab. 3).

After testing many soil samples, we isolated the desired strain, *Enterobacter* sp. DS-S-75, which converted (S)-CHB into (S)-HL with high stereoselectivity. (R)-CHB and (S)-HL were simultaneously produced with excellent optical purities in a high molar yield of 48% from the racemic methyl 4-chloro-3-hydroxybutyrate (CHBM) (Fig. 13). The resting cells of the strain DS–S–75 exhibited no activity for (R)-CHB, indicating that this reaction is carried out with complete stereoselectivity [21].

The cell-free extracts of the strain DS-S-75 showed high enantiospecificity for CHB but not for other halohydrins such as BN, DCP, CPD, and butylene chlorohydrin. Haloacid, chloro acetone and EP were also not degraded. Thus the (S)-CHB dechlorinating enzyme from the strain DS-S-75 is a novel type of dehalogenase, differing from haloalcohol dehalogenase [3, 6–8, 22–25] and halohydrin dehydro-dehalogenase [26]. On the other hand, we have recently found that this *Enterobacter* sp. also had stereoselective carboxylic acid ester degrading activity. The detailed mechanism of this and the applications are now under study.

5.3.2.1 Cultivation and Preparation of *Enterobacter* sp. Resting Cells with High Degradation Activity

The method of how cells with high activity are cultured and obtained is a most important and most critical subject. For preparing cells of *Enterobacter* sp. DS-S-

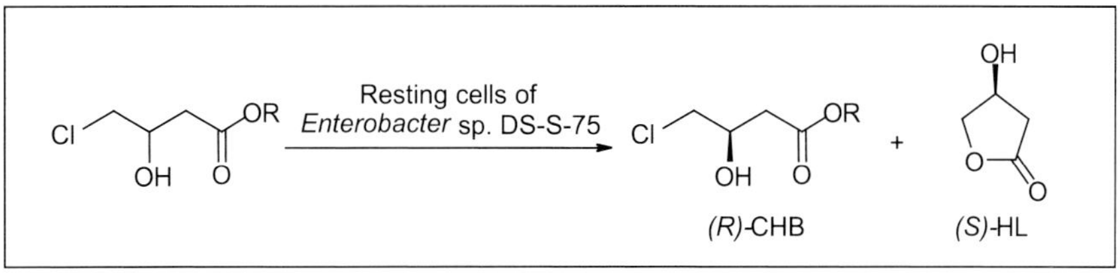

5.3 Production of C4 Chiral Synthetic Units

			CHB		HL	
substrate	specific activity (mU/mg of wet cells)	Residual ratio[a] (%)		Optical purity (% ee)	Conversion ratio[b] (mol %)	Optical purity (% ee)
Methyl CHB	19.5	48		99.5 (R)	48	95.9 (S)
Ethyl CHB	18	42.7		99.8 (R)	45.1	92.4 (S)
i-Propyl CHB	17.1	50		99.4 (R)	47.5	96.8 (S)
Propyl CHB	16.3	49.6		96.6 (R)	47	96.6 (S)
Butyl CHB	13.9	48.9		99.7 (R)	47.1	96.7 (S)

Fig. 13 Stereoselectivity for various esters of CHB with *Enterobacter* sp. DS-S-75. The resolution reaction was carried out in a 500 mL Erlenmeyer flask with 100 mL of 20 mM phosphate buffer (pH 7.2) containing 2% (v/v) of racemic substrate, 1% (w/v) CaCO$_3$ and 1 g of the resting cells (wet weight). The flask was incubated at 30 °C for 24 hours with shaking. a) Remaining CHB; b) conversion to HL in mol-%.

75, the following factors were investigated: 1) medium composition and concentration, 2) growth temperature, and 3) concentration of oxygen.

First of all, we generally used a nutrient medium composed of peptone, yeast extracts, and glycerol when culturing several of the bacteria we had isolated. The yeast extract, namely its type and lot, was found to be a major factor so we needed to select the most stable variety (Fig. 14).

Secondly, it often occurred that the optimal temperature for cell growth and the temperature for highest activity were different. Because the total activity of the resting cells can be shown to be a product of cell amounts and activity per unit volume, both high specific activity and high concentration of the cells were required when culturing the resting cells (Fig. 15).

Thirdly, dissolved oxygen had the strongest influence on the specific activity in our experience. Thus, we investigated the relationship between generation numbers and its specific activity under both excess and poor DO conditions (Fig. 16). Though this strain showed lower specific activity with excess dissolved oxygen, in contrast, a higher specific activity (more than twice) was shown with a low oxygen content. We also examined some relationships between cell growth, or specific activity, and dissolved oxygen. The results indicated that to obtain higher specific activity, it is necessary to culture the cells under low oxygen concentration in the log phase (highest growth rate). This could be reproduced on a scale from several liters to several tens of tons. In this manner, we established a cultivation method for producing resting cells with high specific activity.

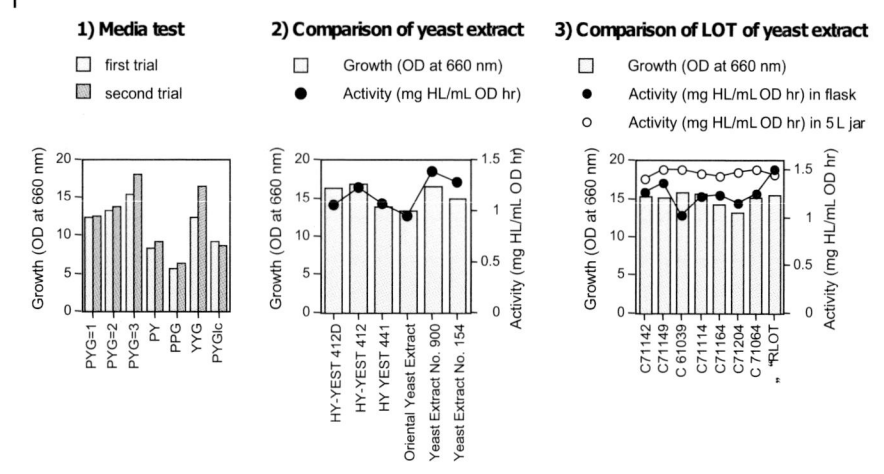

Fig. 14 Selection of cultivation medium. 1) Media composition (type of ingredient and concentration): P, polypeptone (1%); Y, yeast extract (1%); G, glycerol (1%); Glc, glucose (1%); PYG=1, 2, and 3 represents 1%, 2%, and 3% of each, respectively. Cultivation was carried out for 24 hours under aerobic conditions. 2) Various yeast extracts: Medium composition PYG =1. Activity is given as mg of HL formed per mL per OD per hour. 3) Several Lot Nos. of yeast extract No. 900: medium composition; PYG=1. Activity is given as mg of HL formed per mL per OD per hour. Cultivation was carried out in a 500 mL flask and a 5 L jar fermenter.

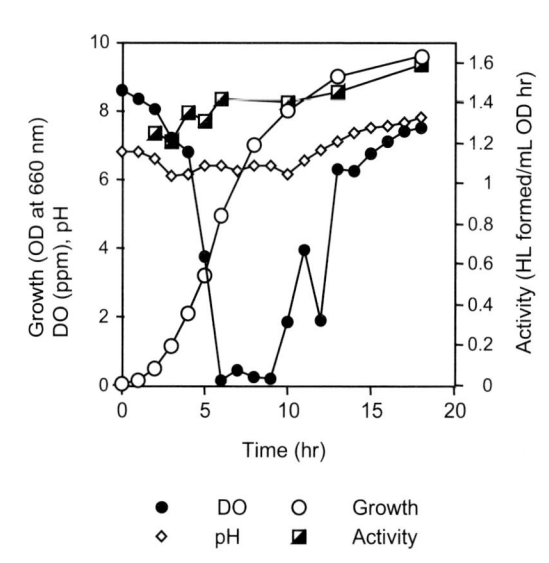

Fig. 15 Cultivation profile of Enterobacter sp. DS-S-75. Cultivation was carried out in 0.6% each of polypeptone, yeast extract, and glycerol in a 5 L jar fermenter under low aerobic conditions at 30 °C.

Fig. 16 Effect of oxygen concentration for specific activity in cultivation. Cultivation was carried out in 0.6% each of polypeptone, yeast extract, and glycerol in a 5 L jar fermenter under low (7% vvm) and high (15% vvm) aerobic conditions at 30 °C. Initial specific activity was taken as 100%.

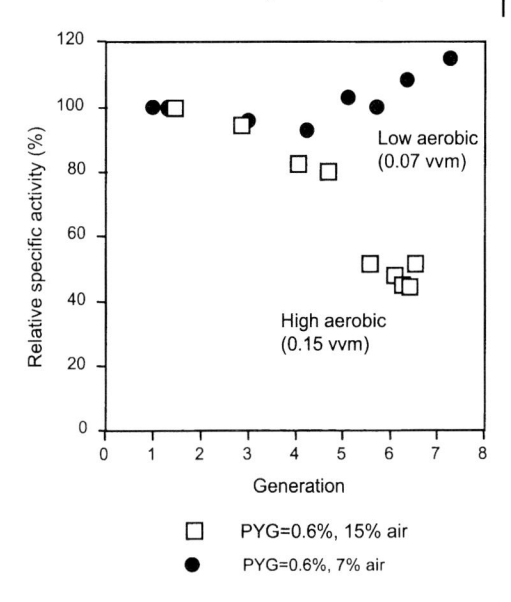

5.3.2.2 Ton-Scale Production of (R)-4-Chloro-3-hydroxy-butyrate (CHB) and (S)-3-Hydroxy-γ-butyrolactone (HL) by the Resting Cells of *Enterobacter* sp.

In order to perform a scale-up to the ton scale, optimal conditions for the activity were investigated with respect to the influence of pH, temperature, and substrate concentration. Optimal pH and temperature were established to be 6.9 and 30 °C, respectively. A concentration above 8% CHB had a negative effect on the activity, and methyl CHB (CHBM) was preferred to ethyl CHB.

The production of (R)-CHBM was carried out in a 5 L jar fermentor with 2.5 L of a reaction solution of 20 mM phosphate buffer (pH 6.7) containing 271 g [8% (w/v)] of CHBM and 90 g (wet weight) of the resting cells prepared in a nutrient medium containing 1% (w/v) each of polypeptone, yeast extract, and glycerol (pH 7.2). The pH was controlled at pH 6.7 with 25% (w/w) aqueous NaOH during the resolution reaction. Fig. 17 shows a profile of the resolution reaction of racemic CHBM with the resting cells of the strain DS-S-75.

During the initial 10 hours of the resolution reaction, degradation of CHBM proceeded strongly, then gradually ceased. As the degradation of CHBM proceeded, the formation of (S)-HL, methanol and chloride ion was detected. The resolution was completely finished in 40 hours. The yield of (R)-CHBM and its optical purity were determined to be 48% and 99.1% ee, respectively. Also, (S)-HL was obtained simultaneously in a conversion yield of 48% with 90.1% ee. When $CaCO_3$ was used as a neutralizing agent instead of aqueous NaOH, (S)-HL with an optical purity of over 95% ee was obtained, suggesting that a scaled-up production of (R)-CHBM and (S)-HL in a one-pot manner is theoretically possible. During this purification, the cells were removed by centrifugation and the resulting supernatant was subjected to an active carbon column treatment to recover the (R)-CHBM and (S)-HL. The elu-

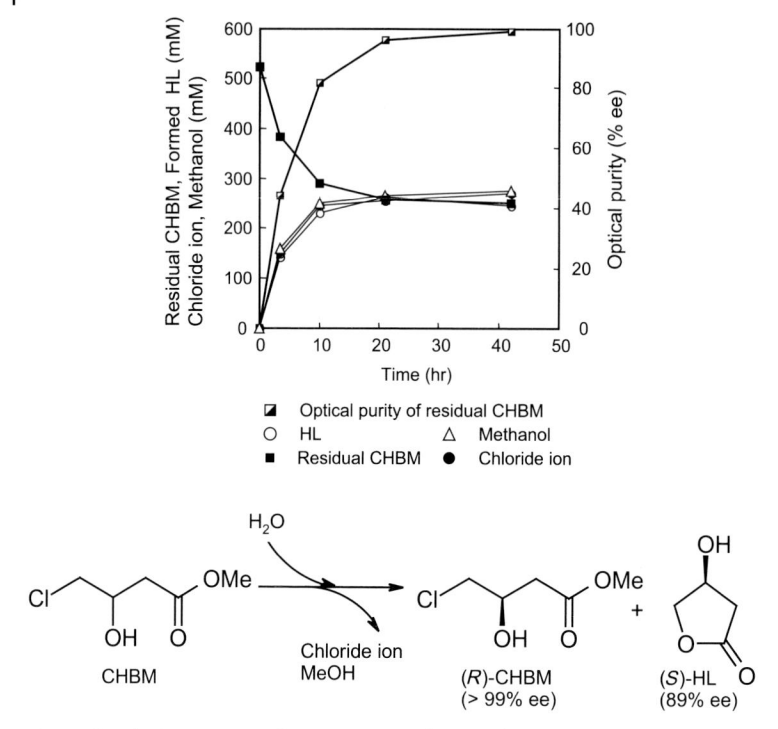

Fig. 17 Resolution reaction of (R,S)-CHBM with *Enterobacter* sp. DS-S-75 under pH control with 25% NaOH aq.

tion was done with acetone, and the resulting (R)-CHBM and (S)-HL were purified by distillation *in vacuo* (67 °C and 110 °C/0.3 mmHg, respectively). 96.1 g of (R)-CHBM and 52.3 g (S)-HL were obtained without loss of ee. The respective specific rotations of the (R)-CHBM and (S)-HL obtained were determined to be $[\alpha]_D^{20}$ +16.1 (c 1.21, CH_3OH) and $[\alpha]_D^{25}$ −68.1 (c 1.2, CH_3CH_2OH).

5.3.2.3 Effect of Base on the Stereoselectivity of *Enterobacter* sp.

As described above, (S)-HL was obtained in high optical purity (95% ee) using $CaCO_3$ as a base, whereas when NaOH solution (25%) was used, lower optical purity was observed, suggesting that stereoselectivity is dependent on the type of base. So, several bases (alkalies) were investigated (Tab. 4). As a result, weak bases such as sodium carbonate and sodium bicarbonate gave higher optical purity of (S)-HL than a strong base such as sodium hydroxide. An ammonium solution (14%) gave the best result not only for optical purity but also for degradation activity. The rate was determined to be twice that of other weak bases. Hence, we decided to use ammonium solution as a base for subsequent work.

Tab. 4 Effect of neutralizing reagent on the CHBM resolution.

Base	(R)-CHBM (% ee)	(S)-HL (% ee)	E value
NaOH 25% (w/w)	87.6	89	49.4
K_2CO_3 10% (w/w)	98.3	95.5	207.1
$KHCO_3$ 20%	98	95.7	209.9
Na_2CO_3 10% (w/w)	98.6	95.5	215.3
$NaHCO_3$ 10%	95.6	95.6	171.5
$(NH_4)_2CO_3$ 10% (w/w)	98.1	95.6	207.2
$CaCO_3$ 5% (w/w)	98.1	95.1	185.4
NH_4OH 7% (w/w)	99.9	95.1	300

E value is measure of stereoselectivity as defined by C.J. Sih (see K. Faber, Biotransformations in Organic Chemistry, 1997, Springer, Berlin).

$$E \text{ value} = \frac{\ln \frac{\text{ee of product}(1-\text{ee of substrate})}{\text{ee of product}+\text{ee of substrate}}}{\ln \frac{\text{ee of product}(1+\text{ee of substrate})}{\text{ee of product}+\text{ee of substrate}}}$$

5.3.2.4 Intelligent Production of (R)-Methyl 4-chloro-3-hydroxybutyrate (CHBM) and (S)-3-Hydroxy-γ-butyrolactone (HL) with a Fed-Batch System of Substrate

In order to produce (R)-CHBM and (S)-HL as C4 chiral building blocks more efficiently on a technical scale, we adapted the fed-batch system with the pH-dependent control used in the production of C3 chirals described in Section 2. Taking advantage of the released chloride ion when (S)-CHBM as a substrate was degraded and dechlorinated, our original fed-batch system was constructed using the computer system described above. Additional substrate according to the amount of consumed ammonium solution. Indeed, we succeeded in an industrial-scale production of (R)-CHBM (>99% ee) and (S)-HL (95% ee). Fig. 18 shows the profile of the resolution reaction of 12% racemate between the 5 L and 15 kL scales. The microbial resolution of racemic CHBM using the resting cells of the strain DS-S-75 made the economic production of highly optically pure (R)-CHBM and (S)-HL in a one-pot reaction possible. Racemic CHBM as a substrate can be easily obtained from EP via BN using petroleum chemicals. More recently, we have succeeded in purification and characterization of this enzyme. It was found to have the catalytic activity of a stereoselective carboxylic acid esterase, and more detailed studies of enzymes for industrial production are now in progress.

5.4 Purification of Chiral Units

Since our chiral products are in an aqueous solution, the purification involves solvent extraction, evaporation of water, and adsorption on charcoal (Fig. 19). Most of our chirals are alcohols, and an azeotropic mixture is formed during distillation.

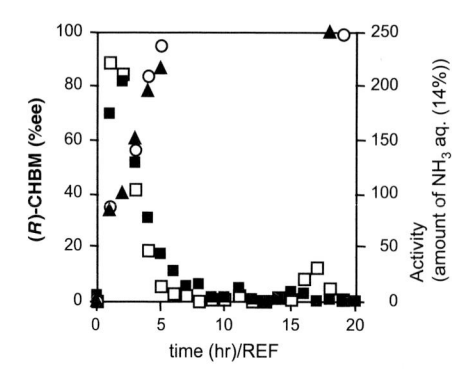

Fig. 18 Comparison of activities between 5 kL and 15 kL scales for the CHBM resolution.

▲ CHBM %ee, 15 kL scale   ○ CHBM %ee, 2 L scale
■ Amount of NH₃ aq. (kg/hr)   □ Amount of NH₃ aq. (kg/hr)

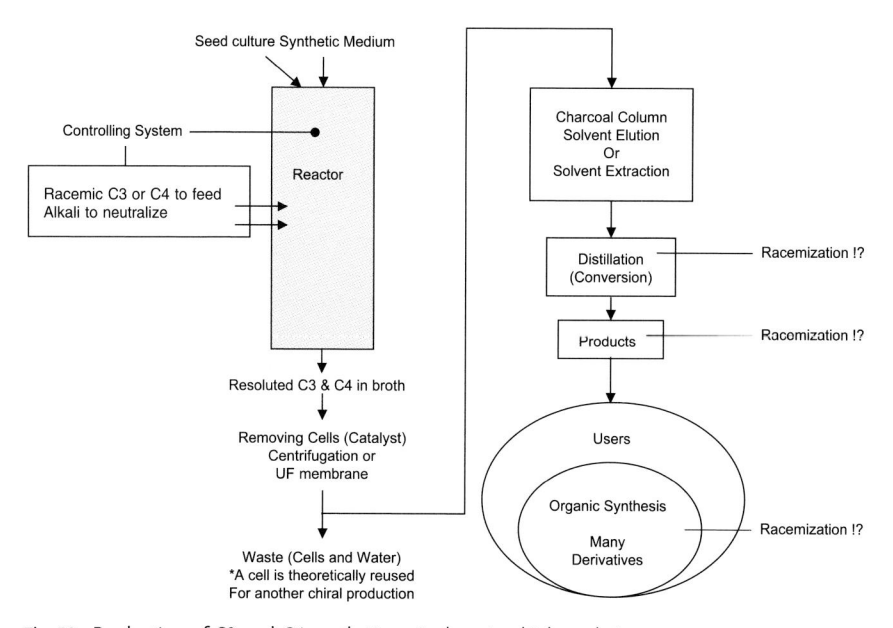

Fig. 19 Production of C3 and C4 synthetic units by microbial resolution.

Therefore the condensed product is treated by adsorption on charcoal followed by desorption with acetone. The acetone solution is distilled, and the chiral units are finally obtained. In some cases, such as C4 chirals, extraction with a solvent such as ethylacetate, butylacetate or butanol can be used. The method chosen depends on the solubility of the chiral compound in water or the solvent.

Fig. 20 Effect of reaction temperature for the racemization in epoxidation from DCP to EP.

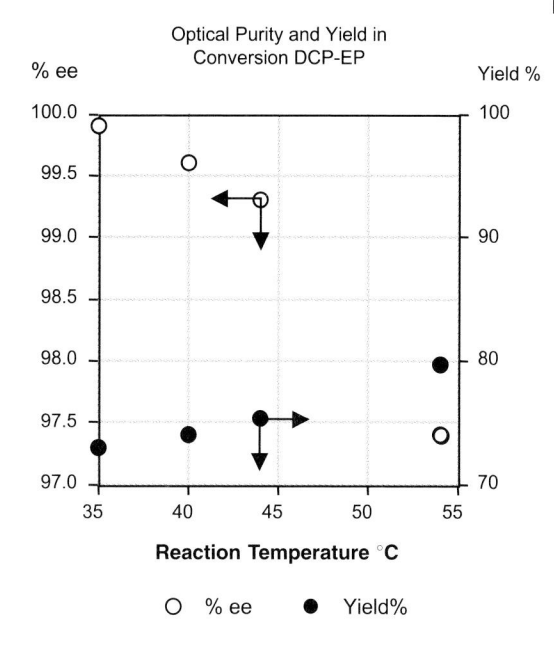

5.5
Racemization

In an industrial plant for EP, 1,3-DCP and DCP are converted into EP with $Ca(OH)_2$/water or CaO/water. The chiral DCP is essentially converted into EP by the same reaction. However, under certain conditions chiral EP racemizes very readily, especially in the presence of Cl^- via an exchange reaction. For example, racemization occurred during epoxidation of DCP to EP when $Ca(OH)_2$ was used as the alkali or when EP was not quickly extracted into the solvent [27], because Cl^- reacted with EP and symmetrical 1,3-DCP was formed, which yields EP more promptly than 2,3-DCP. Less racemization occurs at low temperature (Fig. 20). Furthermore, EP and GLD gradually react with water. Nucleophilic attack on EP usually occurred at the C1 position and partly at the C2 position. For example, pure optically active EP was opened by boiling water/acid, and the % ee values decreased to 88% ee. Epoxides are also unstable because polymerization gradually proceeds. As a consequence, EP and GLD can not be kept for a long time, but these epoxides are useful and reactive synthesis units. CPD is racemized by heating, but the stability is far better than that of EP and GLD, and is considered useful for practical applications.

As pointed out, heating causes racemization of the chiral units. Therefore, the distillation must be carried out at low temperature under a high vacuum, using a thin-film or continuous distillation needed. These limitations and suitable conditions to prevent racemization need to be clarified by optical isomer analysis in real time using high performance GC [28].

5.6
References

1 N. Kasai, T. Suzuki, Y. Furukawa, J. Mol. Catal. B: Enzymatic 1998, 4, 237–252.
2 M. Tokunaga, J.F. Larrow, G. Kakiuchi, N. Jacobsen, Science 1997, 277 936–939.
3 J. Hasegawa, M, Ogura, S. Tsuda, S. Maemoto, H. Kutsuki, T. Ohashi, Agric. Biol. Chem. 1990, 54, 1819–1827.
4 S. Hamaguchi, T. Ohashi, K, Watanabe, Agric. Biol. Chem. 1986, 50, 375–380.
5 N. Kasai, K. Tsujimura, K. Unoura, T. Suzuki, Agric. Biol. Chem. 1990, 54, 3185–3190.
6 N. Kasai, K. Tsujimura, K. Unoura, T. Suzuki, J. Indust. Microbiol. 1992, 9, 97–100.
7 J.J. Baldwin, A.W. Raab, K. Mensler, B.H. Arso, D.E. McClure, J. Org. Chem. 1978, 42, 4876–4878.
8 N. Kasai, K. Tsujimura, K. Unoura, T. Suzuki, J. Indust. Microbiol. 1992, 10, 37–43.
9 S. Kinoshita, Y. Kumoi, A. Ohsima, T. Yosida, N. Kasai, J. Ferment. Bioeng. 1991, 72, 74–78.
10 T. Suzuki, N. Kasai, Bioorg. Med. Chem. Lett. 1991, 1, 343–346.
11 T. Suzuki, N. Kasai, R. Yamamoto, N. Minamiura, J. Ferment. Bioeng. 1992, 73, 443–448.
12 T. Suzuki, N. Kasai, N. Minamiura, Appl. Microbiol. Biotechnol. 1993, 40, 273–278.
13 J.C. Sowden, H.O. Fischer, J. Am. Chem. Soc. 1942, 64, 1291–1293.
14 J.M. Klunder, S.Y. Ko, K.B. Sharpless, J. Org. Chem. 1986, 51, 3710–3712.
15 W. Dougerty, F. Liotta, D. Mondimore, W. Shum, Tetrahedron Lett. 1990, 31, 4389–4390.
16 B. Zhou, A.S. Gopalan, F. van Middlesworth, W.-R. Shieh, C. Sih, J. Am. Chem. Soc. 1983, 105: 5925–5926.
17 B.E. Rossiter, K.B. Sharpless, J. Org. Chem. 1984, 49, 3707–3711.
18 M. Wolberg, W. Hummel, C. Wandrey, M. Muller, Angew. Chem., Int. Ed. Engl. 2000, 39, 4306–4308.
19 T. Nakamura, T. Nagasawa, F. Yu, I. Watanabe, H. Yamada, J. Bacteriol. 1992, 174, 7613–7619.
20 H. Idogaki, N. Kasai, M. Takeuchi, M. Hatada, T. Suzuki, Tetrahedron Asymmetry 2001, 12, 369–373.
21 T. Suzuki, H. Idogaki, N. Kasai, Enzyme Microb. Technol. 1998, 24, 13–20.
22 T. Nakamura, F. Yu, W. Mizunashi, I. Watanabe, Agric. Biol. Chem. 1991, 55, 1931–1933.
23 A.J. van den Wijngaard, D.B. Janssen, B. Witholt, J. Gen. Microbiol. 1989, 135, 2199–2208.
24 A.J. van den Wijngaard, P.T.W. Reuvekamp, D.B. Janssen, J. Bacteriol. 1991, 173, 124–129.
25 H.M.S. Assis, P.J. Sallis, A.T. Bull, D.J. Hardman, Enzyme Microb. Technol. 1998, 22, 568–574.
26 T. Suzuki, N. Kasai, R. Yamamoto, N. Minamiura, Appl. Microbiol. Biotechnol. 1994, 42, 270–279.
27 S. Iriuchijima, A. Keiyu, N. Kojima, Agric. Biol. Chem. 1982, 46, 1593–1597.
28 V. Schurig, J. Chromatogr. 1988, 441, 135–153.

III
Adaptation of Existing Catalysts for Important Building Blocks

1
Synthesis of Unnatural Amino Acids [1]

DAVID J. AGER and SCOTT A. LANEMAN

Abstract

Asymmetric hydrogenation with Knowles' catalyst, [Rh-DIPAMP-COD]BF$_4$ has been used on a large scale to prepare L-dopa. The general method has been taken and applied to a wide variety of unnatural amino acids that are required by the pharmaceutical industry as starting materials for complex drug candidates. The scope and limitations of the catalytic method are well understood which means that the approach can be applied with confidence to a wide variety of substrates.

1.1
Introduction

Our work with Knowles' catalyst (**1**), Rh-DIPAMP, started with the need to prepare unnatural[2] amino acids for the fine chemical and pharmaceutical business. We discussed the advantages and the disadvantages of the various approaches available to us at that time [1–3]. To encapsulate this discussion, the need for general approaches that can be relied upon to make small amounts of materials led us to use resolution or chiral auxiliaries. In some cases, a chiral pool synthesis could be used; that is a natural amino acid could be used as the precursor for the preparation of an unnatural amino acid. For kg amounts, rapid turn-around was still required and asymmetric hydrogenation was usually the method of choice. Some of the problems associated with the use of Knowles' Rh-DIPAMP catalysts with specific systems are mentioned below. On the whole, the methodology could be applied to the preparation of a wide variety of unnatural amino acids. At a large scale, hydrogenation may not be the method of choice as exemplified by the production of L-phenylalanine, which was made by fermentation within Monsanto. For our work on asymmetric hydrogenations to amino acids, we used Knowles' catalyst as it was readily available to us and the scope and limitations of substrate requirements were reasonably well understood.

---

[1] This work was performed at NSC Technologies, 601 East Kensington Road, Mount Prospect, IL 60056, USA.

[2] Unnatural is used as a synonym for non-proteinogenic amino acid. Under this usage all D-amino acids are unnatural.

Asymmetric Catalysis on Industrial Scale: Challenges, Approaches and Solutions
Edited by H. U. Blaser, E. Schmidt
Copyright © 2004 Wiley-VCH Verlag GmbH & Co. KGaA, Weinheim
ISBN: 3-527-30631-5

Fig. 1 Reduction step in the synthesis of L-dopa.

Ar = 3-AcO-4-MeOC$_6$H$_3$

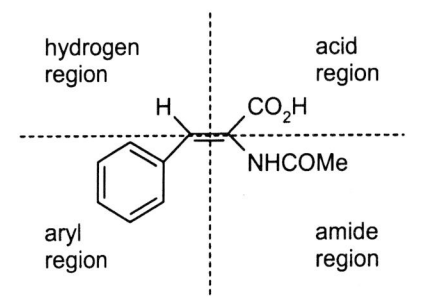

Reaction: Ar-C(=...)(NHAc)(CO$_2$H) + H$_2$ →[Rh(cod)(R,R-dipamp)]$^+$BF$_4^-$ (**1**)→ Ar-CH(NHAc)(CO$_2$H)

Fig. 2 The quadrants of an enamide.

(hydrogen region / acid region / aryl region / amide region; substituents H, CO$_2$H, aryl, NHCOMe)

The literature has many examples of ligands that have been developed to effect asymmetric hydrogenations of a variety of functional groups [4–8]. The reductions of enamides to provide α-amino acids derivatives has been of importance since the pioneering work of Knowles and his colleagues at Monsanto [9, 10]. This approach has been used by Monsanto, Searle and now Egis for the production of L-dopa (Fig. 1) [4]. The chemistry related to L-dopa is discussed in more detail in Chapter I 1, by W.S. Knowles.

1.2
Scope of Enamide Substrates

Although it has some limitations, the Rh-DIPAMP system developed by Knowles has been well studied and this allows for a fair degree of certainty that it can be used with "untried" substrates. For this reason, it has been used to prepare a range of unnatural amino acid derivatives at scale [7, 11]. A model has been reported that allows others to predict whether Knowles' catalyst can be used in a productive manner with their desired enamide. The prediction of the model is based on dividing the enamide substrate into four quadrants (Fig. 2) [7].

If we start at the top left-hand corner and work round anti-clockwise, the scope and limitations of the system can be explained. This quadrant has to be hydrogen. Any substitution results in a drop in the reaction rates and a significant lowering of enantioselectivity. Thus, β-branched α-amino acid derivatives cannot be accessed with Knowles' catalyst. This limitation has been overcome by the use of other catalyst systems such as Burk's DuPHOS [12]. The need for hydrogen at this position also means that a stereoselective synthesis of the enamide has to be undertaken (*vide infra*).

The lower left-hand quadrant can be almost any aryl or alkyl group. For phenyl substituents, electron donating and withdrawing groups are tolerated at any position on the ring [7, 11]. With *ortho*-substituents, however, steric effects can play a role and the hydrogenation may become sluggish, especially if two *ortho*-substituents are present [7, 11, 13]. Heterocycles are tolerated within this quadrant of the enamide [7, 14]. In addition to nitrogen and oxygen examples, sulfur examples, such as thiophene can also be used. In the case of nitrogen heterocycles, it may be necessary to prevent ligation from the nitrogen lone pair into the metal center; this can be achieved by simple protonation [7, 15]. Alkyl groups can also be placed on the enamide rather than aryl, but this can make access to the enamide more challenging.

The lower right-hand quadrant is perhaps the most important, and understanding the requirements at this position has allowed expansion of the original work to other substrates (*vide infra*) [9]. An amide is the most common functionality at this position. The carbonyl oxygen group forms a complex with the rhodium [15]. The nitrogen is just a spacer group and can be replaced by any group of a similar size, such as an oxygen atom. The carbonyl group can also be part of a carbamate rather than an amide. This allows preparation of *N*-protected amino acids. Even a benzylcarbamate (Cbz or Z) can be used as Knowles' catalysis will not perform a hydrogenolysis of this group [13, 16].

The final quadrant, the top right-hand, has to contain an electron withdrawing group. This group activates the carbon–carbon unsaturation. A wide variety of groups are tolerated including a carboxylic acid, ester, amide, nitrile, ketone, and aldehydes [7, 9]. This shows that the catalyst system can be used to prepare a wide range of amino acid derivatives without recourse to a number of transformations on the product unnatural amino acid.

Many of the ligands that have been advocated for analogous reductions provide high enantioselectivity and some can provide good turnover numbers and frequencies [4, 5]. Knowles' catalyst often results in an ee of about 94–95% if the reaction mixture is monitored. However, crystallization of the *N*-acylamino acid product often results in enantioenrichment [10, 11]. In addition, hydrolysis of the amide to provide the amino acid itself also provides an opportunity for enantioenrichment (Fig. 3) [10].

The key step in the preparation of amino acids is the generation of the chiral center by asymmetric hydrogenation. As summarized in Tab. 1 [Rh-COD-(*R,R*)-DI-PAMP)]$^+$BF$_4^-$ (**1**) has demonstrated high catalyst activities for the majority of enamides, with the exception of the *p*-nitrophenylalanine analog (**2d**) [10]. Substrate to catalyst ratios (S/C) of 10 000 are achieved in the reduction of many enamides with **1** while the reduction of **2d** occurs with a ratio of 200. Reduction of the nitro group is not observed. High catalyst activities are observed in the asymmetric hydrogenation of phenylalanine derivatives that contain chloro (**2b**) and fluoro (**2c**) groups in the *para* position [10].

The enantioselectivities did not change upon decrease of the catalyst loading (S/C) from 5 000 to 20 000 in the reduction of the 2-naphthyl enamide (**2a**), but the

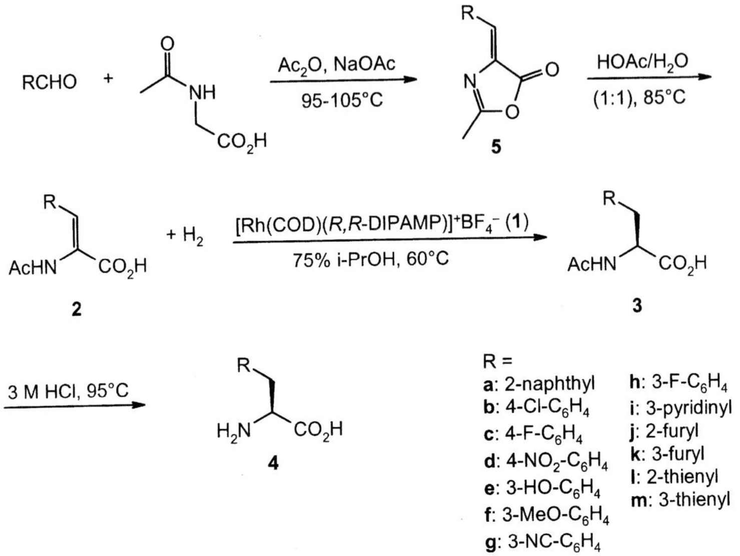
RCHO +
NH
CO2H
Ac2O, NaOAc
95-105°C
R
N
O
O
5
HOAc/H2O
(1:1), 85°C
R
AcHN
CO2H
2
+ H2
[Rh(COD)(R,R-DIPAMP)]+BF4− (1)
75% i-PrOH, 60°C
R
AcHN
CO2H
3
3 M HCl, 95°C
R
H2N
CO2H
4
R =
a: 2-naphthyl
b: 4-Cl-C6H4
c: 4-F-C6H4
d: 4-NO2-C6H4
e: 3-HO-C6H4
f: 3-MeO-C6H4
g: 3-NC-C6H4
h: 3-F-C6H4
i: 3-pyridinyl
j: 2-furyl
k: 3-furyl
l: 2-thienyl
m: 3-thienyl

Fig. 3 Overall sequence for azlactone synthesis, hydrogenation, and hydrolysis for the preparation of amino acids.

Tab. 1 Isolation yields and enantioselectivities on several intermediates and amino acids.

R=	*5* *Yield (%)*	*2* *Yield (%)*	*3* *ee (%)* [a]	*S/C*	*Product (4)*	
					Yield (%)	*ee (%)* [a]
a	72	90	92	5 000–20 000	89	94.7
b	58	95	–	10 000	89	94.6
c	67	94	91	10 000	80	95.4
d	94	96	–	200	90	96.0
e	69 [b]	43 [b]	– [b]	10 000	87	99.3
f	46	84	–	10 000	88	98.7
g	74	87	–	5 000	69	98.7
h	72	81	94	10 000	86	97.2
i	45 [c]	70 [d]	– [d]	1 000	33	77

a) Products are of the *S*-configuration. Enantioselectivities were determined by chiral HPLC.
b) Contains *O*-acetate.
c) Prepared by literature method [18].
d) Isolated as the HBF_4 salt.

catalyst performance is susceptible to trace amounts of oxygen, which can significantly decrease the catalyst activity. High catalytic turnovers are achieved in the asymmetric reduction of enamides that contain methoxy and acetoxy groups in the *meta* position. Hydrolysis of **3e** and **3f** affords L-*m*-tyrosine and L-*O*-methyl-*m*-tyrosine in 99.3% ee and 98.7% ee, respectively [10]. A slightly higher catalyst loading (S/C=5000) is required for complete conversion in the presence of a cyano functionality. L-3-Cyanophenylalanine (**4g**) can be prepared in 98.0% ee without reduction or hydrolysis of the cyano group.

Knowles' catalyst slowly catalyzes the reduction of 3-pyridinyl enamide due to the pyridinyl unit. The asymmetric reduction of the methyl ester analog with [Rh-COD-(R,R)-DIPAMP)]$^+$BF$_4^-$ has been reported to be 99% ee with a substrate to catalyst ratio of 28 [14]. A similar decrease in catalyst activity has been observed with other catalysts [17]. Protonation of the pyridinyl group with a non-coordinating strong acid, such as HBF$_4$, improves the S/C to 1000, but 50% esterification occurs during hydrogenation due to the alcoholic solvent. The enamide **2i** · HBF$_4$ was prepared by ring opening of **5i** with HBF$_4$ instead of acetic acid at room temperature. Hydrolysis of the resultant acid/ester hydrogenation mixture with HCl in water produces 3-pyridinylalanine in 77% ee.

The reduction of Z-protected enamide methyl esters of 2- and 3-furyl, and thienyl with Rh-DIPAMP has been reported to give enantioselectivities of from 85 to 91% at an S/C ratio of 4000 [16].

The properties of the N-acyl amino acid derivative and the amino acid itself can be important if crystallization is employed to ensure high optical purity. However, in the experience of the authors, it is seldom a problem. Only 3-pyridylalanine could not be purified to high optical purity by recrystallization. Use of enzymes to enhance enantiomeric excess as well as allowing the use of mild hydrolysis conditions has been advocated by others [12]. Amino acids are surprisingly stable to acid-induced racemization.

1.3
Preparation of Enamides

As noted above, a stereoselective synthesis of the enamide is important. The azlactone method (Fig. 3) results in the preferential formation of the Z-enamide when an aromatic aldehyde is employed. In addition, this isomer usually precipitates from the reaction mixture and this simplifies purification. When an alkyl aldehyde is used, the ratio of enamide isomers is often 1:1 or close to this. In addition, many of these alkyl examples are not crystalline and physical separations such as chromatography have to be employed. This is obviously a limitation of the methodology when compared with catalysts that employ the DuPHOS ligands, and related ligand families where both isomers can be reduced down to the same enantiomer of the desired amino acid [12].

An alternative approach to the enamides is to use a Wadsworth-Emmons variation of the Wittig reaction (Fig. 4) [13, 16, 19]. As the parameters that control this reaction are well understood, it is possible to control the stereochemical outcome of the reaction in favor of the desired enamide isomer.

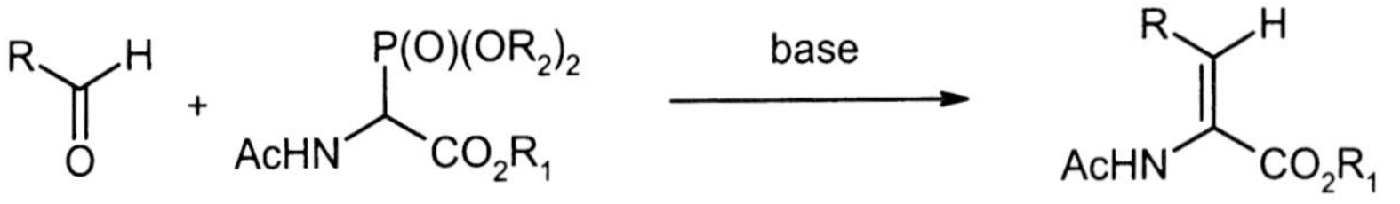

Fig. 4 Wittig-type approach to enamides.

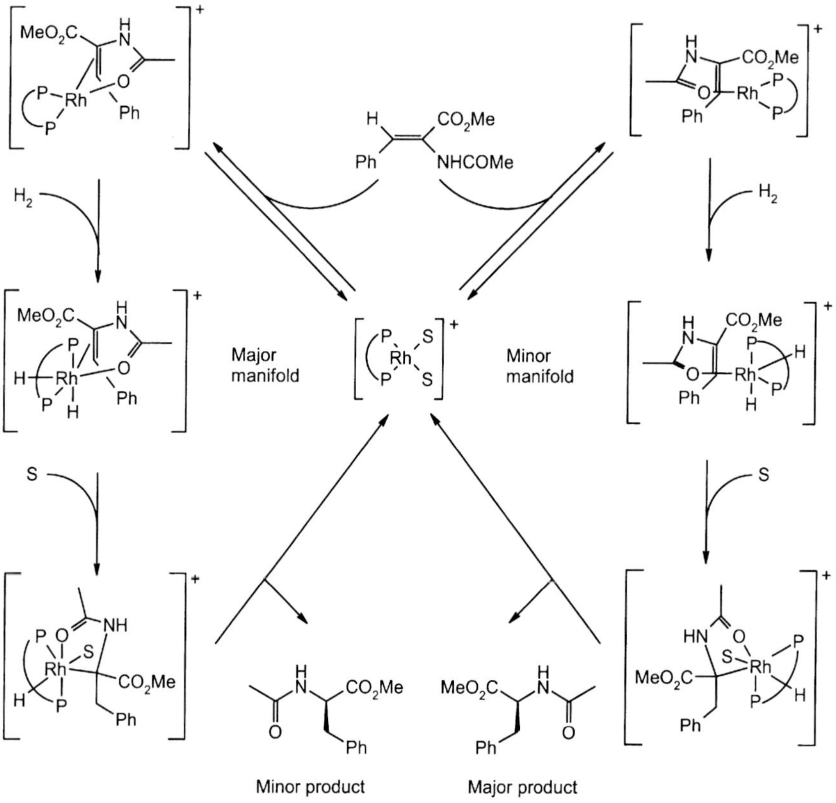

Fig. 5 Mechanism of enamide reductions with Knowles' catalyst.

1.4
Mechanism

The mechanism for the asymmetric hydrogenation of enamides by Knowles' catalyst is well understood due to the work of Halpern (Fig. 5) [20]. The intermediates were identified by spectroscopy. The surprising finding was that two catalytic cycles were possible. The one that contains the lower concentrations of intermediates gives rise to the major product isomer as the reaction rates are faster compared with the cycle that has more detectable intermediates.

1.5
Ligand Synthesis

The catalyst precursor, **1**, is air-stable which simplifies handling operations on a manufacturing scale. Despite these advantages, ligand synthesis is very difficult. After the initial preparation of menthylmethylphenylphosphinate (**6**), the $(R)_P$-isomer is separated by two fractional crystallizations (Fig. 6). The yield of (R,R)-DIPAMP (**8**) is 18% based on **7** [9]. Another disadvantage is that the (R,R)-DIPAMP can racemize at 57 °C, which can be problematic since the last step is performed at 70 °C. Once the ligand is coordinated to rhodium, however, racemization is no longer a problem.

Other routes to (R,R)-DIPAMP (**2**) have been reported [21–23]. At present, the most practical synthesis of DIPAMP involves the formation of a single diastereoisomer of **9** by the combination of $PhP(NEt_2)_2$ and (–)-ephedrine followed by formation of the borane adduct (Fig. 7) [21, 24].

The cost of the ligand and catalyst preparation can be critical in the decision to pursue technical transfer to the manufacturing stage. Fortunately, the high reactivity of $[Rh\text{-COD-}(R,R)\text{-DIPAMP})]^+BF_4^-$ (**13**) in most enamide reductions can offset the high price, low yield synthesis of DIPAMP.

D-Amino acids can be prepared in an analogous manner if (S,S)-DIPAMP is used as the ligand.

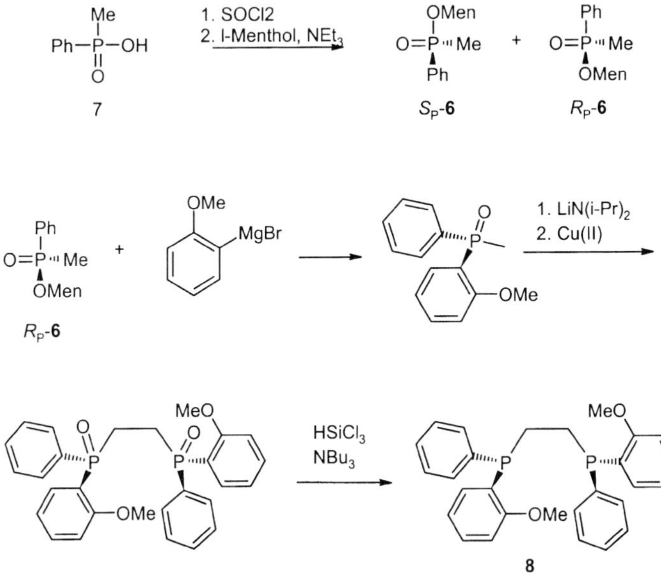

Fig. 6 Synthesis of DIPAMP.

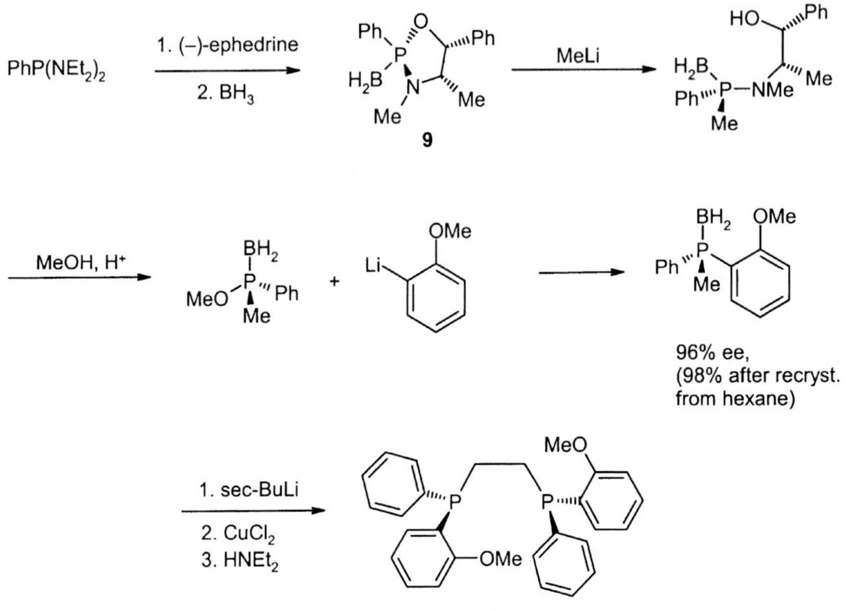

Fig. 7 Alternative synthesis of DIPAMP.

1.6
Other Substrates

Although Knowles' catalyst has been known for 30 years, there are still applications that are coming to light.

The phosphorus analogs of enamide esters, the dimethyl phosphonates can be reduced by Knowles' catalyst to provide the corresponding α-amido phosphonates (Fig. 8) [25].

The reduction of itaconic acid derivatives was found to be selective and was used to access compounds under investigation as HIV protease inhibitors (Fig. 9) [26–28].

Knowles had shown that α-keto esters could be reduced to α-hydroxy esters if the vinyl acetate was employed (Fig. 10) [9]. If the R-group is aryl, the enol ether must have the Z-geometry for reduction to occur [19]. When the substituent R is an alkyl group this requirement is relaxed. Unfortunately, turnover numbers are not high [29].

The enol acetates of α-keto acids can be accessed by a variation of the azlactone synthesis. They can also be reduced by Knowles' catalyst, but ee values are 87–88% and the substrate to catalysis ratio needed for reduction does not allow this to be an economic approach to α-hydroxy acids although the reaction times are reduced compared with the corresponding carboxylate esters [9, 29].

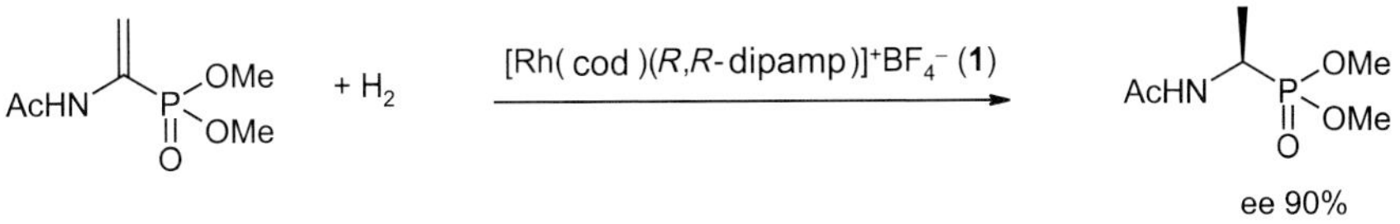

Fig. 8 Synthesis of α-amido phosphonates.

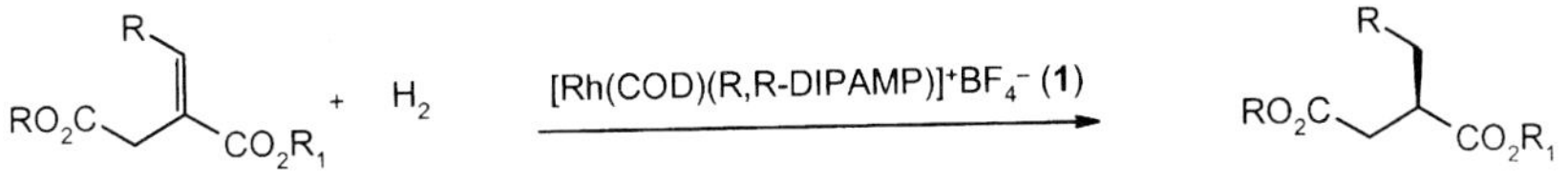

Fig. 9 Reductions of itaconic acid derivatives.

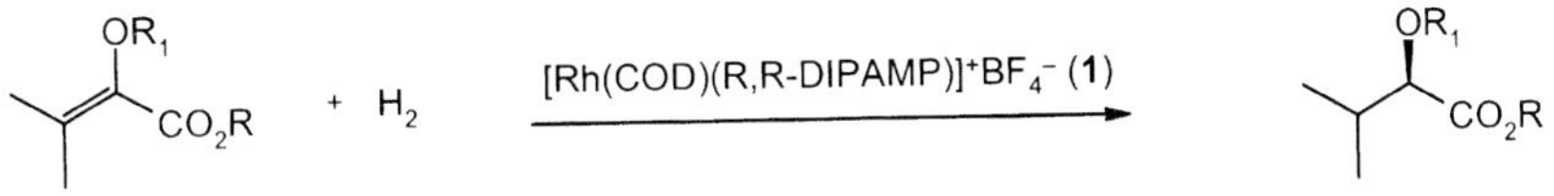

Fig. 10 Reductions of α-keto esters to α-hydroxy acid derivatives.

1.7 Summary

Knowles catalyst has been used to prepare a wide range of α-amino acids at scale. L-Dopa is still produced at scale using the original process developed by Monsanto. Other amino acids are produced at a smaller scale for a variety of pharmaceutical applications.

In addition to enamides, other substrates can be reduced by the catalyst system, including enol acetates of α-keto acids and esters and itaconic esters and acids.

Despite its age, Knowles catalyst will continue to be a work horse in the fine chemical industry and is still used as the reference system for amino acid synthesis. After 25 years, although ee values may be slightly lower than more modern catalysts, the turnover numbers and turnover frequencies continue to make it a stiff competitor for new catalysts systems to better.

1.8
References

1 Handbook of Chiral Chemicals (D.J. AGER, Ed.), Marcel Dekker, New York, 1999.
2 D.J. AGER, I.G. FOTHERINGHAM, N.J. TURNER, Spec. Chem. 2000, 20, 46.
3 D.J. AGER, I. FOTHERINGHAM, W. LIU, D.P. PANTALEONE, P.P. TAYLOR, K. BEHRENDT, Spec. Chem. 1999, 19, 10.
4 S.A. LANEMAN, in Handbook of Chiral Chemicals (D.J. AGER, Ed.), Marcel Dekker, New York, 1999, p. 143.
5 Comprehensive Asymmetric Catalysis (E.N. JACOBSEN, A. PFALTZ, H. YAMAMOTO, Eds.), Springer, Berlin, 1999.
6 H. TAKAYA, T. OHTA, R. NOYORI, in Catalytic Asymmetric Synthesis (I. OJIMA, Ed.), VCH Publishers, Inc., New York, 1993, p. 1.
7 K.E. KOENIG, in Asymmetric Synthesis (J.D. MORRISON, Ed.), Academic Press, Inc., Orlando, FL, 1985, Vol. 5, p. 71.
8 D.J. AGER, M.B. EAST, Asymmetric Synthetic Methodology, CRC Press, Boca Raton, 1995.
9 B.D. VINEYARD, W.S. KNOWLES, M.J. SABACKY, G.L. BACHMAN, D.J.J. WEINKAUFF, Am. Chem. Soc. 1977, 99, 5946.
10 W.S. KNOWLES, Acc. Chem. Res. 1983, 16, 106.
11 S.A. LANEMAN, D.E. FROEN, D.J. AGER, in Catalysis of Organic Reactions (F.E. HERKES, Ed.), Marcel Dekker, New York, 1998, p. 525.
12 M. BURK, in Handbook of Chiral Chemicals (D.J. AGER, Ed.), Marcel Dekker, New York, 1999, p. 339.
13 S.A. LANEMAN, unpublished results.
14 J.J. BOZELL, C.E. VOGT, J.J. GOZUM, Org. Chem. 1991, 56, 2584.
15 J. HALPERN, in Asymmetric Synthesis (J.D. MORRISON, Ed.), Academic Press, Orlando, 1985, Vol. 5, p. 41.
16 T. MASQUELIN, E. BROGER, K. MULLER, R. SCHMID, D. OBRECHT, Helv. Chim. Acta, 1994, 77, 1395.
17 C. DOBLER, H.J. KREUZFELD, M. MICHALIK, H.W. KRAUSE, Tetrahedron Asymmetry 1996, 1, 117.
18 C. CATIVIELA, M.D. DIAZ DE VILLEGAS, J.I. GARCIA, J.A. MAYORAL, E. MELENDEZ, An. Quim., Ser. C 1985, 81, 56.
19 U. SCHMIDT, J. LANGER, B. KIRSCHBAUM, C. BRAUN, Synthesis 1994, 1138.
20 J. HALPERN, in Asymmetric Synthesis (J.D. MORRISON, Ed.), Academic Press: Orlando, 1985, Vol. 5, p. 41.
21 S. JUGE, M. STEPHAN, J.A. LAFFITTE, J.P. GENET, Tetrahedron Lett. 1990, 31, 6357.
22 U. SCHMIDT, B. RIEDL, H. GRIESSER, C. FITZ, Synthesis 1991, 655.
23 S. JUGE, J.P. GENET, Tetrahedron Lett. 1989, 30, 2783.
24 S. JUGE, M. STEPHAN, R. MERDES, J.P. GENET, S.J. HALUT-DESPORTES, J. Chem. Soc., Chem. Commun. 1993, 531.
25 J.J. TALLEY, US Patent 5321153.
26 M.A. SCHMIDT, J.J. TALLEY, in Catalysis of Organic Reactions, Marcel Dekker, New York, 1995, p. 105.
27 J.J. TALLEY, US Patent 4939288.
28 J.J. TALLEY, US Patent 5473092.
29 D.J. AGER, S.A. LANEMAN, P. LEEMING, V. WALL, unpublished results.

2
The Application of DuPHOS Rhodium(I) Catalysts for Commercial Scale Asymmetric Hydrogenation

Christopher J. Cobley, Nicholas B. Johnson, Ian C. Lennon, Raymond McCague, James A. Ramsden, and Antonio Zanotti-Gerosa

Abstract

The introduction of chiral phospholane ligands has proved to be a turning point in the rhodium-catalyzed asymmetric hydrogenation of C=C bonds. A variety of chiral building blocks, such as itaconates and α-amino acids, as well as a number of pharmaceutical intermediates [Candoxatril, Tipranavir intermediate, Cbz-protected 2,7-L,L-diaminosuberic acid, and (S)-phorenol] have been prepared employing this technology. The applications already reported in the literature are briefly reviewed. DuPHOS-rhodium catalysts have been successfully employed in the multi-hundred kilogram production of non-natural α-amino acids. The synthesis of 3-fluorophenylalanine and 3-pyridylalanine are described with experimental details. In addition, advantages and limitations of this synthetic methodology will be considered. The synergistic combination with biocatalysis will also be discussed. Some practical and economic issues related to the industrial use of rhodium chiral catalysts, such as problems of catalyst recycling or separation and the nature of the catalyst precursor, are examined. Activity and selectivity requirements are considered in the light of the application of rhodium asymmetric catalysis to the production of high value, relatively low volume pharmaceutical intermediates.

2.1
Introduction

The successful industrial application of the homogeneous catalytic asymmetric hydrogenation of prochiral olefins depends on the ability of the catalyst systems to offer high activity both in terms of reaction rate and efficient utilization of the catalyst (as the molar substrate to catalyst ratio; S/C). It is also essential for the reaction to deliver the product with high enantiomeric excess and in good yield under conditions appropriate for industrial manufacture. In addition, the catalysts should be accessible in appropriate quantities for commercial manufacture.

A class of catalyst that has proven to be valuable for industrial processes is the rhodium DuPHOS family [1]; an example of which, as its precatalyst [(R,R)-Me-DuPHOS Rh-COD]BF$_4$ **1**, is illustrated in Fig. 1.

Asymmetric Catalysis on Industrial Scale: Challenges, Approaches and Solutions
Edited by H. U. Blaser, E. Schmidt
Copyright © 2004 Wiley-VCH Verlag GmbH & Co. KGaA, Weinheim
ISBN: 3-527-30631-5

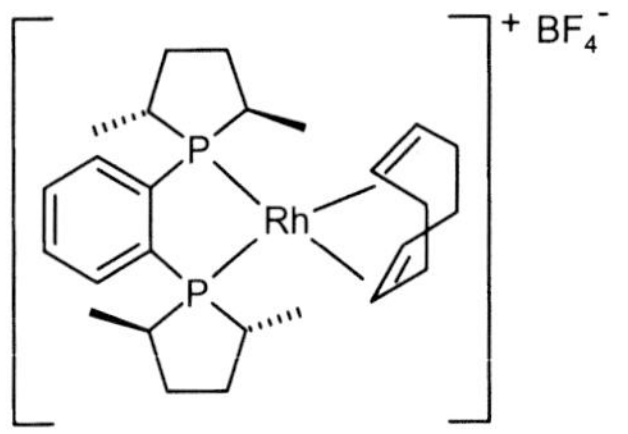

Fig. 1 Structural formula of [(R,R)-Me-DuPHOS Rh-COD]BF$_4$, **1**.

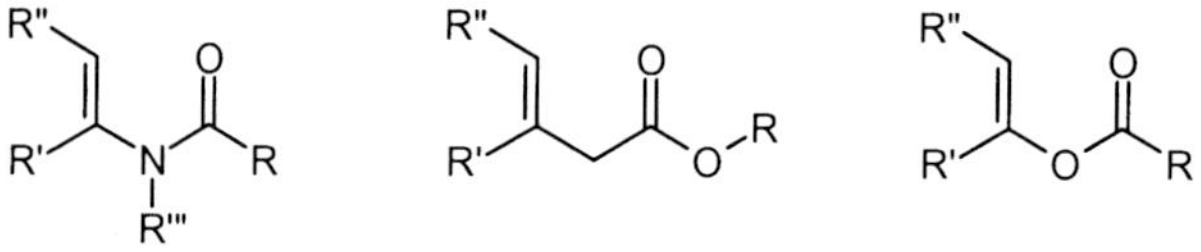

Fig. 2 Examples of hydrogenation substrates for which rhodium DuPHOS catalysts are suitable.

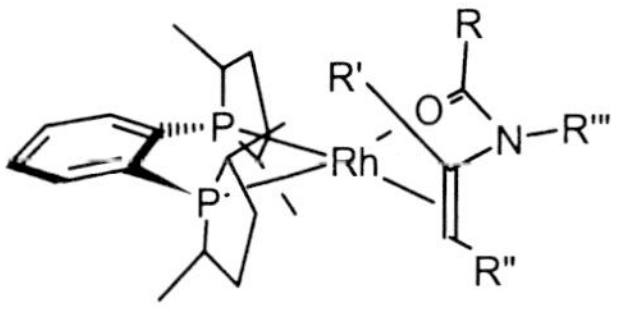

Fig. 3 Proposed substrate-catalyst intermediate **5** [2b].

In order for these rhodium DuPHOS catalysts to achieve the desired reactivity and selectivity, the hydrogenation substrate must contain certain features to facilitate the highly diastereoselective transition state required for the reaction. All the substrates to which rhodium DuPHOS hydrogenation catalysts have been successfully applied thus far possess a donor atom γ to the olefin (Fig. 2). Within the constraint of this geometric requirement a wide array of prochiral olefins have been demonstrated as suitable substrates for asymmetric hydrogenation with rhodium DuPHOS catalysts. Examples include enamides **2** [1, 2], vinylacetic acid derivatives **3** [3], and enol acetates **4** [4].

The secondary binding group γ to the olefin is typically a carbonyl group and is necessary for chelation of the substrate to the rhodium atom to provide an intermediate of the type **5** (Fig. 3). This ordered transition state orientates the substrate so that the hydrogen adds predominantly onto only one face of the olefin, providing high enantiofacial discrimination in the formation of the product.

In this chapter, an example of the synthesis of an α-amino acid, N-Boc-(S)-3-fluorophenylalanine (Section 2.2.1) is provided, highlighting the process requirements for the 50–200 kg scale. Consideration is then given to more general and practical details necessary to provide an economic process, with reference to other examples.

2.2
Large-Scale Manufacture of α-Amino Acids

α-Amino acids find great utility for pharmaceutical development, but one of the challenges is the broad spectrum of functionality required on the α-side chains. Chirotech has demonstrated that the DuPHOS technology is highly versatile and it has been used in the manufacture of numerous non-proteinogenic amino acids [5]. These vary not only by side chain functionality, but also by the associated acid and amine protection. The example of (S)-3-fluorophenylalanine is one that falls into a general scheme as given in Fig. 4 where, following the asymmetric hydrogenation step, the N-acetyl function in the hydrogenation product is removed by treatment with an aminoacylase enzyme.

The preferred method practiced by Chirotech demonstrates the efficient combination of catalytic technologies, where biocatalysis is used in tandem with catalytic asymmetric hydrogenation. The use of an enzymatic N-acetyl group deprotection has three major advantages:

i) The N-acetyl group is removed by the enzyme in a highly enantioselective manner, therefore the product is obtained as essentially a single enantiomer in high yield even if the asymmetric hydrogenation itself is not entirely selective.

ii) Other methods of N-acetyl group removal, such as reflux in hydrochloric acid, can lead to partial racemization of the product. This is more pronounced if a bulky β-substituent, such as naphthyl, is present in the molecule.

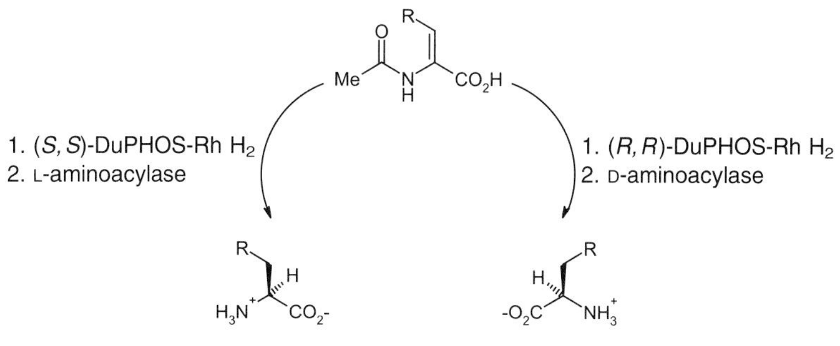

Fig. 4 Combination of asymmetric hydrogenation and biocatalysis.

iii) Using this process the amino acid is obtained in the free amino form and is therefore readily protected without any need to add base to neutralize the HCl salt. This also means that there is no excess acid to remove and leads to a far simpler waste stream management regime.

Both L- and D-aminoacylases have been isolated and cloned by Chirotech [6], but in addition commercial L-aminoacylases, such as Amano L-acylase, are also suitable for the synthesis of (S)-amino acids. By combining these enzymes with the appropriate chemocatalyst both enantiomers of a wide range of novel amino acids have been synthesized in high yield and exceptional enantiomeric purity [7]. In many cases the minor enantiomer is not detected; such selectivity is essential when supplying high purity pharmaceutical intermediates. Moreover, this combination is especially suited to industrial operation because both stages are catalytic and scale-up well. Furthermore, the process is robust, thereby delivering a product with assured quality.

For the asymmetric hydrogenation of a wide range of β-substituted N-acetyl-α-enamides, rhodium Et-DuPHOS catalysts have been reported to provide the best selectivity [5, 8]. However, for this class of compound, more often than not our choice of catalyst would be rhodium Me-DuPHOS, using the (S,S)-ligand for the L-amino acids and the (R,R)-ligand for the D-amino acids. Although not necessarily providing the best selectivity, this catalyst system is highly active and readily available (see Section 2.3.1) and hence provides direct scale-up and a lower overall cost contribution to the process, particularly when considered in combination with the biocatalytic deacylation step.

2.2.1
N-Boc-(S)-3-Fluorophenylalanine

The manufacture of N-Boc-(S)-3-fluorophenylalanine **10** has been carried out on a pilot scale to produce over 200 kg of product [9] (Fig. 5). The substrate was made by an Erlenmeyer reaction [10] between 3-fluorobenzaldehyde **6** and N-acetyl glycine **7**, to provide an intermediate azlactone, which was hydrolyzed to give the hydrogenation substrate **8**. Asymmetric hydrogenation proceeded readily using [(S,S)-Me-DuPHOS Rh-COD]BF$_4$ (S/C 1400) (86% ee). A commercial L-acylase was used to provide facile acetyl group removal and to upgrade the enantiomeric excess from 86% to >99%. N-Boc protection and final crystallization afforded the product in very high chemical and enantiomeric purity (>98% purity and >99% ee).

The route as outlined in Fig. 5 is highly versatile and has formed the basis for many other products. The main differentiation clearly being the choice of aryl aldehyde used in the first step as this dictates what functionality will be present in the final product side chain. Other than making appropriate solvent changes and considerations, such as the physical form of the finished product, these methods have been widely used for producing pilot quantities (10–200 kg) of many derivatives, a selection of which are shown in Fig. 6. For each of these products we have

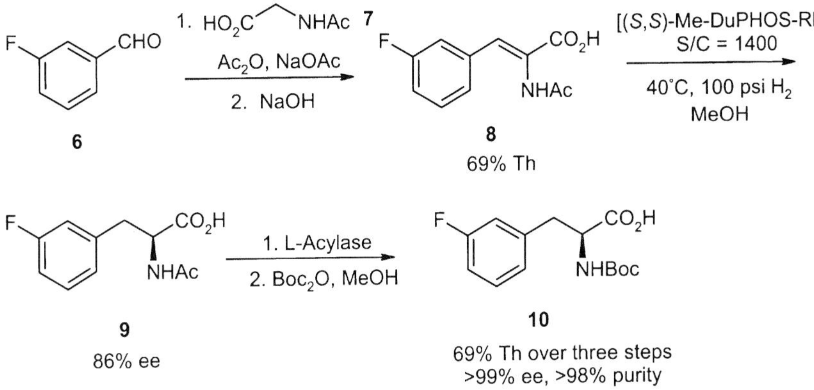

Fig. 5 Route used for large-scale manufacture of N-Boc-(S)-3-fluorophenylalanine.

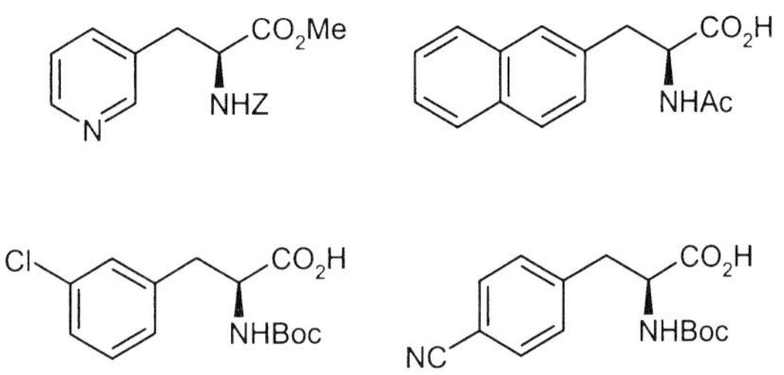

Fig. 6 Examples of α-amino acids manufactured on a 10–200 kg scale.

found that both the asymmetric hydrogenation and the biotransformation have performed exceedingly well in the pilot plant.

2.2.2
Experimental for N-t-Butoxycarbonyl-(S)-3-fluorophenylalanine

2.2.2.1 N-Acetyl Dehydro-3-fluorophenylalanine 8 via an Azlactone Intermediate

Acetic anhydride is added to 3-fluorobenzaldehyde, N-acetyl glycine, and anhydrous sodium acetate in ethyl acetate. The reaction mixture is heated to 80 °C. After completion of the reaction, the mixture is cooled, water is added, and the resultant precipitate is filtered and washed. The crude azlactone intermediate is then treated with aqueous sodium hydroxide followed by acidification. The resultant precipitate is filtered, washed, and dried to provide the N-acetyl dehydro-3-fluorophenylalanine in 70% yield from the aldehyde on a 50 kg scale.

2.2.2.2 N-Acetyl-3-fluorophenylalanine 9

N-Acetyl dehydro-3-fluorophenylalanine is added to methanol in a 150 L autoclave. The vessel is then purged with nitrogen to degas the contents. [(S,S)-Me-DuPHOS Rh-COD)]BF$_4$ is added directly (molar substrate to catalyst ratio: 1400). The vessel is then pressurized with hydrogen (100 psi). The hydrogen uptake of the reaction is monitored and following completion, the reactor is vented and again purged with nitrogen. The crude product is then used directly for the next step without any further purification. The product at this stage is obtained in 86% ee.

2.2.2.3 N-BOC-(S)-3-fluorophenylalanine 10

N-Acetyl-3-fluorophenylalanine is charged into a reactor and following pH adjustment and solvent exchange with water, Amano L-acylase (5% by weight) is added. The reaction is stirred at 35–40 °C until the biotransformation is complete. The mixture is then acidified and filtered. The filtrate is basified and concentrated. The resultant product solution is subjected directly to the Boc protection step.

Methanol is added to the crude aqueous solution of 3-fluorophenylalanine, followed by a solution of di-t-butylcarbonate in methanol. Following this addition, the reaction is acidified and the resultant precipitate filtered and dried to provide the N-BOC-(S)-3-fluorophenylalanine in >99% ee and >98% purity by HPLC analysis (70% yield from N-acetyl-3-fluorophenylalanine).

2.3
Practical and Economic Considerations for Manufacture

Asymmetric hydrogenation using homogeneous catalysts is particularly well suited to the manufacture of chiral intermediates for the pharmaceutical industry where the products are of high value and the volumes are relatively low. When endeavoring to supply customers with such intermediates, the initial volumes are typically 10–50 kg for early clinical trial supply and these are prepared either in a kilo or pilot facility. At this stage the primary concerns are quality of product and speed of delivery and to a lesser extent price. For initial catalyst selection, having high throughput screening techniques available in-house allows rapid identification of a suitable catalyst system from the wide range of available catalyst systems. Following the laboratory route development, where the general applicability of the substrate synthesis and catalytic route has been demonstrated, we typically have found that process scale-up can be achieved rapidly and safely with minimal modification to the laboratory route. For this initial scale-up, where speed and product quality are the main drivers, the catalytic step does not necessarily need to be highly optimized and at this stage S/C 1000–5000 will suffice.

Larger volume product supplies are typically for late-stage clinical trials and then commercial manufacture. These will be in the order of hundreds of kg to multi-tons. At this stage further optimization will be conducted. Typically the catalytic step will be more rigorously defined to obtain the most efficient use of the catalyst, both in terms of selectivity and productivity. The ultimate aim now becomes to reduce the catalyst cost contribution as far as possible in order to provide the optimum process economics. At this stage issues such as vessel residency and reactor space-time yields must also be considered in conjunction with absolute catalyst loading.

2.3.1
Manufacture and Choice of Precatalysts

Chirotech has invested much time and effort in the development of rhodium DuPHOS precatalysts. Robust synthetic routes have been developed for these materials, such that commercial quantities of the (DuPHOS Rh-COD) precatalysts are available [11]. In addition, Chirotech has manufactured and supplied [(R,R)-Me-DuPHOS Rh-COD]BF$_4$ **1** on a multi-kilogram scale, using the route to the ligand outlined in Fig. 7 [12].

The cost of rhodium DuPHOS precatalysts is relatively high, with a large proportion of the cost being associated with the market value of the metal. Therefore, it is essential that the catalysts be utilized appropriately. There has been some debate about COD precatalysts not being economic for use in industrial processes when compared with the NBD precatalyst [13]. A study on the use of COD and NBD precatalysts for the asymmetric hydrogenation of a range of substrates [14] has been carried out. This confirmed that at high catalyst loading (S/C 100) there is a rate difference for some substrates, such as methyl acetamidocinnamate, with the NBD precatalyst producing the active species at a faster rate, but no difference in enantiomeric selectivity is observed. However, at low catalyst loadings the difference between the productivity of the NBD and COD precatalysts became negligible. Moreover, this effect was found to be substrate dependent. With dimethyl

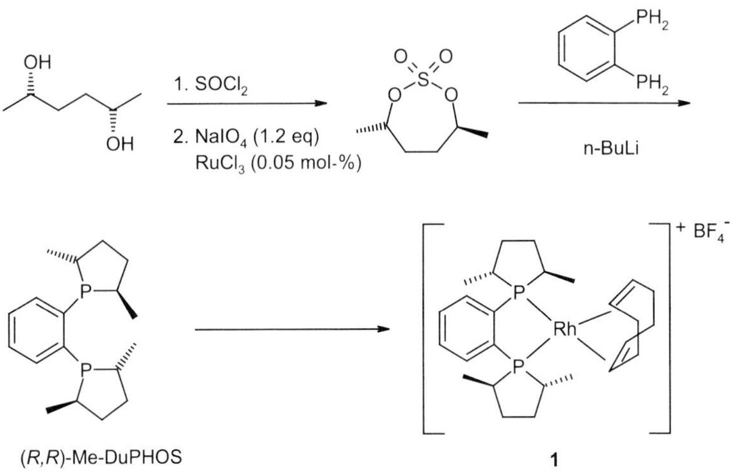

Fig. 7 Manufacture of [(R,R)-Me-DuPHOS Rh-COD]BF$_4$ **1**.

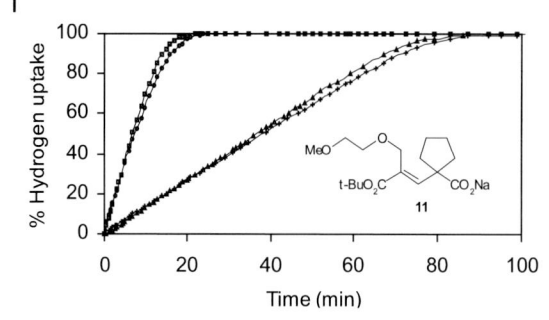

Fig. 8 Hydrogenation of the candoxatril precursor **11**. The first two curves are Rh-DuPHOS COD and NBD precatalysts with efficient stirring. The second two curves are Rh-DuPHOS COD and NBD precatalysts with less efficient stirring.

itaconate and the candoxatril precursor **11** (Fig. 9, Section 2.3.2) no rate difference between the two precatalysts was observed.

It was found that experimental conditions such as rate of stirring have a much more dramatic effect than the choice of precatalyst. For example, hydrogenation of the candoxatril precursor **11** was 4.5 times slower with less efficient stirring (Fig. 8), using either the COD or NBD precatalysts. For this class of reaction, hydrogen mass transfer into solution [15] is the most important single process parameter affecting the overall reaction rate.

2.3.2
Substrate Synthesis, Purity and Catalyst Loading

A significant obstacle in the application of rhodium-catalyzed hydrogenation, in common with many other catalytic processes, is the requirement of a relatively pure substrate. To be commercially viable these expensive catalysts must be used at relatively low catalyst loading. A concomitantly low level of an inhibiting impurity could be sufficient to prohibit catalysis. However, it is not necessary to remove all impurities, only those which adversely effect the catalytic process, such as organophosphorus compounds and chloride ions. Consequently, particular attention must be paid to the synthetic route used to prepare the substrate. For example, synthesis of an olefin via Horner-Emmons chemistry inevitably leaves trace amounts of active phosphorus compounds in the crude substrate. These will bind to rhodium and must be rigorously removed if high substrate to catalyst ratios are to be achieved. Alternative approaches involving synthetic routes such as an Erlenmeyer [10] reaction are much more attractive since potentially problematic impurities are much more readily removed.

Work on the candoxatril precursor **11** [16] gave an insight into the importance of substrate purity for efficient hydrogenation using rhodium DuPHOS catalyst systems. The asymmetric hydrogenation of **11** with rhodium Me-DuPHOS furnished the desired intermediate **12** in excellent enantiomeric excess and yield (Fig. 9).

However, some variability with different batches of sodium salt **11** was observed [16]. The salt was normally made by treatment of the acid with solid NaOH followed by recrystallization [17]. Either residual NaOH or incomplete sodium salt formation

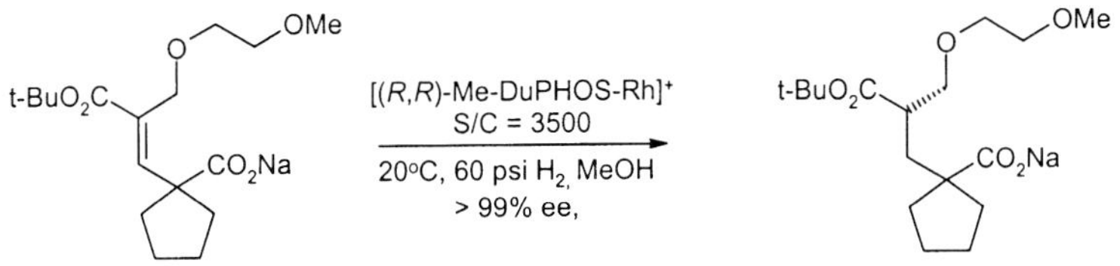

2.3 Practical and Economic Considerations for Manufacture | 277

Fig. 9: Synthesis of an intermediate for Candoxatril.

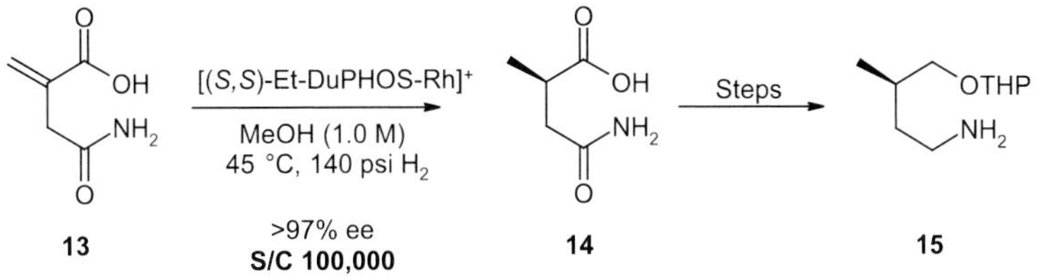

Fig. 10 Asymmetric hydrogenation of 2-methylene succinamic acid.

could be causing the catalyst to be deactivated, as it was previously demonstrated that the free acid was a very poor substrate giving less than 5% conversion over 1.5 h under our standard conditions [16]. Development of an alternative recrystallization protocol making use of soluble base allowed for the reproducible hydrogenation of the sodium salt **11** at S/C 4000. In addition higher S/C ratios (S/C 10000) could be achieved, but at the expense of reactor time and conversion.

Another example where the substrate purity proved to be of great importance was for the asymmetric hydrogenation of 2-methyl succinamic acid **13** to provide a route to the THP ether of 4-amino-2(R)-methylbutan-1-ol **15**. This is a key intermediate used in the synthesis of Takeda's NK1 receptor antagonist TAK-637 [18] and for other pharmaceutical applications. In the original substrate synthesis excess ammonium hydroxide was neutralized using aqueous HCl. For this material the best S/C ratio that could be achieved was 1000 giving complete conversion in 2 hours and 97% ee. Investigation of substrate purity revealed small amounts of NH_4Cl (approx. 1 mol-%) in the product. When HCl was replaced with H_2SO_4 in the synthetic route a remarkable increase in activity was observed. At S/C 1000 the reaction was complete in 4 min and the optimized process used S/C 100000 (21 400 w/w) giving complete reaction in 6 hours and 97% ee [19] (Fig. 10).

For this reaction [(S,S)-Et-DuPHOS Rh-COD]BF_4 was found to be the best pre-catalyst, but good results were also obtained with [Me-DuPHOS Rh-COD]BF_4, [Et-FerroTANE Rh-COD]BF_4 and [i-Pr-FerroTANE Rh-COD]BF_4 [11].

As well as impurities being deleterious we have also found the judicious introduction of additives can be critical for developing economic catalytic processes. This is demonstrated in the asymmetric hydrogenation of itaconic acid derivatives to give chiral succinates **17** (Fig. 11) [3].

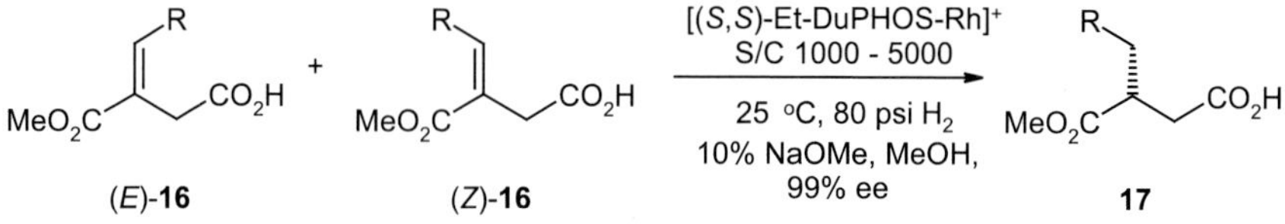

Fig. 11 Asymmetric hydrogenation of itaconate substrates.

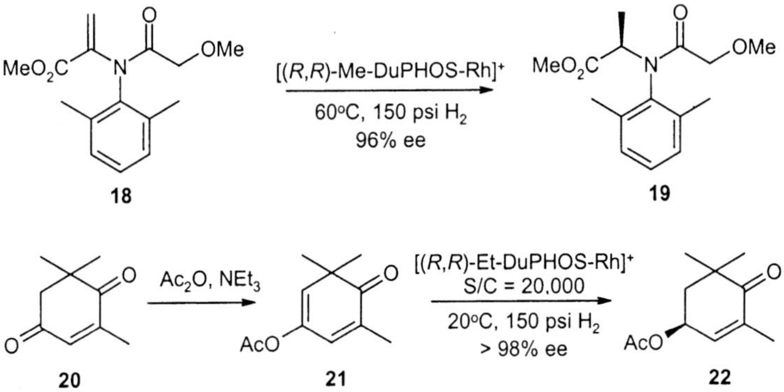

Fig. 12 Asymmetric hydrogenation approaches to (R)-metalaxyl and zeaxanthin.

This reaction is greatly accelerated by the presence of a sub-stoichiometric amount of base. It is proposed that this facilitates the coordination of the resultant carboxylate group to the catalyst. A wide range of functionalities are tolerated as is the presence of E/Z isomers at the prochiral center [3]. On the contrary, itaconate derivatives with ester or amide functionalities on the β-carboxylate and the free α-carboxylic acid are not good substrates for rhodium DuPHOS catalysts, giving low conversion and selectivity. Catalysts based on rhodium FerroTANE are superior for the asymmetric hydrogenation of such substrates [20]. In this case no added base is required, since the high selectivity comes from the preferential secondary binding of the amide or ester group to the metal.

There have been many other independent reports of excellent S/C ratios using rhodium DuPHOS systems. For the synthesis of an (R)-metalaxyl intermediate **19** [21], a turnover number of 50000 has been demonstrated using Me-DuPHOS-Rh. Hoffmann la Roche have reported the Et-DuPHOS-Rh-catalyzed hydrogenation at S/C 10000–20000 of a cyclic enol acetate **21** to provide an intermediate **22** to Zeaxanthin in 98% ee [22] (Fig. 12).

2.3.3
Catalyst Charging

The rhodium DuPHOS precatalysts are very sensitive to oxygen when in solution and should be handled under an inert atmosphere, but the crystalline complexes are stable for long periods of time under nitrogen and can be weighed and handled in air. Indeed, at Chirotech some batches of precatalysts have been used over a 4 year period. All of these precatalysts are stored in Schlenk tubes, under nitrogen, but are regularly opened to weigh out samples for hydrogenation screening work. For small-scale laboratory work the precatalyst is easily introduced into a vessel by injection of a solution. However, when hydrogenation processes are scaled-up the addition of small amounts of precatalyst solution to the reactor requires more consideration. In addition, equipment to add solids directly into larger reactors is often unable to handle a few grams of precatalyst. Many hydrogenation vessels are in fact designed for heterogeneous reactions and integrated charging chambers are designed to deliver a few 100 g up to kilograms of catalyst. When the hydrogenation of the candoxatril precursor **11** was scaled-up to 12 kg, the solid precatalyst (6 g) was dissolved in degassed methanol and the solution was added via a charging port under a positive flow of nitrogen [16]. For our pilot-scale manufacture of a-amino acids (see Section 2.2) the solid precatalyst was added directly through a charging port, with the substrate solution under a blanket of nitrogen. Once the catalyst has been added the nitrogen atmosphere is replaced by hydrogen to commence the reaction. Thus, rhodium DuPHOS precatalysts can be readily charged into large-scale hydrogenation vessels without undue regard for their stability or possible degradation.

2.3.4
Catalyst Removal

There are several methods to remove homogeneous catalyst residues from either the immediate hydrogenation product, or a more suitable downstream product. These include:

i) Distillation of the product [23].
ii) Recrystallization of the product [24].
iii) Adsorption of the catalyst onto a solid support (e.g., carbon, silica, alumina, etc.).
iv) Extraction of the product into either basic or acidic aqueous solution (amino acid, amines, acids, etc.) and back extraction of the catalyst residues with an organic solvent.

Virtually all of the products that have been manufactured by Chirotech have used recrystallization at some point of the process to remove the catalyst (see Section 2.2). This not only removes the catalyst, but also generally raises the enantiomeric excess and enhances the product purity. It should be noted that at S/C 5000 for a substrate such as the candoxatril precursor **11** the catalyst would only contribute 0.006% (60 ppm) rhodium impurity if all of it were retained in the product. Current regulations stipulate that the rhodium content needs to be less than 5 ppm in the final product [25], but in our experience crystallization is a highly effective method to provide final products that are within specification for rhodium content.

Catalyst immobilization is another method that has been used in an attempt to prevent the contamination of products with metal residues and to reuse the valuable catalyst [26]. In virtually all of these methods the immobilized catalyst is of lower activity than the corresponding homogeneous catalyst due to the tethering technology. Furthermore, enantioselectivity and activity may decrease and vary with the number of reuses of the catalyst. In addition, significant leaching of the metal is often observed and given that a greater amount of catalyst may be required in the immobilized case, this can lead to significant contamination. Indeed this may well be above the level of the input of metal for a truly optimized homogeneous process. When [DuPHOS Rh-COD]BF_4 precatalysts were immobilized on alumina [27] and used for the asymmetric hydrogenation of methyl-2-acetamidoacrylate, variable enantiomeric excess values were obtained over four runs (S/C 100, 92–95% ee) [28]. The homogeneous variant operates at S/C 10 000 providing product in >99% ee, reproducibly [14]. At the current state of development of immobilization technology our opinion is that optimization of the loading of the homogeneous catalyst for a single pass will provide the best process economy.

2.4
Conclusions

From the results of our extensive laboratory and pilot plant research we are confident that many asymmetric hydrogenation processes, using rhodium DuPHOS catalysts can be carried out at S/C 5 000–100 000, whilst maintaining high enantiomeric excess [14]. This can be substrate dependent, but in many cases the poor reactivity is due to low levels of impurities. It is important to consider the complete route to a target molecule and not just the asymmetric hydrogenation step. Routes using this technology have significant advantages: they can often be concise, there are many efficient methods to produce olefin substrates with a coordinating group, all of the substrate is converted into product in high enantiomeric excess, and hydrogenation is a clean and readily scaleable process. The cost of the precatalyst should be viewed in the context of the efficiency of the overall process to the target molecule and not in isolation.

Finally, an important trade-off with obtaining high S/C ratios is that of time. In early development it is often more desirable to have reactions complete in a short and reproducible timeframe, rather than to optimize catalyst loading. For larger scale manufacture, especially for markets other than pharmaceuticals, obtaining the best S/C ratio will be of greater importance.

2.5
References

1 M. J. Burk, Acc. Chem. Res. 2000, 33, 363–372.
2 a) M. J. Burk, J. E. Feaster, W. A. Nugent, R. L. Harlow, J. Am. Chem. Soc. 1993, 115, 10125–10138; b) S. K. Armstrong, J. M. Brown, M. J. Burk, Tetrahedron Lett. 1993, 34, 879–882.
3 M. J. Burk, F. Bienewald, M. Harris, A. Zanotti-Gerosa, Angew. Chem., Int. Ed. Engl. 1998, 37, 1931–1933.
4 a) M. J. Burk, C. S. Kalberg, A. Pizzano, J. Am. Chem. Soc. 1998, 120, 4345–4353; b) N. W. Boaz, Tetrahedron Lett. 1998, 39, 5505–5508.
5 M. J. Burk, F. Bienewald, Unnatural α-Amino Acids via Asymmetric Hydrogenation of Enamides, in Transition Metals for Organic Synthesis and Fine Chemicals, Wiley-VCH, Weinheim, Germany, 1998, Vol. 2, pp. 13–25.
6 a) H. S. Toogood, I. N. Taylor, R. C. Brown, S. J. C. Taylor, R. McCague, J. A. Littlechild, Biocatalysis and Biotransformation 2002, 20 (4), 241–249; b) H. S. Toogood, E. J. Hollingsworth, R. C. Brown, I. N. Taylor, S. J. C. Taylor, R. McCague, J. A. Littlechild, Extremophiles 2002, 6, 111–122.
7 S. J. C. Taylor, R. C. Brown, WO 00/23598, 27 Apr 2000.
8 For recent literature examples see: a) S. W. Jones, C. F. Palmer, J. M. Paul, P. D. Tiffin, Tetrahedron Lett. 1999, 40, 1211–1214; b) M. Adamczyk, S. R. Akireddy, R. E. Reddy, Tetrahedron 2002, 58, 6951–6963; c) W. Wang, C. Xiong J., Zhang, V. J., Hruby, Tetrahedron 2002, 58, 3101–3110.
9 Work carried out in collaboration with Robinson Brothers Limited.
10 R. M. Herbst, D. Shemin, Organic Syntheses 1943, Col. Vol. II, 1–3 (A. H. Blatt Ed.).
11 Research quantities (<25 g) of DuPHOS, BPE and FerroTANE and PhanePHOS precatalysts are available from Strem Chemicals Inc., 7 Mulliken Way, Dexter Industrial Park, Newburyport, MA 01950-4098, USA. www.strem.com. Commercial quantities, above 25 g, are available from the Dowpharma business of The Dow Chemical Company.
12 U. Berens, EP 1028967 B1, 26 September 2001.
13 A. Börner, D. Heller, Tetrahedron Lett. 2001, 42, 223–225.
14 a) C. J. Cobley, I. C. Lennon, R. McCague, J. A. Ramsden, A. Zanotti-Gerosa. Tetrahedron Lett. 2001, 42, 7481–7483; b) C. J. Cobley, I. C. Lennon, R. McCague, J. A. Ramsden, A. Zanotti-Gerosa in "Catalysis of Organic Reactions", New York, Marcel Dekker, chapter 24, 2002, p. 329–339.
15 Y. Sun, R. N. Landau, J. Wang, C. LeBlond, D. G. Blackmond, J. Am. Chem. Soc. 1996, 118, 1348–1353.
16 M. J. Burk, F. Bienewald, S. Challenger, A. Derrick, J. A. Ramsden, J. Org. Chem. 1999, 64, 3290–3298.
17 a) S. Challenger, A. Derrick, C. P. Mason, T. V. Silk, Tetrahedron Lett. 1999, 40, 2187–2190; b) M. Bulliard, B. Laboue, J. Lastennet, S. Roussiasse, Org. Process Res. Dev. 2001, 5, 438–441.
18 a) Y. Ikeura, T. Ishimaru, T. Doi, M. Kawada, A. Fujishima, H. Natsugari, J. Chem. Soc., Chem. Commun. 1998, pp. 141–2142; b) H. Natsugari, Y. Ikeura, I. Kamo, T. Ishimaru, Y. Ishichi, A. Fujishima, T. Tanaka, K. Kasahara, M. Kawada, T. Doi, J. Med. Chem. 1999, 42, 3982–3993.
19 R. B. Appell, C. J. Cobley, C. Goralski, I. C. Lennon, C. Praquin, A. C. Sutterer, A. Zanotti-Gerosa, Org. Process Res. Dev. 2003, 7, 407–411.
20 a) U. Berens, M. J. Burk, A. Gerlach, W. Hems, Angew. Chem., Int. Ed. Engl. 2000, 39, 1981–1984; b) C. P. Ashcroft, S. Challenger, A. M. Derrick, R. Storey, N. M. Thomson, Org. Process Res. Dev. 2003, 7, 362–368.
21 H. U. Blaser, F. Spindler, Topics Catal. 1997, 4, 275–282.
22 a) E. A. Broger, Y. Crameri, R. Schmid, T. Siegfried, EP 691325, 10 Jan 1996; b) M. Scalone, R. Schmid, E. A. Broger, W. Burkart, M. Cereghetti, J. Crameri, J. Foricher, M. Henning,

F. Kienze, F. Montavon, G. Schoettel, D. Tesauro, S. Wang, R. Zell, U. Zutter, Proceedings Chiratech '97 Symposium, The Catalyst Group, Spring House USA, 1997.
23 H. U. Blaser, Adv. Synth. Catal. 2002, 344, 17–31.
24 M. J. Burk, N. B. Johnson, B. D. Hewitt, US 6211386, 3 Apr. 2001.
25 See draft guidelines from the European Agency for the Evaluation of Medicinal Products, 27 June 2002. www.emea.eu.int/pdfs/human/swp/444600en.pdf.
26 C. Bianchini, P. Barbaro, Topics Catal. 2002, 19, 17–32.
27 R. Augustine, S. Tanielyan, S. Anderson, H. Yang, J. Chem. Soc., Chem. Commun. 1999, pp. 1257–1258.
28 J. A. M. Brandts, J. G. Donkervoort, C. Ansems, P. H. Berben, A. Gerlach, M. J. Burk in Catalysis of Organic Reactions. New York, Marcel Dekker, 2000, pp. 573–581.

3
Liberties and Constraints in the Development of Asymmetric Hydrogenations on a Technical Scale

JOHN F. MCGARRITY, WALTER BRIEDEN, RUDOLF FUCHS, HANS-PETER METTLER, BEAT SCHMIDT, and OLEG WERBITZKY

Abstract

Lonza's experience in process development and execution on a technical scale in the field of asymmetric hydrogenations is illustrated by four examples, two from internally-generated projects, and two from custom-manufacturing projects. In all cases the design of the hydrogenation step itself, and the appropriate optical purity specifications, are determined by the flexibility in the synthetic strategy, for example the freedom to modify activating substituents, to enhance ligand characteristics, or crystallization conditions to enhance optical purity. This freedom is often reduced in custom-synthesis projects.

In all four examples, a large number of catalyst ligands were screened. In the cases of (d)-(+)-biotin, (an internal project) and 4-Boc-piperazine-2(S)-N-^{t}butylcarboxamide (the custom manufacture of a starting material for a cGMP production) the substrate structure could be modified to optimize the stereoselectivity. In the cases of dextromethorphan (an internal project), and a benzodiazepine carboxylic ester (an intermediate in a custom manufacture cGMP project) moderate enantioselectivities (89% and 88% ee) could be tolerated, as favorable crystallization conditions could be chosen.

In all cases very extensive screening experiments were necessary, and efforts are described to rationalize the experimental programs. Examples are given of scale-up problems and changes, and also of post-piloting improvement of the processes.

3.1
Introduction

Since Lonza carried out its first homogeneous asymmetric hydrogenation on a 17 ton scale in 1994, the development and application of asymmetric hydrogenations has been a major strength in Lonza's technology portfolio. The applications in Lonza have varied from internally-generated projects, for end-products marketed by Lonza, to projects in the custom-manufacturing field, in which the Lonza Exclusive Synthesis department is active. This chapter will illustrate how process de-

Asymmetric Catalysis on Industrial Scale: Challenges, Approaches and Solutions
Edited by H. U. Blaser, E. Schmidt
Copyright © 2004 Wiley-VCH Verlag GmbH & Co. KGaA, Weinheim
ISBN: 3-527-30631-5

velopment was influenced by the different boundary conditions of the different sorts of project.

Process costs will always be a constraint, and there must always be constant effort to reduce these with high yield, high selectivity, high S/C ratio, low pressure, reasonable reaction time, high concentration, simple work-up, low waste, non-hazardous chemicals. However, for projects in the custom-manufacturing area, where speed of development and delivery is essential, the most economical process need not be the first to be implemented. In many cases, especially where the target product is considered to be a starting material, development of a second-generation process can be considered.

Very high enantioselectivity is always desirable, however the possibility of enriching an enantiomer further down the process pathway should always be used. The ability to design the synthesis is primordial in this respect. A further key ability in synthesis design is the possibility to modify substituents on the substrate of the hydrogenation reaction to optimize selectivity or reactivity. This option often does not exist for custom-synthesis projects. A constraint in realization of a technical process may be the unavailability of the chosen catalyst or ligand in the desired quantities, or could be due to a restrictive licensing policy of the patent holder. The effects of several of these conditions among others will be illustrated with examples of process development for two internal Lonza processes; (d)-(+)-biotin, and dextrometorphan, and for two custom-synthesis projects, a chiral piperazinamide (starting material) and a chiral benzodiazepinone (advanced intermediate).

3.2
Asymmetric Hydrogenation in the Lonza Biotin Process

3.2.1
Historical Development

The original Lonza synthesis [1, 2] of (d)-(+)-biotin **1** (outlined in Fig. 1) was designed around the critical diastereoselective hydrogenation of the tetrasubstituted double bond in imidazolinone **2** to the desired 3aS,6aR isomer **3**. This strategy was a natural consequence of the experience gained with an analogous diastereoselective hydrogenation in the synthesis of an intermediate for the Merck antibiotic thienamycine (see Fig. 2).

Despite an extensive catalyst screening, the best diastereoselectivity obtained for the conversion **2–3** (Fig. 3) was modest (70:30) and unfortunately could not be significantly improved by variation of the substituents. However, it was adequate for a first generation process, as isolation conditions could be optimized to provide a 58% yield of **3** with less than 1% of the diastereomer **4**.

Once the complete biotin process had been established as technically feasible, the evident weakness of the hydrogenation step was addressed. In fact the original synthesis design involved the hydrogenation of the corresponding thiophe-

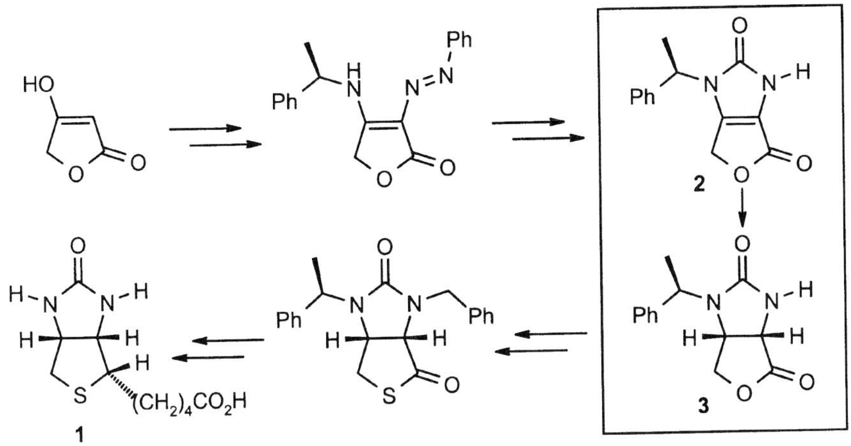

Fig. 1 General scheme of the Lonza synthesis of (d)-(+)-biotin.

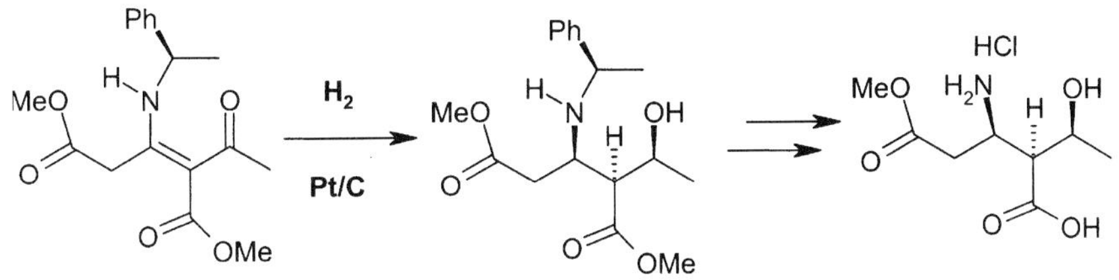

Fig. 2 The critical diastereoselective hydrogenation in the synthesis of a thienomycine intermediate.

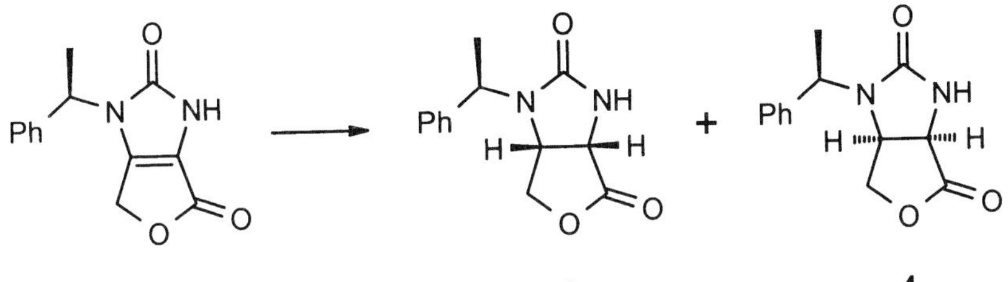

Fig. 3 The critical diastereoselective hydrogenation in the Lonza (d)-(+)-biotin.

nones, however all noble metal heterogeneous catalysts tested were inactive, presumably due to poisoning by sulfur. Encouraging results were obtained however with Wilkinson's catalyst, therefore the obvious solution lay with enantioselective homogeneous catalysis. It was also hoped that substituents other than (R) 1-phenylethyl, or none at all, could be incorporated on N-1 and N-3, to diminish the raw material costs and possibly simplify the later stages of the synthesis.

3.2.2
Laboratory Screening Experiments

3.2.2.1 Ligand Screening

A very definite constraint in efficiently developing a process with homogeneous enantioselective catalysts in 1993 was the complete lack of experience in the use of such systems in Lonza. Rather than go through the laborious and time-consuming apprenticeship in the technology, Lonza decided to outsource the initial screening to the expert Ciba-Geigy central research group (which has since been devolved and mutated into the Solvias company). As fortune would have it, the Ciba-Geigy group had, among an extensive series of standard catalysts, several which were undergoing in-house development in their Josiphos series. As this chapter makes clear, these catalysts almost became objects of infatuation in the Lonza asymmetric hydrogenation programs.

The initial catalyst screening was carried out with the substrate **2** from the original synthesis [3, 4]. The structures of the diastereoisomers formed are given in Fig. 3. Selected results from the initial screening are detailed in Tab. 1. The structures of the ligands mentioned in this chapter are given in Figs. 4 and 5.

Summary of screening results:

- The only ruthenium and iridium catalysts tried ([Ru$_2$Cl$_4$(S-BINAP)$_2$]NEt$_3$ in methanol and [Ir(COD)py(Pcy$_3$)]PF$_6$ in dichloromethane were completely inactive.
- Charged rhodium catalysts particularly in methanol were less active than neutral rhodium catalysts in toluene.
- Of the (at that time) standard ligands tested, only MOD-DIOP gave promising results.
- The (at that time) development Josiphos ligands provided the most promising results, and one, (R)-(S)-PPF-P(^{t}Bu)$_2$ was sensational.

Tab. 1 Initial catalyst screening for the hydrogenation of **2** to **3**.

M	Ligand	S/C	Solvent	p[H$_2$] (bar)	T (°C)	Time (h)	Conv.[a] (%)	D.r.[a] 3:4
Rh+	(4R,5R)-DIOP	250	THF/MeCl$_2$	60	60	18	100	67:33
Rh+	(4R,5R)-MOD-DIOP	250	Tol/MeOH (1:1)	70	60	19	73	83:17
Rh(0)	(4R,5R)-MOD-DIOP	500	Toluene	35	70	3	100	88:12
Rh+	(R,R)-NORPHOS	250	EtOH	35	70	16	7	65:35
Rh(0)	(2S,4S)-BDPP	500	Tol/MeOH (4:1)	35	70	19	51	75:25
Rh(0)	(R)-(S)-BPPFOAc	500	Toluene	20	790	2.5	88	56:44
Rh+	(R)-(S)-BPPFOAc	250	MeOH	40	40	68	55.5	60:40
Rh(0)	(R)-(S)-PPF-P(^{t}Bu)$_2$	500	Toluene	50	70	18	>98	>98:2

a) Abbreviations: Conv.=conversion, D.r.=diastereomer ratio.

Josiphos Ligands (R)-(S)-R₂PF-PR*₂

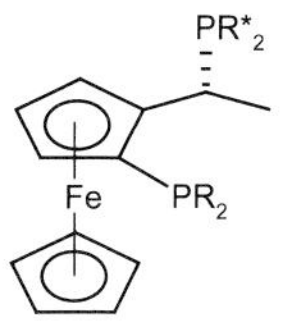

1) (R)-(S)-PPF- P(^{t}Bu)₂
2) (R)-(S)-PPF-P(phobyl)₂
3) (R)-(S)-(Cy)₂PF-P(^{t}Bu)₂
4) (R)-(S)-(Et)₂PF-P(^{t}Bu)₂
5) (R)-(S)-PPF-P(Cy)₂
6) (R)-(S)-PPF-P(^{i}Pr)₂
7) (R)-(S)-PPF-P(4-MeOXyl)₂
8) (R)-(S)-(4-MeOXyl)₂PF-P(^{t}Bu)₂
9) (R)-(S)-(t-Bu)₂PF-P(Ph)₂
10) (R)-(S)-PPF-P(^{i}Pr)₂
11) (R)-(S)-(Cy)₂PF-P(Cy)₂
12) (R)-(S)-(^{i}Pr)₂ PF-P(^{t}Bu)₂
13) (R)-(S)-(4-MeO-3,5-diethylphenyl)₂PF-P(^{t}Bu)₂

Fig. 4 Structures of the Josiphos ligands used in the Lonza asymmetric hydrogenation programs.

3.2.2.2 Substrate Screening

As Lonza had complete freedom to modify its own process for (d)-(+)-biotin, and selective introduction of substituents at N-1 (through choice of starting material) and N-3 (through functionalization of **2**) was simple, the effect of substituents on the hydrogenation was briefly studied. The results for a selection of the N-1- and N-3-substituted analogues investigated in Fig. 6 are summarized in Tab. 2 and Tab. 3, respectively.

The conclusions were:

- N-3 substituents lower activity and selectivity, particularly for bulky ligands.
- The enantioselectivity of the hydrogenation of the N-1 benzyl-substituted analogue with both (R)-(S)-PPF P^t(Bu)₂ and (4R,5R)-MOD-DIOP is very impressive. None of the undesired 3aR,6aS enantiomer could be detected with the NMR method used.
- The 2-(R)-o-methoxyphenylethyl substituent in the N-1 position diminished the activity, and had varying effects on the selectivity.
- The largest N-1 substituent, 2(R)-napthylethyl enhanced the selectivity, but diminished the activity of the catalyst.

The consequences for the synthetic strategy were as follows.

The very high enantioselectivity of the hydrogenation of the 1-benzyl intermediate, even with the conventional MOD-DIOP ligand was probably the most impressive result. Development of the process with the N-1 benzyl substituent would have made the later intermediates of the biotin process identical to those in the existent commercial process, which could have expanded the market opportunities

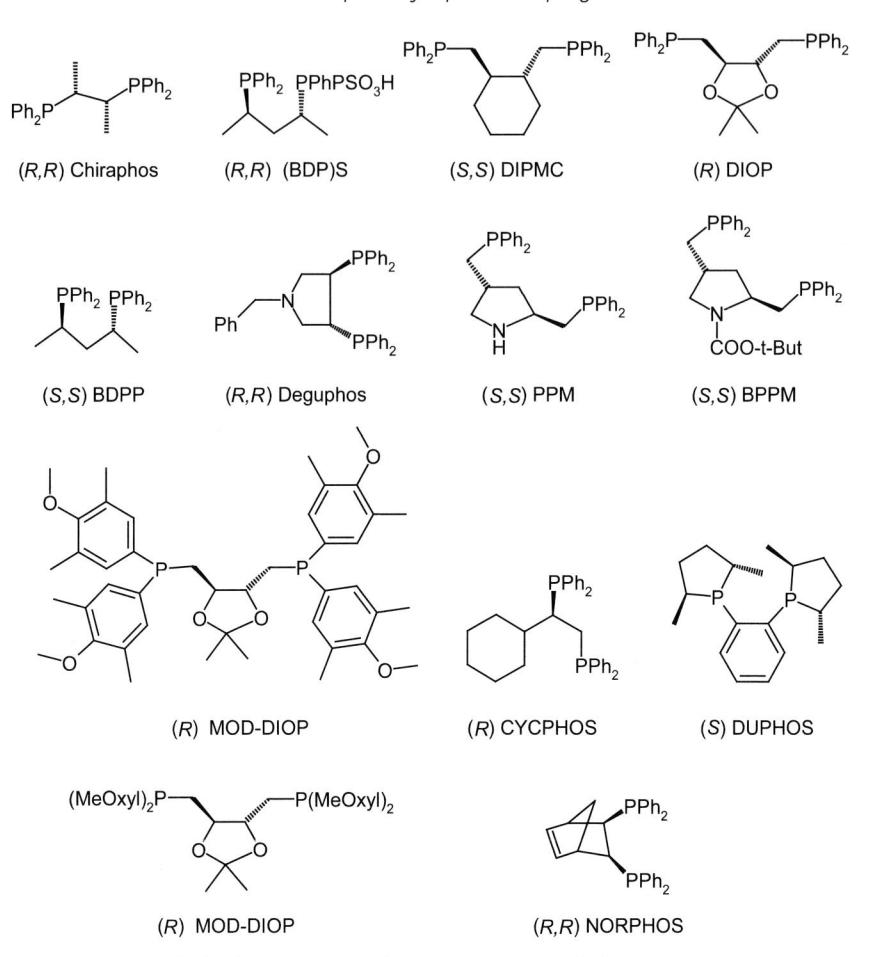

Fig. 5 Structures of other ligands used in the Lonza asymmetric hydrogenation programs.

for Lonza. However a major drawback was that the deprotective removal of the benzyl group in the last step is much more difficult than that of a phenylethyl substituent. Furthermore, the phenylethyl-substituted intermediates showed better crystallization properties than the benzyl analogues. Therefore it was decided to proceed at least in the short term with the existing substituent series.

A double diastereoselection effect was noticed, the effect depending on the ligand used. For example hydrogenation of the (R)-phenylethyl-substituted **2** with (R)-(S)-PPF PCy$_2$ gave a ratio of **3** (3aS,6aR):**4** (3aR,6aS) of 94:6, whereas when the (S)-phenethyl analogue was hydrogenated with the same ligand, the ratio of 3aS,6aR:3aR,6aS isomers was 36:64. A disappointment was the complete inactivity of the thiophenone analogue of **2** to hydrogenation in the presence of rhodium catalysts with the most active ligands given above.

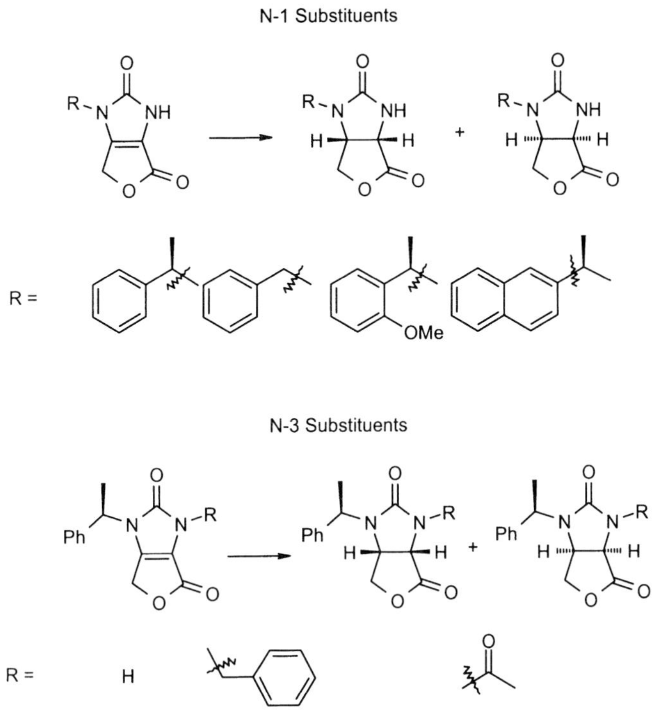

N-1 Substituents

R = (phenyl, m-tolyl, o-methoxyphenyl, 2-naphthyl, isopropyl group structures)

N-3 Substituents

R = H, benzyl, acetyl

Fig. 6 Effects of N-1 and N-3 substituents on the asymmetric hydrogenation of Furoimidazolones.

Tab. 2 Effect of N-1 substituents on the asymmetric hydrogenation of analogues of 2.

R Ligand	(R) 2-Phenylethyl		Benzyl		(R) 2-(2-Mephenyl)ethyl		(R) 2-2-Naphthylethyl	
	Act[a]	Sel	Act	Sel	Act	Sel	Act	Sel
(4R,5R)-DIOP	1.0	4:1	1.0	1:1	1.0	3:1	0.3	6:1
(4R,5R)-MOD-DIOP	10.0	85:15	5	>95:5	0.1	92:8	1.0	99:1
(R)-(S)-BBPF-OAc	2.0	70:30	1.0	76:23	0.2	83:17	1.5	99:1
(R)-(S)-PPF-P(^{t}Bu)$_2$	7.0	99:1	7.0	>95:5	1.0	99:1		

a) Abbreviations: Act=activity relative to (4R,5R)-DIOP=1, [Rh (0), S/C 1000:1, 30 bar H$_2$, 70°C, toluene, 60 h for complete conversion]. Sel.=diastereomer ratio.

Tab. 3 Effect of N-3 substituents on the asymmetric hydrogenation of analogues of **2**.

Substituent	H		Benzyl		Acetyl	
Ligand	Act[a]	Sel	Act	Sel	Act	Sel
(4R,5R)-DIOP	1.0	4:1	0.1	3:1	0.7	2:1
(4R,5R)-MOD-DIOP	10.0	85:15	0.05	60:40	1.0	68:32
(R)-(S)-BPPFOAc	2.0	70:30	0.2	56:44	0.4	40:60
(R)-(S)-PPF P(^{t}Bu)$_2$	7.0	99:1	0.03	–	1.0	57:43

a) Abbreviations: Act=activity relative to (4R,5R)-DIOP=1, [Rh(0), S/C 1000:1, 30 bar H$_2$, 70°C, toluene, 60 h for complete conversion]. Sel=diastereomer ratio.

3.2.2.3 Parameter Screening and Optimization

An extensive series (120) of parameter optimization experiments were carried out with the complex of [Rh(COD)Cl]$_2$ with (R)-(S)-PPF P^t(Bu)$_2$. The substrate was recrystallized (content >98.5%) to ensure consistency of results.

The diastereoisomer ratio (d.r.) was >99:1 for the following conditions:

Temperature: 50–100°C
Pressure: 7–50 bar
Concentration: 5–30%
S/C: 500–5000

Although Lonza disposes of high pressure hydrogenation reactors it was clearly advantageous to develop the hydrogenation for standard reactors qualified to 10 bar pressure. Several experiments were run whereby the catalyst solution was reused for further hydrogenations at an S/C ratio of 1000. It was found that after one recycle the performance was undiminished, however both the activity and selectivity decreased regularly between the second and fifth recycle. The effects of solvent on activity, selectivity, and isolated yield are given in Tab. 4.

Although the selectivity is identical for the solvents toluene, EtOAc, and ^{i}PrOH, and the activities are by and large similar, toluene was the solvent of choice due to the ease and efficiency of product isolation. In fact product **3** crystallized out during the reaction, and simple filtration after cooling to room temperature gave

Tab. 4 Solvent effects on the asymmetric hydrogenation of **2** with Rh(0)/(R)-(S)-PPF P^{t}Bu$_2$.

S/C	Solvent	°C	Time (h)	Conversion (%)	D.r.	Yield (%)
5000	Toluene	90°C	6	99.8	>99:1	96.6
5000	EtOAc	90°C	7	99.8	>99:1	91.1
5000	i-PrOH	90°C	7.5	99.6	>99:1	86
5000	CH$_2$Cl$_2$	50°C	6	6	>99:1	4.3

the yield quoted above. The optimized laboratory procedure for hydrogenation of 30 g of **2** was: S/C 3500, in 100 mL toluene, at 100 °C, under 10 bar H_2, with stirring at 400 rpm for 6 hours, yielded 29.1 g of **3** content 99.4% (content **2** 0.3%), d.r. >99:1, 96.3% yield.

3.2.3
Scale-up and Production

In a pilot campaign, altogether 17,392 kg of **3** were produced in 182 batches, each run in a 630 L hydrogenation reactor. The average yield was 92.2%, the average content was 99.0%, with <0.5% of the isomer **4**. In all 1% yield was lost in 19 of the total 182 batches due to incomplete conversion. This was almost certainly caused by incursion of oxygen, probably in the H_2 or N_2 lines. All attempts to revitalize a reaction that had stopped, through addition of fresh catalyst, were unsuccessful. The toluene solvent was successfully recycled after distillation under N_2.

The optimized parameters from the lab experiments were only slightly modified as follows:

- *Concentration*: 23 kg **2** per 100 L. Lower concentrations decreased the productivity and also the yield, due to dissolution of the product, higher concentrations reduced the conversion, as the product precipitation during the hydrogenation made the suspension too thick for efficient stirring.
- *S/C ratio*: Ratio **2**: [Rh(COD)Cl]$_2$ 8771:1; Ratio **2**: (R)-(S)-PPF P(^{t}Bu)$_2$ 5000:1. As the Josiphos was the more expensive component, it was used in a less than stoichiometric amount relative to the rhodium complex.
- *Hydrogen pressure*: 9 bar. The hydrogen pressure in the Lonza lines is 10 bar. The hydrogenation also takes place at 5 bar, however the reaction time then exceeds 10 hours.
- In a production campaign 12,879 kg of **3** were produced, standard batch size 1500 kg in a 10 m^3 reactor vessel. The average yield was 95.3%, average content >99%. Use of an oxygen-removing catalyst in the hydrogen and nitrogen lines ensured that conversion was complete in all batches.

In order to minimize the risk of a failed batch on the production scale, the S/C ratio was lowered to 4200:1. The TOF (Turn Over Frequency) was 700 per hour.

3.2.4
Preparation of Josiphos (R)-(S)-PPF-P(^{t}Bu)$_2$ on a Technical Scale

A significant constraint in the use of enantioselective ligands for technical asymmetric hydrogenation is their availability not only in the physical state in the quantities needed, but also to be free of patent or prohibitive license restrictions. The availability of the Josiphos ligands has been published in a recent review [5]. However at the time of our original research, the desired (R)-(S)-PPFP(^{t}Bu)$_2$ ligand was only available in sub-gram quantities. For the pilot campaign we prepared 8.0 kg, and for the production campaign we prepared 20 kg. For the first

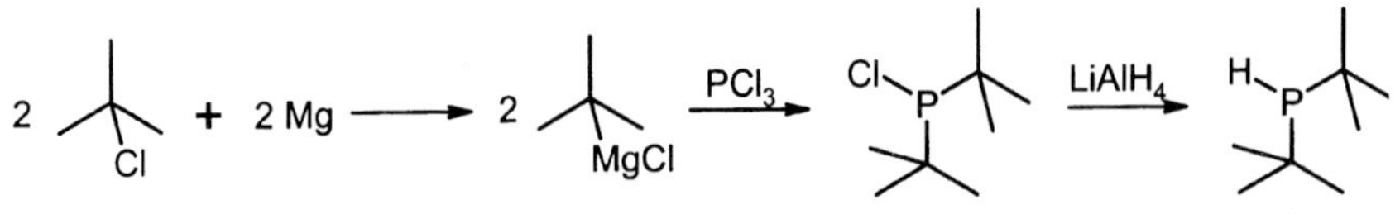

Fig. 7 Synthesis of di-^{t}butylphosphine.

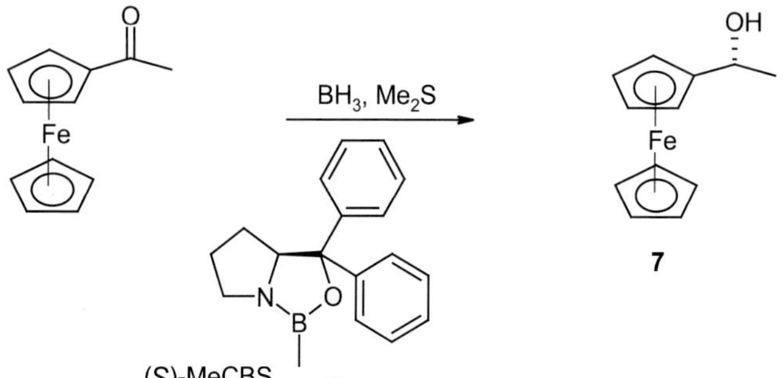

Fig. 8 Enantioselective synthesis of [1(R)-hydroxyethyl]ferrocene catalyzed by (S)-MeCBS.

production the standard lipase-catalyzed route [6] was used to generate (R)-(S)-PPFA. For both campaigns the necessary "exotic", pyrophoric, toxic and noxious HP(^{t}Bu)$_2$ **6** was not available and had to be prepared in-house. The ability to prepare such a reagent according to the route in Fig. 7 was a major contributing factor in advancing the project.

For the second production campaign of (R)-(S)-PPFP(^{t}Bu)$_2$, the (R)-hydroxyethylferrocene **7** was formed directly by borane reduction of acetylferrocene using (S)-MeCBS-oxazaborolidine catalysis [7] (see Fig. 8). Typically in laboratory studies, after conversion to the acetate with acetic anhydride a 91% yield with 88% ee was obtained. On a production scale 149 kg of crude **7** were prepared in 14 batches in a 500 L reactor in 88% yield, and 86% ee.

3.2.5
Conclusions

The diastereoselectivity reached, even on a production scale, was remarkable. Although the chiral phenylethyl substituent certainly enhanced the selectivity of the hydrogenation, the results with the prochiral benzyl-substituted analogue were also very impressive. A licensing agreement was rapidly reached between Lonza and Ciba-Geigy for the commercial production. As a result of turbulence in the market price of (d)-(+)-biotin, Lonza did not continue with its production after the first campaign.

3.3
The Lonza Dextrometorphan Process

3.3.1
Introduction

Dextromethorphan **9** is an important bronchodilating agent, which is on the market as an antitussive drug. The product is an alkaloid of the morphinan group, and it has three contiguous chiral centers. Fig. 9 shows the classical synthetic pathway for its preparation. The C=N bond of the critical hexahydroisoquinoline intermediate **10** is reduced by NaBH$_4$, leading to the formation of the racemic octahydroisoquinoline **11**. Finally after the resolution of the racemic mixture with mandelic acid, the morphinan skeleton is formed from the (S)-octahydroisoquinoline **12** by a stereoselective Grewe type cyclization. The absolute configuration of the chiral center of the isoquinoline induces the formation of the newly built chiral centers in the morphinan skeleton.

The classical resolution of the racemic octahydroisoquinoline via crystallization of a diastereomeric salt is very time and volume consuming, in particular because of the necessity of recycling the resolving agent and the undesired enantiomer. Several attempts have been made to develop an asymmetric hydrogenation route leading to **12**, using an enantioselective hydrogenation of the C=C double bond of the corresponding N-acyl enamides derived from **10** with Ru(II) complexes [8, 9]. By use of this methodology, the N-acylated octahydroisoquinoline derivatives are obtained in excellent optical yields. The disadvantage of this strategy is that only the Z-configurated enamides are hydrogenated, whereas the acylation of the hexahydroisoquinolines always leads to the formation of mixtures of the Z- and E-isomers and the latter has either to be separated or isomerized before hydrogenation.

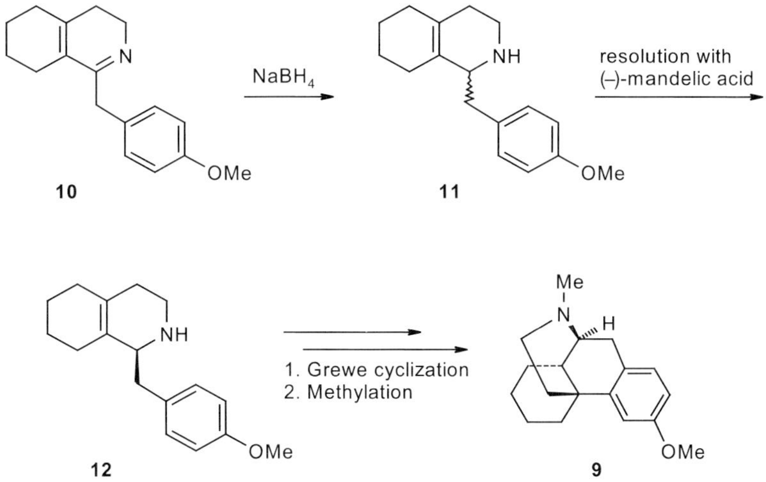

Fig. 9 Industrial process for the production of dextrometorphan.

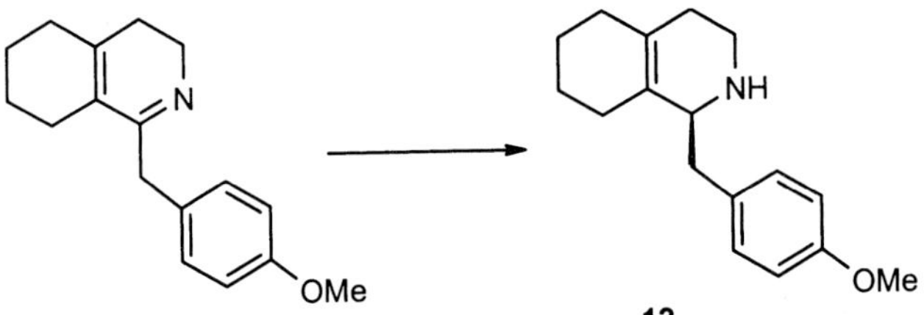

Fig. 10 Enantioselective hydrogenation of hexahydroquinoline **10**.

These additional unit operations make the economic advantage of this strategy questionable. Therefore a direct enantioselective hydrogenation of **10** to **12** (Fig. 10) was sought [10].

3.3.2
Imine Hydrogenation

3.3.2.1 Background

Although many highly enantioselective chiral catalysts and reagents are known for the hydrogenation of C=C and C=O bonds, the reduction of a C=N bond is still problematic [11, 12]. Despite the remarkably successful development of the (S)-metolachlor process by the (at that time) Ciba-Geigy group [13], the enantioselective hydrogenation of other substrates remains difficult. This can be due to several of the following reasons:

- *Catalyst activity*: many of the classical homogeneous hydrogenation catalysts have a low activity for the hydrogenation of imines. Some of the reactions require high pressures and temperatures (which may lead to lower selectivities).
- *Basic system*: both imines and amines are basic compounds; this can lead to deprotonation of the metal hydrides formed as intermediates in the catalytic cycle and finally cause a decomposition of the active catalyst.
- *Product inhibition*: amines are potential ligands for the transition metals of the catalyst; this can lead to product inhibition of the catalytic cycle.
- *Substrate stability*: many imines are not very stable. They can be hydrolyzed or form oligomers and trimers, and the formation of syn/anti or enamide isomers is also possible.

At the beginning of our investigations, we set the minimal objectives for the catalyst performance which would allow a future industrial application of the process. The following targets had to be reached: enantioselectivity >80%, S/C ratio >1500, TOF_{av} >200 per hour.

3.3.2.2 Screening Experiments

It rapidly became clear that iridium is the metal of choice. Using the BDPP$_{sulf}$ ligand we obtained the following results: ee=24.3%, S/C=100, TOF$_{av}$=20. An initial extended screening with over 40 different ligands was carried out with neutral iridium(I) catalysts, at room temperature, 70 bar pressure in a methanol/toluene mixture. Fig. 11 shows some of the most significant results obtained. Whereas most of the classical diphosphine ligands had a poor performance, the experiments with the Josiphos catalysts looked quite promising. As is evident from Section 3.2.4 the mode of preparation of these ligands permits an efficient fine tuning of the electronic and steric properties of the two phosphino groups, and so this possibility for ligand optimization was exploited.

It was found that the reproducibility and activity of pre-formed cationic [Ir(COD)ferrocenyl diphosphine] complexes were superior to those of the neutral systems. A series of hydrogenations was carried out in a bi-phasic toluene/water system (see below). As anticipated, the activity and selectivity of the catalysts is strongly influenced by the nature of the phosphorous substituents of the ferrocenyl diphosphine. These effects are summarized in Tab. 5. The final level of 89% enantioselectivity was achieved with the *tert*-butyl/4-methoxy-3,5-dimethylphenyl derivative.

Ligand	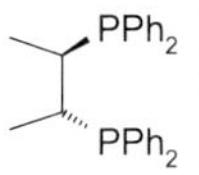 PPh$_2$ / PPh$_2$	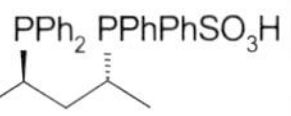 PPh$_2$ PPhPhSO$_3$H	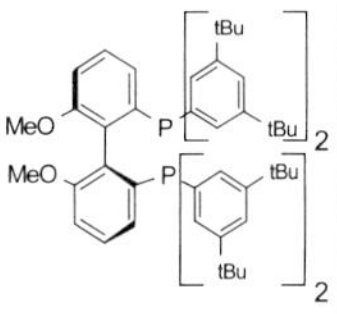 (MeO/tBu biaryl diphosphine)	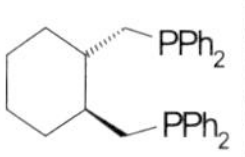 PPh$_2$ / PPh$_2$	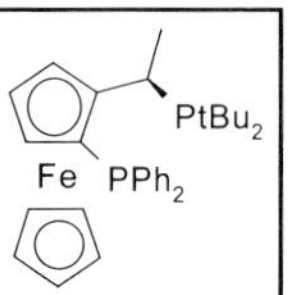 Fe PPh$_2$ / PtBu$_2$
S/C	100	100	200	100	400
% ee	38,4 (R)	24,3 (S)	13 (S)	18,9 (S)	68 (S)

Fig. 11 Effect of ligand types on the enantioselective hydrogenation of **10**.

Tab. 5 Effect of substituents on Josiphos ligands on the enantioselectivity and reactivity in the hydrogenation of **10**.

Ligand	S/C	Rel. activity	ee (%)
(R)-(S)-Et$_2$PF-P(^{t}Bu)$_2$	200	1	13.7 (R)
(R)-(S)-^{t}Bu$_2$PF-P(^{t}Bu)$_2$	200	1.7	14.9 (S)
(R)-(S)-PPF-P(^{t}Bu)$_2$	200	2.3	35.5 (S)
(R)-(S)-^{i}Pr$_2$PF P(^{t}Bu)$_2$	200	4.0	54.4 (R)
(R)-(S)-PPF-(^{t}Bu)$_2$	1500	125	85.5 (S)
(R)-(S)-4-(MeOXyl)$_2$PF-P(^{t}Bu)$_2$	2000	200	89.2 (S)

3.3.2.3 Optimization Phase: Development of an Economic Process

The hexahydroisoquinoline **10** has a very limited stability as a free base, as it has the tendency to undergo disproportionation reactions to the racemic octahydro and aromatic compounds, and it is also easily oxidized. These instability problems necessitate fresh preparation prior to every hydrogenation reaction on a lab and technical scale, which would strongly limit the flexibility of the production process. Fortunately we found that the hexahydroisoquinoline **10** forms a solid monophosphate salt, which is perfectly stable. This product also has the advantage that it is very easy to handle, whereas the free hexahydroisoquinoline is a sticky and viscous oil. In addition the isolation of the phosphate is a purification step leading reproducibly to a substrate of high quality, a very important point for the development of highly efficient catalytic processes. The phosphate salt of the imine **10** cannot be directly hydrogenated but needs to be deprotonated to the free base. We found that the *in situ* neutralization with NaOH in a biphasic toluene/water system worked well and that the hydrogenation actually takes place in the water phase containing sodium phosphate. The catalyst is temperature-sensitive and temperatures above 22 °C give a fast initial reaction but incomplete conversion due to catalyst decomposition. Temperatures below 15 °C are not desirable as the reaction rate is significantly lowered and full conversion is difficult to achieve due to the high amount of solids formed. The type and concentration of coordinating anions present in the reaction solution has a strong influence on the rate and to a certain extent on the enantioselectivity of the hydrogenation. The addition of a halide generally leads to a strong increase of the activity of the catalysts. The effect decreases in the order iodide > bromide > chloride. The halides are typically added in the form of their tetraalkylammonium salts. The effect of additives on the selectivity is dependent not only on the halide, but also on the nature of the ligand, and the hydrogen pressure. Fig. 12 shows an example of this complicated effect.

As even the optimized hydrogenation conditions gave rise to not more than 89% enantioselectivity it became very important to find an efficient purification procedure which would allow the enrichment of the desired (–)-enantiomer. A rapid screening of several weak and strong acids led to an acceptable solution. The acetates of both the racemate **11** as well as the (–)-enantiomer **12** are readily crystalline salts but the acetate of the desired (S)-enantiomer is less soluble in toluene at 0 °C than that of the racemate by a factor of 10. Thus it could be crystallized in overall yield of 84%, content 99.1%, 98.9% ee. Finally it was noticed that traces of iridium (70–200 ppm) in the acetate salt hindered the following reaction in the synthetic sequence, by leading to decomposition of the formic acid (used as formylation reagent). A treatment of the octahydroisoquinoline solution with charcoal prior to precipitation of the acetate salt lowers the Ir content to 20–40 ppm. This new quality can then be formylated without problems under the standard conditions.

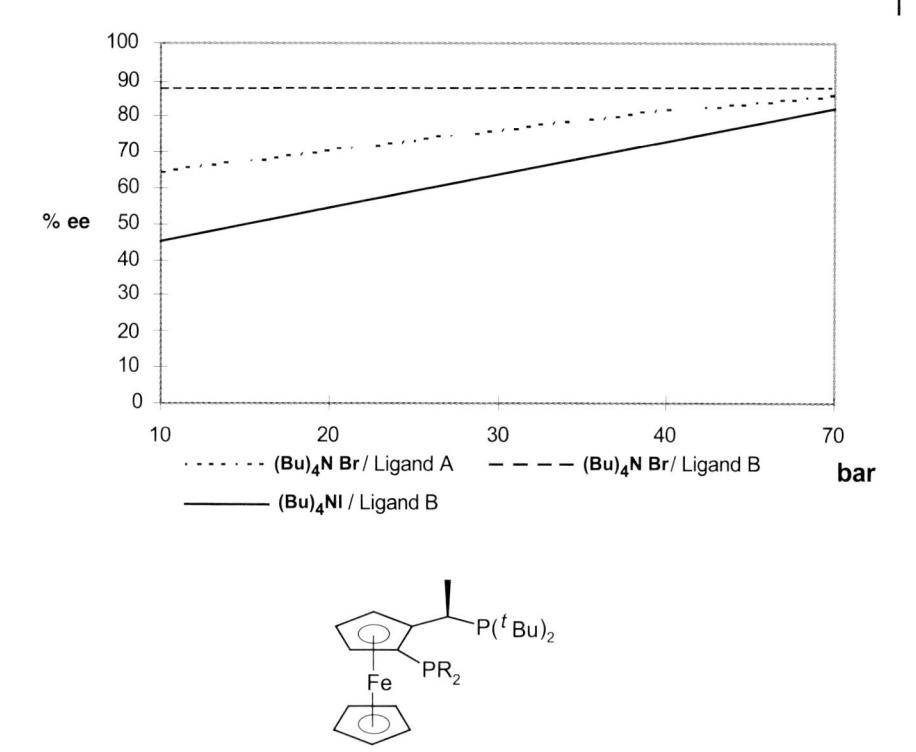

Fig. 12 Influence of hydrogen pressure, and additives on the selectivity of the hydrogenation of **10** in the presence of Josiphos ligands.

Ligand A: R = Ph
Ligand B: R = 4-OMeXyl

3.3.3
Scale-up

A scale-up of the laboratory process into a 100 L BUSS type loop reactor at 70 bar was not trivial. Altogether 28 batches each of 5 kg **10** (as the phosphate salt) were hydrogenated. The first batches reacted considerably more slowly than expected, gave incomplete conversions, and high variations from batch to batch. Two major problems were identified:

- The reaction system contained considerable amounts of solids (mostly sodium phosphates) which led to partial blockage and formation of crusts especially in the heat exchange tubes.
- As a consequence of this, the heat transfer was reduced and a high temperature gradient of up to 5 °C between inlet and outlet of the heat exchanger was observed.

Several changes to the initial conditions were tested in the laboratory and directly applied to the BUSS reactor:

- Increasing the excess of NaOH from 1.1 equiv. to 1.6 equiv. had only little influence on reaction rate and selectivity in the laboratory reactor but gave rise to a finer suspension.
- Increasing the amount of water led to increased dissolution of the solids.
- A change of the additive from NBu$_4$Cl to NBu$_4$Br was shown to give a significantly faster reaction without affecting the selectivity at high pressure.

The combination of these changes together with increasing the amount of catalyst then allowed us to attain complete conversion with good enantioselectivity and a TOF of 170 per hour.

The finalized stable process conditions, S/C 1100, temperature 20 °C, pressure 70 bar, reaction time 6 hours, led to a conversion of 91.5–92.9% and a crude product with 80.4–81.4% ee, which could be crystallized as the acetate with a content of >99%, an ee of 98.1–98.4% and a yield of 75.2–76.5% based on **10**.

In order to find conditions suitable for transfer to another Lonza plant, the performance of our process was further elaborated in a 30 L standard hydrogenation reactor. This reactor behaved very similarly to the laboratory autoclave and at 30 bar hydrogen pressure and an S/C ratio of 1500 good enantioselectivities and high conversions were possible. We could even conduct the reaction at 10 bar pressure, to deliver the acetate with content >99% and 98.3% ee with a TOF of 210 per hour, making the process amenable to the standard large scale hydrogenation reactors available at Lonza.

Owing to strategic reasons, Lonza exited the dextrometorphan business before the process could be transferred. Subsequently a further study of the hydrogenation of **10**, using iridium complexes with chiral amidophosphine-phosphonite ligands, was published by another group [14]. An enantioselectivity of 86% ee was reached, but the chemoselectivity was low.

3.3.4
Conclusions

An interesting point that emerges clearly from this study is that only one experiment with any catalyst system is insufficient to form a definite conclusion concerning its utility. For example, with reference to Fig. 12 the performance of the two ligands (R)-(S)-4-(MeOXyl)$_2$PF-P(^{t}Bu)$_2$ and (R)-(S)-PPF-P(^{t}Bu)$_2$ at hydrogenation pressures of 70 bar in the presence of bromide are similar, whereas at lower pressure the former is clearly superior. In the presence of iodide, the performance of the former ligand is strongly dependent on pressure, whereas in the presence of bromide it is pressure-insensitive.

3.4
4-Boc-piperazine-2(S)-N-ᵗbutylcarboxamide

3.4.1
Introduction

4-Boc-piperazine-2(S)-N-ᵗbutylcarboxamide **13** is an intermediate for the production of the orally active HIV protease inhibitor Crixivan® from Merck & Co. As **13** was designated as a starting material for the Crixivan® process, Lonza had complete freedom in choosing a synthetic route. Several approaches to **13** were investigated [2], of which the enantioselective hydrogenation routes will be discussed here. The development of the Lonza route was subject to the pressure of time, so that Lonza would qualify as a supplier for the NDA filing. The ongoing results of this research were communicated freely to, and discussed with Merck.

3.4.2
Initial Studies

Initial investigations showed that the direct hydrogenation of pyrazine carboxylic acid derivatives **14–16** was possible [15] (Fig. 13). This is one of the few examples known today for a homogeneous heteroarene hydrogenation [16]. With [Rh(NBD)Cl]$_2$ and a variety of standard ligands, for example Chiraphos, Norphos, Duphos, Deguphos, PPm, BDPP, BINAP, BPPFOAc no reaction was observed. Similarly no reaction could be observed with Ir catalysts. The positive results are given in Tab. 6.

The impracticality of the conditions, particularly the high S/C ratio necessary to achieve a reasonable conversion, aligned with the difficulty in enrichment of the desired enantiomer by crystallization led to the abandon of this variant.

The second approach, hydrogenation of tetrahydropyrazine derivatives was much more fruitful, as the substrate reactivity could be tuned by modification of the substituents on N-1 and N-4 [17]. Initially the esters were synthesized by partial hydrogenation of the corresponding pyrazine derivatives followed by monofunctionalization at N-1 (see Fig. 14). The corresponding amides could only be prepared through trapping the intermediate tetrahydropyrazine as the less reactive

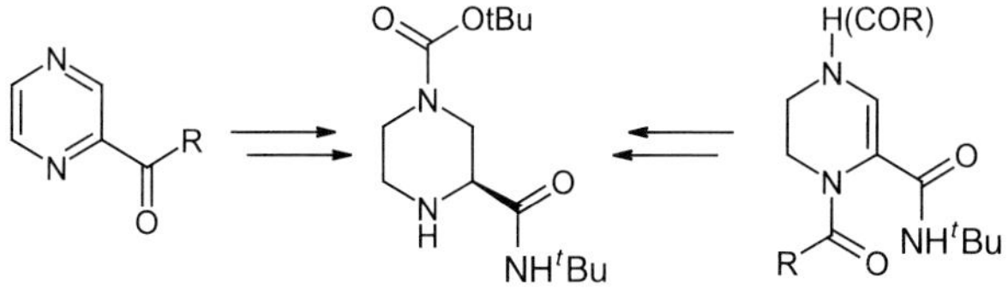

14 R= OMe
15 R= Ot-ᵗBu
16 R= NHt-ᵗBu

Fig. 13 Enantioselective hydrogenation of pyrazine and tetrahydropyrazine derivatives.

Tab. 6 Direct enantioselective hydrogenation of pyrazine carboxylic acid derivatives.

Substrate	Ligand	Conversion (%)	ee (%)
14	(R)-(S)-PPF-P(^{t}Bu)$_2$	95	4
15	(R)-(S)-PPF-P(^{t}Bu)$_2$	95	67
16	(R)-(S)-PPF-P(^{t}Bu)$_2$	59	58
16	(R)-(S)-PPF-PCy$_2$	80	77
16	(4R,5R)-DIOP	75	3

Conditions: [Rh(NBD)Cl]$_2$, S/C 20, MeOH, 70 °C, 50 bar, 20 hours.

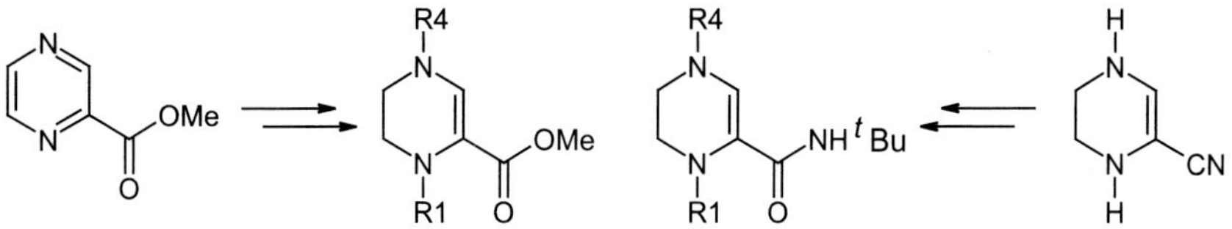

Fig. 14 Synthetic routes to tetrahydropyrazine carboxylic acid derivatives.

Tab. 7 Enantioselective hydrogenation of tetrahydropyrazine carboxylic esters **13**.

R1	R4	Metal	Ligand	S/C	Temp (°C)	Conversion (%)	ee (%)
H	H	Rh	(R)-(S)-PPF-P(^{t}Bu)$_2$	25	70	95	40
Ac	H	Rh	(R)-(S)-PPF-P(^{t}Bu)$_2$	25	70	95	80
Ac	Ac	Rh	(R)-(S)-PPF-P(^{t}Bu)$_2$	25	70	95	96
Ac	Ac	Rh	(+)-Chiraphos	25	130	90	80
Ac	Ac	Ru	R-BINAP	50	50	90	95
Ac	Ac	Rh	(R)-(S)-PPF-P(^{t}Bu)$_2$	250	50	95	90

[Rh(COD)Cl]$_2$, MeOH, 50 bar H$_2$, 24 hours.

N-1 acetyl derivative. A much more powerful synthesis of the *tert*-butylamides was the Ritter reaction of the corresponding nitrile, which allowed direct production of N-1-acylated tetrahydropyrazines [18]. These in turn could be selectively further functionalized at N-4 so that selective protective group removal was possible, giving access to the final N-4 Boc substitution pattern [2]. Some examples of the results obtained in the original screening are given in Tab. 7.

The positive effect of acyl substituents in the N-1 and N-4 positions is striking, both for the ester experiments above and also for the corresponding experiments with the amides **14**. The ruthenium BINAP system performed similarly to the rhodium (R)-(S)-PPF-P(^{t}Bu)$_2$ system, but as the (R)-(S)-PPF-P(^{t}Bu)$_2$ ligand had

Tab. 8 Enantioselective hydrogenation of tetrahydropyrazine carboxylic amides **14**.

R1	R4	S/C	Temp. (°C)	Time (h)	Pressure (bar)	Conversion (%)	ee (%)
Ac	Ac	500	90	4	50	94	94
C_2H_5CO	C_2H_5CO	1000	90	22	50	91	87
C_2H_5CO	$COCF_3$	400	90	17	50	95	88
C_2H_5CO	$CO_2{}^tBu$	475	90	20	20	50	90
Ac	CO_2Me	1000	90	20	50	97	84
Ac	CHO	480	90	1	50	99	97
C_2H_5CO	CHO	2000	90	2	50	99	96

$Rh(COD)_2BF_4$, (R)-(S)-PPF-P^{t}Bu$_2$, MeOH.

been produced in house, and as the licensing situation was less complicated, further studies and development were carried out with it. It was also remarked that cationic complexes of $Rh(COD)_2BF_4$ were significantly more active with the (R)-(S)-PPF-P(^{t}Bu)$_2$ ligand than were the neutral complexes, so they were used for further development. A selection of the results of experiments with the tetrahydropyrazine carboxylic amides **14** is given in Tab. 8.

The conclusions were as follows:

- The activity is superior with a propionyl substituent at N-1 than acetyl.
- The presence of a Boc (CO_2-^{t}Bu) group at N-4 diminishes the activity.
- The presence of a formyl group at N-4 increases the activity and selectivity.
- The selectivity was also high in EtOAc as solvent, however the activity was much superior in methanol.

The conditions given in the last entry in Tab. 7 were chosen for the pilot campaign.

3.4.3
Scale-up

A systematic parameter study indicated that the selectivity was unaffected by:

Pressure:	10–50 bar hydrogen
Temperature:	80–110 °C
Mole ratio Rh:ligand:	1.0 to 1.3
Concentration:	160–400 g/L, above this concentration the reaction was slower.

A total of 333 kg 4-formyl-1-propionyl-1,4,5,6-tetrahydropyrazine-2-carboxylic acid *tert*-butylamide was hydrogenated at 10 bar pressure in 4 batches in a 630 L stainless steel hydrogenation reactor in a yield of 96% and 95% ee. The only identified problem in the first batch was incomplete conversion. This could be remedied by further addition of catalyst. Finally an S/C ratio of 1000:1 was used. The TOF was

300 per hour. This result would most certainly be improved with further piloting. The product was hydrolyzed to remove the two protective groups and reacted with Boc anhydride to form the target 9 in 63% overall yield, average content 99.5%, average 99.6% ee.

3.4.4
Further Improvements in Synthetic Strategy

A drawback of the above strategy is that both protective groups are removed simultaneously, and so the introduction of the Boc group at N-4 is associated with selectivity problems. This problem can be avoided by the use of the trifluoroacetyl-protecting group at N-1, as it can be selectively removed with base in the presence of Boc or formyl groups in N-4 [2]. A selection of results of the hydrogenation of N-1 trifluoroacetyl derivatives (Fig. 15) is given in Tab. 9.
The conclusions were as follows:

- The trifluoroacetyl group increases the reactivity relative to the acetyl group.
- The Boc group at N-4 diminishes the reactivity relative to the acetyl group.
- Addition of 1 equiv. acid increases the reactivity dramatically.

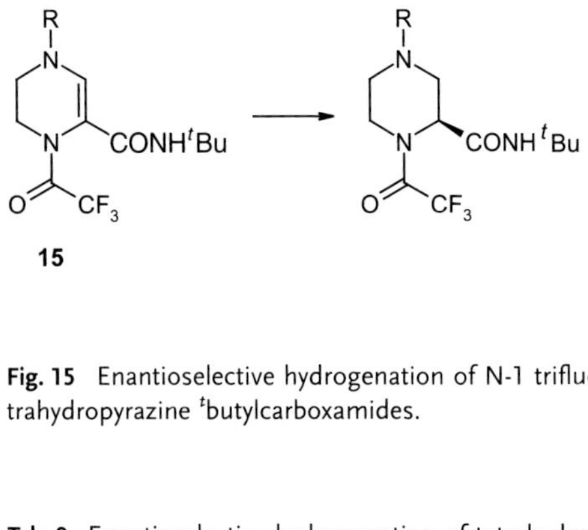

Fig. 15 Enantioselective hydrogenation of N-1 trifluouroacetyltetrahydropyrazine ^{t}butylcarboxamides.

Tab. 9 Enantioselective hydrogenation of tetrahydropyrazine carboxylic amides **15**.

R	Ligand	S/C	Acid	Solvent	Temp. (°C)	Time (h)	Conv. (%)	ee (%)
CHO	(R)-(S)-PPF-P(^{t}Bu)$_2$	600	–	MeOH	90	12	95	96
CHO	(R)-(S)-PPF-P(^{t}Bu)$_2$	820	–	MeOH	90	12	73	42
Boc	(R)-(S)-PPF-P(^{t}Bu)$_2$	200 a)	–	EtOAc	90	12	95	94
H	(R)-(S)-PPF-P(^{t}Bu)$_2$	500 a)	–	THF	116	6	91	79
H	(R)-(S)-PPF-P(^{t}Bu)$_2$	500 a)	H$_2$SO$_4$	THF	40	6	99	90
H	(R)-(S)-(^{t}Bu)$_2$PF-P(^{t}Bu)$_2$	1500 a)	H$_2$SO$_4$	THF	40	6	99	90

Rh(COD)$_2$BF$_4$,
a) Rh(COD)$_2$OAc, pressure 10 bar.

A major advantage of protonation of the N-4-unsubstituted **15** was that, after neutralization with ammonia, the desired enantiomer crystallized out in excellent yield and >99% ee.

3.4.5
Conclusions

Altogether 500 screening experiments were run in the laboratory. Although systematic conclusions could be made concerning the effect of substituents on reactivity and selectivity, no rational predictions of the most advantageous ligand could be made. Independently the Merck group developed two syntheses of **9** where the hydrogenation of N-4 Boc-substituted tetrahydropyrazines with Cbz [19] or formyl [20] substituents at N-1 were carried out with excellent enantioselectivity. The best conditions were found with 2 mol-% Rh-BINAP at 70 bar pressure and 7 mol-% at 100 bar pressure, respectively. Unfortunately, the preferred Lonza route to **9** was economically not competitive due to the unexpected high cost of trifluoroacetic acid.

3.5
Intermediates for SB-214857 (Lotrafiban)

3.5.1
Introduction

SB-214857 **16** is a potent GPIIb/IIIa receptor antagonist developed at Glaxo SmithKline pharmaceuticals (GSK) as a platelet aggregation inhibitor. The key intermediate, an optically active benzodiazepine carboxylic acid **17**, was produced by lipase-catalyzed enantioselective hydrolysis of the racemic methyl ester **18**, with subsequent racemization of the remaining (*R*)-ester by treatment with base [21, 22]. The racemic methyl ester was generated by catalytic hydrogenation of unsaturated precursor **19** (see Fig. 16). An alternative method of establishing the (2*S*)-acetic acid side-chain by incorporation of L-aspartic acid was also investigated by GSK [23].

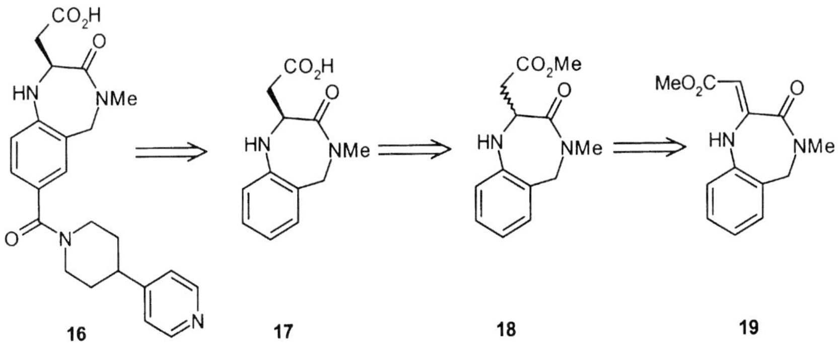

Fig. 16 Retrosynthetic analysis of **16** (SB-214857).

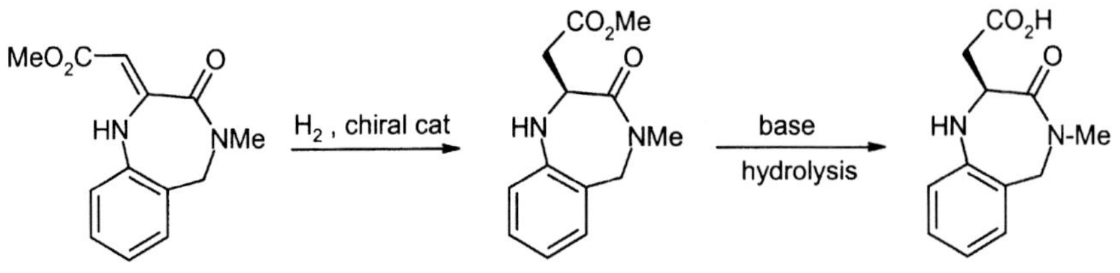

Fig. 17 Enantioselective hydrogenation of benzodiazepinone-substituted unsaturated carboxylic esters.

The Lonza approach, elaborated in discussion with GSK, was to carry out an enantioselective hydrogenation of **19**, and subsequently hydrolyze the (2S)-ester **20** (Fig. 17). A clear objective was to establish a process which could be validated within a time-frame of 3–6 months.

Although the possibility exists to modify the electronic distribution in the double bond through substitution at N-1, it was decided to avoid introducing extra steps, and to carry out the initial screening without substitution. An unexpected but welcome advantage of this approach was revealed by the much reduced solubility (by a constant factor of 10 over the temperature range –10 °C to +50 °C) of the racemate **18** relative to the (2S)-ester **20** in methanol. The technical simplicity of the crystallization of **18** to leave **20** in high optical purity in solution, allied with the very satisfactory enantioselectivities attained in the reaction itself, was decisive in fixing this strategy.

3.5.2
Initial Screening of Catalysts

In the initial screening, use of cationic Rh-complexes [Rh(COD)$_2$]BF$_4$ and diphosphines, with a substrate/catalyst ratio of between 20 and 50, led to a facile hydrogenation. Excellent conversions to **20** could be obtained at industrially attractive conditions (room temperature, <10 bar, <12 hours) using a variety of chiral diphosphines. A selection of the results is outlined in Tab. 10.

Surprisingly an experiment with [Rh(COD)Cl]$_2$ as metal complex and (R)-(S)-PPF-P(^{t}Bu)$_2$ as ligand led to negligible conversion. Furthermore, the use of other cationic complexes with the counter anions hexafluorophosphate, acetate or trifluoroacetate led to mediocre enantioselectivities, and so these variations were not pursued. From the ligands tested, the most promising was the member of the Josiphos family (R)-(S)-(Cy)$_2$PF-P(^{t}Bu)$_2$ followed by BDDP and (R)-(S)-PPF-P(^{t}Bu)$_2$. It was striking that the two related ligands (R)-(S)-PPF-P(^{t}Bu)$_2$ and (R)-(S)-PPF-P(phobyl)$_2$ (where phobyl is the bulky bicyclo[3.3.1]nonane substituent) show similar activities but inverse enantioselectivities. Some minor solvent effects were noticed on the activity and selectivity of the catalysts, however the preferential crystallization of the racemate in methanol as mentioned above, defined methanol as the solvent of choice. A number of cationic Ir-complexes were also tested for the

Tab. 10 Enantioselective hydrogenation of **19** (S/C = 20–50) using [Rh(COD)$_2$]BF$_4$.

Conditions	Ligand	Conversion (%)	ee (%)
A	(R,R)-Chiraphos	4	–
A	(R,R)-Deguphos	22	16 (R)
A	(S,S)-DIPMC	87	10 (R)
A	(R)-DIOP	98.6	54 (S)
A	(S,S)-BPPM	98.6	2 (R)
A	(S,S)-BDPP	99	78 (S)
A	(R)-(S)-(MeOXyl)PF-P(^{t}Bu)$_2$	85	22 (R)
A	(R)-(S)-PPF-P(4-MeOXyl)$_2$	99.6	20 (S)
A	(R,S)-BPPF-OAc	94	8 (R)
A	(R)-(S)-(Cy)$_2$PF-P(^{t}Bu)$_2$	98.7	86 (S)
A	(R)-(S)-(Cy)$_2$PF-P(Cy)$_2$	99	10 (R)
A	(R,S)-PPFA	No reaction	–
A	(R)-(S)-(4-MeOXyl)$_2$PF-P(^{t}Bu)$_2$	10	–
B	(S,S)-Me-Duphos	18	–
B	(R)-Binap	95	62 (R)
B	(R)-(S)-PPF-P(^{t}Bu)$_2$	99	76 (S)
A	(R)-(S)-PPF-P(^{t}Bu)$_2$	97	62 (S)
A	(R)-(S)-(Et)$_2$PF-P(^{t}Bu)$_2$	98	68 (S)
B	(R)-(S)-PPF-P(phobyl)$_2$	96.8	86 (R)

Conditions: A, MeOH, 23 °C, 20 hours, 6 bar; B, THF, 23–50 °C, 1–3 hours, 6 bar.

asymmetric hydrogenation of **19**, and although, enantioselectivities of >90/10 could be obtained in certain cases, the reaction was always accompanied by hydrolytic cleavage of the enamine functionality of **19** in substantial amounts up to 30%. For this reason the use of Ir-complexes was abandoned. Further screening with cationic Rh-complexes [Rh(COD)$_2$]BF$_4$ was carried out at industrially relevant S/C ratios of 500–2000, with the most successful ligand (R)-(S)-(Cy)$_2$PF-P(^{t}Bu)$_2$, accompanied by some other Josiphos ligands for comparison. The results are summarized in Table 11. At this phase of the experimentation, the ligand of choice was clearly defined, the conversion under economically attractive conditions was very good, and the enantioselectivity was satisfactory, given the ease of isolation.

Tab. 11 Catalytic enantioselective hydrogenation of **19** (S/C > 310) using [Rh(COD)$_2$]BF$_4$.

Conditions	Ligand	S/C	Conversion (%)	ee (%)
A	(R)-(S)-PPF-P(^{t}Bu)$_2$	400	99	25 (S)
A	(R)-(S)-PPF-P(^{t}Bu)$_2$	820	99	26 (S)
A	(R)-(S)-(Et)$_2$PF-P(^{t}Bu)$_2$	310	97.7	69 (S)
A	(R)-(S)-(^{i}Pr)$_2$PF-P(^{t}Bu)$_2$	310	97.3	80 (S)
A	(R)-(S)-(^{i}Pr)$_2$PF-P(^{t}Bu)$_2$	1000	99.4	71.6 (S)
A	(R)-(S)-(Cy)$_2$PF-P(^{t}Bu)$_2$	1000	94.2	88 (S)
A	(R)-(S)-(Cy)$_2$PF-P(^{t}Bu)$_2$	2175	98.9	88 (S)

3.5.3
Statistically Evaluated Screening of Experimental Conditions

For the next phase of experimentation 7 g of the required ligand, which was not commercially available, were synthesized from [1-(R)-(dimethylamino)ethyl]ferrocene using the established protocol for Josiphos ligands [5, 6]. During this time an initial screening of conditions was carried out using the analogous (R)-(S)-(^{i}Pr)$_2$PF-P(^{t}Bu)$_2$ ligand. Eight experiments were carried out to screen the substrate/catalyst ratio (range 1700–2300), temperature (15–45 °C), pressure (4–10 bar), methanol concentration (800–1800 g/mol **4**) and stirring rate (550–950 rpm). The Stavex® software and a Plackett-Burman-Plan (screening) were used. The reaction was followed by HPLC samples were taken every half hour.

The statistical evaluation done by the Stavex® program identified only temperature as a critical parameter. Temperatures in the range of 40 °C were necessary to complete conversion in reasonable times. The highest stereoselectivity and conversion were observed with the following conditions: 45 °C; 4 bar pressure; an S/C ratio of 2300; 800 g MeOH/mol; stirring rate 900 rpm; reaction time 5 hours. The desired enantiomer was obtained in 85% yield. A repetition with the newly prepared ligand (R)-(S)-(Cy)$_2$PF-P(^{t}Bu)$_2$ provided an ee of 87%. Interestingly, when the pressure was increased to 10 bar and the stirring rate dropped to 500 rpm, the enantioselectivity was considerably diminished from 88% to 37%.

On the basis of these results new ranges of the parameters for the optimization work (and scope of operation) were set: substrate/catalyst ratio (range 1700–2300), temperature (25–45 °C), pressure (4–10 bar), methanol concentration (800–1200 g/mol **19**). The stirring rate was not included for the subsequent statistical design. The vertex centroid plan (quadr.; D-opt., 19 runs) recommended by Stavex was used. The goal of the optimization was to refine conditions in order to ensure a robust and technically feasible process, and the measure of performance was the yield of the desired optically active **20**.

A series of 19 experiments were run each for a reaction time of 3 hours, after which period they were worked up, in all cases the racemate crystallized out, complicating the analysis. The statistical analysis of the data gave a good model for the conversion (see Fig. 18) but was not satisfactory for the enantioselectivity, and consequently not for the yield. This is believed to be at least partially due to the inhomogeneity of the reaction mixtures leading to imprecision of the analytical measurements.

The trends mentioned above could however be confirmed. The most significant factor determining the conversion is the temperature. High pressure combined with high temperature leads to a fast reaction but is detrimental to the enantioselectivity. The concentration of the reaction solution and the S/C ratio are less important within the ranges investigated. The optimal parameters proposed by the Stavex® program for optimum yield with a reaction time of 3 hours were essentially those in the best experiment: temperature 45 °C, pressure 4 bar, S/C 2300, MeOH 800 g/mol **19**. An additional experiment run under these conditions yielded, after 98% conversion, **20** in 87.3% yield, 99.7% ee, content 98.6% with

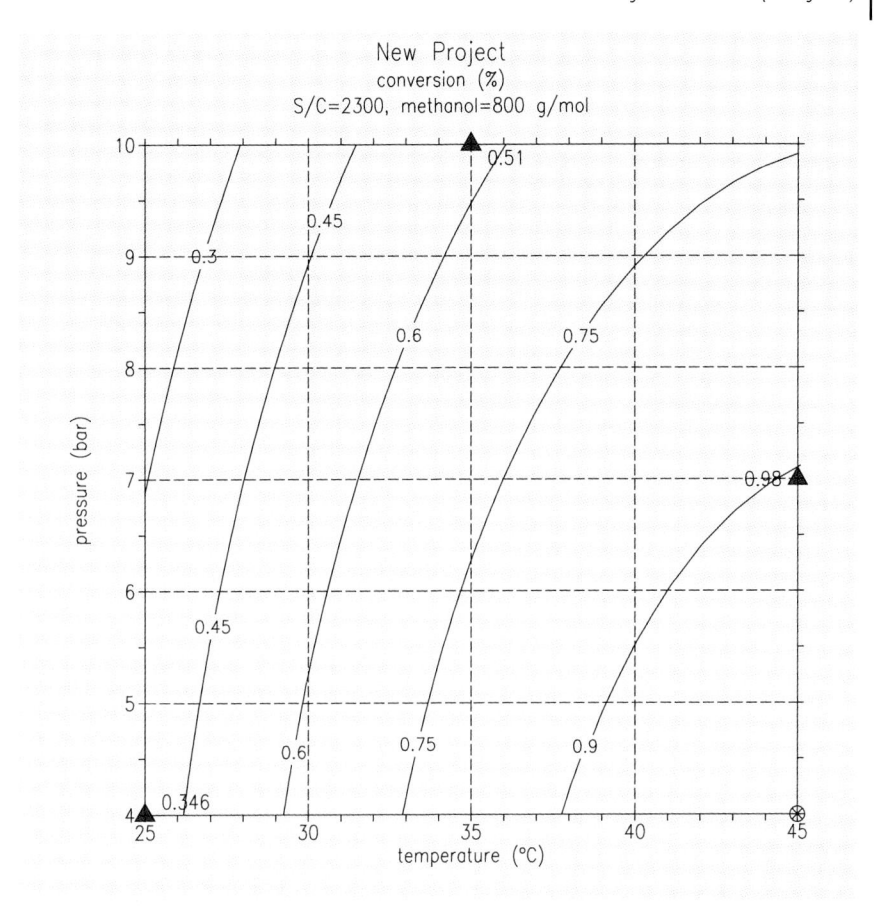

Fig. 18 Conversion of **19** to **20** as function of temperature and pressure (at S/C 2300 and 800 g MeOH/mol).

0.5% **19**, and highest unknown impurity 0.1 area-%. Of course the further hydrolysis step **20**–**17** was also optimized during the course of this study, leading to a straightforward and economical process.

As the proposed optimal parameters were at the extremes of the ranges tested, it was foreseen in the next phase of process development, i.e., establishment of qualified ranges for cGMP production, to examine higher temperatures, lower pressures and higher S/C ratios.

At this time GSK decided to suspend further experimentation.

3.6
References

1 J. F. McGarrity, L. Tenud, EP 273270, 1988.
2 W. Brieden, Proceedings of the Chiral USA'97 Symposium, p. 45.
3 M. Eyer, R. Fuchs, J. F. McGarrity, EP 602653, 1994.
4 J. F. McGarrity, F. Spindler, R. Fuchs, M. Eyer, EP 624587, 1994.
5 H.-U. Blaser, W. Brieden, B. Pugin, F. Spindler, M. Studer, A. Togni, Top. Catalysis 2002, 19, 3–16.
6 A. Togni, C. Breutel, A. Schnyder, F. Spindler, H. Landert, J. Am. Chem. Soc. 1994, 116, 4062–4066.
7 W. Brieden, USP 5760264, 1998.
8 M. Kitamura, Y. Hsiao, Y. Ohta, M. Tsukamoto, T. Ohta, H. Takayama, R. Noyori, J. Org. Chem. 1994, 59, 297–310.
9 B. Heiser, E. A. Broger, Y. Crameri, Tetrahedron Asymmetry 1991, 2, 51–62
10 O. Werbitzky,. WO 97 03052 (1997).
11 H.-U. Blaser, F. Spindler, in Comprehensive Asymmetric Catalysis (E. N. Jacobsen, A. Pfaltz, H. Yamomoto, Eds.), Springer, Berlin, 1999, p. 351.
12 T. Ohkuma, M. Kitamura, R. Noyori, in Catalytic Asymmetric Synthesis, (I. Ojima, Ed.), Wiley-VCH, New York, 2000, p. 83.
13 H.-U. Blaser, H. P. Buser, K. Coers, R. Hanreich, H. P. Jalett, E. Jelsch, B. Pugin, H. D. Schneider, F. Spindler, A. Wegmann, Chimia 1999, pp. 275–280.
14 E. Broger, M. Scalone, R. Schmid, Tetrahedron Asymmetry 1998, 9, 4043–4054.
15 R. Fuchs, EP 803502, 1996.
16 M. Studer, C. Wedermeyer-Exl, F. Spindler, H.-U. Blaser, Monatschr. Chemie 2000, 131, 1335–1344.
17 R. Fuchs, J.-P. Roduit, EP 744401, 1995.
18 W. Brieden, J.-P. Roduit, R. Fuchs, WO 9622981, 1996.
19 K. Rossen, S. A. Weissmann, J. Sager, R. A. Reamer, D. A. Askin, R. P. Volante, P. J. Rieder, Tetrahedron Lett. 1995, 36, 6419–6422.
20 K. Rossen, P. J. Pye, L. M. DiMichele, R. P. Volante, P. J. Rieder, Tetrahedron Lett. 1998, 39, 6823–6826.
21 I. P. Andrews, R. J. Atkins, N. F. Badham, R. K. Bellingham, G. F. Breen, J. S. Carey, S. K. Etridge, J. F. Hayes, N. Hussain, D. O. Morgan, A. C. Share, S. A. C. Smith, T. C. Walsgrove, A. S. Wells, Tetrahedron Lett. 2001, 42, 4915–4917.
22 A. S. Wells, WO 9829561.
23 J.-B. Clement, J. F. Hayes, H. M. Sheldrake, P. W. Sheldrake, A. S. Wells, Synlett 2001, 9, 1423–1427.

4
Large-Scale Applications of Biocatalysis in the Asymmetric Synthesis of Laboratory Chemicals

Roland Wohlgemuth

Abstract

Although biocatalysts were used in organic synthesis early on in the history of Fluka, if chosen at all, biocatalysis was applied only for those products where all other synthetic methods worked poorly or failed. In the last 15 years, biocatalysts have been used more and more at Fluka and today biocatalysis has become an established manufacturing tool for the synthesis of chiral compounds, metabolites, and natural products. The combination of biotechnology and organic chemistry has become a synthetic strategy even at the planning level.

The large-scale resolution of racemic 2-alkanols to enantiomerically pure (R–)- and (S+)-alkanols by lipase shows environmental, health and safety advantages of biocatalysis compared with nonenzymatic syntheses. The biocatalytic Baeyer-Villiger oxidation of racemic bicyclo[3.2.0]-hept-2-en-6-one to enantiomerically pure 2-oxabicyclo[3.3.0]-oct-6-en-3-one and 3-oxabicyclo[3.3.0]-oct-6-en-2-one by CHMO represents an asymmetric Baeyer-Villiger oxidation.

Prerequisites for the large-scale applications are a rapid production technology for the biocatalysts, techniques for the stabilization of the biocatalyst and the use of the stable biocatalysts in appropriate reactors.

4.1
Introduction

Asymmetric synthesis of chiral molecules by any type of technology has always been a central theme in organic chemistry and is, furthermore, of vital and fundamental interest for understanding chiral biology in the life sciences. In addition chirality is a key factor in the interaction of man-made new molecules with biopolymers from nature and therefore the mode of action of each enantiomer in a racemic mixture is of enormous practical importance.

The spontaneous separation of the two enantiomers in a racemic mixture sodium/ammonium tartrate tetrahydrate into enantiomorphic crystals and their subsequent manual separation by Louis Pasteur [1] in 1848 started the road to modern stereochemistry. The resolution of racemic mixtures by chemical and biocatalytical methods has been and continues to be a major source of new single enantiomers. In contrast to the separation of enantiomers in a racemic mixture, symmetry-breaking synthetic operations are powerful synthetic tools, providing a direct and quantitative access to the required single enantiomer. This can be achieved by the use of a chiral auxiliary or a chiral catalyst, which are able to perform the symmetry-breaking operation by differentiating two faces of an achiral structure or by differentiating two enantiotopic functional groups in an achiral or meso molecule. A traditional way of producing such chiral molecules is by using the abundant renewable resources from nature (the chiral pool) and performing classical synthetic organic reactions making sure that the handedness of the molecules, which has been created by the natural bio-catalytic pathways, is preserved under reaction conditions. The advances in the stereochemistry of organic compounds have been accompanied by progress in understanding the structure and function of nature's catalysts [2–5].

The number and types of biocatalytic reactions is steadily increasing [6] and a growing number of processes are transferred from the laboratory to the pilot plant and manufacturing units [7]. This is because of the increasing importance of chiral, enantiomerically pure laboratory chemicals in such diverse areas as the development of efficient and safe new pharmaceuticals, the early introduction of chirality into synthetic schemes in organic chemistry by using chiral building blocks or as chiral auxiliaries, as standards and auxiliaries for the creation of chiral molecular interactions in analytical technologies and also for applications concerning food, agro- and environmental institutions [8]. In addition, the quality improvements of reagents and the elimination of unwanted side effects have led to a continuous need to provide molecules with the proper chirality [9]. It is therefore not surprising that the manufacturers of laboratory chemicals have also built up expertise in the preparation of these products.

4.2
History of Applied Biocatalysis at Fluka

Hydrolases were in the first catalogue after the company was founded in 1950 but, not surprisingly, the chiral molecules originated mainly from the chiral pool. The first biocatalytic reactions were developed with kidney acylases and later with esterases and lipases, in the beginning mainly animal-derived biocatalysts [10]. The set-up of in-house biocatalyst production from microbial and plant sources as well as the construction of a new biotechnology laboratory with ten fermenters of up to 300 L total volume, allow the development and production of improved biocatalysts and for them to be applied in the asymmetric synthesis of laboratory chemicals. There are today more than 100 biocatalytic processes in routine production and a project management team is handling custom biotransformations.

4.2.1
Resolution of Racemic Mixtures

Although the resolution of racemic mixtures is a rather old technique, it is still of great interest, because technology platforms can be developed for specific classes of racemic compounds.

4.2.1.1 Resolution of Racemic Amino Acids

The resolution of *N*-acetyl-D,L-amino acids to prepare non-natural L- and D-amino acids was the beginning of applied biocatalysis, and aminobutyric and di-amino-butyric acids as well as beta-D,L-phenylalanines can be resolved. A series of microbial acylases from *Streptomyces, Alcaligenes, Comamonas* and *Pseudomonas* species were produced for these applications. Immobilized acylase on Eupergit C or the use of membrane reactors allow the facile production of such chiral amino acids.

4.2.1.2 Resolution of Racemic Alcohols

In the case of the resolution of alcohols, lipases are used to prepare both enantiomers of 2-alkanols, phenylethanols and phenylpropanols, furylethanols, binaphthols and cyclo-pentene-1,3-diols. Both enantioselective acylations with irreversible acyl donors in organic solvents as well as enantioselective hydrolyses in aqueous buffer solutions are in use to achieve the resolutions. Immobilized lipases for process improvements have been prepared and are available.

4.2.2
Decomposition Reactions of Educts and Side Products

The oxidation of *N,N*-dimethylalkylamines with hydrogen peroxide to *N,N*-dimethyl-alkylamin-*N*-oxides leaves some hydrogen peroxide as unreacted educt, which needs to be removed to a level of less than 0.03%. This removal can be achieved nicely by the use of catalase after the completion of the reaction.

4.2.3
Introduction and Removal of Protecting Groups

The need to have orthogonally stable blocking functions that can be selectively removed under mild conditions excludes the use of many classical chemical protecting groups. Enzymatic protecting group techniques offer viable alternatives if the proper combination of enzyme and protecting group is available [11]. The introduction of an acyl group in primary versus secondary alcohols, e.g., in glycerol, can be achieved with lipase from *Candida antarctica*.

If the removal of a protecting group by chemical methods fails in a lengthy synthesis, this can be quite costly and therefore the availability of biocatalysts on the shelf is very useful. The removal of the acetoxy-group in (*S*)-2-acetoxy octanoic acid *tert*-butylester is catalyzed by lipase from wheat germ. Thereby the corresponding (S)-2-hydroxy octanoic acid *tert*-butylester is obtained with excellent enantiomeric purity ($S:R = 98.9:1$), while all chemical procedures gave side reactions and esterases did not work at all or very slowly.

4.2.4
Development and Production of Biocatalysts

The search for the ideal biocatalyst [12] and the subsequent development and production of robust and reproducible biocatalysts are key success factors for establishing viable large-scale applications. In order to fully exploit the capabilities and potential of biocatalysis for catalyzing a wide variety of chemical reactions, it is important to enlarge the collection of available biocatalysts. The practical application of biocatalysts depends on adequate bioinformatics tools, stable biocatalysts-in-the-bottle and efficient screening technology [13]. In order to interface the biocatalysts with chemical reaction steps, modules of biocatalyst classes are useful and can be put together in collections [6]. The large-scale production of biocatalysts is facilitated by the construction of recombinant strains over-expressing active enzyme and the subsequent fermentation process development. As an example of the large-scale availability of Baeyer-Villiger enzymes [12], we have scaled-up the fermentation of a recombinant *E. coli* strain expressing the cyclohexanone-monooxygenase (CHMO) gene from *Acinetobacter* sp. Enzyme isolation via cell harvesting, homogenization, centrifugation, and ammonium sulfate fractionation with 50% and 80% enabled us to produce the isolated CHMO at 12 U/mL with excellent stability. Stabilization for storage has been achieved by 20 mM KNa-phosphate buffer pH 7, 80% ammonium sulfate, and 3.5 mM DTE. Stability under process conditions is another biocatalyst property of practical importance and immobilization techniques such as covalent attachment on Eupergit C or sol-gelentrapment of biocatalysts are used routinely in our applications.

4.2.5
Complex Reaction Paths to Natural Products

Although a huge number of enzymes are known [15, 16] and can be isolated, there is a universe of biocatalysts in nature and by design which has not yet been explored and remains to be discovered. The pathways of natural product biosynthesis explored in the last century contain a wealth of stereochemical knowledge about complex secondary metabolites and is still of great interest [17]. As an example of this type of asymmetric synthesis the production of nonactin with *Streptomyces griseus* is shown in Fig. 1 [18]. In this case, the biocatalytic machinery of the secondary metabolism of whole cells is used to accumulate the natural product, which is then further processed to the purified nonactin. It is interesting to note that nonactin is composed of building blocks of nonactic acid, which have opposite configurations.

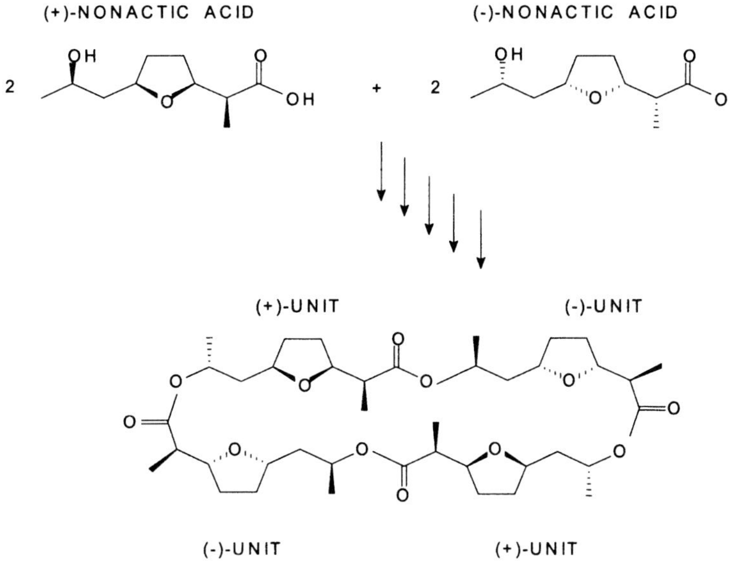

Fig. 1 Asymmetric synthesis of (+)- and (–)-nonactic acid and cyclic (+)(–)(+)(–)-linkage of (+)-nonactic acid and (–)-nonactic acid in the synthesis of nonactin.

4.2.6
Regio- and Stereoselective Synthesis

The framework construction of products from building blocks with concomitant development of functionalities has been and continues to be a major synthetic approach in organic chemistry. It is therefore not surprising that the interest in developing regio- and stereoselective reactions making use of biocatalysts is growing rapidly, because the construction approach demands exact stereochemical control over separate fragments and that is what enzymes have been optimized to do by nature.

The aldol reaction is one of the most important reactions for the construction of C–C bonds and therefore the search for asymmetric methods progresses along all branches of inorganic and organic catalysis as well as biocatalysis [19, 20]. Practical concerns for large-scale applications make commercially available, inexpensive small-molecule catalysts such as L-proline very attractive [21], not only for aldol reactions, but also for mechanistically related addition reactions such as Mannich- and Michael-type reactions. The large-scale availability of a variety of microbial aldolases as well as the necessary donors such as dihydroxyacetone phosphate give access to a wide range of potential targets that can be synthesized by enzymatic aldol-reactions [22]. The scope of preparative applications for transketolases is rapidly expanding because of their high enantioselectivity [23].

An area where regio- and stereoselective synthesis with biocatalysts is particularly attractive because of highly specific reactions are carbohydrate synthesis and glycosylation of natural products [24, 25]. We have prepared a series of glycosyl transferases with different bond-specificities such as beta-1,4- and alpha-1,3-galactosyl transferases in stable and usable form for carbohydrate synthesis and the glycosylation of natural products. Thereby the purification needs to take into account that glycosidases have to be removed in order to obtain the maximum yield of product per unit of glycosyltransferase.

4.3
Reasons for Large-Scale Application of Biocatalysts for the Synthesis of Laboratory Chemicals

The reasons for applying biocatalysts are mainly that existing synthetic tools do not work when preparing a compound or that there are significant method problems. If complicated protection-deprotection schemes have to be used in addition to the reaction, the costs of the process might become too high, and additional energy and waste disposal costs will add to these expenses. Direct biocatalytic steps have the potential to reduce the number of reaction steps and also to yield favorable environment, health and safety risk analyses. Taken together, the above-mentioned reasons can lead to decreased production costs or to the basic feasibility of the synthesis under given conditions.

4.4
Examples of Large-Scale Application of Biocatalysis

At Fluka, we have, for safety reasons, decided to split the large-scale application of biocatalysts into two zones, which have different requirements with respect to reactors, instrumentation, and building infrastructure. One zone allows working under sterile, but not explosion-safety conditions, and a separate zone has infrastructure and reactors with explosion-safety, but not sterile conditions. The advantage of this configuration is that the biocatalyst production in fermenters can be coupled with the application of the biocatalyst in the normal chemical production whereby existing chemical reactors can be used. This translates directly into enormous cost-savings. In addition, standard work-up procedures from organic chemistry can be used. Large-scale at Fluka is defined either as the maximum volume of the reactor or as the maximum amount of product achievable in a continuous process.

4.4.1
Matrix Approach: Product Groups Produced with Specific Enzyme Classes

In order to structure the technology platform, various product groups produced with the same enzyme class are continuously enlarged based on prior experience. In Fig. 2, a series of chiral products prepared with hydrolases are shown and they are usually obtained with very high enantiomeric purities. Lipases and esterases have been the main contributors to the breakthrough of enzymatic methods in or-

Fig. 2 Products made at Fluka by lipase applications.

ganic chemistry and have been assembled into three kits together with an irreversible acyl donor kit. The advantage of the matrix approach with biocatalysts and product groups produced with these biocatalysts is the knowledge base in continuous process improvement and growth in two dimensions.

4.4.2
(S)-2-Octanol

As an example of an enantiospecific acylation in organic solvent with the irreversible acyl donor vinyl laurate (see Fig. 3), only the (R)-enantiomer is acylated, while the (S)-2-octanol is obtained directly in 69% yield, a purity >99% and with an excellent enantiomeric ratio $S:R > 99.5:0.5$ (as determined by chiral GC-analysis). The selectivity of the forward reaction, catalyzed by a lipase in MTBE, is thereby maximized, because the use of the enolester precludes the reverse reaction. Batches of about 70 kg are produced routinely and the technology can be easily transferred to larger sizes.

Yield	69%
Purity	> 99 %
Enantiomeric Ratio	S:R > 99.5:0.5

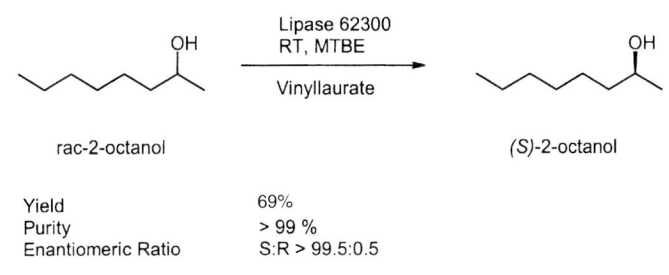

Fig. 3 Lipase applications: (S+)-2-octanol enantiomer.

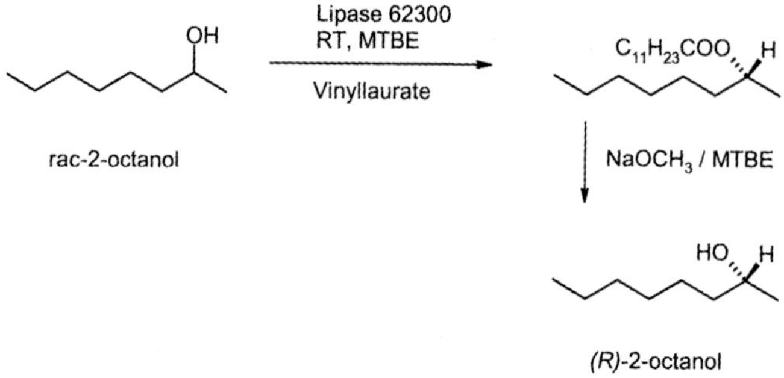

316 | 4 Large-Scale Applications of Biocatalysis in the Asymmetric Synthesis of Laboratory Chemicals

rac-2-octanol → (Lipase 62300, RT, MTBE, Vinyllaurate) → $C_{11}H_{23}COO$-R intermediate → (NaOCH$_3$ / MTBE) → (R)-2-octanol

Yield	77 %
Purity	> 99 %
Enantiomeric Ratio	R:S > 99.5:0.5

Fig. 4 Lipase applications: (R–)-2-octanol enantiomer.

4.4.3
(R)-2-Octanol

The opposite enantiomer (R)-2-octanol is obtained in 77% yield after removal of the acyl group, as shown in Fig. 4. The purity of > 99% and the enantiomeric ratio of $R:S > 99.5:0.5$ (determined by chiral GC) is also excellent. The acylated (R)-2-octanol is obtained as the first intermediate in the above-mentioned lipase-catalyzed reaction in MTBE and has to be separated from unreacted (S)-2-octanol.

4.4.4
Comparison of Classical Resolution with the Biocatalytic Procedure

The practical advantages of the biocatalytic versus the classical procedure at large-scale consist of a) a reduced number of steps in the biocatalytic reaction (1 step) versus the classical organic resolution (5 steps), b) reduced production costs compared with the resolution with brucine, c) a reduced reactor volume of only 20%, d) improved environmental, health and safety conditions, because no toxic agent is involved, and e) expanded volume capacity, because no investment in additional equipment is necessary with a 5-fold scale-up.

4.4.5
2-Oxabicyclo[3.3.0]-oct-6-en-3-one

The (1S,5R)-2-oxabicyclo-octenone (see Fig. 5) is formed with 96% ee from one ketone enantiomer of racemic bicyclo[3.2.0]hept-2-en-6-one, which is synthesized in a cycloaddition reaction from cyclopentadiene and dichloroacetylchloride with subsequent Zn-reduction. The key limitation to the biocatalytic Baeyer-Villiger process is product inhibition [26]. In order to overcome product inhibition problems, *in situ* product removal was required.

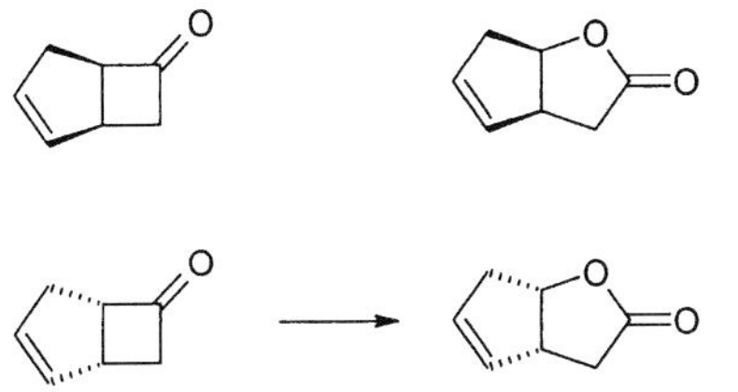

Fig. 5 Biocatalytic BV-oxidation of bicyclo[3.2.0]-hept-2-en-6-one with *E. coli* pQR209.

(+)-*(1R,5S)*-2-oxa-bicyclo [3.3.0]-oct-6-en-3-one

(–)-*(1S,5R)*-2-oxa-bicyclo [3.3.0]-oct-6-en-3-one (96% ee)

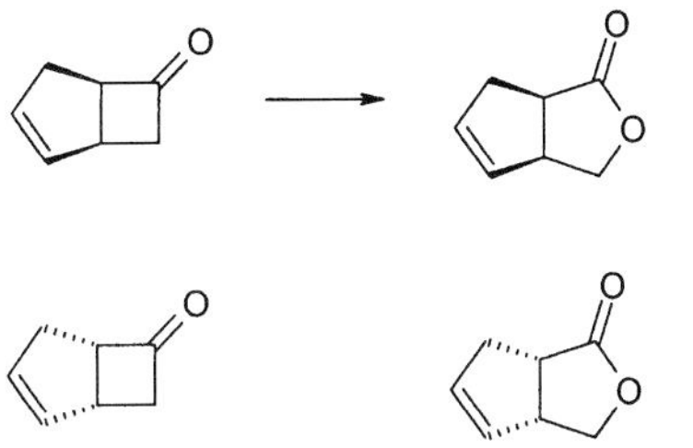

(–)-*(1R,5S)*-3-oxa-bicyclo [3.3.0]-oct-6-en-2-one (> 98% ee)

(+)-*(1S,5R)*-3-oxa-bicyclo [3.3.0]-oct-6-en-2-one

Fig. 6 Biocatalytic BV-oxidation of bicyclo[3.2.0]-hept-2-en-6-one with *E. coli* pQR209.

4.4.6
3-Oxabicyclo[3.3.0]-oct-6-en-2-one

In addition, the Baeyer-Villiger oxidation of racemic bicyclo[3.2.0]-hept-2-en-6-one with *E. coli* pQR209 gives about 50% of nearly enantiomerically pure (1R,5S)-3-oxabicyclo octenone as shown in Fig. 6; this regioisomer cannot be made selectively by chemical Baeyer-Villiger oxidation. In both cases, a yield of product on substrate of 85 to 90% has been obtained. Applying a ketone feed rate which keeps the reactant concentration below the inhibitory concentration was essential.

4.4.7
Comparison of the Classical Procedure with the Biocatalytic Baeyer-Villiger Reaction

A comparison between the classical chemical oxidation system with MCPBA and the biocatalytic Baeyer-Villiger reaction clearly shows that MCPBA has lower product selectivity. Even when the formation of 34% epoxylactone and 37% epoxyketone can be inhibited by using zeolite-catalysts [27], thereby avoiding double bond oxidation as a side reaction, the product is not chiral as is the case with the biocatalytic Baeyer-Villiger reaction. The ratio of 2- to 3-lactone depends on the oxidation system as well as on the conditions, but the 2- and 3-lactone can easily be separated to give the pure chiral products [26]. The replacement of peracids or hydrogen peroxide/catalyst-mixtures by biocatalysts has distinct advantages with respect to process safety and is useful for the introduction of chirality in prochiral ketones or by divergent oxidation of racemic ketones to regioisomers of chiral lactones.

4.5
Discussion and Outlook

The large-scale application of biocatalysis requires an integrated approach of biology, chemistry, and engineering, taking advantage of the progress in each of these disciplines. The development technology for the biocatalyst is essential as well as the process of going from the first sample to the first kg amounts of the biocatalyst. Efficient isolation procedures for the final product are important to take full advantage of the product yield in the biocatalytic reaction. Novel asymmetric methods for the synthesis of chiral compounds can open new opportunities for the preparation of the chiral pharmaceuticals of the future. The growth in chiral therapeutics and their chiral intermediates continues and the production of single enantiomers is increasingly important [8]. The availability of different classes of biocatalysts has enabled a broad variety of chiral compounds to be synthesized. Progress in microbiology, biochemistry, and fermentation technology has provided access to a variety of enzymes and microbial cells as tools for organic synthesis. The introduction of biocatalytic reactions into a whole production process can harmonize improvements in economic efficiency with the solution to environmental, health and safety problems [28]. The use of biocatalysts is on the verge of significant growth for industrial synthetic chemistry [29–32], for convenient solutions to certain intractable synthetic problems [33], and for the expansion of the repertoire of chemistry by modular biocatalysts [34].

4.6
References

1 L. PASTEUR, Ann. Phys. 1848, 24, 442.
2 E. FISCHER, Berichte der deutschen chemischen Gesellschaft 1894, 27, 2892, 3229.
3 V. PRELOG, Chirality in Chemistry, Nobel Lecture, December 12, 1975. http://www.nobel.se/chemistry/laureates/1975/.
4 E.L. ELIEL, S.H. WILEN, Stereochemistry of Organic Compounds, John Wiley and Sons, 1994.
5 R. NOYORI, Asymmetric Catalysis in Organic Synthesis, Wiley, New York, 1994.
6 K. DRAUZ, H. WALDMANN, Enzyme Catalysis in Organic Synthesis: A Comprehensive Handbook, second, completely revised and enlarged edn., Wiley-VCH, Weinheim 2002. Vol. I–III.
7 A. LIESE, K. SEELBACH, C. WANDREY, Industrial Biotransformations, Wiley-VCH, Weinheim 2000.
8 A.M. ROUHI, Chem. Eng. News 2002, 80, 43.
9 Food and Drug Administration, Chirality 1992, 4, 338.
10 R. WOHLGEMUTH, Chimia 1999, 53, 543.
11 E. NÄGELE, M. SCHELHAAS, N. KUDER, H. WALDMANN, J. Am. Chem. Soc. 1998, 120, 6889.
12 S.G. BURTON, D.A. COWAN, J.W. WOODLEY, Nature Biotechnol. 2002, 20, 37.
13 Y. ASANO, J. Biotechnol. 2002, 94, 65.
14 V. ALPHAND, G. CARREA, R. WOHLGEMUTH, R. FURSTOSS, J.M. WOODLEY, Trends Biotechnol. 2003, 21, 318.
15 I. SCHOMBURG, A. CHANG, D. SCHOMBURG, Nucl. Acids Res. 2002, 30, 47. http://www.brenda.uni-koeln.de.
16 A. BAIROCH, The ENZYME database in 2000, Nucl. Acids Res. 2000, 28, 304. http://us.expasy.org/enzyme/.
17 R. BENTLEY, Crit. Rev. Biotechnol. 1999, 19, 1.

18 R.J. Walczak, A.J. Woo, W.R. Strohl, N.D. Priestley, FEMS Microbiol. Lett. 2000, 183, 171.
19 T.D. Machajewski, C.H. Wong, Angew. Chem. 2000, 39, 1352.
20 W.D. Fessner, V. Helaine, Curr. Opin. Biotechnol. 2002, 12, 574.
21 K. Sakthivel, W. Notz, T. Bui, C.F. Barbas III, J. Am. Chem. Soc. 2002, 123, 5260.
22 R. Wohlgemuth, INBIO Europe, Scientific Update, Amsterdam, 2000.
23 N.J. Turner, Curr. Opin. Biotechnol. 2000, 11, 527.
24 Z. Guo, P.G. Wang, Appl. Biochem. Biotechnol. 1997, 68, 1.
25 S. Riva, Curr. Opin. Chem. Biol. 2001, 5, 106.
26 S.D. Doig, P.J. Avenell, P.A. Bird, P. Gallati, K.S. Lander, G.J. Lye, R. Wohlgemuth, J.M. Woodley, Biotechnol. Progr. 2002, 18, 1039–1046.
27 A. Corma, L. Nemeth, M. Renz, S. Valencia, Nature 2001, 412, 423.
28 J. Ogawa, S. Shimizu, Curr. Opin. Biotechnol. 2002, 13, 1.
29 A. Schmid, J.S. Dordick, B. Hauer, A. Kiener, M. Wubbolts, B. Witholt, Nature 2001, 409, 258.
30 R.N. Patel, Curr. Opin. Biotechnol. 2001, 12, 587.
31 G.W. Huisman, D. Gray, Curr. Opin. Biotechnol. 2002, 13, 1.
32 A. Zaks, Curr. Opin. Chem. Biol. 2001, 5, 130.
33 K.M. Koeller, C.H. Wong, Nature 2001, 409, 232.
34 C. Khosla, P.B. Harbury, Nature 2001, 409, 247.

IV
Processes for New Chemical Entities (NCE)

1
Development of an Efficient Synthesis of Chiral 2-Hydroxy Acids
Junhua Tao and Kevin McGee

Abstract

While many methods have been reported for the synthesis of chiral 2-hydroxy acids, few have proven to be reliable toward the synthesis of the title compound in terms of overall yield and enantioselectivity. Herein we describe a continuous enzymatic process for an efficient synthesis of (R)-3-(4-fluorophenyl)-2-hydroxy propionic acid at a multi-kilogram scale with a high space-time yield (560 g/L/d) using a membrane reactor. The product was generated in excellent enantiomeric excess (ee > 99.9%) and good overall yield (68–72%). This process can also be adapted to the synthesis of a variety of chiral 2-hydroxy acids with high yield and stereoselectivity.

1.1
Introduction

(R)-3-(4-fluorophenyl)-2-hydroxy propionic acid **1** is a building block for the synthesis of *Rupintrivir*, a rhinovirus protease inhibitor currently in human clinical trials to treat the common cold (Fig. 1) [1, 2]. Retrosynthetically, *Rupintrivir* was prepared from four fragments: the lactam derivative P_1, the chiral 2-hydroxy acid P_2 (compound **1**), the valine derivative P_3, and an isoxazole acid chloride P_4 (Fig. 1). In this chapter the preparation of **1** using a biocatalytic reduction performed in a membrane reactor will be discussed in detail.

To our knowledge, no methods have been described for the preparation of **1**, although many have been reported in the literature for the synthesis of related chiral 2 hydroxy acids [3]. Among them, only a few are concise and practical for performance at a large scale (Fig. 2). For example, using oxynitrilases a-hydroxy acids could be prepared with optical purity higher than 90–95% ee from an aldehyde and a cyanide equivalent upon hydrolysis if $n=0$ or $n>1$, where n denotes the number of carbons between the aldehyde and the arene (path a, Fig. 2) [4]. However, the stereoselectivity is usually quite low for substrates where $n=1$, which is needed for the preparation of **1**. Extensive studies have been reported to synthesize chiral a-hydroxy acids from ketoesters using asymmetric hydrogenation

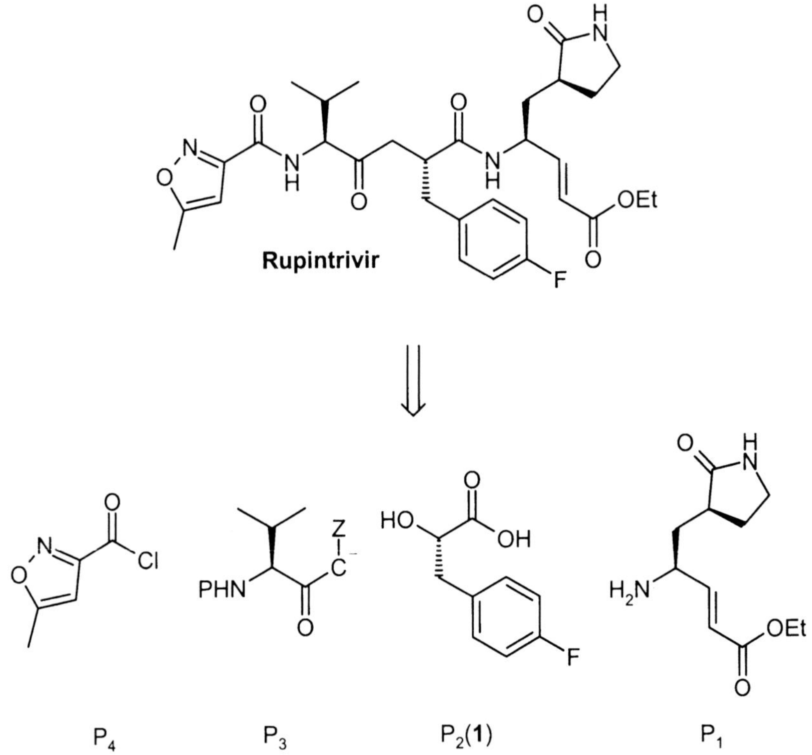
Rupintrivir
OEt
F
Cl
Z
PHN
HO
OH
H₂N
OEt
P4
P3
P2(1)
P1

Fig. 1

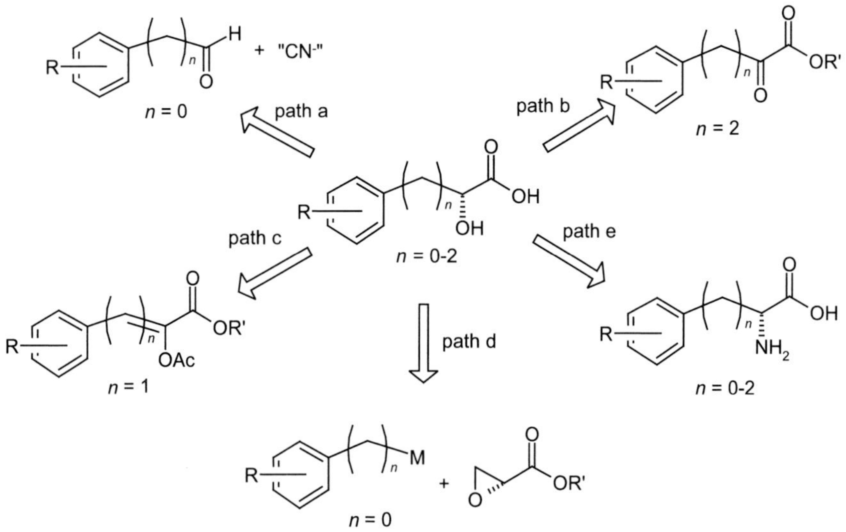
+ "CN-"
n = 0
path a
path b
n = 2
path c
n = 0-2
path e
OAc
n = 1
path d
NH2
n = 0-2
M
OR'
n = 0

Fig. 2

(path b, Fig. 2). In this strategy excellent enantioselectivity could be obtained by a careful selection of organometallics and chiral ligands if $n=2$ [5]. Unfortunately, in our hands the stereoselectivity is far from ideal if $n=1$ under a variety of conditions. For example, excellent ee values can be obtained using Noyori's Ru(II)-BINAP system for substrates where $n=0$. If $n=1$, however, the highest ee value achieved is 88% with less than 20% conversion under high pressure (100 atm) after 48 hours [6]. As for the Rh(I)-NORPHOS system, the ee obtained is 0% if $n=1$ while the same conditions lead to >95% ee for substrates where $n=2$ [6]. Borane-mediated asymmetric reductions have also been investigated for the same purpose [7]. However, most of them require the use of a stoichiometric amount of expensive reducing agents and/or substrates which are not readily available. For example, the corresponding a-keto acid is not stable, undergoing spontaneous decarboxylation if $n=1$ [6]. It was also reported that Corey's CBS reagent can not be used directly for chiral synthesis of 2-hydroxy acids [8]. In addition, enzymatic reduction methods have been reported to prepare compounds where $n=2$ [8]. Only a few were targeted at substrates where $n=1$ and all of them were performed only on small scales using a stoichiometric amount of cofactor NADH. Alternatively, hydroxy acids could be obtained from an enol acetate using asymmetric hydrogenation [9] (path c, Fig. 2) or ring-opening of glycidic acid derivatives by a metallated nucleophilic species [10] (path d, Fig. 2). In our hands, both approaches suffer from either low yields or poor stereocontrol. In the first campaign for scaling-up the preparation of **1**, the only pathway we could rely on utilizes amino acids as starting materials (path e, Fig. 2) [11].

Using this method, the preparation of **1** could be accomplished starting from a protected (R)-4-fluorophenylalanine **2** (Fig. 3). Upon deprotection and diazotization, the desired product could be obtained in a modest yield of 50% for two steps via the isolated intermediate **3**. However, the starting amino acid **2** is fairly expensive and the optical purity of the product **1** varied from a low 78% to a high 97% in scaled-up runs, indicating that the process is not robust. Both problems render this method impractical for production at a large scale.

These issues motivated us to seek an alternative route to preparing the desired optically active a-hydroxy acid. Herein we describe an efficient and concise synthesis of **1** at a multi-kilogram scale using a continuous enzyme membrane reactor.

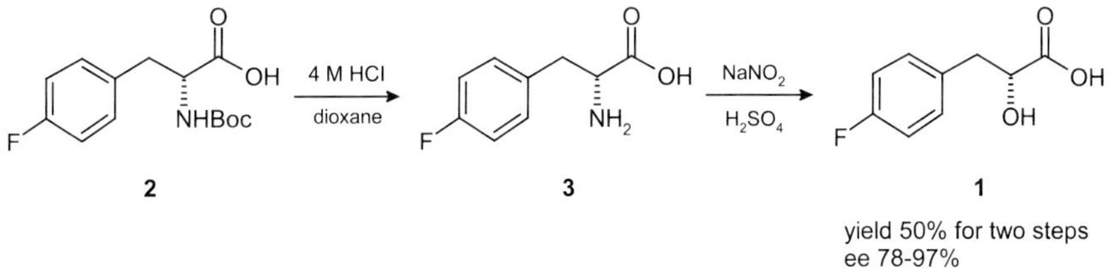

Fig. 3

1.2
Results and Discussion

The route began with 4-fluorobenzaldehyde **4** and hydantoin **5**, which are inexpensive and readily available (Fig. 4). Upon condensation and saponification using a modified literature protocol [12], α-keto acid salt **6** was obtained as a white solid in one step in good yields (77–82%). By this method, an overall quantity of 23 kg of **6** was prepared.

Alternatively, **6** can be prepared from 4-fluorobenzaldehyde and N-acetyl glycine **7** (Fig. 5). However, the product obtained is the α-keto acid **8** with lower stability than its salt. A direct conversion of this acid into its salt **6** was proven to be problematic suffering from low yields [6].

The key step was an aqueous enzymatic reduction using D-lactate dehydrogenase (D-LDH) and formate dehydrogenase (FDH) (Fig. 5) [8e, 13]. Mechanistically, the keto acid salt **6** is stereoselectively reduced to the corresponding α-hydroxy acid in the presence of D-LDH by NADH. The cofactor itself is oxidized to NAD in the process. Subsequently, in the presence of FDH, NAD is reduced back to NADH by ammonium formate, which was oxidized to CO_2 and NH_3. In this fashion the expensive cofactor NAD is regenerated by FDH and only a catalytic amount of NAD is required. Overall the ketone group in **6** is reduced stereoselectively to the hydroxy group in the presence of two enzymes with byproducts being CO_2 and NH_3. The optical purity of the product was determined by converting **1** into the methyl ester **9** (Fig. 6). At a small scale using a batch process, the desired 2-hydroxy acid could be obtained in good yields (80–90%) and excellent enantiomeric excess (>99.9%).

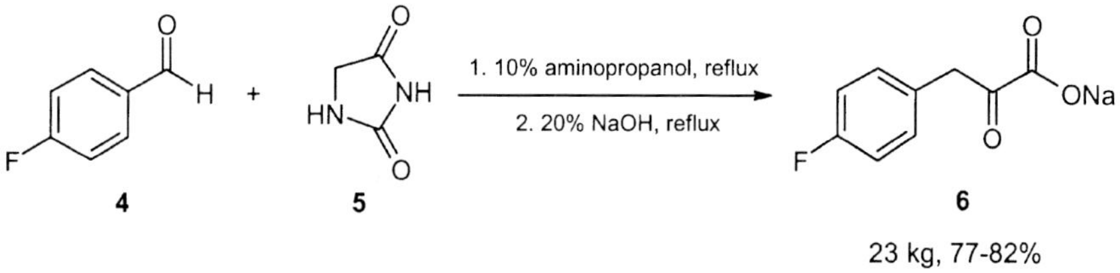

Fig. 4

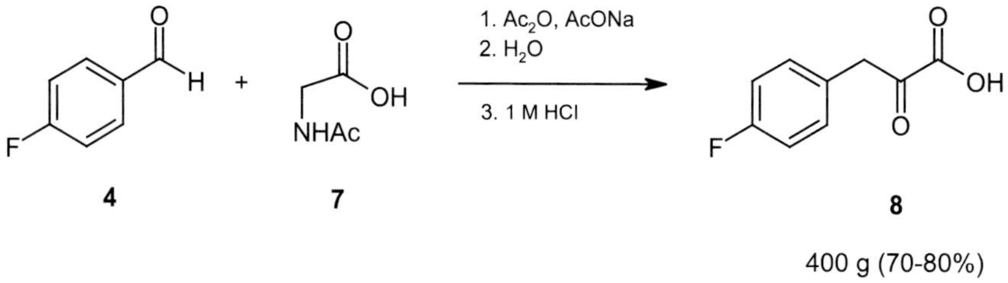

Fig. 5

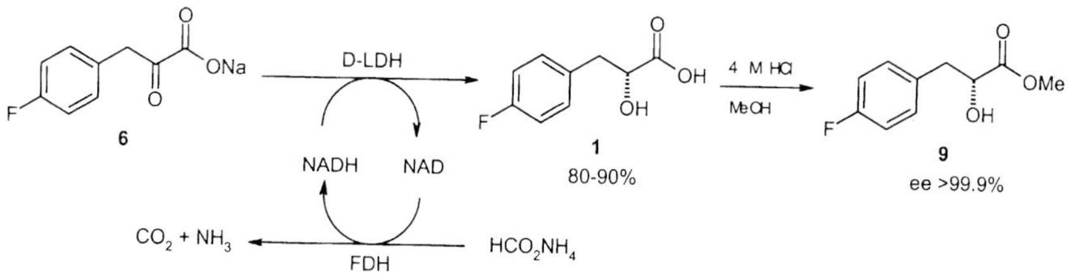

Fig. 6

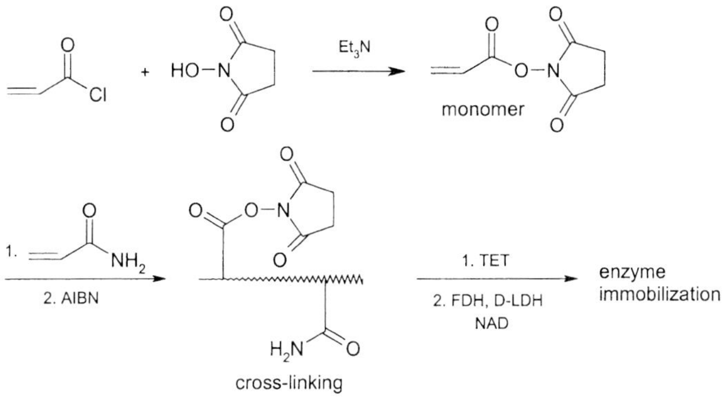

Fig. 7

In order to scale-up the method, however, both D-LDH and FDH have to be recycled to make the process economically feasible. While the starting material **6** could be prepared readily in large scale and only a catalytic amount of NAD is needed for the reaction, neither of the commercially available enzymes is inexpensive. Initial recycling efforts were directed to a batch process using either membrane-enclosed enzyme catalysis or enzyme immobilization methods [13] (Fig. 7). In our hands, these reactor systems were not ideal for scaling-up.

After a period of trial and error, we found that a continuous membrane reactor, shown schematically in Fig. 8, would allow recycling of both D-LDH and FDH and could fit both the critical path time line of product delivery and the cost requirement [14].

The key part of the reactor is a nanofiltration membrane unit (**a**), which allows the permeation of small molecules but not macromolecules such as enzymes. In operation, the reactor is initially charged with D-LDH and FDH before the start of the reaction. An aqueous mixture, which consists of **6**, ammonium formate, and a catalytic amount of NAD, is then continuously fed into the reactor by a peristaltic pump (**b**). After passing a check valve (**c**), the substrate solution is mixed with enzymes inside the reactor by a circulation pump (**d**). The product is collected continuously as an effluent from the filtration membrane unit. In this fashion, both enzymes are retained inside the reactor by the membrane leading to high turnover.

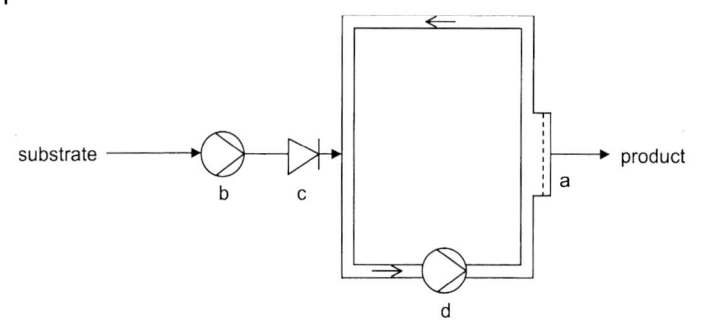

1 Development of an Efficient Synthesis of Chiral 2-Hydroxy Acids

Fig. 8 Schematic drawing of the enzyme membrane reactor.

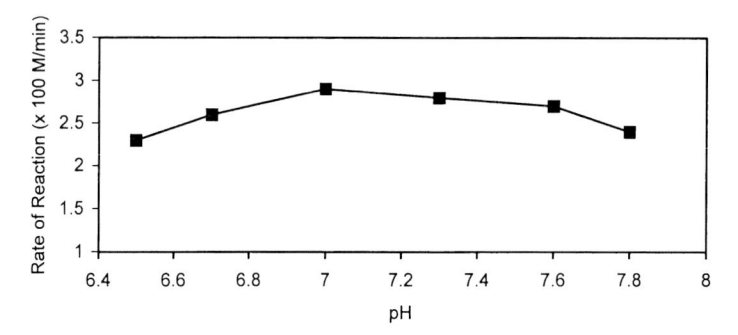

Fig. 9 Effect of pH on reaction rate.

Once a prototype reactor had been designed, optimum conditions were obtained by studying reaction kinetics. D-LDHs from *Leuconostoc mesenteroides* and *Staphylococcus epidermidis* showed the same activity toward the substrate 6 and the former was chosen for subsequent studies solely for economic reasons. FDH from *Candida boidinii* was preferred over the one from yeast (Sigma F8649) since this preparation is not only more reactive but also significantly cheaper. To maximize the throughput and reduce the working volume of the process, a saturated aqueous substrate solution was used in all studies. The concentration was calculated to be at 44 g/L or 16.7 M. As for ammonium formate, four equivalents were used to achieve a high reaction rate. Below that level, the reactivity of the system goes down proportionally. However, no significant advantage was achieved at a higher concentration. The optimum concentration of the cofactor NAD was found to be at 0.167 M or 1% with respect to the substrate. In general, the enzymatic reduction attains the highest reactivity at a pH range between 7.0 and 7.3, which was chosen for subsequent investigations (Fig. 9). To reach this pH range inside the reactor and obtain a conversion above 90%, the substrate solution was maintained at pH 6.3 before being fed into the reactor.

The concentration ratio of D-LDH to FDH also has an important effect on the reactivity and the optimum ratio was found to be around 20 (Fig. 10). Below that

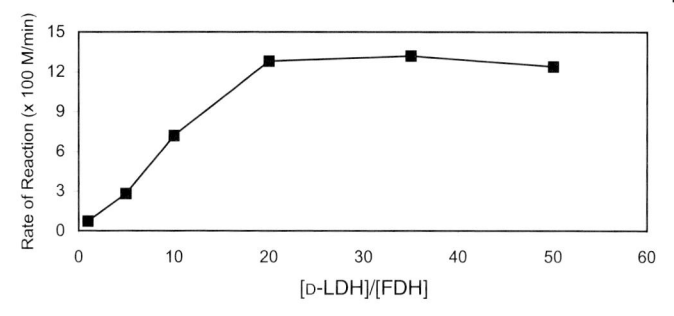

Fig. 10 Effect of concentration ratio of enzymes on reaction rate.

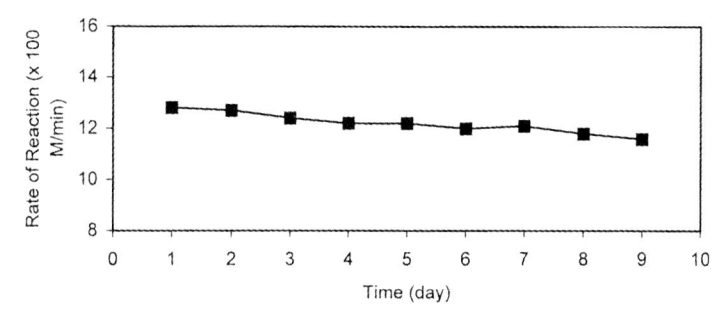

Fig. 11 Enzyme deactivation in the membrane reactor.

level, the reaction slows down significantly. Above that level, a slight gain in the rate of reaction occurs at the expense of a large excess of D-LDH. Moreover, the reactivity starts to level off when the ratio is above 30. In the final material production, a concentration of 400 U/mL was used for D-LDH and 20 U/mL for FDH.

Enzyme deactivation was a key factor in determining the overall cost of the process. In a period of nine days without adding fresh D-LDH and FDH, the rate of reaction decreases slowly and the enzymes lose their activity at a rate of only about 1% per day (Fig. 11). During actual production, the reactor was charged with new enzymes periodically to get a high conversion of above 90%.

In the full-scale production of the chiral hydroxy acid, a continuous membrane reactor with a volume of 2.2 L was used to meet the time line for product delivery. Under the optimum conditions, a substrate solution with pH=6.3 was fed into the reactor at a rate of 12.0 mL/min resulting in a residence time of about 3 hours. The average conversion was maintained above 90%. Upon work-up, the desired product was obtained in a yield of up to 88% with ee values >99.9%. Using this process, a total of 14.5 kg of the desired chiral 2-hydroxy acid **1** was prepared within a period of four weeks with a productivity of ca. 560 g/L/d (grams per liter per day). The cost of goods of the current process is significantly lower than that of the diazotization route.

In contrast to the asymmetric synthesis using organometallics described above which are highly sensitive to substrates, the current enzymatic reduction approach

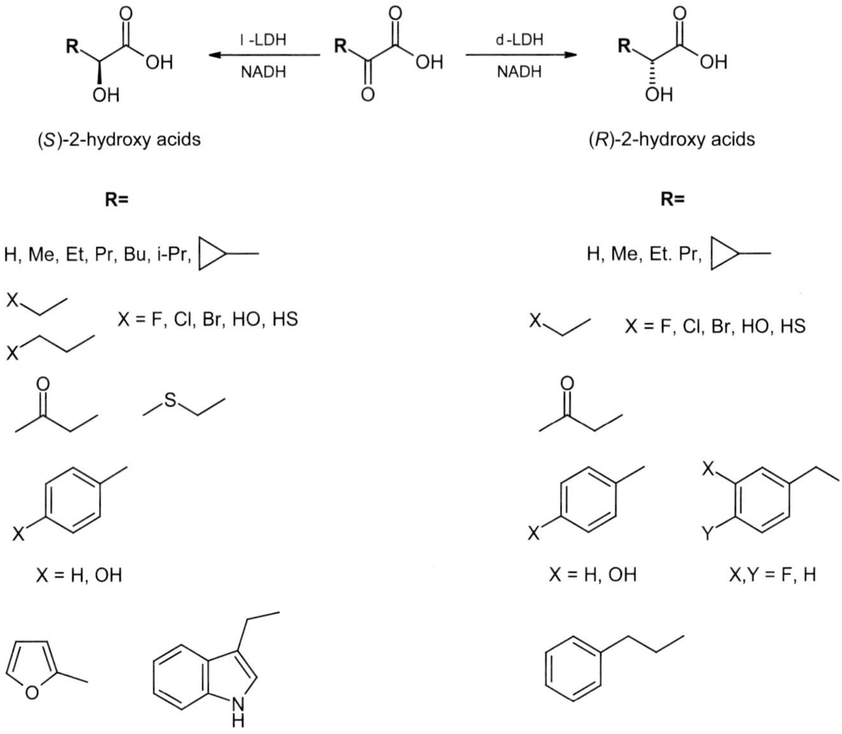

1 Development of an Efficient Synthesis of Chiral 2-Hydroxy Acids

Fig. 12

can not only be used to prepare (R)-2-hydroxy acids in high yield and ee for substrates where n=1, it can also be used to prepare a variety of chiral 2-hydroxy acids with high efficiency (Fig. 12) [6, 8, 13]. By changing D-LDH to L-LDH, (S)-2-hydroxy acids can also be synthesized. It should be noted that L-LDH is significantly cheaper than D-LDH. Using the same approach, several other chiral hydroxy acids have been prepared as chiral drug intermediates [6].

1.3
Conclusion

An efficient and practical process has been described for the synthesis of (R)-3-(4-fluorophenyl)-2-hydroxy propionic acid at a multi-kilogram scale with good overall yields (68–72% for two steps), excellent stereoselectivity (>99.9% ee), and significant cost savings. The key to the process is the use of a continuous membrane reactor which was simple in concept, low cost in design, and provided high space-time yields. This method has a broad substrate spectrum in contrast to asymmetric chemical catalysis and a variety of chiral 2-hydroxy acids can be prepared. Moreover, by alternating D-LDH with L-LDH, both enantiomers can be synthesized.

1.4
Experimental Section

1.4.1
General Remarks

Commercially available solvents and reagents were used without further purification. d-LDHs from *Leuconostoc mesenteroides* and *Staphylococcus epidermidis* were available from Roche Diagnostics and Sigma-Aldrich, respectively. FDH from *Candida boidinii* was purchased from Jülich Fine Chemicals and the yeast preparation from Sigma-Aldrich. NAD was a product of Roche Diagnostics. The enzyme reactor was constructed as a thermostatic loop using Viton tubing with an ultrafiltration membrane (cut-off: 10 000 Da) as a separation unit. ^{1}H NMR spectra were recorded at 300 MHz in d_6 DMSO, D_2O or $CDCl_3$. Reversed phase HPLC was performed on a Phenomenex Prodigy 3 µ ODS (4.6×100) column and chiral HPLC was performed on a Chiralpak AS column. The elemental analyses were carried out by Atlantic Microlab, Inc.

1.4.2
Sodium 3-(4-Fluorophenyl)-2-oxo-propionate (6)

4-Fluorobenzaldehyde **4** (3.72 kg, 30 mol), hydantoin **5** (3.00 kg, 30 moles), 1-amino-2-propanol (225 g, 3.0 mol) and water (7.5 L) were added to a 50 L reactor equipped with a temperature probe, reflux condenser, agitator, and cooling coils. The resulting mixture was heated and refluxed for approximately 10 hours. The reaction was monitored by ^{1}H NMR and was deemed complete upon disappearance of the hydantoin **5** proton signal at δ 3.9 (s, 2H, CH_2) and the appearance of condensed intermediate olefin proton δ 6.4 (s, 2H, CH) resulting in a bright yellow slurry. Aqueous sodium hydroxide (6.00 kg in 30.0 L) was then added and reflux continued until completion as shown by HPLC leading to a transparent orange solution. The mixture was cooled to 20±5 °C and sodium chloride (3.51 kg, 60.0 mol) was added with agitation. The pH of the mixture was adjusted to 8.0 using concentrated HCl and the suspension was stirred for 4 hours until a pale yellow slurry was obtained. The solids were filtered off and purified via slurrying in methanol (30.0 L) followed by filtration. Upon drying under house vacuum at ambient temperature for four days, 5.47 kg of solids (**6**) were obtained with a yield of 82% and HPLC purity above 80%. ^{1}H NMR (D_2O): δ 4.72 (s, 2H), 7.02–7.19 (m, 4H). Anal. Calc. for $C_9H_6O_3FNa \cdot H_2O$: C, 48.66; H, 3.63. Found: C, 48.64, H, 3.74.

1.4.3
(R)-3-(4-Fluorophenyl)-2-hydroxy propionic acid (1)

To a 22 L reactor equipped with agitator and gas diffuser was added EDTA (3.35 g, 9.0 mmol), mercaptoethanol (1.41 g, 18 mmol), ammonium formate (908 g,

14.4 mol), and sterile water (18.0 L), which was degassed prior to addition of the keto acid salt **6** (800 g, 3.6 mol). The suspension was stirred until all solids dissolved. The resulting solution was filtered through a 0.2 μm filter and transferred into a clean 22 L reactor equipped with an argon purge and overhead stirrer with argon degassing maintained throughout the remaining operations. NAD (23.88 g, 36 mmol) was added and the pH adjusted to 6.3 by addition of 1 N HCl. This substrate solution was then fed into a membrane reactor with an ultrafiltration membrane for enzymatic reduction. The reactor was previously filled with an aqueous mixture of enzymes (D-LDH, 400 U/mL and FDH, 20 U/mL). An appropriate feed rate was used to maintain a conversion of 90% or greater when sampled by HPLC. The circulation rate was kept between 15 to 30 times that of the feed rate. Internal pressure was regulated by a permeate control valve. The aqueous effluent solution thus obtained was adjusted to pH 3.0 with 2 N HCl and extracted with MTBE (3×5.0 L). The organic layer was evaporated to obtain 972 g of acid **1** as an off-white solid in a yield of 88% and UV purity >90% (reversed phase HPLC). ^{1}H NMR (CDCl$_3$): δ 2.95 (dd, 1H, J=8.0 Hz, 14.0 Hz), 3.15 (dd, 1H, J=8.0 Hz, 14.0 Hz), 4.50 (dd, 1H, J=8.0 Hz, 4.0 Hz), 6.97 (t, 2H, J=8.0 Hz), 7.00–7.25 (m, 4H). To check optical purity, **1** was converted into **9** by a standard esterification procedure and the enantiomeric excess of **9** was found to be >99.9% by chiral HPLC. In this fashion a total of 14.5 kg of **1** were produced over a period of two weeks.

1.4.4
Reactor Set-up and Preparation

All pumps, temperature probes, and pH meters were calibrated prior to use. Filter membranes were tested under pressure and fully saturated prior to use. All water used was sterile and filtered (0.2 micron). With the effluent valve closed, the reactor was filled with a known amount of water. This volume was used to calculate the enzyme amounts required and was ca. 2.2 L. A solution of 12 L of 0.02% (v/v) peracetic acid in water was prepared. With the circulation pump running, the effluent valve was opened and the peracetic solution was fed in at a rate of ≈25 mL/min until all was consumed. This was to sterilize the assembled system and the final pH reading was 3.7. The system was then flushed with water at a rate of ≈25 mL/min until 25 L of water were consumed and the internal pH reading rose to ≈6.0. Water (12.0 L), EDTA as the disodium salt (2.23 g, 6.9 mmol), mercaptoethanol (938 mg, 12 mmol), and ammonium formate (606 g, 9.61 mol) were added to a 22 L reactor equipped with an agitator and argon sparge tube. The mixture was stirred until these compounds were dissolved. After the solution was thoroughly degassed, it was filtered through a 0.2 micron filter into a similarly fitted clean 22 L reactor. The solution was adjusted to pH=6.3 before use with either 1 N H$_2$SO$_4$ or 1 N NaOH and fed into the reactor at a rate of ≈25 mL/min until consumed. At this point the reactor was ready for enzyme loading with the internal pH reading 6.6.

1.4.5
Enzyme Loading and Replenishment

Water (1.0 L), EDTA as disodium salt (186 mg, 0.5 mmol), mercaptoethanol (78 mg, 1.0 mmol), ammonium formate (50.45 g, 800 mmol) were added to a 2 L reactor fitted with an agitator. The mixture was stirred to effect solution and the pH was adjusted to 7.0±0.1 with either 1 N H_2SO_4 or 1 N NaOH. This was then filtered through a 0.2 micron filter and transferred into a clean flask. Formate dehydrogenase (44 000 units, 20 U/mL) and D-lactic dehydrogenase (880 000 units, 400 U/mL) were dissolved in 500 mL of the solution. The mixture was again filtered through a 0.2 micron filter and then fed into the constructed membrane reactor. The remaining portion of solution without enzymes was used to rinse the mixing flask and feed lines. The reactor was now ready for conversion of the sodium 3-(4-fluorophenyl)-2-oxo-propionate **6** to (R)-3-(4-fluorophenyl)-2-hydroxy propionic acid **1**. Conversion was monitored by HPLC and when deemed to have dropped below 90%, the D-LDH was replenished by dissolving it into solution and feeding it into the reactor as described above.

1.5
Acknowledgment

The authors would like to thank Dr. Srin Babu, Dr. Naresh Nayyar, Steven Lee, James Stein in CRD and Dr. Barbara Potts, Christine Albizati, Jason Ewanicki in ARD for experimental support. We also thank Dr. Andreas Liese at the Institute of Biotechnology, Jülich, Germany for helpful suggestions in designing a membrane reactor. Finally, we thank Dr. Kim Albizati and Dr. Van Martin for fruitful discussions in preparing this manuscript.

1.6
References

1 L. S. Zalman, M. A. Brothers, P. S. Dragovich, R. Zhou, T. J. Prins, S. T. Worland, A. K. Patick, Antimicrob. Agents Chemother. 2000, 44, 1236.

2 P. S. Dragovich, T. J. Prins, R. Zhou, S. E. Webber, J. T. Marakovits, S. A. Fuhrman, A. K. Patick, D. A. Matthews, C. A. Lee, C. E. Ford, B. J. Burke, P. A. Rejto, T. F. Hendrickson, T. Tuntland, E. L. Brown, J. W. Meador III, R. A. Ferre, J. E. V. Harr, M. B. Kosa, S. T. Worland, J. Med. Chem. 1999, 42, 1213.

3 For recent leads see a) H. Yu, C. E. Ballard, B. Wang, Tetrahedron Lett. 2001, 42, 1835; b) D. A. Evans, S. W. Tregay, C. S. Burgey, N. A. Paras, T. Vojkovsky, J. Am. Chem. Soc. 2000, 122, 7936; c) F. F. Huerta, Y. R. S. Laxmi, J.-E. Bäckvall, Org. Lett. 2000, 2, 1037.

4 a) D. H. Dao, M. Okamura, T. Akasaka, Y. Kawai, K. Hida, A. Ohno, Tetrahedron Asymmetry 1998, 9, 2725; b) W. Adam, M. Lazarus, C. R. Saha-Möller, P. Schreier, Tetrahedron Asymmetry 1998, 9, 351; c) Y. Hashimoto, E. Kobayashi, T. Endo, M. Nishiyama, S. Horinouchi, Biosci. Biotech. Biochem. 1996, 60, 1279; d) F. Effenberger, Angew. Chem., Int. Ed. Engl. 1994, 33, 1555.

5 a) C. LeBlond, J. Wang, J. Liu, A.T. Andrews, Y.-K. Sun, J. Am. Chem. Soc. 1999, 121, 4920; b) X. Zuo, H. Liu, M. Liu, Tetrahedron Lett. 1998, 39, 1941; c) H.U. Blaser, H.-P. Jalett, F. Spindler, J. Mol. Catal. A 1996, 107, 85; d) M. Kitamura, T. Ohkuma, S. Inoue, N. Sayo, H. Kumobayashi, S. Akutagawa, T. Ohta, H. Takaya, R. Noyori, J. Am. Chem. Soc. 1988, 110, 629.
6 J. Tao, unpublished results.
7 Z. Wang, B. La, J. Fortunak, X.-J. Meng, G.W. Kabalka, Tetrahedron Lett. 1998, 39, 5501 and references therein.
8 a) K. Ishihara, M. Nishitani, H. Yamaguchi, N. Nakajima, T. Ohshima, K. Nakamura, J. Ferment. Bioeng. 1997, 3, 268; b) E. Schmidt, O. Ghisalba, D. Gygax, G. Sedelmeier, J. Biotech. 1992, 24, 315; c) H.K.W. Kallwass, Enzyme Microb. Technol. 1992, 14, 28; d) M.J. Kim, J.Y. Kim, J. Chem. Soc. Chem. Commun. 1991, 5, 326; e) M.-J. Kim, G.M. Whitesides, J. Am. Chem. Soc. 1988, 110, 2959.
9 M.J. Burk, C.S. Kalberg, A. Pizzano, J. Am. Chem. Soc. 1998, 120, 4345.
10 a) M. Larchevêque, Y. Petit, Tetrahedron Lett. 1987, 28, 1993; b) K.B. Sharpless, J.M. Chong, Tetrahedron Lett. 1985, 26, 4683.
11 F.J. Urban, B.S. Moore, J. Heterocyclic Chem. 1992, 29, 431.
12 G. Billek, Org. Syntheses 1973, Coll. Vol. 5, 627.
13 H.K. Chenault, G.M. Whitesides, Appl. Biochem. Biotech. 1987, 14, 147.
14 a) J. Wöltinger, A.S. Bommarius, K. Drauz, C. Wandrey, Org. Process. Res. Dev. 2001, 5, 241; b) G. Goetz, P. Iwan, B. Hauer, M. Breuer, M. Pohl, Biotechnol. Bioeng. 2001, 74, 317–325; c) W. Neuhauser, M. Steininger, D. Haltrich, K. Kulbe, B. Nidetzky, Biotechnol. Bioeng. 1998, 60, 277; d) U. Kragl, W. Kruse, W. Hummel, C. Wandrey, Biotechnol. Bioeng. 1996, 52, 309; e) E. Schmidt, O. Ghisalba, D. Gygax, G. Sedelmeier, J. Biotech. 1992, 24, 315.

2
Factors Influencing the Application of Literature Methods Toward the Preparation of a Chiral *trans*-Cyclopropane Carboxylic Acid Intermediate During Development of a Melatonin Agonist

Ambarish K. Singh, J. Siva Prasad, and Edward J. Delaney

Abstract

From numerous methods that have been published in recent years, chemists can now prepare chiral cyclopropanes in high enantiomeric excess without resorting to low-yielding, classical resolution techniques. Thus, industrial chemists have improved tools with which to access these molecules. Nonetheless, the choice of specific technologies is based on a myriad of factors. Important among those factors is the stage of development and the trade off that exists between the need to make material rapidly and the requirement to develop a long-term process. The application of a number of asymmetric technologies over the evolution of development of a cyclopropane-containing drug candidate is presented.

2.1
Introduction

Advances in synthetic methodology reported in the chemical literature are frequently translated into practical applications within industrial pharmaceutical research labs, and the nature of this translation is multifaceted. Medicinal chemists constantly look for novel structural elements and new methods by which they can efficiently place specific functionality into three-dimensional space. Alternatively, new methodology is applied to increase diversity within the pool of small molecules that guide medicinal chemists toward potential drug candidates. As specific drug candidates are brought forward for development, the presence of specific functional elements possessing asymmetry presents a very different challenge to the process research chemist and the nature of that challenge varies with each stage of development.

Molecules incorporating a cyclopropane unit have received strong interest within the pharmaceutical industry since they provide chemists with the capability to rigidly arrange pendant groups in specific three-dimensional orientations [1]. Similarly, a wide variety of cyclopropane-containing natural products are known, many of which have become targets of academic total synthesis groups. As a re-

Asymmetric Catalysis on Industrial Scale: Challenges, Approaches and Solutions
Edited by H. U. Blaser, E. Schmidt
Copyright © 2004 Wiley-VCH Verlag GmbH & Co. KGaA, Weinheim
ISBN: 3-527-30631-5

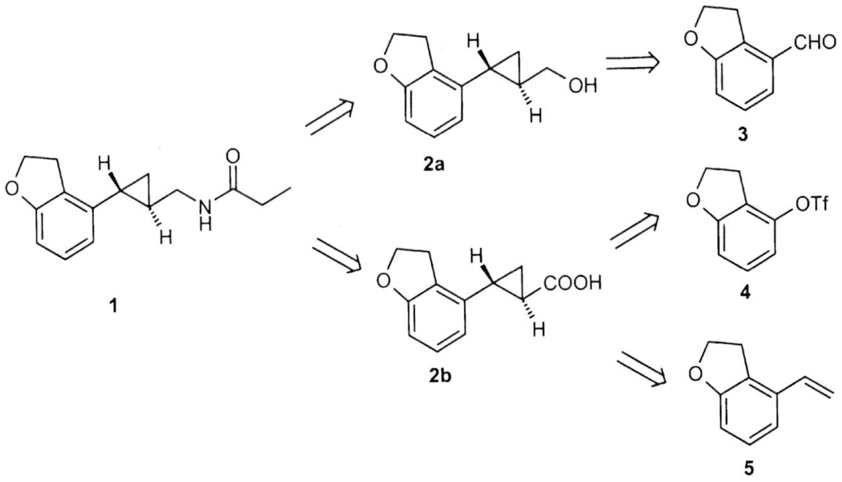

Fig. 1 Retrosynthetic pathways to melatonin agonist drug candidate **1**.

sult, numerous chemical approaches have been published that now allow ready access to chiral as well as achiral cyclopropane derivatives [2].

In the course of developing the melatonin agonist **1** (Fig. 1), potentially indicated for the treatment of sleep disorders, we evaluated approaches to the chiral cyclopropane intermediates **2a** and **2b** based on both classical resolution and asymmetric induction. Conceptually, these intermediates could be derived from dihydrobenzofurans **3**, **4** and **5**; each of these was used at one stage or another during our research and development work.

In this account, we describe how several documented methodologies for the preparation of chiral cyclopropane derivatives were applied across various stages of development and we present insights into the considerations that ultimately drove the selection of the process technology for long-term commercial use.

2.2
Chemical Approach Employed During Preclinical Research and Development (Route A)

During drug discovery, screening and candidate optimization, an expedient approach to target molecules is preferred to prepare small amounts of material for rapid assessment of *in vitro* activity. Accordingly, the initial preparation of 50 g of our drug candidate relied upon a direct adaptation of a literature procedure that employed a chiral auxiliary (Fig. 2) [3].

Thus, allylation of commercially available ester **6** and subsequent Claisen rearrangement provided an approximately 3:1 mixture of the 2- and 4-allyl derivatives. Clean-up was achieved by selective hydrolysis of the ester of the minor, undesired 4-isomer, and extractive removal of the resulting carboxylate with base.

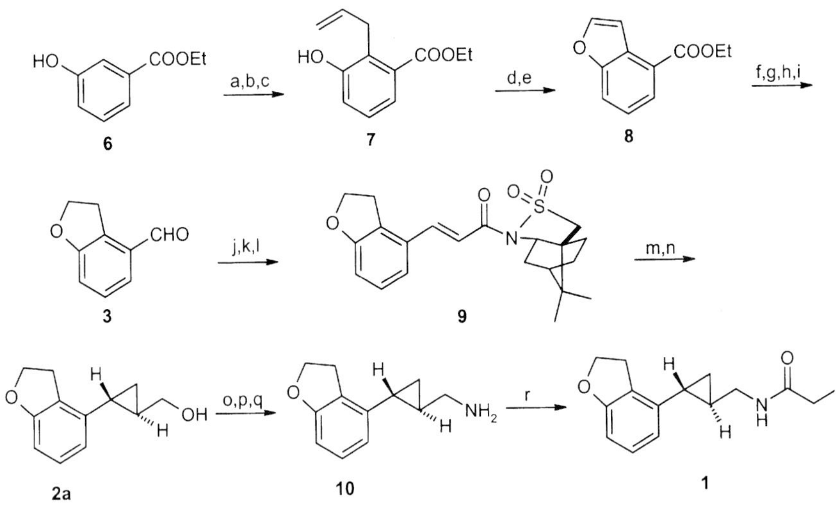

Fig. 2 Preparation of **1** via chiral sultam amide **9**. Route A: (a) allylbromide, K$_2$CO$_3$; (b) 220 °C; (c) KOH; (d) OsO$_4$, NaIO$_4$; (e) TsOH, benzene; (f) NaOH, EtOH, heat; (g) H$_2$, Pd/C; (h) LAH; (i) Swern oxidation; (j) malonic acid, pyrrolidine, pyridine; (k) SOCl$_2$; (l) (1S)-(–)-2,10-camphorsultam sodium salt; (m) CH$_2$N$_2$, Pd(OAc)$_2$; (n) LAH; (o) Swern oxidation; (p) NH$_2$OH·HCl, NaOH; (q) AlH$_3$; (r) propionyl chloride, Et$_3$N.

The remaining 2-isomer **7** was converted into benzofuran **8** by treatment with osmium tetroxide and sodium periodate to give an intermediate hemiacetal, which was dehydrated in the presence of acid. Hydrolysis of **8** to the carboxylic acid, hydride reduction, and Swern oxidation provided the key intermediate **3**. After two carbon homologation with malonic acid, the chiral auxiliary was appended by conversion into the acid chloride and treatment with the sodium salt of (–)camphorsultam to afford amide **9**. Asymmetric cyclopropanation was then accomplished by treatment of **9** with diazomethane in the presence of palladium acetate. Following purification of crude material possessing >80% ee, a 67% isolated yield of cyclopropane intermediate was obtained possessing >99% ee. The chiral auxiliary was subsequently removed by reduction with LAH to give **2a**, which was converted into amine **10** by successive applications of Swern oxidation, oxime formation, and reduction with aluminum hydride. Finally, acylation of **10** with propionyl chloride provided **1** in an overall yield of 3% from ester **6** in 18 chemical steps. This synthesis was carried out relatively quickly with the aid of numerous intermediate chromatographic clean-ups, and it performed well for the purpose of making early supplies for pre-clinical evaluation. At such a point in the typical drug discovery and development cycle, it is not unusual for a candidate such as **1** to be one of several from which a single entity is chosen for Phase I safety studies in man. Therefore, a highly expedient route is called for, even if it is low yielding.

2.3
Chemical Approaches Employed to Support Early Clinical Development (Route B)

In making supplies for early toxicological evaluation and Phase I clinical studies, somewhat larger quantities (0.5 to 2 kg) of the chosen drug candidate are needed as rapidly as possible. At this point, a safe, scaleable, and reasonably efficient synthesis is called for, particularly since early steps may require the processing of many kilograms of intermediates. However, a dynamic tension always exists at this stage between the need to apply efforts and resources to make material and the need to develop process knowledge for long-term use. Since at this stage process chemistry functions are presented with numerous candidates, each possessing a high risk of future dropout, intermediate processes which are practical, but which may not meet criteria for long-term use, are generally employed. Higher costs and an intermediate level of efficiency are reasonable tradeoffs in exchange for speed to enable early clinical evaluation of the candidate.

In the case of the melatonin agonist **1** a classical resolution was employed to prepare intermediate **2b**, as shown in Fig. 3 (Route B). The remaining steps were then safely scaled up to prepare 1.1 kg of **1** in our kilo-lab. Route B began with re-

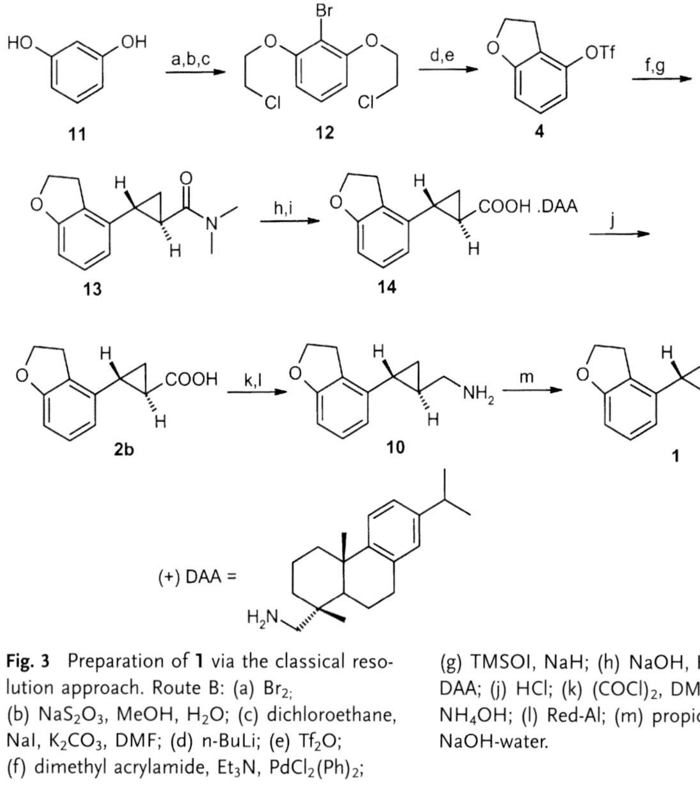

Fig. 3 Preparation of **1** via the classical resolution approach. Route B: (a) Br$_2$; (b) NaS$_2$O$_3$, MeOH, H$_2$O; (c) dichloroethane, NaI, K$_2$CO$_3$, DMF; (d) n-BuLi; (e) Tf$_2$O; (f) dimethyl acrylamide, Et$_3$N, PdCl$_2$(Ph)$_2$; (g) TMSOI, NaH; (h) NaOH, H$_2$O-EtOH; (i) DAA; (j) HCl; (k) (COCl)$_2$, DMF, CH$_2$Cl$_2$ then NH$_4$OH; (l) Red-Al; (m) propionyl chloride, NaOH-water.

sorcinol (**11**). Perbromination, reduction and double O-alkylation were performed to access intermediate **12**. Selective metal–halogen exchange was followed by intramolecular cyclization and triflation to give dihydrobenzofuran **4**. Heck coupling [4] with N,N-dimethylacrylamide, followed by Corey-Kim cyclopropanation [5] gave **13**, which was hydrolyzed and then resolved using (+)-dehydroabietylamine to afford salt **14** in a 98:2 diastereomeric ratio. The enantiomerically enriched acid **2b** was liberated and converted into amine **10** via amidation and reduction. Acylation of **10** with propionyl chloride then provided drug candidate **1**. As inefficient as this route appears, it involved fewer steps (13) and gave a higher overall yield (5%) from a readily available starting material (resorcinol). As such, it represents a good example of an intermediate synthesis to meet a short-term need.

2.4
Development of Process Technology to Support Phase II/III Clinical Studies and Future Commercialization (Routes C and D)

As Route B preparation of our drug candidate was ongoing, research efforts commenced to design procedures capable of producing 100 kg or more of our drug candidate, while at the same time serving as the foundation for a manufacturing process. We recognized olefin derivative **5** to be an important target, since it could be accessed readily from a preferred starting material, dihydronaphthol (**15**, Fig. 4). The latter material is prepared by Birch reduction of α-naphthol and was already being used in a commercial process to prepare the beta-blocker drug nadolol. As such, the infrastructure was already in place to make **15** in high volume at low cost.

Our research focus on this target was rewarded with the development of a highly efficient procedure by which we could access our desired olefin intermediate **5**. Dihydronaphthol (**15**) was cleanly converted into triol **16** by ozonolysis fol-

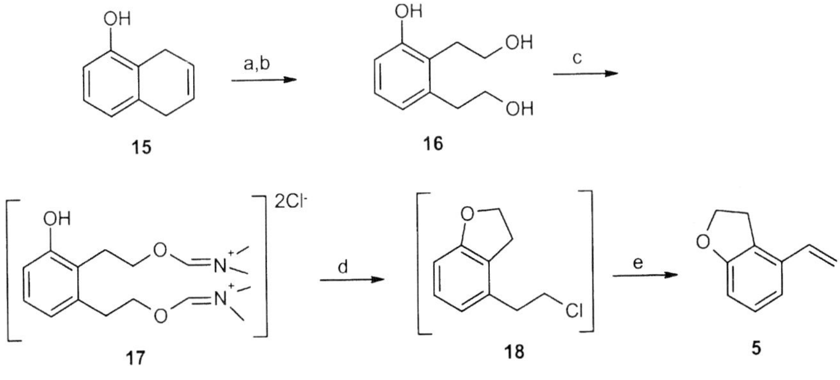

Fig. 4 Preparation of olefin **5**. (a) Ozone, MeOH; (b) NaBH$_4$; (c) Vilsmeier reagent, MeCN; (d) TEA; (e) NaOH, KI, tetrabutylammonium hydroxide.

lowed by *in situ* reduction with sodium borohydride. While ozonide intermediates formed during ozonolysis reactions have a well-deserved reputation for uncontrolled exothermic decomposition, the engineering technology exists to perform these reactions safely by conducting them continuously. Thus, ozone is generated as needed and the dangerous ozonide formed is limited to only minute quantities at any given time. Combined with appropriate controls the reaction can be performed safely and efficiently. Moreover, while our company lacked direct experience with this technology, it is not uncommon for pharmaceutical companies to partner with specialty chemical companies who have the appropriate equipment and expertise to prepare such substances as intermediates.

Imidate chemistry published by Barrett and co-workers [6] around that time was subsequently found to effect clean conversion of triol **16** into dihydrobenzofuran **18**. Thus, **16** was converted into the bisimidate **17**, which quickly cyclized to **18** upon treatment with triethylamine. Under phase transfer reaction conditions, **18** was cleanly converted into our desired olefin **5**. From this chemistry, a process was designed which afforded a 75% yield of **5** from dihydronaphthol in four steps with a minimum number of unit operations and only two intermediate isolations [7].

Having identified **5** as a preferred intermediate, we chose to focus on two approaches to reach the desired chiral cyclopropane derivative: the preparation of chiral epoxide **19**, and direct asymmetric cyclopropanation to afford **2b** (Fig. 5). While the latter appeared more direct, several issues needed to be resolved before scale-up and thus it required more extensive development work. Therefore, initial efforts focused on developing a process based on epoxide **19** and subsequent conversion into **20** via a modified Wadsworth-Emmons procedure [8]. Two approaches were explored to convert **5** into **19**.

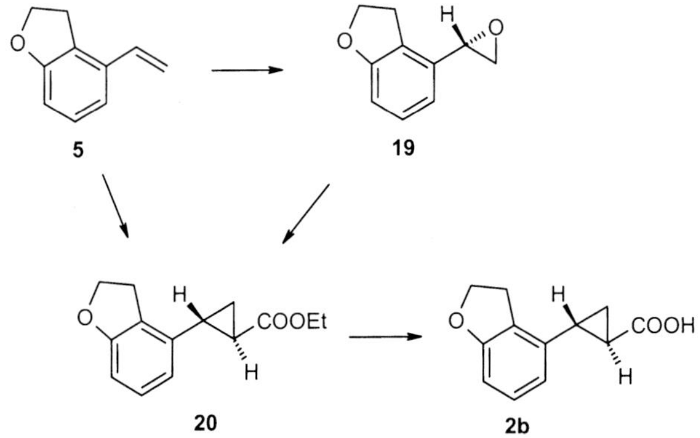

Fig. 5 Strategies based on using olefin **5** as the starting material.

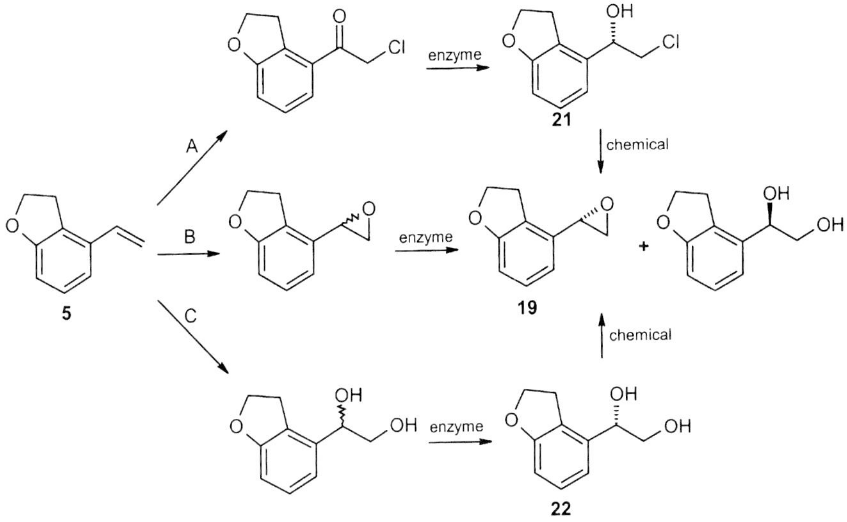

Fig. 6 Enzymatic approaches to epoxide **19**.

2.4.1
Enzymatic Approaches to Convert 5 into 19

Processes based on enzymatic transformations often have an advantage of being environmentally friendly and low cost. We examined three enzymatic approaches as outlined in Fig. 6.

Although representing viable alternatives, each suffered certain drawbacks. Approach A provided moderate enantiomeric excess and involved several chemical steps surrounding the enzymatic reduction of halohydrin substrate **21** [9]. In approach B, a short number of steps existed, but yields would be limited to no more than 50% [10]. The third approach (C) [11] involved a kinetic resolution involving stereoinversion of undesired diol to produce **22**. This approach appeared more attractive, but the overall number of transformations and work-up steps were substantial. Beyond these issues, low aqueous solubility of substrates and low titer of the product necessitated the handling of large volumes of broth and the subsequent need for large volumes of solvent. Thus, the enzymatic approaches were not selected based on high cost, low throughput, and increased waste.

2.4.2
Chemical Approaches to Convert 5 into 19

Asymmetric epoxidation using the catalysts reported by Jacobsen and co-workers [12] have been studied extensively (Fig. 7). Under the best conditions, the desired epoxide was obtained in 72% ee and 80% yield, necessitating an upgrade of the enantiomeric homogeneity in a later intermediate [13]. Sodium hypochlorite (household bleach)

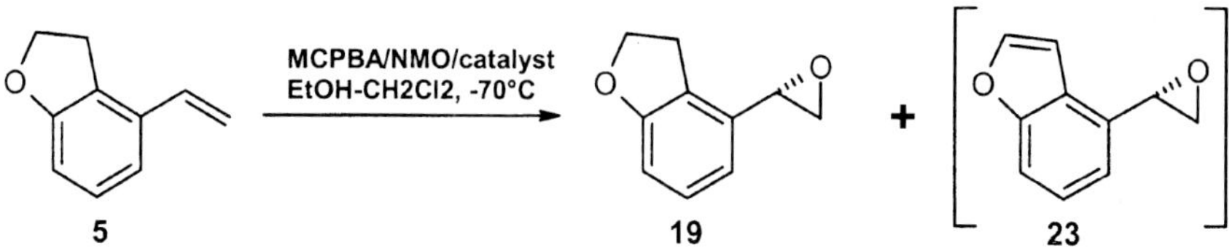

Fig. 7 Epoxidation of **5** under Jacobsen's conditions.

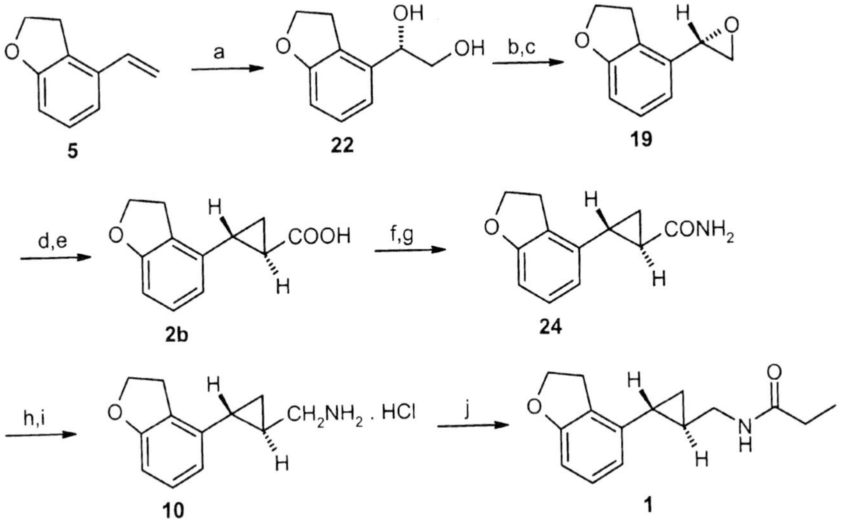

Fig. 8 Preparation of **1** via the epoxide route. Route C: (a) K_2OsO_4, $K_3Fe(CN)_6$, $(DHQ)_2$-PHAL, ^tBuOH-water; (b) TMSCl, trimethylorthoacetate, THF; (c) KO^tBu; (d) NaO^tBu, triethylphosphonoacetate, DME; (e) NaOH, water; (f) Vilsmeier reagent, EtOAc; (g) NH_4OH; (h) Red-Al, toluene (i) HCl, MeOH; (j) propionyl chloride, NaOH, water.

was studied as a safer alternative to *meta*-chloroperbenzoic acid (MCPBA) as the oxidant in consideration of the handling hazards of the latter (shock sensitivity and deflagration potential) [14]. However, enantiomeric purity and yield were significantly lower with bleach. In addition, oxidation of **19** to the corresponding benzofuran existed as a troublesome side reaction, leading to **23** as an impurity. This impurity undergoes analogous transformations as **19**, and could only be purged to acceptable levels via multiple crystallizations at the penultimate step.

In contrast to the Jacobsen conditions, application of the Sharpless asymmetric dihydroxylation procedure [13, 15] on our substrate **5** gave much better overall performance. Chiral diol **22** could be prepared with 98–99% ee in isolated yields of ≈ 90% (Fig. 8). Moreover, since this diol was crystalline, it provided a convenient point for intermediate clean-up.

Conversion of the diol to epoxide **19** followed the recommended procedure of Sharpless [16] with minor modifications. No loss of enantiomeric purity was observed in this process. The epoxide was converted into **2b** by using a modification of Wadsworth-Emmons protocol in high yield, again with no loss of enantiomeric purity [17]. Enhancement of enantiomeric purity of **2b** was not required, as additional enrichment occurred in subsequent steps. Acid chloride formation followed by amidation afforded **24**, which was reduced with Red-Al to the amine **10**, which was isolated as its HCl salt. Acylation with propionyl chloride under Schotten-Baumann conditions afforded **1** in an overall yield of 24% (>99.8% ee) from dihydronaphthol **15**. Using this route, we prepared 22 kg of diol **22**, which afforded 11 kg of drug substance. This quantity supported all requirements through the Phase II development.

2.5
Definition of Process Technology for the Optimal Route (Route D)

While the Sharpless asymmetric dihydroxylation procedure proved effective in making the desired chiral cyclopropane acid intermediate from **5**, several unit operations were required to make the key intermediate **2b**. Overall efficiency, however, is the key criterion for developing a long-term, cost-effective manufacturing process. Asymmetric cyclopropanation [18] certainly offered the most direct approach starting from **5**, and so a number of catalyst systems known to effect this type of reaction were screened. Three catalytic systems were examined in detail for potential, as outlined in Tab. 1.

A key factor in our case was to limit the amount of *cis*-isomer formed as a by-product. Thus, based on overall performance the Nishiyama catalyst **27** was chosen for further development. Implementation of this chemistry did require, however, several modifications to allow for scale-up. For example:

i) Owing to the hazardous nature of ethyl diazoacetate (EDA), bulk quantities cannot be shipped. A thorough examination of the hazards associated with preparing and handling of EDA was conducted by our in-house laboratory [19]. While a procedure for preparing EDA has been published [20], we took on additional development to ensure safe preparation of EDA solutions on a large scale. From the parameters we studied (pH, reaction temperature, and reaction solvent; Fig. 9), a process was defined that presented no safety issues, and the EDA/toluene solutions generated were employed directly in the cyclopropanation step. The procedure developed provided EDA in 85% yield and was demonstrated at a 50 kg scale.

ii) While available commercially in smaller amounts, an efficient and high yielding procedure to prepare large amounts of the Nishiyama catalyst **27** was needed. Thus, preparations of the ligand and catalyst were developed [21]. Procedures involving column chromatography purification were replaced with an isolation that provided the catalyst as a free-flowing powder. Procedures developed were demonstrated at a kilogram scale.

iii) Literature procedures for asymmetric cyclopropanation [24] commonly employ a large excess of the olefin reactant, and keep EDA as the limiting reagent in

Tab. 1 Comparison of catalysts for carrying out asymmetric cyclopropanation on **5**.

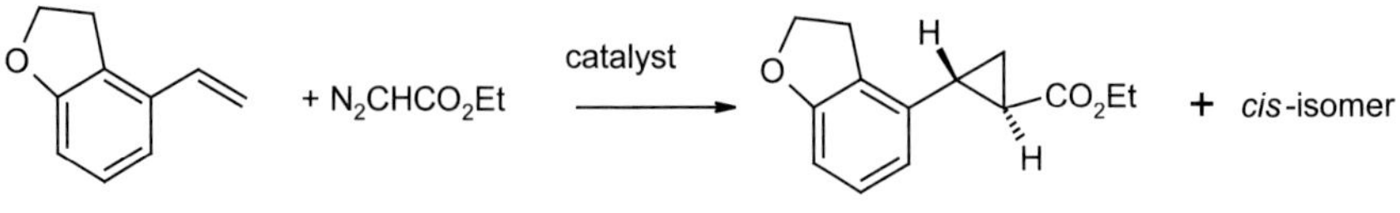

	Evans [22]	Katsuki [23]	Nishiyama [24]
Catalyst	25	26	27
Catalyst load (equiv.)	0.1%	10%	2%
% ee of desired *trans* product	99	84	98–99
% *cis* impurity formed	26	19	8
% conversion	>99	72	95

Structures of Catalysts:

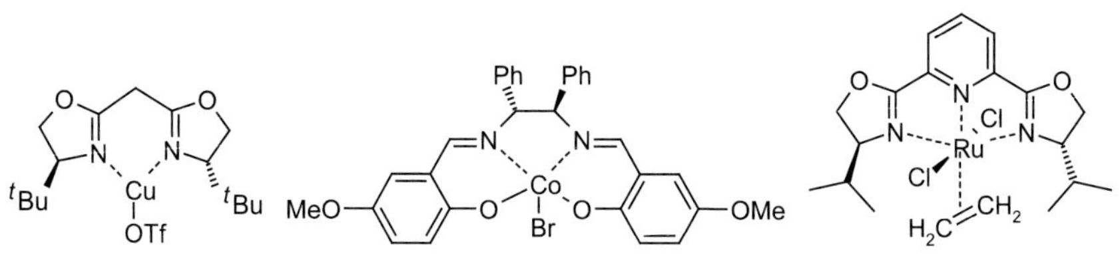

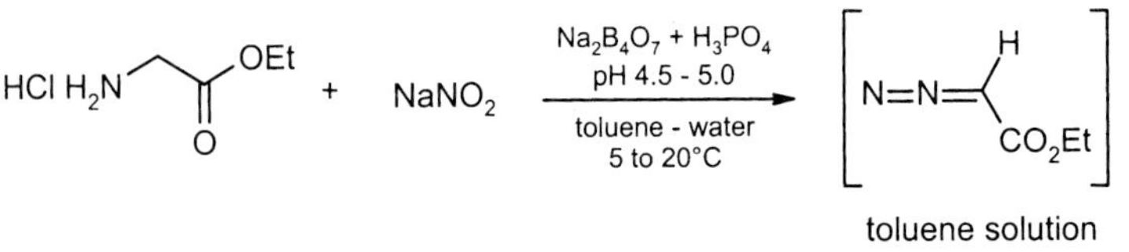

Fig. 9 Safe preparation of ethyldiazoacetate.

order to minimize EDA dimerization to fumarate and maleate esters. Because our olefin intermediate **5** is relatively expensive, we needed to optimize the reaction conditions in such a way so that it could be used as the limiting reactant. An experimental design protocol indicated that the main effects to achieve full conversion were slow addition and higher equivalents of EDA. Thus, addition of 2.5 equivalents of EDA over a 16 hour period was employed to get complete conversion of **5** into cyclopropyl ester while minimizing fumarate and maleate formation.

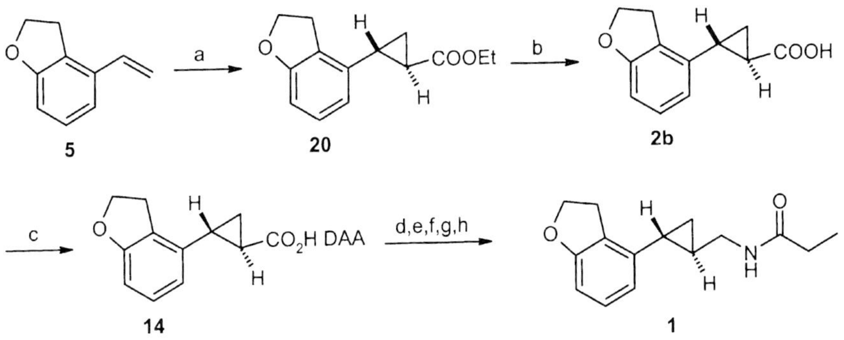

Fig. 10 Preparation of **1** via the asymmetric cyclopropanation route. Route D: (a) EDA, 2 mol-% Nishiyama catalyst 27, toluene; (b) NaOH, TBAH; (c) dehydroabeityl amine, MTBE; (d) Vilsmeier reagent, EtOAc; (e) NH$_4$OH; (f) LAH, toluene; (g) HCl, MeOH; (h) propionyl chloride, NaOH, water.

iv) For the isolation of the desired ester **20** arising from the cyclopropanation, we required the means to remove undesired byproducts such as the *cis*-isomer (8%), traces of enantiomer (2–4%), and fumarate and maleate esters present. Fortuitously, the ester hydrolysis step showed a significantly higher rate for the *trans*-isomer relative to that for *cis*. Thus, following a selective hydrolysis step, most of the *cis*-ester initially present could be readily removed via extraction, improving the *trans*:*cis* ratio to 97%. Additionally, further upgrade in enantiomeric excess and *trans*:*cis* ratio was accomplished following isolation of the (+)-dehydroabietylamine salt (as used to advantage in the racemic Route B) providing material of >99% ee, and >99% overall purity by HPLC analysis. Overall isolated yield of this salt starting from the olefin intermediate **5** was 74%, unoptimized for product recovery (an additional 12–18% is typically present in the crystallization mother liquor). This chemistry was successfully scaled to prepare 250 kg of the cyclopropyl acid **14** (Fig. 10).

With the direct cyclopropanation step in hand, the free acid **2b** was liberated from **14** and taken through the amidation, reduction, and acylation steps (as described in Route C, Fig. 8) to afford the melatonin agonist **1** in 28% overall yield in 12 steps from dihydronaphthol **15** (Route D, Fig. 10). This sequence of procedures formed the basis of a manufacturing process. A comparison of key features relative to those from prior routes is given in Tab. 2.

2.6
Summary and Conclusions

The selection of process technology for commercial use is based on a host of considerations, such as availability of starting materials, length of synthetic route, process intensity (kg of intermediates per kilo of drug), presence and controllability of reac-

Tab. 2 A comparison of four routes for the preparation of **1**.

	Route A (chiral sultam amide)	Route B (resolution with DAA)	Route C (epoxide)	Route D (asymmetric cyclopropanation)
Starting material	6	11	15	15
Chemical steps	18	13	13	12
Process intensity	97	108	15	12
Safety/ environmental	CH_2N_2, OsO_4, $NaIO_4$, benzene, pyridine	Dimethylacryl- amide, pyridine, dichloroethane	Ozone, osmium reagent	Ozone, ethyldiazoacetate
Special equipment	chromatography	cryogenic	none	none
Overall yield	3%	5%	24%	28%
Drug prepared	50 g	1.1 kg	11 kg	8 kg

tion hazards, number of unit operations, the need for special equipment, and overall cost (Tab. 2). Decisions in industrial process research and development are thus framed from a complex set of opportunities, needs and tradeoffs. Exemplary of the earliest stages of development, Routes A and B were considered interim solutions. In evaluating other routes, ready access to **5** in a minimal number of steps from commercially available dihydronaphthol narrowed the options we were to explore to asymmetric dihydroxylation and direct cyclopropanation (Routes C and D). As work proceeded, the superior step yield and % ee afforded by the Sharpless asymmetric dihydroxylation procedure were ultimately seen as being outweighed by the greater number of steps and unit operations required in comparison with the greater overall yield potential of direct asymmetric cyclopropanation.

It is both tempting and convenient to judge the value of a chemical technology by its ultimate use in commercial manufacturing processes. In the end, however, it is important to recognize that the choice of technology frequently depends upon specific factors and contexts unique to the drug candidate, rather than the inherent nature of the underlying chemical technology itself. Furthermore, while expediency and speed are more important than cost efficiency during the earlier stages of development, there is still a need to develop and demonstrate a long-term process strategy for each candidate well before substantial quantities are required for late-stage development. When one considers that candidate dropout rates between initial clinical safety assessment and marketing approval are still about 90% across the pharmaceutical industry, one can appreciate that the majority of processes developed for long-term commercial use never actually reach a production stage. The process work described above for melatonin agonist **1** was just such a case.

Pharmaceutical development is ultimately a game of numbers, as all candidates are viewed as potential winners until discouraging data from clinical trials block the majority from moving forward. Since it is impossible to predict the winners, it is important to recognize that even when candidates approaching late-stage development fail, they've played an important role in keeping you in the game. In addition, knowledge gained and technology developed throughout the evolution of such projects is often applied toward research and development of new candidates. When viewed from that broader perspective, the more important fact demonstrated in this account is that a wide variety of approaches to synthetic targets can and do play important roles fueling the engines of drug discovery and process development.

2.7
Acknowledgement

The authors acknowledge the following individuals who were responsible for the advancement of all of the chemistry described in this account: Gerard Crispino, Rajendra Deshpande, Rita Fox, Jollie Godfrey, Animesh Goswami, Jason Hamm, David Kacsur, Donald Kienzler, Atul Kotnis, Daniel Kuehner, Meena Rao, James Simpson, Shankar Swaminathan, Michael Totleben, Truc Vu, Wen-sen Li, Jason Zhu, John Thornton, and Ming Yang.

2.8
References

1 a) J. Salaun, M.S. Baird, Curr. Medic. Chem. 1995, 2(1), 511; b) J. Salaun, Top. Curr. Chem. 2000, 207, 1.
2 a) W.A. Donaldson, Tetrahedron 2001, 57, 8589 and references therein; b) R. Faust, Angew. Chem., Int. Ed. Engl. 2001, 40, 2251.
3 a) J. Vallgaarda, U. Hacksell, Tetrahedron Lett. 1991, 32, 5625; b) J. Vallgaarda, U. Appelberg, I. Csoregh, U. Hacksell, J. Chem. Soc., Perkin Trans. I 1994, 461.
4 I.P. Beletskaya, A.V. Cheprakov, Chem. Rev. 2000, 100(8), 3009.
5 E.J. Corey, M. Chaykovsky, J. Am. Chem. Soc. 1965, 87, 1353.
6 A.G.M. Barrett, D.C. Braddock, R.A. James, N. Koike, P.A. Procopiou, J. Org. Chem. 1998, 63, 6273.
7 M. Rao, M. Yang, D. Kuehner, J. Grosso, R. Deshpande, Org. Process Res. Dev. 2003, 7, 547.
8 W.S. Wadsworth, W.D. Emmons, J. Am. Chem. Soc. 1961, 83, 1733.
9 A. Goswami, K.D. Mirfakhrae, M.J. Totleben, S. Swaminathan, R.N. Patel, J. Indust. Microbiol. Technol. 2001, 26, 259.
10 A. Goswami, M.J. Totleben, A.K. Singh, R.N. Patel, Tetrahedron Asymmetry 1999, 10, 317.
11 A. Goswami, K.D. Mirfakhrae, R.N. Patel, Tetrahedron Asymmetry 1999, 10, 4239.
12 a) M. Palucki, P.J. Pospisil, W. Zhang, E.N. Jacobsen, J. Am. Chem. Soc. 1994, 116, 9333; b) E.N. Jacobsen, M.H. Wu, Comprehensive Asymmetric Catalysis, Springer, New York, 1999, 649; c) W. Zhang, E.N. Jacobsen, J. Org. Chem. 1991, 56, 2296.
13 J.S. Prasad, T. Vu, M.J. Totleben, G.A. Crispino, O.J. Kaesur, S. Swaminathan, J.E. Thornton, A. Fritz, A.K. Singh, Org. Process Res. Dev., in press.
14 P. Brougham, M.S. Cooper, D.A. Cummerson, H. Heaney, N. Thompson, Synthesis 1987, 1015.
15 a) K.B. Sharpless, W. Amberg, Y.L. Bennani, G.A. Crispino, J. Hartung, K.-S. Jeong, H.-L. Kwong, K. Morikawa, Z.M. Wang, D. Yu, X.-L. Zhang, J. Org. Chem. 1992, 57, 2768. For leading

references in this area see: b) R. A. Johnson, K. B. Sharpless, Catalytic Asymmetric Synthesis, VCH, New York, 1993, 103; c) T. Katsuki, V. S. Martin, Org. React. 1996, 48, 1; d) T. Katsuki, Transition Metals for Organic Synthesis, Wiley-VCH, New York, 1998, vol. 2, 261; e) T. Katsuki, Comprehensive Asymmetric Catalysis, Springer, New York, 1999, II, 621; f) I. E. Marko, J. S. Svendsen, Comprehensive Asymmetric Catalysis, Springer, New York, 1999, vol. 2, 713.

16 H. C. Kolb, K. B. Sharpless, Tetrahedron 1992, 48, 10515.

17 A. K. Singh, M. N. Rao, J. H. Simpson, W.-S. Li, J. E. Thornton, D. Keuhner, D. Kacsur, Org. Process Res. Dev. 2002, 6, 618.

18 a) M.-H. Xu, G.-Q. Lin, Youji Huaxue 2000, 20, 475; b) M. P. Doyle, M. N. Protopopova, Tetrahedron 1998, 54, 7919; c) M. P. Doyle, Catalytic Asymmetric Synthesis, VCH, New York, 1993, 63; d) M. P. Doyle, Pure Appl. Chem. 1998, 70, 1123; e) A. B. Charette, H. Lebel, Comprehensive Asymmetric Catalysis, Springer, New York, 1999, 581; (f) H.-U. Reissig, Angew. Chem., Int. Ed. Engl. 1996, 35, 971.

19 EDA is friction-sensitive, starts self-heating at 80 °C, and its decomposition energy has been measured to be about 114 kJ/mol (about 5 times above our safety limit).

20 N. E. Searle, Org. Synth., Coll. Vol. 4, 424.

21 M. J. Totleben, J. S. Prasad, J. H. Simpson, S. H. Chan, D. J. Vanyo, D. E. Kuehner, R. Deshpande, G. A. Kodersha, J. Org. Chem. 2001, 66, 1057.

22 D. A. Evans, K. A. Woerpel, M. M. Hinman, M. M. Faul, J. Am. Chem. Soc. 1991, 113, 726.

23 T. Fukuda, T. Katsuki, Tetrahedron 1997, 53, 7201.

24 a) H. Nishiyama, K. Itoh, H. Matsumoto, S.-B. Park, K. J. Itoh, J. Am. Chem. Soc. 1994, 116, 2223; b) S.-B. Park, K. Murata, H. Matsumoto, H. Nishiyama, Tetrahedron Asymmetry 1995, 6, 2487.

3
Hetero Diels-Alder-Biocatalysis Approach for the Synthesis of (S)-3-[2-{(Methylsulfonyl)oxy}ethoxy]-4-(triphenylmethoxy)-1-butanol Methanesulfonate:
Successful Application of an Enzyme Resolution Process

Jean-Claude Caille, Jim Lalonde, Yiming Yao, and C.K. Govindan

Abstract

A cost effective and easily scaled-up process has been developed for the synthesis of (S)-3-[2-{(methylsulfonyl)oxy}ethoxy]-4-(triphenylmethoxy)-1-butanol methanesulfonate, a key intermediate used in the synthesis of a protein kinase C inhibitor drug through a combination of hetero-Diels-Alder and biocatalytic reactions. The Diels-Alder reaction between ethyl glyoxylate and butadiene was used to make racemic 2-ethoxycarbonyl-3,6-dihydro-2H-pyran. Treatment of the racemic ester with *Bacillus lentus* protease resulted in the selective hydrolysis of the (R)-enantiomer and yielded (S)-2-ethoxycarbonyl-3,6-dihydro-2H-pyran in excellent optical purity, which was reduced to (S)-3,6-dihydro-2H-pyran-2-yl methanol. Tritylation of this alcohol, followed by reductive ozonolysis and mesylation afforded the product in 10–15% overall yield with excellent optical and chemical purity. Details of the process development work done on each step are given.

3.1
Introduction

The title compound **1** (Fig. 1) is one of the two key intermediates used by Eli Lilly & Co. in the synthesis of **2** (LY 333531), a protein kinase C inhibitor [1]. Compound **2** is in clinical trials for the treatment of retinopathy and nephropathy in patients with diabetes mellitus. A recent paper from Lilly describes [2] the development of a process to prepare **1** and **2** in multi-kilogram quantities. In this process, an acyclic bis-indolylmaleimide is reacted with **1**, and the product obtained is converted into **2** in several steps as shown in Fig. 1.

The method chosen by Eli Lilly to prepare **1** in large quantities is shown in Fig. 2. Although this synthesis is amenable to scale-up, it uses rather expensive starting materials and reagents such as (R)-chloro-2,3-propanediol or (R)-glycidol, vinyl magnesium bromide, and potassium *tert*-butoxide. An evaluation of the previously published synthetic methods indicated to us that a lower-cost synthesis of **1** could be developed by the use of alternative raw materials.

Asymmetric Catalysis on Industrial Scale: Challenges, Approaches and Solutions
Edited by H.U. Blaser, E. Schmidt
Copyright © 2004 Wiley-VCH Verlag GmbH & Co. KGaA, Weinheim
ISBN: 3-527-30631-5

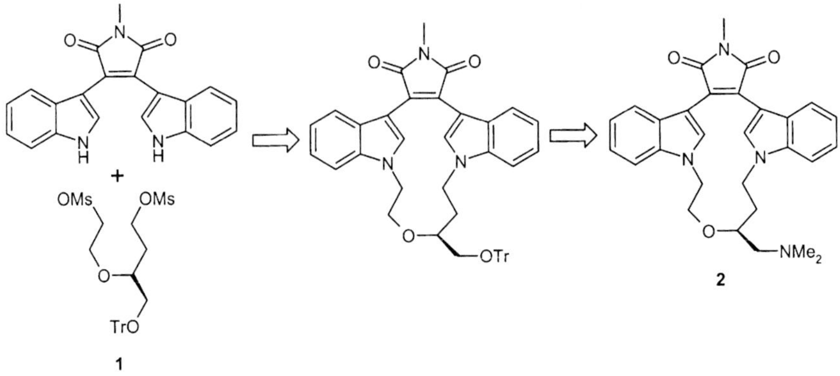

Fig. 1 Synthesis of LY 333531.

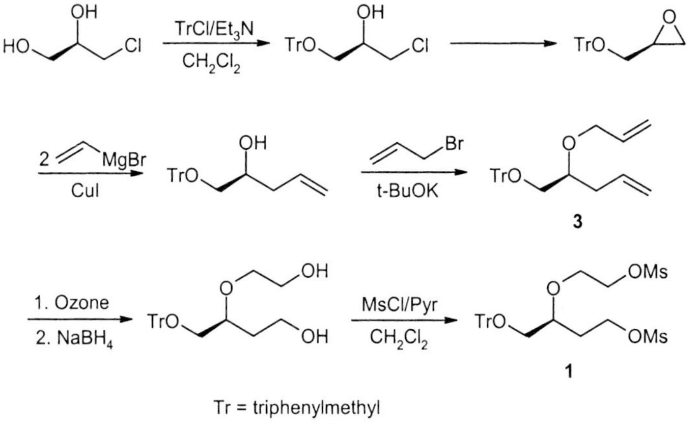

Fig. 2 Lilly synthesis of **1**.

3.2
The Hetero Diels-Alder-Biocatalysis Strategy

A retro-synthetic analysis of **1** as shown in Fig. 3 indicates that 3,6-dihydro-2H-pyran derivatives could serve as key intermediates in the synthesis of **1** and that hetero Diels-Alder reactions could be used to make these 3,6-dihydro-2H-pyran intermediate [3, 4].

Our first approach, which looked very attractive, was based on a catalytic asymmetric hetero Diels-Alder reaction between butadiene and trityloxyacetaldehyde (Fig. 3). This reaction would have been highly atom economic, and if it could be done with a chiral catalyst, would lead directly to 2-(S)-trityloxymethyl-3,6-dihydro-2H-pyran **6** (Fig. 4) that could be used to make **1**. There is a precedent in the literature for such a hetero Diels-Alder reaction, where benzyloxyacetaldehyde reacts

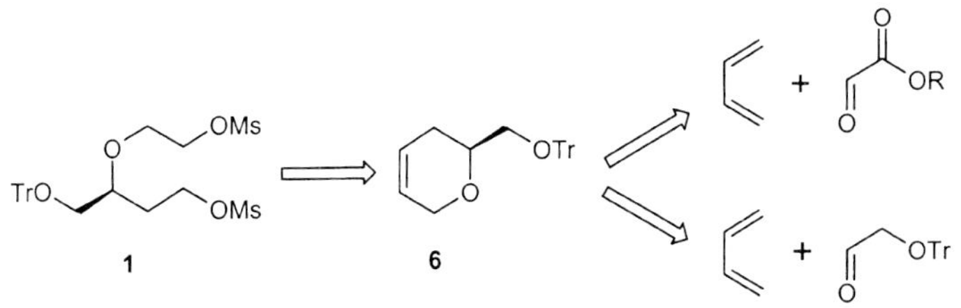

Fig. 3 Retro-synthetic analysis of **1**.

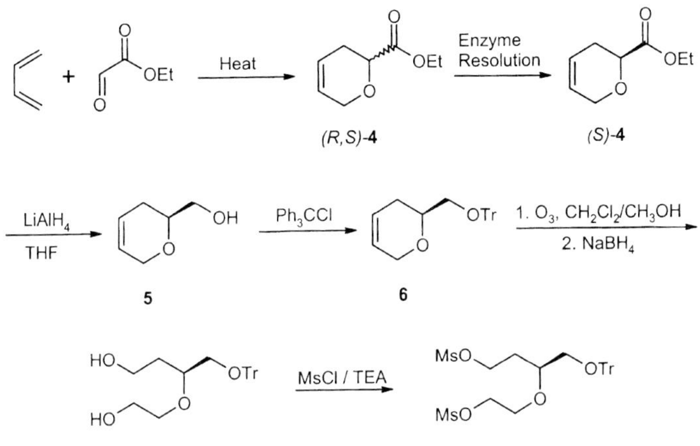

Fig. 4 Proposed synthesis of **1**.

with Danishefsky's diene in the presence of Cu(II)-bis(oxazoline) complexes to yield the corresponding dihydropyranone in 75% yield and 85% ee [5]. When butadiene was allowed to react with trityloxyacetaldehyde in the presence of metal complexes we obtained either a complex mixture as the product or low chemical and optical yield.

Encouraged by successful results reported by Johannsen and Jorgensen [6], who have shown that butadiene reacts slowly with isopropyl glyoxylate in the presence of 2,2-isopropylidene-bis[(4S)-4-tert-butyl-2-oxazoline] and copper triflate to produce (S)-2-isopropyloxycarbonyl-3,6-dihydro-2H-pyran in 55% yield and 87% ee, we studied the reaction of butadiene with esters of glyoxylic acid under a variety of conditions, by varying the solvent, catalyst, temperature, and pressure. In spite of significant efforts, we failed to come up with conditions applicable to large-scale production. Only forced reaction conditions based on high-pressure chemistry allowed us to obtain acceptable chemical yields and enantiomeric excess [7].

Since our catalytic approaches failed, we next attempted a chiral auxiliary approach. The glyoxyloyl function was incorporated into (2R)-bornane-10,2-sultam, a well known chiral auxiliary. However, the reaction of this sultam with butadiene did not occur readily and required pressures of up to 10 kbar [8]. Therefore, this reaction could not be readily scaled-up.

We, therefore, concentrated our efforts on developing a process based on the enzymatic resolution of 2-alkoxycarbonyl-3,6-dihydropyrans. Interest in the catalytic kinetic resolution of racemates has revived recently with the realization by organic chemists that early stage resolutions of chiral intermediates can lead to highly efficient syntheses [9]. An effective strategy for the rapid development of efficient and inexpensive processes is to devise syntheses that incorporate the resolution of chiral intermediates using commercially available bulk enzymes. The largest class of these commercial enzymes is the hydrolases [10], proteases and lipases produced on the ton scale for use in detergents and food processing. In the case of enzymatic ester hydrolysis, this strategy involves selection of a racemic ester intermediate which has low or moderate water solubility, a chiral center close to the reacting center and substantially different groups at the chiral center, but with low or moderate steric hindrance. By subjecting the selected racemate to a library of commercially available enzymes under screening conditions one can very quickly identify useful enzymes. By selecting those enzymes, which convert the undesired enantiomer into the acid, one can develop resolutions in which the unreacted isomer in high optical purity is recovered. Such subtractive resolutions give much higher optical purities than resolutions in which the reaction product is the desired enantiomer [11].

Our proposed synthesis of **1** using a hetero Diels-Alder-biocatalysis approach is shown in Fig. 4. In this study, we report the details of the development of a commercially viable process to make **1** based on the enzyme resolution of (R,S)-**4** [12].

3.3
Hetero Diels-Alder Reaction:
Synthesis of 2-Ethoxycarbonyl-3,6-dihydro-2H-pyran, (R,S)-4

The reaction between butadiene and ethyl glyoxylate was chosen for development because these two raw materials are commercially available and relatively inexpensive. According to the literature, the reaction between butadiene and alkyl glyoxylate ethyl hemiacetal affords 2-alkoxycarbonyl-3,6-dihydro-2H-pyran in 41–46% yield [13]. Ethyl glyoxylate is commercially available as a 50% solution in toluene, and all our experiments were carried out with this solution. To carry out the Diels-Alder reaction, ethyl glyoxylate solution and hydroquinone were charged to an autoclave and heated to 150–170 °C. An excess of butadiene was then pumped into the reaction mixture over 1–2 hours. Temperature was a critical factor in obtaining good conversions. Below 155 °C, conversion of ethyl glyoxylate to the product was lower, and up to 20% of ethyl glyoxylate was recovered. After the reaction the product was separated from the byproduct polymers by distillation. The yield of 2-ethoxycarbonyl-3,6-dihydro-2H-pyran ranged from 55 to 60% and the purity ranged from 90 to 95%. It should be noted that the purity of the racemic ester was not a critical factor that affected the enzyme resolution described in detail below. Even racemic ester of 85–90% purity could be used for the enzymatic resolution.

3.4 Enzymatic Resolution of (*R*,*S*)-4

Using the strategy described above, we set out to identify an enzyme that can be used for the hydrolysis of (*R*)-2-ethoxycarbonyl-3,6-dihydropyran (Fig. 5) on a large scale. To be economical, the enzyme should be commercially available, relatively inexpensive, and highly active. We anticipated that the conformational rigidity of the unsaturated dihydropyran ring will enhance enantio-differentiation and a commercially available enzyme would selectively hydrolyze the undesired (*R*)-ester to the acid, leaving the (*S*)-ester unreacted. Tab. 1 lists the results of our initial screening.

Examination of the results in Tab. 1 indicates that several enzymes are efficient and selective in the hydrolysis of the (*R*,*S*)-**4**, however most lipases preferentially hydrolyzed the (*S*)-ester while several proteases exhibited the desired (*R*)-ester selectivity. Based on these results, we focused all further efforts on the screening and optimization of inexpensive alkaline proteases. The corresponding butyl ester was also screened as a substrate. While several selective hydrolases were identified, the rate of hydrolysis was about three-fold less than that of ethyl ester (see Experimental), presumably due to the decrease in water solubility of the butyl ester.

3.4.1 Secondary Screen and *E* Determination

A more thorough analysis of the enantioselectivity of several commercially available alkaline proteases was performed by determining the enantioselectivity factor *E* of these enzymes in the hydrolysis of (*R*,*S*)-**4**. In order to determine the enantioselectivity factor *E* accurately (the kinetic ratios for the conversion of the two enantiomers in a first-order kinetic resolution) [11], accurate measurement of the optical purity of the product acid and unreacted ester is necessary. Separation of the carboxylic acid enantiomers by gas chromatography was not possible due to decomposition of the acid on the column, so a derivatization method was developed to convert the acid into the corresponding methyl ester. Treatment of the extracted acid and ester with trimethylsilyldiazomethane resulted in the conversion of the acid into the methyl

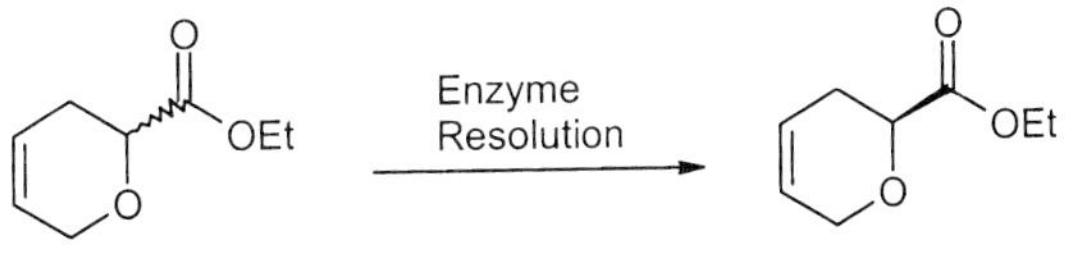

(*R*,*S*)-**4** (*S*)-**4**

Fig. 5 Resolution of (*R*,*S*)-**4**.

Tab. 1 Results for hydrolase library screening at 2 hours.

Entry	Enzyme	Configuration/% ee of remaining ester	Relative conversion
1	Pig liver esterase	–	Full
2	Burkholderia cepacia lipase	R/27.4	High
3	Porcine pancreatic lipase	S/57.1	High
4	Candida rugosa lipase	–	Full
5	alpha-Chymotrypsin	S/28.8	~50
6	Penicillin acylase (E. coli)	R/5.0	Low
7	Aspergillus niger lipase	R/100.0	High
8	Rhizomucor miehei lipase	–	High
9	ChiroCLEC™-CR	–	High
10	Subtilisin Carlsberg	S/100.0	High
11	Candida antarctica "A" lipase	R/>98	High
12	Candida lypolytica lipase	R/41.3	Low
13	Candida antarctica "B" lipase	R/2.1	High
14	Thermomyces lanuginosus lipase	S/–	Full
15	Bacillus stearothermophilus protease	S/77.0	High
16	ChiroCLEC™-BL (slurry)	S/91.4	~50
17	ChiroCLEC™-PC (slurry)	R/92.2	High
18	ChiroCLEC™-EC (slurry)	S/17.5	High
19	Rhizopus delemar lipase	R/97.7	High
20	Rhizopus niveus lipase	R/76.2	High
21	Alcaligenes species lipase	R/5.9	High
22	Aspergillus oryzae protease	S/41.9	High
23	Candida rugosa esterase	R/–	High
24	Papain	R/10.7	High
25	Aspergillus melleus protease	S/26.2	~75
26	PeptiCLEC-TR (slurry)	0	~75
27	Pseudomonas aeruginosa lipase	R/3.3	High
28	Bacillus lentus protease-III	S/100.0	High
29	Control	0	0

ester with only a small amount of racemization (the highest optical purity observed was 89% for the acid). This allowed calculation of a minimum E value for the different enzymes. Of these, Tab. 1 entries 10, 16, and 28 were found to be selective, with ChiroCLEC BL and Bacillus lentus protease-III showing the highest selectivity for the required hydrolysis of the (R)-ester (E value > 25).

3.4.2
Protease Screen

Four additional commercial proteases were examined for the hydrolysis of (R,S)-4 including three variants of Bacillus lentus proteases engineered for stability (Tab. 2). Of these Bacillus lentus protease-I gave equivalent selectivity to that of Ba-

Tab. 2 Selectivity and rate of proteases in (R,S)-**4** resolution.

Protease[a]	Enzyme loading	Conversion (%) after 24 h	E
Bacillus lentus-I	0.3 mL	55.07	41.7
Fungal protease	100 mg	56.45	1.5
Bacillus lentus-II	0.3 mL	39.22	27.5
BL CLEC	100 mg	30.00 (20 min)	28.4
Bacillus lentus-III	0.3 mL	39.32	36.6

Reaction conditions: room temperature, 10% substrate loading, 0.3 M phosphate buffer with pH=7.0.
a) I, II, and III represent proteases from different commercial sources.

cillus lentus protease-III. Altus' commercial protease cross-linked enzyme crystal based on *Bacillus licheniformis* protease, ChiroCLEC©-BL and a variant *Bacillus lentus* protease-I gave slightly lower selectivity than *Bacillus lentus* protease-III. Fungal protease was inferior to all of the *Bacillus lentus* protease variants in both activity and selectivity. Based on these results and cost considerations, *Bacillus lentus* protease-III was selected for further optimization.

3.4.3
Optimization of the Resolution of (R,S)-4

3.4.3.1 Buffer pH and Concentration

The effect of pH on the reaction rate and enantioselectivity using *Bacillus lentus* protease-III was studied and the results are summarized in Tab. 3. As with most ester resolutions using alkaline proteases, at high pH the hydrolysis is more rapid, however the selectivity is somewhat lower. We also observed that base-catalyzed hydrolysis is significant when the pH is over 9.0 (<1%), which has a negative effect on selectivity. Based on initial rate measurements, the rate of hydrolysis at pH 8 was four-fold higher than that at pH 7. Thus there is a substantial decrease in rate of hydrolysis with decreasing pH.

The effect of buffer strength on the rate of hydrolysis was also examined. From the pH/selectivity determination, it was clear that pH control is essential to obtain good selectivity, especially during scale-up. The sensitivity of the process to deviations in pH could be decreased by the use of very concentrated buffers. From the results in Tab. 4, it is apparent that the rate of hydrolysis also decreases at very high buffer concentrations, possibly due to the decreased solubility of the substrate at high salt concentrations or inactivation of the enzyme. Based on these results, a buffer strength of 0.3 M was selected for scale-up of the process.

Tab. 3 Effect of buffer pH on conversion at fixed enzyme/substrate ratio.

Reaction pH	Conversion (%) after 16 min	E
7.0	39.32	36.6
8.0	50.52	31.5
9.0	54.13	27.3

Reaction conditions: room temperature, 1.0 g substrate, 8.7 mL 0.3 M phosphate buffer, 0.3 mL *Bacillus lentus* protease-III.

Tab. 4 Effect of buffer concentration on conversion at fixed enzyme/substrate ratio.

Buffer concentration (M)	Conversion (%) after 16 min
0.1	27.0
0.3	27.2
1.0	7.0

Reaction conditions: room temperature, 1.0 g substrate, 8.7 mL phosphate buffer with pH = 7.0, 0.3 mL *Bacillus lentus* protease-III.

3.4.3.2 Optimization of Enzyme Loading and Other Parameters

The effect of enzyme loading on rate of the reaction and selectivity was examined at a pH of 7.8. In several relatively large-scale trials, we found that 0.10 to 0.15 mL of ~5% enzyme solution per gram of substrate was sufficient to give a product with an ee of 99.5% within 8 hours. The results of a more systematic analysis of the effects of substrate concentration are shown in Tab. 5.

The effects of temperature, co-solvents and substrate were investigated in an effort to decrease enzyme loading further (see Tab. 6). A study of temperature effects showed that, at higher temperatures, the activity of *Bacillus lentus* protease increased dramatically, but selectivity decreased. Increasing the solubility of the ester in the reaction media by the addition of water-miscible co-solvents such as acetonitrile (results not shown) had the effect of slowing the reaction. Water-immiscible solvents such as toluene also lowered the reaction rate. Examination of the effects of substrate loading was more productive (Tab. 7). The absolute rate of hydrolysis decreases with increasing substrate concentration. Thus, at a fixed ratio of protein catalyst to substrate, the rate of conversion decreases with increasing ester concentration. The mechanism of this effect is unclear. Since in each case the aqueous phase is saturated with the ester substrate, it may result from autolysis or enzyme inactivation from high concentrations of the hydrolysis product.

The results presented above clearly indicate that the resolution of (R,S)-4 can be easily carried out using *Bacillus lentus* proteases. As has been mentioned before we chose *Bacillus lentus* protease-III for further development because of its efficacy and lower cost. Commercial formulations of this enzyme are stable and avail-

Tab. 5 Effect of enzyme loading and ester concentration on reaction time.

Enzyme [a] (mL)	Substrate (g)	mL Enzyme/g substrate	Ester concn. (%)	Reaction time (h)	ee of ester (%)
1.8	6.0	0.3	25	6	99.9
7.5	25	0.3	25	6	99.9
1.0	5.0	0.2	25	2.5	98.5
0.75	5.0	0.15	25	6	99.7
0.5	5.0	0.1	25	6	99.7
3.0	10.0	0.3	50	16	99.5

Temperature was maintained at 25 °C and reaction allowed to proceed until the desired optical purity was reached.
a) Protein concentration is approximately 5% in the enzyme solution.

Tab. 6 Temperature and toluene effect.

Temperature (°C)	Conversion (%) after 1 h	ee Substrate (%)	E
45	52.0	89.5	31.4
25	42.5	66.2	36.7
25 Toluene (20%)	25.7	31.5	29.0

Reaction conditions: 0.3 mL *B. lentus* protease-I, 10% substrate loading, 0.3 M phosphate buffer with pH = 7.0.

Tab. 7 Effect of substrate concentration on conversion at fixed enzyme/substrate ratio.

Substrate loading (%)	Conversion (%)	% ee Substrate after 2 h
10	37	57.1
25	23	27.1
50	16	18.9

Reaction conditions: room temperature, 0.3 M phosphate buffer with pH = 7.0, enzyme loading 0.3 mL/g of substrate.

able as a solution. Having identified optimum conditions for the reaction we concentrated our efforts on increasing the volume efficiency of the process. Initial small-scale experiments were carried out using 1 N solutions of sodium hydroxide to control the pH. This significantly reduced the productivity of the process. In large-scale experiments it was shown that even 10 N sodium hydroxide solution could be used, significantly reducing the volume of aqueous base required to control pH. It was also observed that in the range of 25–30 °C the reaction was over in 2–4 hours. Agitation rate of the reaction mixture was also found to have an effect on reaction time and conversion.

The progress of the reaction can be followed by gas chromatography using chiral columns that separates the (R)- and (S)-isomers of 4 well. However, in a manufacturing plant such analyses can take a significant amount of time. The consumption of base during pH control can also be used to follow the conversion.

The equipment we used to conduct the reaction in the laboratory and in the pilot plant were quite simple. Any reactor equipped with a variable speed stirrer, temperature, and pH controllers can be used for the reaction. It was observed that slow agitation which is sufficient to provide good pH control is best. Vigorous agitation was found to lead to lower conversions, probably because of the denaturation of the enzyme.

The product is recovered by simple extraction. Since the next step in our synthesis is a lithium aluminum hydride reduction, it was necessary for us to dry the product completely. However, it has not been found necessary to purify the product further by distillation. Typical yields we have obtained in the enzymatic resolution step in large-scale runs have varied from 38 to 41%.

3.5
Pilot Plant Trials

The enzymatic resolution of (R,S)-2-ethoxycarbonyl-3,6-dihydropyran has been carried out repeatedly in a pilot plant on a 100–125 kg scale. The conditions of pH, agitation, concentration, and enzyme/substrate ratio have been further optimized, but for the most part conditions developed during the laboratory optimization studies have been found to work well. This technology is currently being used to produce ton quantities of (S)-4. It has to be pointed out that no major difficulties were encountered during the scale-up. The work-up of the product in this process, unlike in many other enzyme-catalyzed processes, is quite simple and volume efficiencies are better than some chemical reactions. The recovery and recycling of the enzyme is not needed since it is commercially available and relatively inexpensive.

3.6
Attempted Resolution of 3,6-Dihydro-2H-pyran-2-ylmethanol, (R,S)-5

As an alternative to hydrolysis of the racemic ester in water, we investigated the possibility of resolving the alcohol (R,S)-5 obtained by the reduction of (R,S)-4, by enzymatic *trans* esterification in organic solvents as shown in Fig. 6. High rates and selectivities of enzymatic acylation of unhindered alcohols can be achieved in an organic solvent using vinyl acetate as an acyl donor [14].

The racemic alcohol was prepared by lithium aluminum hydride reduction of (R,S)-4. An abbreviated library of commercially available enzymes (entries 1–11 in Tab. 1) was used to screen the vinyl acetate acylation. Enzymes in entries 2, 3, 4, 8, and 11 catalyzed the acylation with a relatively fast rate (>20% conversion in

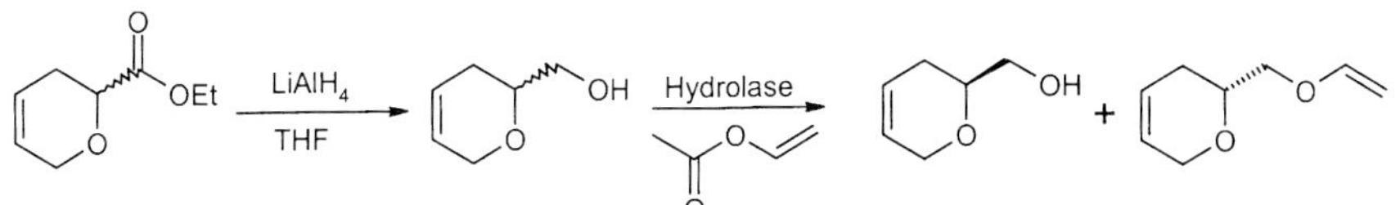

Fig. 6 Resolution of alcohol intermediate, (R,S)-**5**.

2 hours) however none did so with an *E* value exceeding 3. These results confirmed our initial selection of (R,S)-**4** as the best substrate for enzymatic resolution.

3.7 Experimental

3.7.1 Hydrolase Library Screening – Hydrolysis of (R,S)-4

Twenty-eight commercially available hydrolases were screened (ChiroScreen™ EH, Altus Biologics, Cambridge, MA, USA) to identify an enzyme that has the desired characteristics. Initial screenings were performed on a small scale in a phosphate buffer (1 mL) at pH 7 using 20 µL of ester and 100 mg of the enzyme. The reaction mixtures were stirred by a magnetic stirrer and 100 µL aliquots were removed at 1 hour, 2 hours and 6 hours. The aliquots were added to 1 mL of methyl *tert*-butyl ether (MTBE), dried over MgSO$_4$ and analyzed by gas chromatography. Optical purity and a rough indication of conversion were used to screen the candidate enzymes for rate and the desired selectivity.

3.7.2 Gram Scale Runs

To a 20 mL flask was added 2.5 g of (R,S)-2-ethoxycarbonyl-3,6-dihydro-2H-pyran, followed by 7 mL of 0.2 M pH 7.5 phosphate buffer and 2 mL of *Bacillus lentus* protease-III solution (approximately 5% solution of the protein). The biphasic solution was stirred at room temperature (23 °C) using a magnetic stirrer. The pH was checked at 0.5 hour intervals and readjusted to 7.5 by the drop-wise addition of 1 N NaOH (approximately 7 mL were required over the complete reaction). The progress of the reaction was monitored by chiral gas chromatography. After 5 hours, the enantiomeric purity of the unreacted ester was >99% and the reaction was stopped by the addition of 10 mL of MTBE. The pH of the aqueous phase was adjusted to 8.5 and the mixture was transferred into a separatory funnel. The aqueous phase was extracted twice with 20 mL of MTBE and the combined organic layers were extracted once with saturated sodium bicarbonate (10 mL), followed by saturated sodium chloride solution (10 mL), and the organic layer was dried over 2 g of anhydrous MgSO$_4$. The solvent was removed under reduced pressure (water aspirator) using a rotary evaporator (bath $T=50\,°C$) to give 1.2 g (48% yield) of a fragrant clear yellow liquid.

3.7.3
Butyl Ester

The butyl ester was hydrolyzed enantioselectively by the *Bacillus lentus* protease-III, albeit at a slower rate. Racemic butyl ester (1.07 g) was added to 10 mL of pH 7.5 phosphate buffer and 1 mL of *Bacillus lentus* protease-III solution. After 6.5 hours, the optical purity of the unreacted ester was only 72%. After stirring for an additional 14 hours the optical purity reached 99.9%.

3.7.4
Determination of Enantiomeric Purity by Gas Chromatography

Aliquots (200 µL) of the well stirred reaction mixture were removed periodically and added to a 2 mL vial containing 1.5 mL of MTBE. Hydrochloric acid (20 µL of 1 N) was added, the vial was shaken and then the solvent was dried over anhydrous MgSO$_4$ and the solution was transferred into a vial. Five drops of methanol and 1 drop of trimethylsilyldiazomethane solution (2 M in hexane) were added to convert the carboxylic acid product into the methyl ester for GC analysis. A J&W Scientific Cyclodex B column (0.25 micron, 30 meter×0.25 mm) was used. GC conditions: 110 °C for 10 min then 1 °C/min for 5 min. Retention times: (S)-methyl ester 8.35, (R)-methyl ester 8.65, (S)-ethyl ester 11.35, and the (R)-ethyl ester at 11.64 min.

The butyl ester was analyzed using the same column and the temperature program: 10 min at 110 °C, the temperature of the oven was increased at a rate of 1 °C/min to 115 °C, held for 15 min and then increased to 150 °C at a rate of 1 °C/min. The two enantiomers of the butyl ester eluted at 28.5 and 29.1 min.

3.7.5
Resolution of (R,S)-4

(R,S)-2-ethoxycarbonyl-3,6-dihydro-2H-pyran (400 g) and phosphate buffer (1080 mL of pH=7.8, 0.3 M) were charged to a 2 L reactor. The biphasic mixture was stirred using an overhead mechanical stirrer at room temperature. The pH of the mixture was adjusted to 7.8 by the careful addition of 32% NaOH. A ~5% solution of *Bacillus lentus* protease-III (120 mL) was added to initiate the reaction. The pH of the mixture was maintained at 7.8 during the reaction using a Radiometer Auto titration apparatus (Radiometer VIT 90, Copenhagen, Denmark) with 32% NaOH as the titrant. Conversion was followed by the consumption of base and by periodically analyzing aliquots from the mixture. After 2 hours, the ee of the unreacted ester reached 99.7%. A total 206 mL of 32% NaOH were added during the hydrolysis. After 2 hours, the agitation was stopped. The mixture was extracted with three portions of MTBE (400 mL, and 2×200 mL). The combined organic phases were washed with 100 mL of saturated brine and MTBE was removed under reduced pressure using a rotary evaporator. Trace amounts of water were removed by the addition and evaporation of 20 mL toluene. The product was filtered to remove a small amount of NaCl. The yield of (S)-**4** was 163 g (41%).

3.7.6
Reduction of (S)-4

The reduction of (S)-**4** to (S)-3,6-dihydro-2H-pyran-2-yl methanol **5** was carried out readily by adding the ester slowly to solutions of lithium aluminum hydride (LAH) in THF or adding LAH solution to the ester. The reaction was carried out at 0–10 °C to reduce racemization. The ee of the alcohol **5** obtained was generally close to that of the starting ester. The yield of the product in this step was 85–90%.

3.7.7
Tritylation of 5

The tritylation of **5** was performed in pyridine using standard procedures. Reaction mixtures were worked-up by water quenching followed by filtration and recrystallization of the crude product. Recrystallization of the crude product was done from 2-propanol containing small amounts of triethylamine. The addition of triethylamine was necessary to avoid the formation of trityl isopropyl ether by solvolysis of the product when heated in 2-propanol. The yield of the product in this step was 75–80%. Impurities such as ethyl glyoxylate diethyl acetal carried over from previous steps were removed during the isolation of **6**.

3.7.8
Reductive Ozonolysis of (S)-2-Trityloxymethyl-3,6-dihydro-2H-pyran 6

Ozonolysis of **6** to produce **7** was studied at temperatures ranging from –78 to –0 °C in mixtures of dichloromethane and methanol. The intermediates that formed were found to be stable below –30 °C. Diol **7** was obtained in high yields when the reaction mixture was quenched with aqueous sodium borohydride. If the reaction mixture was warmed to room temperature before the quench, a complex mixture of products formed as indicated by the NMR spectrum. Similarly, only polymeric products were formed when the ozonolysis was carried out without methanol even at –78 °C. When ozonolysis was done at –5 to 0 °C in the presence of methanol, formation of trityl methyl ether was noticed, a result similar to which has been observed before [2]. Diol **7** was obtained in near quantitative crude yield when ozonolysis was carried out below –30 °C, followed by the addition of the reaction mixture to cold, aqueous sodium borohydride solution. The product was used without further purification in the next step.

3.7.9
Preparation of 1

The conversion of **7** to **1** in high yields has previously been described by the Lilly group. The reaction of the diol with methane sulfonyl chloride was carried out in dichloromethane in the presence of triethylamine. The crude product was then recrystallized from a mixture of ethyl acetate and heptane. The reaction can also be carried out in ethyl acetate, eliminating the use of dichloromethane. Care should be exercised to remove all traces of acid from **1** since the trityl group can cleave in the presence of acids, causing the decomposition of the product. The yield of the product in this step was 85–90%. The chemical purity was >99% and ee was >99.5%.

3.8
Summary

We have developed a commercially viable process for the synthesis of **1** through a combination of chemical and biocatalytic reactions. Process chemists often shy away from enzyme-catalyzed reactions because of the perception that these reactions are difficult to carry out. In many cases this is true since such reactions can have very low productivity and work-up of the reaction mixture (for example, filtration of the enzyme after the reaction) can be very difficult. However, in this work we have shown that such processes can also be extremely efficient, if carefully chosen. One can often find an enzyme that can be used just like a chemical reagent. Since we started out with low cost raw materials and the resolution was done early in the synthesis, our process is very cost-competitive. In spite of the fact that our overall yields are lower than that reported in the previous synthesis, our costs are lower since we have been able to eliminate the use of several expensive reagents and simplify the overall process by the use of biocatalysis. We have also been able to scale-up the process without difficulties. Further efforts to improve this process will be concentrated on the recovery and recycling of (R)-3,6-dihydro-2H-pyran-2-carboxylic acid. Recovery and reuse of the undesired enantiomer is an area that warrants significant attention since in most resolutions, whether chemical or enzymatic, half of the starting material is wasted. Efficient methods of recycling the undesired enantiomer can significantly reduce the cost of the final product.

Enantiomerically enriched hetero Diels-Alder cycloadducts have found applications in the synthesis of other pharmaceutical intermediates such as compactin and mevinolin [4, 5]. The strategy described in this work should be applicable to these and similar compounds.

3.9
References

1. a) J.-C. Caille, US Patent 6300106 B1, 2001; b) J.-C. Caille, C. K. Govindan, H. Junga, J. Lalonde, Y. Yao, Org. Process Res. Dev. 2002, 6, 471–476.
2. M. M. Faul, L. L. Winneroski, C. A. Krumrich, K. A. Sullivan, J. R. Gillig, D. A. Neel, C. J. Rito, M. R. Jirousek, J. Org. Chem. 1998, 63, 1961–1973.
3. For a review on hetero Diels-Alder reactions see: M. D. Bednarski, J. P. Lissikatos, in Comprehensive Organic Synthesis, Vol. 2 (B. M. Trost, I. Fleming, Eds.), Pergamon Press Oxford, UK, 1991, p. 661.
4. a) For a review on asymmetric hetero Diels-Alder reactions see: H. Waldman, Synthesis 1994, 535–551; b) K. Mikami, Y. Motoyama, M. Terada, J. Am. Chem. Soc. 1994, 116, 2812 and references therein.
5. A. K. Gosh, P. Mathivanan, J. Cappiello, Tetrahedron Lett. 1997, 38, 2427–2430.
6. M. Johannsen, K. A. Jorgensen, J. Org. Chem. 1995, 60, 573.
7. J. Jurczak, personal communication.
8. a) T. Bauer, C. Chapuis, J. Kozak, J. Jurczak, Helv. Chim. Acta 1989, 72, 482; b) T. Bauer, C. Chapuis, J. Kozak, J. Jurczak, J. Chem. Soc., Chem. Commun. 1990, 1178.
9. J. M. Keith, J. F. Larrow, E. N. Jacobsen, Adv. Synth. Catal. 2001, 343, 5–26.
10. U. T. Bornscheuer, R. J. Kazlaauskas, Hydrolases in Organic Synthesis; Wiley-VCH, Weinheim, 1999.
11. Recovery of the unreacted enantiomer affords much higher optical purities than that obtainable for the product of the reaction for a given yield. For a discussion and definition of the E value see C. S. Chen, C. J. Sih, Y. Fujimoto, G. Girdaukas, J. Am. Chem. Soc. 1982, 104, 7294.
12. For a detailed discussion and experimental procedures of the complete process see reference 1 b.
13. A. Y. Abuzov, E. M. Klimov, E. I. Klimova, Dokl. Akad. Nauk SSSR 1962, 142, 341–343.
14. Y. F. Wang, J. J. Lalonde, M. M. Momongan, D. E. Bergbreiter, C. H. Wong, J. Am. Chem. Soc. 1988, 110, 7200–7205.

4
Multi-Kilo Resolution of XU305, a Key Intermediate to the Platelet Glycoprotein IIb/IIIa Receptor Antagonist Roxifiban via Kinetic and Dynamic Enzymatic Resolution

Jaan A. Pesti and Luigi Anzalone [1]

Abstract

We describe the process development for the preparation of (R)-XU305, a key intermediate of roxifiban. This process evolved from a small-scale enzymatic kinetic resolution into a pilot plant-scale preparation. The combination of an unusual method of racemization of (S)-XU305 as a thioester permitted a dynamic enzymatic resolution to proceed, removing the need to recycle the less-reactive isomer in order to obtain acceptable overall yields.

4.1
Introduction

A significant trend of the pharmaceutical field over the past 20 years has been the shift of emphasis from racemic to chiral entities [2]. In addition to the large differences in biological activity displayed by enantiomers, other reasons include: i) the continuing improvements in the efficacy of asymmetric syntheses [3], ii) the entry of new starting materials into the chiral pool [4], and iii) the increase in the chemical art of resolution by both diastereoisomeric salt crystallization and enzymatic resolution [5, 6]. More apparent to the patient has been the increasing reluctance of physicians to prescribe formulations that contain the inactive or less active enantiomer and the recent policy of the Food and Drug Administration (FDA) and other regulatory agencies to steer the industry toward the production of single-isomer drugs [7]. Related to this last point is the FDA's encouragement of "racemic switches" in order to remove the remaining marketed racemates [2]. This trend has transformed formerly exotic purification techniques, such as preparative chromatographic resolution and simulated moving bed separation, into procedures that now resolve (or separate) kilograms of enantiomerically pure active pharmaceutical intermediates (API) daily. Unfortunately, nearly all enantio-purification methods come with a handicap. Since the "wrong" isomer is removed at this point, 50% of the input can be a dead loss. Nevertheless, many nascent pharmaceutical processes at first do sacrifice the undesired enantiomer. As the chemistry is honed, the unwanted isomer may be recycled, most ideally via its racemization

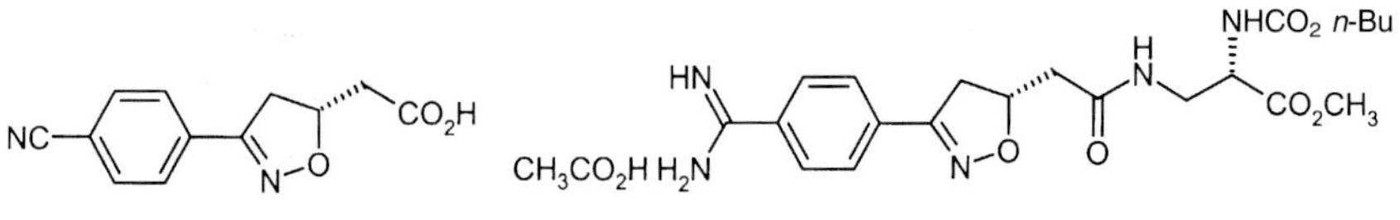

Fig. 1 Structures of the key intermediate (R)-XU305 (**1**) and of roxifiban (**2**).

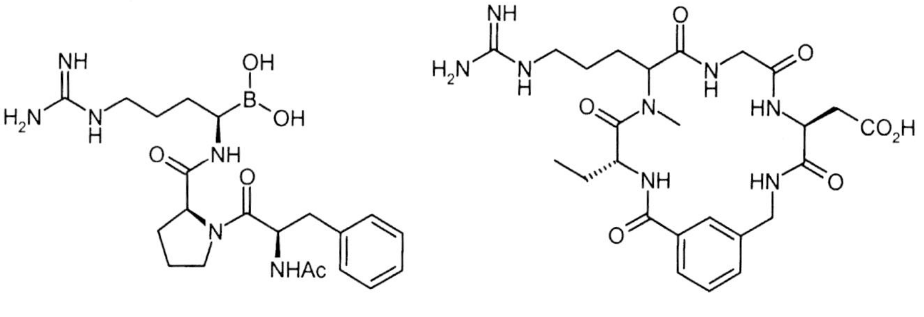

Fig. 2 Structures of the boropeptide thrombin inhibitor DuP 714 (**3**) and the cyclic pentapeptide DMP 728 (**4**).

during the resolution to create a dynamic resolution [8]. This creates the theoretical possibility of a 100% conversion into the desired enantiomer in a single step. We wish to describe research that experienced such an evolution, which allowed our group to prepare multi-kilogram quantities of (R)-XU305 (**1**), the key intermediate in the synthesis of roxifiban (**2**) (Fig. 1).

A brief introduction to the discovery of roxifiban will set the context of this work. About 20 years ago, our company [9] initiated research to develop a replacement for the anti-coagulant therapy: Coumadin®. While its indications include a variety of cardiovascular problems, the narrow therapeutic index required careful patient monitoring to avoid necrosis or hemorrhage. This research led to the boropeptide thrombin inhibitor DuP 714 (**3**) [10] and subsequently the cyclic pentapeptide DMP 728 (**4**) [11] (Fig. 2). DMP 728 targeted the receptor glycoprotein IIb/IIIa, which otherwise would bind fibrinogen once activated, a prelude to platelet aggregation. However, peptides have disadvantages, especially for oral administration, and further structure activity research suggested that 3,5-disubstituted isoxazolines would function as good core structures that would allow the rapid preparation of a wide array of candidates for anti-thrombotic screening. This premise was correct and rapidly led to the discovery of roxifiban [12].

Roxifiban is a potent and selective prodrug antagonist of our target receptor in that it disrupts the binding of fibrinogen to GPIIb/IIIa. This results in a potentially powerful weapon to prevent and treat a wide variety of thrombotic diseases resulting from undesired platelet adhesion, possibly including stroke, acute myocardial infarction, transient ischemic attack, and unstable angina. The success of clopidogrel, eptifibatide, abciximab, tirofiban hydrochloride, and aspirin indicate that antiplatelet therapies are well established. However, none of these agents possess all of the key attributes desired by clinicians, namely selectivity for the platelet, freedom of adverse effects, activity versus all agonists, rapid onset of action, and oral activity. Initial screening indicated that roxifiban could exceed the ability of the existing therapies to meet these criteria. It is unlike other known IIb/IIIa antagonists in that it will bind to both activated and resting platelets. This was expected to result in increased antiplatelet activity and damp the typical peak/valley pharmacokinetic profile due to the store of roxifiban available on the unactivated platelets. The possibility that roxifiban could be an important weapon against cardiovascular disease gave us a reason to devote considerable resources to develop a safe, efficient, and inexpensive large-scale process. We will discuss the evolution of the XU305 (**1**) process from a laboratory kinetic resolution to a pilot plant dynamic resolution.

4.2
Scale-up of the Kinetic Enzymatic Resolution of 9 to 1

The discovery group fulfilled the first preparations of roxifiban in small amounts for initial screening studies. The desirable pharmaceutical possibilities of this molecule were soon appreciated and our process R&D group rapidly set out to prepare kilogram quantities of API to support further Phase I and II project needs. We found that we could closely adapt the existing synthesis but had to modify most of the reaction conditions and some of the starting materials due to cost, safety, and scalability factors [13]. Our final synthesis appears in Fig. 3. While the sidechain's stereogenic center of **2** would be easily introduced via inexpensive L-asparagine (**5**) (after formylation, Hofmann reaction and methylation), the isoxazoline chirality was unlikely to be available from commercially available starting materials. Fortunately, the power of the enzyme, Amano lipase PS-30 (*Pseudomonas cepacia*) to resolve the 5 position of an isoxazoline similar to **9** was realized early [12]. Subsequently this reasonably priced catalyst has fulfilled the project needs for establishing this stereogenic center at all stages of development.

The first preparations of roxifiban were limited in scale to about 100 g and on such a scale the question of the recycle of the less-reactive enantiomer (S)-**9** was unimportant. Since we expected annual API requests to exceed 100 kg within 1–2 years, a means to recycle this enantiomer would soon be required. First, a good preparation of the isoxazoline substrate (R/S)-**9** was necessary.

Vinyl acetic acid, as well as its methyl and 1-butyl esters, was examined by the discovery chemists as dipolarophiles in the [3+2] cycloaddition to form analogues of **9**. The yields of the isoxazolines were variable, presumably due to the propensity of these vinylic compounds to polymerize. A better alternative was isobutyl ester (**8**) that is a stable compound and commercially available in large quantities. When this reactant was tested in our group, it regiospecifically produced **9** in consistent 90–95% yields. The accessibility of the two precursors **7** and **8**, and the efficiency of the cycloaddition fulfilled our goals of a practical preparation of (R/S)-**9**.

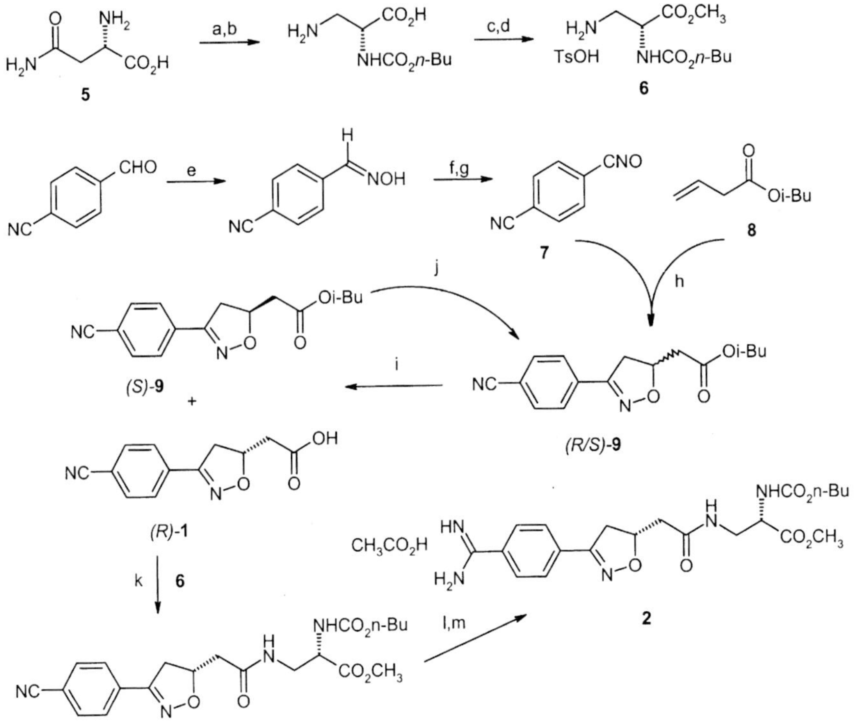

Fig. 3 Plant-scale process for the preparation of roxifiban. a) BuOCOCl, 80%; b) PhI(OAc)$_2$, 75%; c) SOCl$_2$, MeOH; d) tosic acid (82% over 2 steps); e) H$_2$NOH, MeOH, 97%; f) NCS, DMF; g) Et$_3$N, 90%; h) [3+2] (90% over 3 steps); i) Amano PS-30, buffer, Triton X-100®, 36%; j) KOtBu, toluene, 85%; k) SOCl$_2$, Hunig's base, CH$_3$CN, 82%; l) H$_2$NOH, MeOH, 95%; m) Ac$_2$O/HOAc, H$_2$, Pd/C, 59%.

4.2.1
Development of the Resolution of (R,S)-9

At first, a protocol similar to that developed by the discovery group was adopted in order to determine whether (R,S)-9 would undergo enantioselective hydrolysis. A mixture of (R,S)-9 and Amano PS-30 lipase was stirred overnight in dimethyl sulfoxide (DMSO) in phosphate buffer at 38–40 °C. Analysis by chiral HPLC indicated that hydrolysis was selective for the (R)-enantiomer.

While it is a powerful solvent, DMSO is undesirable for industrial use. Instead, the surfactant Triton X-100® can be an effective replacement for DMSO in enzymatic hydrolyses [14], and in our case, worked very well. Triton X-100® forms micelles when mixed with water and would solubilize low concentrations of ester for hydrolysis. However, increasing the surfactant charge did not accelerate hydrolysis. In an attempt to reduce the somewhat long reaction time, we screened the enzymes Amano PS Lipase XIII, *Pseudomonas fluorescens* and *Candida cylindracea* versus Amano PS-30. PS-30 was still considered the best choice of enzyme when expense, commercial availability, and activity were assessed, and was retained as our preferred lipase.

The preparation was now increased to 50 g of (R,S)-**9**. The elimination of DMSO provided a convenient bonus in that unreacted (S)-**9** was not appreciably soluble in the aqueous reaction mixture and was recovered with the other insoluble components by filtration. Subsequently, (R)-**1** could be isolated as a solid upon acidification of the filtrate. The chemical and optical purity of the crude (R)-**1** could be efficiently raised by recrystallization from ethanol. These results solved our work-up concerns and allowed us to scale-up to the kilo-lab.

The resolution of 1.0 kg of (R,S)-**9** in the kilo-lab required 48 hours instead of the 25–35 hours observed on a smaller scale, but 94% of the available yield of (R)-**9** was obtained in 96% enantiomeric excess (ee$_p$). With the exception that the filtration was slower than expected, all proceeded well and we moved into the plant using 50 and 200 gallon glass-lined vessels to resolve 7.90 kg of (R,S)-**9**. After 48 hours, the (R)-ester had decreased to a satisfactory level of 1.7% and the isolation/purification was conducted as described above to produce 2.42 kg of (R)-**1** in 75% available yield (98.5 wt.-%, >99.9% ee). We were delighted to note that our process development work had anticipated all potential problems; while this had been our first plant-scale biotransformation of any sort, the chemistry and subsequent work-up had proceeded without incident.

Over the years, we have scaled this resolution up to 40 kg of isobutyl ester **9** in 300 gallon vessels without problems. Although the larger scale runs are somewhat limited by the slow filtrations, yields have remained reproducible. We consider this preparation a mature and robust process and have devoted our further energy to other aspects of the roxifiban synthesis.

4.2.2
Racemization of 9

Once the (R)-**1** carboxylic acid had been collected, the remaining ester (S)-**9** could be isolated and purified in high yield. An acceptable commercial process would require the racemization and recycling of this isomer through additional rounds of resolution. A series of experiments were conducted to define the best racemization conditions (Tab. 1).

There are several key conclusions derived from this study:

i) Ammonia and other amines do cause racemization only when methanol is the solvent. Other solvents including the higher alcohols promote little or no racemization.
ii) Inorganic bases such as potassium or sodium carbonate cause racemization over 16–18 hours in a 1:1 mixture of acetonitrile/methanol. In N,N-dimethylacetamide only 8% of the ester (S)-**9** racemized over the same period.
iii) Barium hydroxide octahydrate caused racemization of the substrate with concomitant conversion of the nitrile moiety to the amide.
iv) Attempts to use aqueous systems failed, thus excluding the most likely reaction conditions for attaining a dynamic resolution.

Tab. 1 Racemization of (S)-9 using 1 mole of various bases.

Conditions	Time (h)	% Racemized[a]	Observations
NH_3/MeOH	48	100	some transesterification
NH_3/EtOH	48	trace	
Et_2NH/IPAc	48	14	
Et_2NH/MeOH	8	48	some transesterification
K_2CO_3/[(1:1 CH_3CN/MeOH)]	16	100	
K_2CO_3/DMAC	18	6	
$NaHCO_3$/[(1:1 CH_3CN/MeOH)]	16	100	some transesterification
NaOMe/MeOH	0.2	100	
NaOMe/THF	0	100	20% hydrolysis
$Ba(OH)_2 \cdot 8H_2O$/MeOH	18	100	amide formation
NaOH/H_2O	18	(–)	decomposition
KO^tBu/^tBuOH	0.2	100	
KO^tBu/toluene	0.2	100	
Na_2HPO_4/MeOH/H_2O	16	0	

a) 100% racemized indicates an ee of 0%.

v) Racemization was very rapid in the presence of alkoxides. A catalytic amount of potassium *tert*-butoxide (as low as 5 mol-%) was sufficient for complete loss of optical activity within minutes [12].

As a result of this study, potassium *tert*-butoxide in toluene was selected as the racemization conditions. The speedy reaction with alkoxides was significant since the substrate was not stable in the presence of strong bases for extended periods. After a brief treatment of the (S)-9 solution with butoxide, it was neutralized with an equimolar amount of HCl, washed with brine and concentrated to ~1/4 of its original volume. Heptane was added to precipitate the racemic ester in ~90% recovery. This process was used for the kilo-lab campaign and when that scale of operation proceeded flawlessly, we transferred it to the plant. In general, the filter cake [up to 120 kg of wet cake containing 35–40 kg of (S)-9] from the resolution work-up was extracted with toluene while still on the filter, washed with brine, and dried by azeotropic distillation. Racemization was rapid and clean at 40 °C after the addition of 1 mol-% of 1 M potassium *t*-butoxide in *t*-butanol. After cooling and water addition, the mixture was quenched with 1 N hydrochloric acid, and the organic phase was separated and concentrated. After dilution with heptane, the product crystallized out to afford 80–90% recovery of (R,S)-9 (~95 wt.-% purity, 0–% ee_p). The racemization process was generally very reproducible and required only one reactor [13].

4.2.2.1 Mechanism of Isoxazoline (9) Racemization

Intriguingly, since bases as mild as sodium bicarbonate were able to accomplish complete racemization, it seemed the stereogenic position of **9** was significantly

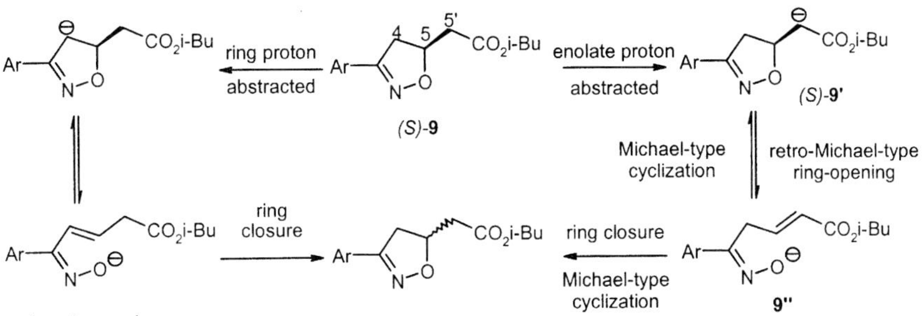

Fig. 4 A proposed mechanism for the base-induced racemization of (S)-**9**.

more acidic than initially predicted. In order to improve racemization conditions that would be tolerant of enzyme activity, we sought to define this mechanism further [15].

In general, deprotonation of unactivated 3-aryl-2-isoxazoles has been hypothesized and sometimes demonstrated to occur at the 4-position instead of the 5′ carbon (Fig. 4) [16]. However, we believe the 5′ position (α to the ester) is the logical initial site for deprotonation since an electron-withdrawing group is adjacent [17]. Anions at either the 4 or 5′ position would presumably be capable of leading to racemization via ring-opened structures since the putative oxime anion could reform the ring from either side of the α,β- or β,γ-unsaturated ester. Both are reasonable explanations for the observed racemization and the key to discovering the mechanism is to determine which proton is abstracted initially [18]. When **9** was stirred in d_4-methanol in the presence of catalytic sodium methoxide, the side-chain 5′ carbon was >95% deuterated after 26 hours while none was detected on the ring carbon [19]. The simplest explanation for this facile racemization would be a Michael/retro-Michael-type equilibrium between the enolate (S)-**9′** and the enone/oxime anion **9″** [20]. Racemization via a Michael/retro-Michael equilibrium is infrequently cited and even less often demonstrated [21], but it seems to occur when **9** is in the presence of some bases [22].

Besides eliminating the need to recycle (S)-**9** in separate steps, there are other advantages to converting this kinetic resolution into a dynamic resolution. If both enantiomers of **9** are substrates for an enzyme, enantioselectivity will remain constant only with the continuous racemization of **1** [24, 25]. Otherwise under non-racemizing conditions, the ee_p would remain high only at low conversions; once the more-reactive enantiomer supply is exhausted, the ee_p would typically begin to deteriorate as the now-predominant, less-reactive enantiomer begins to hydrolyze.

4.3
Process Development of the Dynamic Enzymatic Resolution of Thioester 10b

While we were pursuing racemization conditions for (S)-9, we also screened a variety of aqueous systems as potential dynamic resolutions. Without exception, this was futile. Either no (S)-enantiomer isomerized to replace the fast-reacting enantiomer, the high pH needed for racemization deactivated the enzyme, or the substrate decomposed. Minor resolution (up to 55% ee$_p$) was possible if the alkyl group contained electron-withdrawing groups such as chlorine, but in general, such molecules did not seem to be good substrates for the lipase. In light of the numerous unsuccessful attempts, it began to appear unlikely that we would attain a dynamic resolution and avoid the inconvenient stepwise (S)-9 racemization/recycle protocol. In spite of that, this sequence of resolution, (S)-9 isolation, purification, and repeated resolution fulfilled our requirements of a safe, cost-effective and reproducible process (70% overall yield at plant scale) for eventual advancement to commercialization. However, we ultimately still desired a better alternative to this tedious recycling procedure.

Drueckhammer and his associates supplied the answer we sought when they reported the first example of the enzymatic dynamic resolution of substrates containing thioesters [26]. For certain propionates, the presence of a thio substituent next to a thioester sufficiently lowered the pK_a of the α-proton to permit racemization to occur under enzyme-friendly reaction conditions (Fig. 5).

They later discovered it was possible to forego the second thio function (the thiophenol) by careful selection of the thioester, the enzyme, and reaction conditions [27]. This considerably broadened the scope of the new reaction, but their first communication suggested a new path for our process. The use of thioesters as substrates in enzymatic resolution is not frequent but it is still well precedented [26–28]. In particular, an example of its use in large-scale pharmaceutical manufacture is our company's preparation of the antihypertensive Captopril [29].

Realistically, we did not have the option of incorporating a second thio substituent to enhance the acidity of the substrate. Rather, we expected the twin effects of the thioester's electronic influence and the resonance stabilization of the Michael/retro-Michael tandem to enable the formation of the enolate under mildly basic conditions. The electronic effect of the thioester was supported by a calculation, indicating that the pK_a of the α-proton of the thioester would lower (by 1.5 pK_a units) in comparison with 9 [30]. It was not clear if this would be enough to permit racemization under reaction conditions that would support sufficient enzyme activity but relatively little work would be required to test the hypothesis.

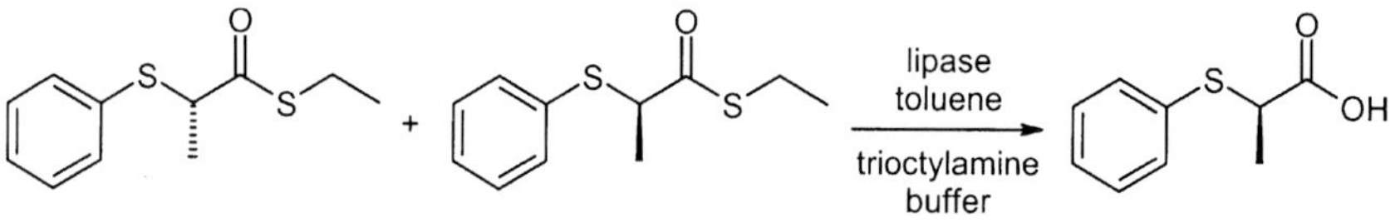

Fig. 5 First example of thioester dynamic enzymatic resolution.

Another factor to consider is that the site of enzyme-catalyzed hydrolysis would not be adjacent to the stereocenter but rather β to it. While there are many examples where reaction enantioselectivity is attenuated by the distance of the stereogenic carbon from the enzymatic action site [31], resolution of centers up to five bonds distant from the site of enzymatic action is still possible [32]. We hoped that we would be able to reproduce the example of the enzymatic kinetic resolution of **9** but with a thioester, inasmuch as the resolution of **9** had been excellent and the resolved center was located two bonds from the ester.

Finally, it would be necessary to develop an excellent preparation of the thioester substrate since it appeared likely that any route to it would be longer than the analogous preparation of **9**. While it might be possible to establish a dynamic resolution, the advantage would evaporate if the preparation of the thioester was not highly efficient.

4.3.1
Identification of Efficient Reaction Conditions for Dynamic Enzymatic Resolution

We decided a reasonable approach would be to prepare most of the thioesters of **9** incorporating R groups of C_4 and lower. These would be screened against a small library of commercially available esterases and lipases that we had found useful in other work. A convenient laboratory means for rapidly preparing a series of thioesters from a single substrate is the reaction of acyl chlorides with the copper salts of thiols (Fig. 6) [33]. We could prepare (R,S)-**1** by hydrolysis of **9** with lithium hydroxide in aqueous methanolic THF (97% yield) [34]. This was converted into the acyl chloride with thionyl chloride and subsequently reacted with the corresponding copper mercaptan of the desired thioester. The copper salt was itself easily, if odoriferously, prepared in 77 to 99% crude yields by heating 2.5 equivalents of the requisite thiol with copper(II) oxide overnight and filtration (Tab. 2) [35, 36].

Our screening technique consisted of preparing 1 mL mixtures of the thioester **10**, enzyme and Triton X-100® in phosphate buffer at 40 °C. Any reaction that might occur was monitored by chiral HPLC [39] which indicated the presence of hydrolysis, and if so, whether it was stereoselective. While the medium was basic, we were not overly concerned whether racemization occurred at this point; our goal was to find an enzyme/substrate match that resolved in high yield and ee_p.

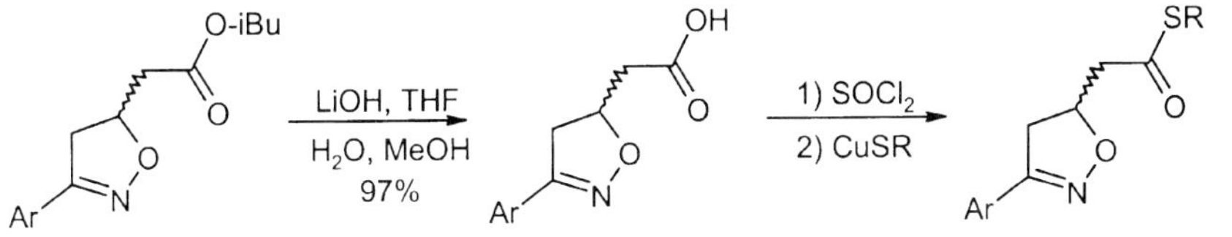

Fig. 6 Laboratory-scale preparation of thioesters.

Ar = 4-cyanobenzene

10a R = ethyl
10b R = 1-propyl
10c R = 1-butyl

Tab. 2 Yields for the preparation of thioesters (**10**) for screening.

Thioester	Yield[a] (%)	Thioester	Yield[a] (%)	Thioester	Yield[a] (%)	Thioester	Yield[a] (%)
Phenyl	92	ethyl	73	1-propyl	86	2-propyl	71[b]
1-Butyl	>99	2-butyl	94	isobutyl	91	tert-butyl	89%

a) Reported as crude yields.
b) Prepared via the silylated mercaptan/AlCl$_3$ method described later.

The thioesters were screened against commercially-available enzymes from the following sources: *Pseudomonas cepacia, Candida cylindracea, Rhizopus delemar, Candida lipolytica, Geotrichum candidum, Mucor* sp.*, Penicillium roqueforti, Pseudomonas* sp.*, Rhizopus oryzae, Rhizopus niveus, Aspergillus niger, Penicillium cyclopium, Porcine pancreas* type II, and *Aspergillus oryzae*. Perhaps not too surprisingly, only the enzyme that resolved **9**, the lipase from *Pseudomonas cepacia* (Amano PS-30), displayed reasonable enantioselectivity and reaction rate for some thioesters, simplifying our choice of enzyme [40]. This was a welcome discovery, as we already possessed large quantities of Amano PS-30 in our warehouse. We were aware of the peculiarities of its handling and were confident of future supply and price, all important factors for our upcoming plant-scale preparations as this catalyst would be the highest cost item.

This study revealed that only the linear (non-branched) alkyl thioesters **10a,b**, and **c** (ethyl, 1-propyl, and 1-butyl) were hydrolyzed efficiently (≤24 hours) with appreciable enantioselectivity (>85% ee$_p$) for some enzymes [41]. The presence of any α or β branching led to, at best, low conversion and poor ee$_p$. This is in marked contrast to the ester **9** [13], which of course possesses a β methyl group and indicated that the demands of the enzyme's active site are stricter for thioesters as compared with esters. The reaction rate and ee$_p$ for all three linear alkyl thioesters were so good that we decided to forego further optimization of the choice of substrate at that time. We felt that the choice of enzyme and thioester class was sufficient for the first pilot-plant run.

Concurrent with the above research, we were considering reaction conditions for racemization. An initial test experiment based on Prof. Drueckhammer's conditions [26] convinced us that the search for acceptable racemizing conditions would not be arduous. A chiral thioester was prepared from (*R*)-**1** with thionyl chloride followed by 2-butyl mercaptan (94% yield). When this was mixed with toluene, 1,8-diazabicyclo[5.4.0]undec-1-ene and Triton X-100® in phosphate buffer at pH 9.2, the heterogeneous mixture racemized completely in only 25 min. While the substrate had also degraded, such rapid racemization convinced us that we would be able to discover optimized conditions in a reasonable time period.

A screening set of racemizations was conducted on (*R*)-1-propyl thioester **10b**. Simply stirring the heterogeneous mixture of thioester, PS-30 lipase, and Triton X-100® in 40 °C phosphate buffer at pH 9.25 produced no detectable racemization

Tab. 3 Racemization conditions [a] for thioester **10b**.

Amine (equiv.)	Toluene added	Reaction period	% Racemized[b]
Trioctylamine (0.2)	(–)	145 h	9
Trioctylamine (0.2)	10% v/v	8 h	49
Trioctylamine (2.0)	(–)	161 h	15
Triethylamine (2.0)	(–)	141 h	47
Triethylamine (2.0)	10% v/v	100 min	52
Trimethylamine (0.2)	(–)	50 h	7
Trimethylamine (1.0)	(–)	146 h	50
Trimethylamine (2.0)	(–)	91 h	50
Trimethylamine (2.0)	10% v/v	24 h	49
Trimethylamine (3.0)	(–)	83 h	50
None[c]	(–)	97 h	0

a) Racemization conditions: 0.2 N NaH$_2$PO$_4$, pH = 9.2, 40 °C, Triton X-100®.
b) A value of 50% racemization represents one half-life or 50% ee.
c) PS-30 lipase also present.

over four days. Higher pH values will significantly reduce the activity of PS-30 lipase so we turned to the known propensity of amines and/or toluene to facilitate racemization [26–28 d–f]. These results are listed in Tab. 3.

Several significant points were established from this set of experiments:

i) The use of trioctylamine, frequently selected to assist racemization during dynamic resolutions, was ineffectual until toluene was added. Once added, the racemization half-life [42] dropped to only 8 hours [43].
ii) The addition of toluene was a general trend, 10% v/v (1.5 mL per gram of thioester) dramatically lowered the half-life for any amine charged [45].
iii) Triethylamine and trimethylamine were also good selections when combined with toluene, however only the choice of ≥2 equivalents of trimethylamine also possessed a reasonable racemization half-life in the absence of toluene: 83 hours. Should the use of toluene be precluded later, we might still establish acceptable racemization conditions using trimethylamine in single-phase systems.
iv) Ester hydrolysis was undetected or minimal (<1%) in all cases, suggesting that non-enzymatic hydrolysis would not degrade the ee$_p$.

With all of our initial questions answered, we were ready to combine these findings to seek dynamic enzymatic resolution.

Our first attempt at dynamic resolution provided satisfactory proof that **10a** would simultaneously racemize and resolve (Fig. 7). The crude ee$_p$ of 91% was raised by ethanol recrystallization to a 67% overall yield of (R)-**1** of 97.2% ee$_p$, convincing evidence of dynamic resolution. Obviously, both enantiomers must have reacted to form the (R)-acid. The presence of a 4% HPLC area impurity (the corresponding amide) in the crystallized product indicated that the reaction conditions still required further refinement.

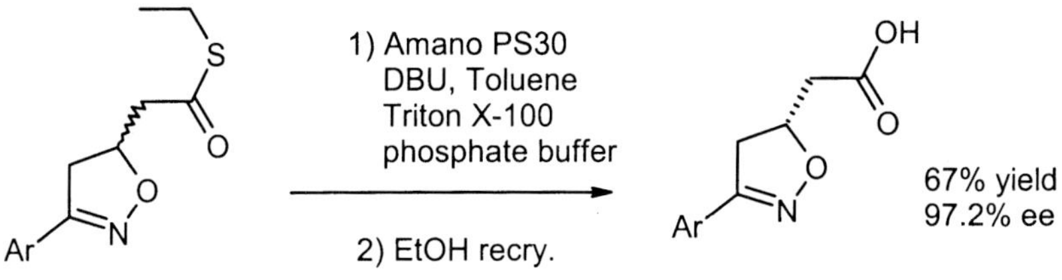

4 Multi-Kilo Resolution of XU305, a Key Intermediate to the Platelet Glycoprotein IIb/IIIa Receptor

Ar = 4-cyanobenzene

Fig. 7 First successful isoxazoline thioester dynamic resolution.

Tab. 4 Search for optimum dynamic resolution conditions using Amano PS-30 lipase.

Thioester	Amine (equiv.)	Reaction period (h)	Conversion (%)	ee_p (R)
10b	trimethylamine (0.2)	65	61	96.5
10b	trimethylamine (1.0)	48	>99	95.6
10b	trimethylamine (2.0)	20	>99	97.6
10b	trimethylamine (3.0)	20	>99	97.3
10b	trimethylamine (2.0)[a]	70	94	96.8
10a	trimethylamine (2.0)	16	>99	96.3
10c	trimethylamine (2.0)	48	>99	97.3
10b	trioctylamine (0.2)[a]	168	87	97.2
10b	triethylamine (1.1)	66	>99	96.2
10b	none	7	23	96.5

a) Reaction medium also contains 10% v/v toluene.

With this validation, we now embarked on a third phase of experimentation. We sought to combine the established choices of enzyme (Amano PS-30) and class of thioesters with proper racemization conditions to optimize the resolution. Key experiments are listed in Tab. 4.

While the addition of toluene to form a biphasic reaction system had dramatically improved racemization rates, we were surprised to find that its combination with Amano PS-30 lipase led to incomplete conversion, although the enantioselectivity was still very good. This observation indicated that in spite of the significantly faster rate of racemization induced by the presence of toluene, the lipase was unacceptably inhibited and/or inactivated under these reaction conditions.

Fortunately, our racemization work had established the ability of trimethylamine to produce acceptable rates of racemization in the absence of toluene. The use of two equivalents of the amine supplied the answer and provided the last factor for establishing a successful, large-scale enzymatic dynamic resolution. In the laboratory, this combination produced >99% conversion to (R)-1 of 97.6% ee_p after 22.1 hours, more than adequate for our objectives.

The power of this resolution is best gauged by measuring the "enantiomeric ratio" E [24, 46], which quantifies the inherent ability of the enzyme to differentiate between the two enantiomers. This could be easily done under non-racemizing conditions [47] and the derived value of 74 predicted a maximum ee_p of 97.3% [48]. This was in good agreement with the measured value of 97.6% when conducted as a dynamic resolution. Usually, any $E > 30$ is considered preparatively useful, suggesting this reaction was efficient.

Furthermore, we had measured a racemization half-life of 90.5 hours for **10b** under our optimized racemization conditions. We had also measured the solubility of this thioester to be only 7.0×10^{-3} g/mL after 5 hours of reaction. The combination of these two observations would reasonably suggest that at least several half-lives would be required to approach reaction completion, much too long to be acceptable for large-scale work. To our delight, complete resolution nearly always occurred in much less time: 20–50 hours. As this is a heterogeneous reaction, measurement of good reaction rate data is problematic, being dependent upon particle size, efficiency of agitation, vessel size, and related physical factors [49]. Nevertheless, this fast and enantioselective reaction indicated that any solubilized **10b** is expeditiously hydrolyzed to (R)-**1**. A further indication of a rapid enantioselective reaction is that the ratio of (S)- to (R)-**10b** gradually increased to 7–8:1 once the conversion was ~80% complete, suggesting a significantly higher enzyme-catalyzed hydrolysis rate as compared with the racemization rate. Otherwise, had the rates been comparable, the enantiomers would have soon equilibrated. This is noteworthy since typical dynamic resolutions require racemization rates to be at least similar to that of the enzymatic reaction of the fast-reacting enantiomer in order to achieve acceptable ee_p values [25]. The combination of these examples demonstrate that if E is high and the reaction conditions properly selected, the ee_p will not degrade even when the desired rate order of racemization and enzymatic hydrolysis is reversed.

4.3.2
Preparation of 1-Propyl Thioester (10b)

The achievement of fast, clean, high-yield dynamic resolution conditions still left one challenge: the requirement that an efficient preparation of the thioester **10b** be found. We had selected the 1-propyl thioester as the developmental candidate since while all three linear alkyl thioesters had resolved similarly during developmental work, the stench of the 1-propanethiol was the least offensive; a practical consideration for working at bulk scale. The synthesis we had used to produce material for process development, hydrolysis of **9** followed by chlorination with thionyl chloride and reaction with copper mercaptans, was simple and high yielding but its length negated the advantages we had established by the dynamic resolution. Copper propyl mercaptan was not commercially available and was a disagreeable reagent in its own right. The handling of reactive or odorous compounds as liquids is already sufficiently difficult on a large scale. The need to manipulate them as solids is unacceptable for a viable process. It was obvious that an alternative preparation was demanded, it would need to be shorter, and it would need to address the problems of manipulating a highly odoriferous liquid.

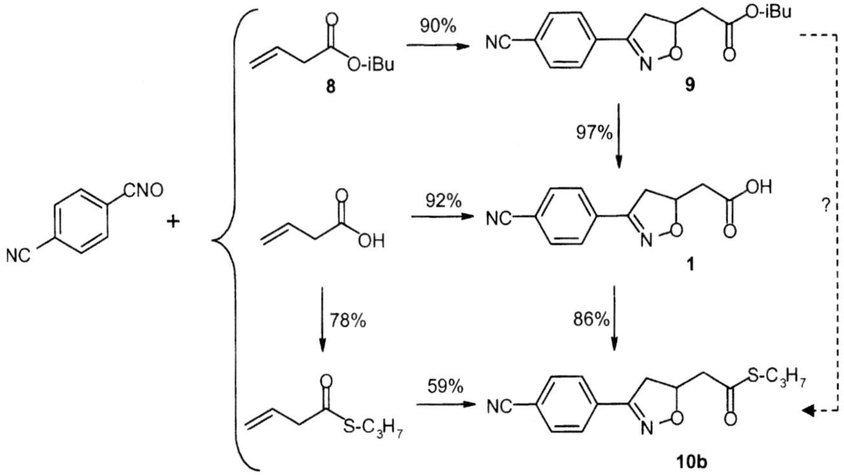

Fig. 8 Routes to thioester **10b**.

After consideration of the moderate complexity of our target **10b**, it was clear that the best hope for multi-kilo preparations would still incorporate the useful [3+2] cycloaddition of an aryl nitrile oxide with a properly functionalized propylene to quickly establish the 2,5-disubstituted isoxazoline ring. Once that is assumed, there remained three conceptual routes to **10b** (Fig. 8).

Racemic **1** could be obtained in excellent yield by the cycloaddition with vinyl acetic acid, but we did not discover any efficient means to convert this directly into **10b** that didn't involve mercaptan salts. The thioester could be produced directly by reaction with the propyl thioester of vinyl acetic acid, however its preparation similarly demanded copper mercaptan. Our previous work had created an excellent process for preparing **9**. This chemistry had by now been demonstrated numerous times in our pilot plant. It was a well-established large-scale process and we could avoid repeating scale-up work related to reagent sourcing, thermochemistry, engineering, and analytical if we could efficiently transesterify this directly to **10b**. This became our primary goal.

There are remarkably few transformations of esters to thioesters. The use of bis(dimethylaluminum) thiolates produced an unoptimized 50% yield of **10b** but was unattractive since the reagent required a separate preparation and it partially reacted with the aryl nitrile [53]. Nor did simply mixing propyl mercaptan with **9** under enzymatic resolution conditions provide any equilibrium between **9** and **10b** that might have allowed a dynamic resolution. The problem was ultimately solved by the use of silylated thiols and aluminum chloride (Fig. 9).

Mukaiyama and co-workers reported this direct conversion of methyl and ethyl esters into thioesters in 1974 [54]. Although the dearth of references to this reaction since then suggests that the reaction has been forgotten by the chemical community [55] and it featured only a limited selection of esters, it was well suited for

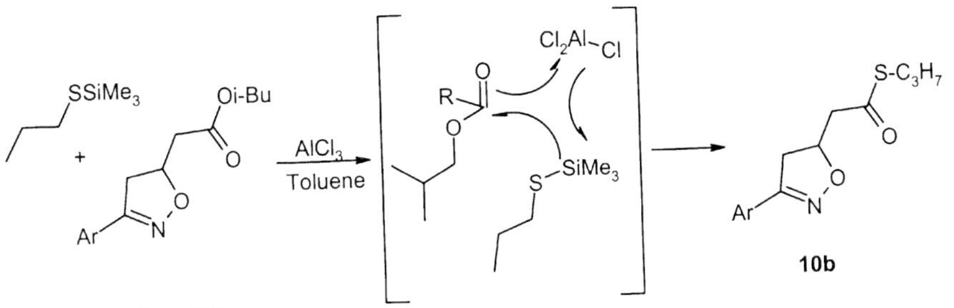

Fig. 9 Direct conversion of esters to thioesters.

our use. We were easily able to prepare a variety of thioesters from **9** in this manner [56] and found the preparation of **10b** went particularly well (91% crystallized yield). Despite the need for a second vessel to prepare the silylated propyl mercaptan, the reaction adapted well into the pilot plant and became our preferred route to multi-kilogram quantities of **10b**.

4.3.3
Scale-up of the Dynamic Enzymatic Resolution Chemistry into the Plant

The work was now sufficiently mature to attempt multi-kilogram preparations. Conducting these reactions first in the kilo-lab would provide a reasonable balance of safety and learning since the use (preparation of **10b**) and the release (dynamic resolution of **10b**) of propyl mercaptan might still have unpredictable consequences at large scale. The conversion of 1.57 kg of **9** into 1.88 kg of **10b** (99.0 wt.-% purity) proceeded facilely in 93% yield and the subsequent dynamic resolution produced 1.31 kg of (R)-1 in 88.7% yield (99.6 wt.-%, >99.5% ee$_p$). These results were excellent and provided experience for entering the pilot plant.

Our first pilot plant run was designed to prepare 45 kg of thioester. As safety was particularly important with the use of mercaptans, our two reaction vessels (200- and 300-gallon glass-lined reactors) would be vented through a scrubber containing bleach and sodium hydroxide solution to control emissions. Another unexpected consideration was the late-stage replacement of n-butyllithium for n-hexyllithium as the base. n-Hexyllithium is preferred since the conjugate acid, hexane, is safer to handle as compared with the volatile butane, but this had become necessary due to a shortage of hexyllithium. At this scale, we could not vent outside the quantity of butane we would form from the use of n-butyllithium, but instead it was directed to our thermal oxidizer to be burned. The rate of natural gas uptake to the burners would also provide a handy means of measuring the butane produced and in turn the endpoint of the reaction.

The preparation of the silylated mercaptan went smoothly; 24.5 kg of 1-propanethiol was reacted with 2.5 M n-butyllithium followed by chlorotrimethylsilane in THF/heptane as in our established procedure. As we had calculated, a 6 hour sparge of nitrogen through this 30 °C solution eliminated all of the butane. This solution was transferred via a cartridge filter to the larger vessel that already contained 50.0 kg of **9** in THF. Addition of the aluminum chloride at this point required careful planning. It is a reactive solid and we wanted to minimize operator exposure. Our engineers designed a solids-charging adapter for safe delivery in portions without exposing the reaction or the operators. The mixture was subsequently heated to 65 °C and the first sampling indicated that the transesterification was complete (**10b**:**9** LC ratio of 162:1). The reaction mixture was worked-up with water, toluene, and brine to produce a solution from which **10b** was crystallized by the solvent exchange of toluene for heptane. After filtration and drying, 44.9 kg (98.6 wt.-%) of product in 91% yield was recovered.

Attention to the subsequent dynamic resolution step was critical due to the possibility of traces of propane mercaptan escaping the reactor or scrubbers. The resolution was carried out in a 300-gallon glass-lined reactor vented through a 50-gallon-stirred vessel charged with bleach and 30% aqueous sodium hydroxide. Rather than use a typical scrubber, this well-stirred 50-gallon kettle would give more assurance that all of the vented thiol would come into contact with bleach. As regards design, the set-up only differed from the kilo-lab preparation in that a pH stat device was not used. Rather, the pH was adjusted every 2–6 hours by the addition of base. Water, sodium phosphate monobasic hydrate, and aqueous 24% trimethylamine were charged and the pH was adjusted to 9.1 with more base. Triton X-100®, 44.9 kg of **10b** and 4.5 kg of Amano PS-30 lipase were subsequently charged and the mixture warmed to 40 °C. After 41 hours, HPLC analysis indicated an (R)-**1**:**10b** ratio of 96:4, sufficient to initiate work-up. This was conducted in a similar manner to the kinetic resolution process: filtration, acidification of the filtrate with phosphoric acid, filtration of the (R)-**1** solids, and eventually recrystallization (95.5% recovery) from ethanol to produce 28.4 kg (>99.9 wt.-%) in 80.4% overall yield. The ee$_p$ was only 90.7% at the end of the reaction but was raised to >99.9% after recrystallization. These results at multi-kilogram scale fulfilled our goal of transforming the kinetic resolution of **9** into a dynamic resolution and successfully executing the chemistry on a plant scale.

4.4
Conclusions

A kinetic resolution of an ester requiring separate steps to recover and recycle the less reactive enantiomer has been transformed into a dynamic resolution of the corresponding thioester. While the thioester route requires an additional synthetic step, the efficiency of this step renders the dynamic resolution route to (R)-**1** shorter. The development of an alternative route provides our company with additional options for commercial-scale roxifiban preparation.

4.5 Acknowledgements

We thank the following colleagues for their contributions to the work described in this chapter. A fuller assessment of the extent of individual participation may be obtained by reviewing the publications authored by some of our colleagues that shared in this work: Pat N. Confalone, Jianguo Yin, Lin-hua Zhang, Philip Ma, Robert E. Waltermire, Joseph Fortunak, Edward Gorko, Charlotte Silverman, John Blackwell, J.C. Chung, Mike Hrytsak, Mary Cooke, Lakisha Powell, Dennis Potter, Charles Ray, Bill Boucher, Walt Conway, Joseph Renai, Dawn Blackburn, Elton Watson, Fred Maccherone, Jeff Koskey, and Robert Wethman. We especially thank the following for their contributions to the preparation and review of this manuscript: Ahmed Abdel-Magid, Robert DiCosimo, Rafael Shapiro, and Peter Mattei.

4.6 References

1. Current address: Johnson & Johnson, Pharmaceutical Research & Development, L.L.C., Spring House, PA 19477, USA.
2. a) S.C. STINSON, Chem. Eng. News 2000, 79, Oct 1, 2001, 79–97; b) reference 3, Chapter 1.2.
3. D.J. AGER, M.B. EAST, Asymmetric Synthetic Methodology, CRC Press, Boca Raton, 1995.
4. a) Reference 3, pp. 12–13; b) N.G. ANDERSON, Practical Process Research & Development, Academic, San Diego, 2000, pp. 330–332.
5. U.T. BORNSCHEUER, R.J. KAZLAUSKAS, Hydrolases in Organic Synthesis, Wiley-VCH, Weinheim, 1999.
6. The following review is an excellent industrial prospective on these three methods: S. KOTHA, Tetrahedron 1994, 50, 3639–3662.
7. J. CROSBY in Introduction and J.J. BLUMENSTEIN in Chiral Drugs: Regulatory Aspects (A.N. COLLINS, G.N. SHELDRAKE, J. CROSBY, Eds.), Chirality in Industry II, Development in the Commercial Manufacture and Applications of Optically Active Compounds, Wiley & Sons, Chichester, 1997, chaps. 1 and 2.
8. a) R.S. WARD, Tetrahedron: Asymmetry 1995, 6, 1475–1490; b) R. NOYORI, M. TOKUNAGA, M. KITAMURE, Bull. Chem. Soc. Jpn. 1995, 68, 36–56; c) K. FABER, Chem. Eur. J. 2001, 7, 5004–5010; d) S. CADDICK, K. JENKINS, Chem. Soc. Rev. 1996, 25, 447–456; e) M.T.E. GIHANI, J.M.J. WILLIAMS, Curr. Opin. Chem. Biol. 1999, 3, 11–15.
9. This work was completed during the period that our group was part of the DuPont Pharmaceutical Company and its precursor, the DuPont-Merck Pharmaceutical Company.
10. J. WITYAK, R.A. EARL, M.M. ABELMAN, Y.B. BETHEL, B.N. FISHER, G.S. KAUFFMAN, C.A. KETTNER, P. MA, J.L. MCMILLAN, L.J. MERSINGER, J. PESTI, M.E. PIERCE, F.W. RANKIN, R.J. CHORVAT, P.N. CONFALONE, J. Org. Chem. 1995, 60, 3717–3722.
11. a) S. JACKSON, W.F. DEGRADO, A. DWIVEDI, A. PARTHASARATHY, A. HIGLEY, J. KRYWKO, A. ROCKWELL, J. MARKWALDER, R. WELLS, R. WEXLER, S. MOUSA, R. HARLOW, J. Am. Chem. Soc. 1994, 116, 3220–3230; b) C.-B XUE, W.F. DEGRADO, J. Org. Chem. 1995, 60, 946–952.
12. C.-B. XUE, J. WITYAK, T.M. SIELECKI, D.J. PINTO, D.G. BATT, G.A. CAIN, M. SWORIN, A.L. ROCKWELL, J.J. RODERCIK, S. WANG, M.J. ORWAT, W.E. FRIETZE, L.L. BOSTROM, J. LIU, C.A. HIGLEY, F.W. RANKIN, A.E. TOBIN, G. EMMETT, G.K. LALKA, J.Y. SZE, S.V. DI MEO, S.A. MOU-

sa, M. J. Thoolen, A. L. Racanelli, E. A. Hausner, T. M. Reilly, W. F. DeGrado, R. R. Wexler, R. E. Olson, J. Med. Chem. 1997, 40, 2064–2084.

13 a) L.-H. Zhang, L. Anzalone, P. Ma, G. S. Kauffman, L. Storace, R. Ward, Tetrahedron Lett. 1996, 37, 4455–4458; b) L.-H. Zhang, J. C. Chung, T. D. Costello, I. Valvis, P. Ma, G. S. Kauffman, R. Ward, J. Org. Chem. 1997, 62, 2466–2470.

14 D. L. Hughes, J. J. Bergan, J. S. Amato, M. Bhupathy, J. L. Leaser, J. M. McNamara, D. R. Sidler, P. J. Reider, E. J. J. Grabowski, J. Org. Chem. 1990, 55, 6252–6259.

15 A recent review of racemization lists only a few examples that proceed via similar mechanisms to that we propose: E. J. Ebbers, G. J. A. Ariaans, J. P. M. Houbiers, A. Bruggink, B. Zwanenburg, Tetrahedron 1997, 53, 9417–9476.

16 a) P. Grünanger, P. Vita-Finzi, in Isoxazoles (E. C. Taylor, A. Weissberger, Eds.); The Chemistry of Heterocyclic Compounds, Wiley, New York, 1991, Vol. 49/Part 1, pp. 544–573; b) V. Jäger, W. Schwab, Tetrahedron Lett. 1978, 19, 3129–3132; c) W. Schwab, V. Jäger, Angew. Chem., Int. Ed. Engl. 1981, 20, 603–605; d) T. G. Burrowes, W. R. Jackson, S. Faulks, I. Sharp, Aust. J. Chem. 1977, 30, 1855–1858; e) A. W. Chuchołowski, S. Uhlendorf, Tetrahedron Lett. 1990, 31, 1949–1952.

17 However, see reference 16e for a different opinion for initial deprotonation of an aryl isoxazoline that is activated at the 5-position.

18 Unsuccessful direct means of detecting the hypothesized intermediates: we have been unable to trap the hypothesized oxime anion intermediate with either methyl iodide or chlorotrimethylsilane in the presence of base. We did not observe any characteristic infrared signals using IR probes.

19 In comparison, the thioester 10b has a significantly lower pK_a. It was 45% deuterated at the side-chain carbon after only 60 s and none at the ring carbon when the experiment was conducted as for 9. Further details are available in the Supporting Information for reference 47.

20 Similar base induced cleavages of tetrahydro-2-furanacetic acid derivatives are known: a) F. Brion, Tetrahedron Lett. 1982, 23, 5299–5302; b) B. A. Keay, R. Rodrigo, Can. J. Chem. 1983, 61, 637–639; c) J. R. Stille, R. H. Grubbs, J. Org. Chem. 1989, 54, 434–444.

21 For examples of Michael/retro-Michael equilibria with the capacity for racemization: a) W.-C. Shieh, J. A. Carlson, J. Org. Chem. 1994, 59, 5463–5465; b) S. M. Reddy, H. M. Walborshy, J. Org. Chem. 1986, 51, 2605–2607; c) M. Yamagishi, Y. Tamada, K. Oszki, T. Da-te, K. Okamura, M. Suzuki, K. Matsumoto, J. Org. Chem. 1992, 57, 1568–1571; d) J. R. Harrison, P. O'Brien, D. W. Porter, N. M. Smith, J. Chem. Soc., Perkin Trans. 1 1999, 3623–3631; e) P. Slosse, C. Hootele, Tetrahedron 1981, 37, 4287–4294; f) C. Morley, D. W. Knight, A. C. Share, J. Chem. Soc., Perkin Trans. 1 1994, 2903–2907; g) E. Kiehlmann, E. P. M. Li, J. Nat. Prod. 1995, 58, 450–455; h) J. P. Michael, D. S. Gravestock, Afr. J. Chem. 1998, 51, 146–157; i) L. Y. Foo, Phytochemistry 1987, 26, 813–817; j) A. Kolarovic, D. Berkeš, P. Baran, F. Povazanec, Tetrahedron Lett. 2001, 42, 2579–2582; k) K. Tan, R. Alvarez, M. Nour, C. Cavé, A. Chiaroni, C. Riche, J. d'Angelo, Tetrahedron Lett. 2001, 42, 5021–5023. Another related example is: l) E. Vázquez, A. Galindo, D. Gnecco, S. Bernès, J. L. Terán, R. G. Enríquez, Tetrahedron Asymmetry 2001, 12, 3209–3211. For a review of this and other tandem reactions, see: T.-L. Ho, Tandem Organic Reactions, Wiley-Interscience, New York, 1992.

22 The more we examined the racemization of 9 and 10b, the more interesting aspects emerged. When 10b was stirred under the resolution conditions (excluding lipase) for 66 h in a system where 97–98% of all acidic protons were composed of deuterium, the ee of the isolated 10b was 77.0% (23% racemization) and the side-chain's α-ester methylene carbon was 14% deuterated (by ^{1}HNMR). As this is opposite to that intuitively expected, it may be that the racemization mechanism is assisted by an unusual ex-

ample of Cram's "guided tour" mechanism. This occurs when a base removes a proton from the stereogenic carbon and "guides" it to the opposite side of the atom without entirely removing it from the molecule [23]. This mechanism is often facilitated by trialkylamines, although it is not known to occur under highly associating conditions, such as in this example.

23 a) D.J. CRAM in Fundamentals of Carbanion Chemistry (A.T. BLOMQUIST, Ed.), Organic Chemistry; Academic: New York, 1965, Vol. 4, pp. 86–105, 188–193; b) E. BUNCEL in Carbanions: Mechanistic and Isotopic Aspects (C. EABORN, N.B. CHAPMAN, Eds.), Reaction Mechanisms in Organic Chemistry; Elsevier, Amsterdam, 1975, Vol. 9, pp. 54–58, 86–88; c) E.M. KAISER, D.W. SLOCUM in Carbanions (S.P. MCMANUS, Ed.), Organic Reactive Intermediates, Academic, New York, 1973, pp. 338–341.

24 C.S. CHEN, Y. FUJIMOTO, G. GIRDAUKAS, C. SIH, J. Am. Chem. Soc. 1982, 104, 7294–7299.

25 G. FULLING, C. SIH, J. Am. Chem. Soc. 1987, 109, 2845–2846.

26 D.S. TAN, M.M. GUNTER, D.G. DRUECKHAMMER, J. Am. Chem. Soc. 1995, 117, 9093–9094.

27 P.-J. UM, D.G. DRUECKHAMMER, J. Am. Chem. Soc. 1998, 120, 5605–5610.

28 a) D. BIANCHI, P. CESTI, J. Org. Chem. 1990, 55, 5657–5659; b) I. KUMAR, R.S. JOLLY, Org. Lett. 1999, 1, 207–209; c) C.-S. CHANG, S.-W. TSAI, C.-N. LIN, Tetrahedron Asymmetry 1998, 9, 2799–2807; d) C.-S. CHANG, S.-W. TSAI, J. KUO, Biotechnol. Bioeng. 1999, 64, 120–126; e) C.-S. CHANG, S.-W. TSAI, Biochem. Eng. J. 1999, 3, 239–242; f) C.-N. LIN, S.-W. TSAI, Biotechnol. Bioeng. 2000, 69, 31–38; g) S. IRIUCHIJIMA, N. KOJIMA, J. Chem. Soc., Chem. Commun. 1981, 185–186; h) H. FRYKMAN, N. ÖHRNER, T. NORIN, K. HULT, Tetrahedron Lett. 1993, 34, 1367–1370; i) R.N. PATEL, J.M. HOWELL, C.G. MCNAMEE, K.F. FORTNEY, L.J. SZARKA, Biotechnol. Appl. Biochem. 1992, 16, 34–47; j) C.-Y. CHEN, Y.-C. CHENG, S.-W. TSAI, J. Chem. Technol. Biotechnol. 2002, 77, 699–705.

29 R.N. PATEL, Biocatalysis of Synthesis of Chiral Pharmaceutical Intermediates, in Encyclopedia of Microbiology, 2000, 1, Academic Press, 2nd edn., 430–444.

30 J. WERNER, J. BROOKWELL, N. TEA, Beaker, Version 2, Brooks & Cole Publishing, Pacific Grove, CA, copyright 1987–1989.

31 Reference 14 and references 2–4 therein.

32 a) T. LEE, J.B. JONES, J. Am. Chem. Soc. 1996, 118, 502–508 and references therein; b) X. YANG, A.R. REINHOLD, R.L. ROSATI, K.K.-C. LIU, Org. Lett. 2000, 2, 4025–4027; c) S. HORIUCHI, H. TAKIKAWA, K. MORI, Bioorg. Med. Chem. 1999, 7, 723–726; d) E. HEDENSTROM, B.-V. NGUYEN, L.A. SILKS III, Tetrahedron Asymmetry 2002, 13, 835–844 and references therein; e) L.E. JANES, A. CIMPOIA, R.J. KAZLAUSKAS, J. Org. Chem. 1999, 64, 9019–9029; f) S. HORIUCHI, H. TAKIKAWA, K. MORI, Bioorg. Med. Chem. 1999, 7, 723–726.

33 H.-U. REISSIG, B. SCHERER, Tetrahedron Lett. 1980, 21, 4259–4262.

34 Acid hydrolysis with HCl in THF required 3 days for complete reaction in only 63% yield versus 2 hours with base.

35 R. ADAMS, W. REIFSCHNEIDER, A. FERRETTI, Org. Synthesis Coll. Vol. 5, Wiley & Sons, New York, 1973, pp. 107–110.

36 Alternatively, the acyl chloride could either be reacted with mercaptan and cobalt(II) chloride [37] or thioacetamide followed by alkyl bromide [38] to produce the corresponding thioester. While these avoid the inconvenient preparation of the copper salt, yields were lower due to side-product formation.

37 S. AHMAD, J. IQBAL, Tetrahedron Lett. 1986, 27, 3791–3794.

38 T. TAKIDO, M. TORIYAMA, K. ITABASHI, Synthesis 1988, 404–406.

39 250×4.6 mm Chiralcel OJ column, ethanol/hexane/trifluoroacetic acid 20/85/0.25, 0.90 mL/min, 280 nm.

40 Lipase AK was the second choice, the only other enzyme to produce greater than trace quantities of (R)-1. However, when this experiment was conducted on larger scale with pH control, the hydrolysis was very slow in comparison with PS-30; 19% completion after one day.

41 We decided to forego testing of the shortest "linear" alkyl thioester: methyl thioes-

ter due to the intense smell, low boiling point of methanethiol, and since the ester analogue was poorly resolved.

42 Racemization half-life is defined as the amount of time for the ee_p to drop by half (i.e., from 100% ee_p to 50% or from 50% to 25%).

43 The difference in the effect of the trialkylamines may be a function of the difference in aqueous solubility. Trioctylamine's solubility is low: 7.5 mg/L [44]. Trimethylamine will dissolve in water at any ratio.

44 A. V. Nikolaev, G. M. Grishin, A. A. Kolesnikov, G. I. Pogadaev, Izv. Sib. Otd. Akad. Nauk SSSR, Ser. Khim. Nauk 1969, 2, 145–148.

45 The amine is still necessary, in spite of the significant effect of added toluene. When toluene was present without amine and the pH still adjusted to ∼9.5 by NaOH, no racemization was detected.

46 H. Stecher, K. Faber, Synthesis 1997, 1–16.

47 J. A. Pesti, J. Yin, L. H. Zhang, L. Anzalone, J. Am. Chem. Soc. 2001, 123, 11075–11076.

48 $ee_p = (E-1)/(E+1)$.

49 There is not much confidence in the ability to determine E for incompletely dissolved substrates. The complications introduced by mass transfer limitations suggest that any E derived for such systems is useful only for comparison with its peers. Measurements of E for heterogeneous systems are discussed in references 50–52.

50 A. J. J. Straathof, A. Wolff, J. J. Heijnen, J. Mol. Cat. B: Enzymatic 1998, 5, 55–61.

51 Wolff, V. von Asperen, A. J. J. Straathof, J. J. Heijnen, Biotechnol. Prog. 1999, 15, 216–227.

52 A. J. J. Straathof, J. A. Jongejan, Enzyme Microb. Technol. 1997, 21, 559–571.

53 a) E. J. Corey, D. J. Beames, J. Am. Chem. Soc. 1973, 95, 5829–5831; b) R. P Hatch, S. M. Weinreb, J. Org. Chem. 1977, 42, 3960–3961; c) T. Cohen, R.E. Gapinski, Tetrahedron Lett. 1978, 19, 4319–4322; d) S. Warwel, B. Ahlfaenger, Chem. Ztg. 1977, 101, 103.

54 T. Mukaiyama, T. Takeda, K. Atsumi, Chem. Lett. 1974, 187–188.

55 A Science Citation search revealed only two references to this initial work from 1974 to 1999: a) N. Krause, Chem. Ber. 1990, 123, 2173–2180; b) V. A. Vigante, Y. A. Ozols, G. Y. Dubur, Khim. Geterotsikl. Soedin. 1991, 7, 953.

56 J. A. Pesti, J. Yin, J. Chung, Synth. Commun. 1999, 2, 3811–3820.

5
Protease-Catalyzed Preparation of (S)-2-[(tert-Butylsulfonyl)-methyl]-hydrocinnamic Acid for Renin Inhibitor RO0425892

Beat Wirz, Stephan Doswald, Ernst Kupfer, Wolfgang Wostl, Thomas Weisbrod, and Heinrich Estermann

Abstract

An enzyme-catalyzed reaction for the large-scale preparation of (S)-2-[(tert-butylsulfonyl)-methyl]hydrocinnamic acid (S)-3, a chiral building block in the synthesis of renin inhibitor RO0425892 (1, remikiren) (Fig. 1), is described. The corresponding racemic ethyl ester substrate 2 is emulsified at elevated temperature in 20–30% concentration in an aqueous buffer and hydrolyzed enantioselectively ($E > 100$) using cheap commercial Subtilisin Carlsberg. The desired acid (S)-3 is separated from the remaining antipodal ester (R)-2 by repetitive extraction at alkaline and acidic pH to give the product in >99% ee and 42% yield. Awkward emulsion problems encountered with these highly concentrated reaction mixtures made the extractive work-up the most critical issue and suggested the application of a disk separator. The development of the reaction from process research to the pilot-scale is described.

5.1
Introduction

The renin-angiotensin system (RAS) is a multiregulated proteolytic cascade that produces the potent pressor and aldosteronogenic peptide angiotensin II (Ang II). In a first step the aspartic proteinase renin cleaves selectively its protein substrate angiotensinogen to release the decapeptide angiotensin I, which in turn is processed by angiotensin converting the enzyme into the octapeptide Ang II. The blocking of the initial step in the enzymatic cascade of the RAS is a promising concept for the treatment of hypertension.

Intensive work worldwide on the design of renin inhibitors has yielded several classes of potent compounds. Most of these inhibitors are transition-state mimetics of the P_1–P'_1 scissile bond (Fig. 2) combined with peptide residues or analogues thereof [1]. Much work has been focused on the minimization of the peptide character of renin inhibitors to overcome the drawbacks of substrate-analogous peptides, such as the instability towards enzymatic cleavage, biliary excretion, and low oral bioavailability. In the course of our studies on renin inhibitors at Roche, we synthesized a series of mimetics incorporating the P_4–P_3 binding elements [2, 3].

Asymmetric Catalysis on Industrial Scale: Challenges, Approaches and Solutions
Edited by H. U. Blaser, E. Schmidt
Copyright © 2004 Wiley-VCH Verlag GmbH & Co. KGaA, Weinheim
ISBN: 3-527-30631-5

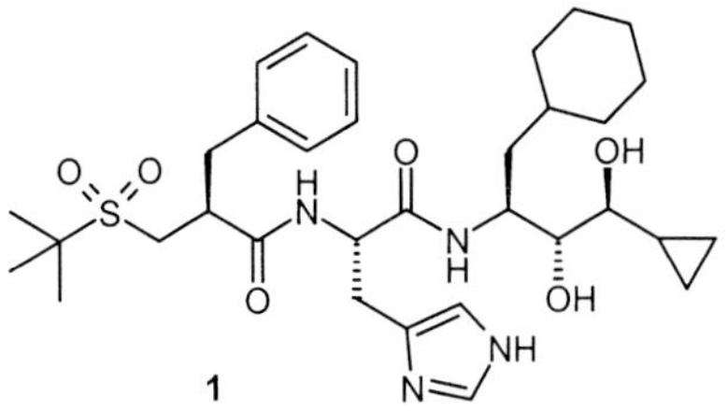

Fig. 1 Structure of the renin inhibitor RO0425892 (**1**, remikiren).

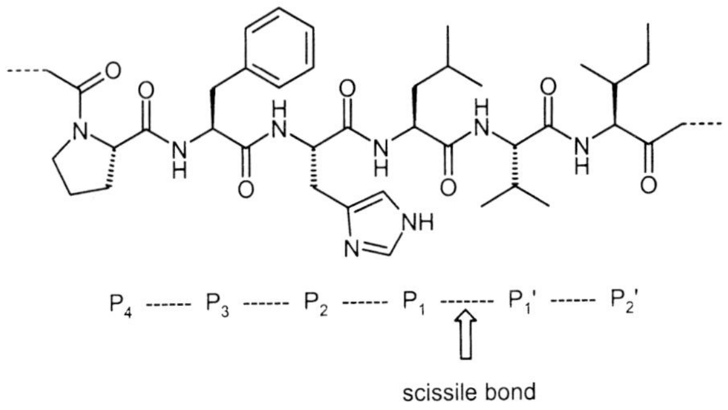

Fig. 2 Substrate side-chain positions (P_4–P_2') for angiotensinogen are illustrated using the notation of Schlechter and Berger.

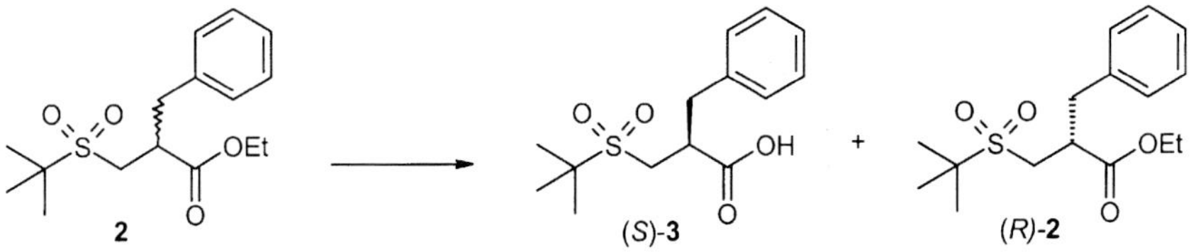

Fig. 3 Enantioselective hydrolysis of ethyl sulfopropionate **2**.

As remikiren was selected for clinical development, bulk quantities had to be prepared. One building block of remikiren, acid (*S*)-**3**, was prepared by an enantioselective enzymatic hydrolysis (Fig. 3) as one of the key steps [4, 5]. This chapter describes the development of this enzymatic step from the lab- to pilot-scale. Process research for the enzymatic step started in March 1989 and the first batches were piloted in February 1992 on the 120–150 kg scale in a campaign for the production of 150 kg of remikiren.

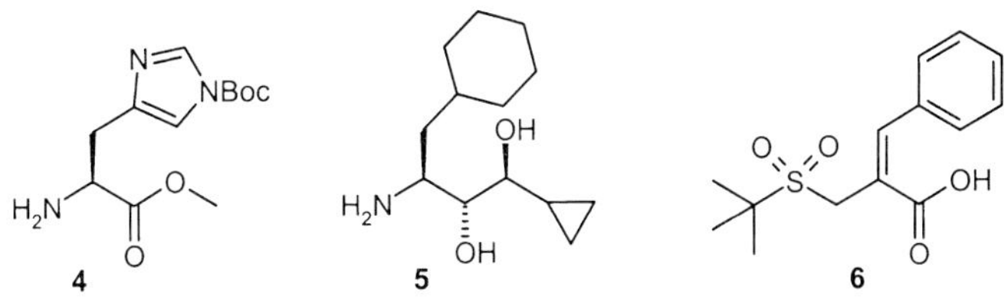

Fig. 4 Structures of compounds 4, 5, 6.

5.2
The Process Research Synthesis

5.2.1
The Synthetic Concept

The structure of remikiren suggested the assembly of three chiral building blocks, namely sulfone (S)-3, (S)-histidine ester 4 and aminediol 5 (see the structures in Fig. 4).

The coupling procedures as well as the sequence of the coupling steps were developed in Process Research by P. Vogt, L. Knight, R. Knorr, M. Schlageter, and J.-P. Gaertner (work not published). According to their protocol, the sulfone-building block (S)-3 was coupled via an activated ester derivative with histidine methylester 4 in essentially quantitative yield. Subsequently, the final coupling step with the aminediol 5 was achieved with the Boc-protected coupling product under activation at the carboxylate moiety. This chapter describes the development of the large-scale preparation of (S)-2-benzyl-3-(tert-butylsulfonyl)-propionic acid (3) by means of an extremely cheap detergent protease. Its precursor, racemic ethyl sulfopropionate 2, was prepared according to a modified published procedure [6].

5.2.2
Alternative Concepts

Various methods for the preparation of enantiomerically pure P_4–P_3 mimetics had been published before this work. Among these were the classical co-crystallization with chiral amines [7, 8], the derivatization as diastereoisomeric amides [6, 9, 10], the synthesis of chiral intermediates following "Evans" methodology [10–12], or the use of lipase-catalyzed reactions [13], and the asymmetric hydrogenation of unsaturated derivatives [8, 14].

The last strategy was followed in-house by R. Schmid, E. Broger and co-workers [15]. The asymmetric hydrogenation of the triethylammonium salt of acid 6 in methanol with a Ru-(p-Tol-BIPHEMP) catalyst produced (S)-3 in 88% ee (97% ee in 85% yield after crystallization). However, the procedure had not yet been established for the large scale when bulk amounts of (S)-3 were requested.

A classical chemoenzymatic approach for the kinetic resolution of racemic N-protected phenylalanines and analogues thereof was based on the enantioselective hydrolysis of their esters by means of serine proteases, namely α-chymotrypsin or Subtilisins BPN′ and Carlsberg generating the corresponding L-acids [16–26]. The same configurational selectivity has been observed also for 2-benzyl succinate esters with α-chymotrypsin [27]. For remikiren we decided to produce the chiral phenylalanine analogue building block (S)-3 with the aforementioned protease approach employing ethyl ester 2 as the substrate. Near the end of this work, the resolution of a series of 2-benzyl-3-sulfonamidopropionic acid esters similar to 2 was reported [21, 22, 28] using Subtilisin Carlsberg.

5.2.3
Enzymatic Approaches

5.2.3.1 Racemic Resolution of Ethyl Sulfopropionate 2 with α-Chymotrypsin

In the early drug discovery phase of the project, α-chymotrypsin was employed as the catalyst according to a procedure reported by Heinz Stadler and co-workers [2, 3]. An analogous procedure using a substrate concentration of 1.25% (w/v) together with an $s/e=77$ [29] at a temperature of 29–30 °C and 20 mM calcium hydroxide solution as the titrating agent was applied repeatedly in order to meet supply requests for research purposes and preclinical evaluations. The assessment of the absolute configuration was based on the results of Cohen and Milovanovic [27]. When remikiren became a drug candidate and larger amounts of (S)-3 were required a more economical, technically feasible procedure was envisaged. The main issues were to have a cheaper bulk enzyme combined with a higher substrate concentration (cf. Section 5.3).

5.2.3.2 Aminolysis of Ethyl Sulfopropionate 2 with Histidine Methylester 4

A very straightforward approach as compared with the existing one (Fig. 3) would have been the direct enzyme-catalyzed peptide formation (cf. Chen et al. [30]) by enantioselective aminolysis of ester 2 with histidine methylester 4 or even racemic histidine ester, as it would resolve the objectives of resolution and coupling in one step. Orientating experiments in which 20 proteases adsorbed on porous glass beads (SIKUG 041/02/120/A, Schott) were in contact with EtOH solutions of 2 and 4 with various water contents, however, did not reveal any reaction.

5.3
Enzymatic Resolution of Ethyl Sulfopropionate 2 with Subtilisin Carlsberg

5.3.1
Process Research

5.3.1.1 Enzyme Screening

The enzyme screening in order to find a substitute for α-chymotrypsin started in March 1989. Since the ester substrate **2** can be considered as a phenylalanine analogue, other proteases recommended themselves as catalysts. Proteases are widely used in many industrial applications such as laundry, food and leather processing and others [31], and therefore, many are available in bulk amounts at comparatively low prices. In a screening of various commercial bulk proteases from MKC [32], Novo [33], Amano Pharmaceuticals and Sigma for the enantioselective hydrolysis of **2**, several enzymes exhibited excellent enantioselectivity. Among them was Subtilisin Carlsberg, an endoproteinase of the serine type that is known to tolerate various organic solvents in high concentrations [21, 22, 25]. Thus, the selective enzymes were re-tested at an unphysiologically high but technically more relevant substrate concentration of 6% combined with elevated temperature (cf. Section 5.3.1.2). The most active biocatalysts were Alcalase (Novo) and Optimase (MKC), both fermentation products of *Bacillus licheniformis* – the major component of which is Subtilisin Carlsberg (Tab. 1). It catalyzes the hydrolysis of proteins and amino acid esters [34] and has been widely used as a detergent additive.

Subtilisin Carlsberg exhibited virtually absolute specificity ($E > 200$ [35]) for the (S)-ester that corresponds to the L-form of a natural amino acid and provided the desired acid (S)-**2** in high enantiomeric and chemical purity. Various commercial enzyme preparations from *Bacillus licheniformis* were available: the cheaper solid preparations Optimase M 440 from MKC [32] and Alcalase 2.0 T from Novo Nordisk [33] as well as the more expensive liquid forms Alcalase 2.5 L and Optimase L 660.

Tab. 1 Best results for the enzyme screening.

Enzyme (amount)	Conversion (%)	Time (h)	ee for (S)-3 (%)
Alcalase 2.0 T (100 mg)	48.1	21.3	>99
Optimase M 440 (100 mg)	47.9	21.6	>99
Savinase 6.0 T (150 mg)	46.5	22.7	>99

Conditions: **2** (1.6 g) was emulsified at 36 °C in 5 mM $CaCl_2$ (25 mL) and the pH adjusted to 7.5. The reaction was started by the addition of the enzyme and the pH kept constant by the controlled addition of 1.0 M NaOH under vigorous stirring. Work-up: repeated extraction with ethyl acetate at pH 7.5 and 2.2.

5.3.1.2 Parameter Optimization

Owing to the high selectivity and activity observed with Subtilisin Carlsberg already seen under the normal reaction conditions of the enzyme screening, the optimization of the procedure concentrated on minimizing the reaction volume while keeping the chemical purity and enantiomeric excess of (S)-3 above 99% and work-up simple. Reduction of the reaction volume was crucial for lowering production costs and could be achieved either by increasing the initial concentration of the starting material in the reaction medium and/or by increasing the concentration of the titrating agent used for maintaining the pH during hydrolysis. The following optimization experiments were carried out with Optimase M 440.

In the original process [2, 3] the crystalline substrate had been applied as a DMSO or ethanol solution. Thus, the substrate concentration could not be increased significantly without increasing the co-solvent concentration, which in turn would affect the enzyme activity. Elevating the temperature to 36 °C or above could solve the problem. At this temperature the suspended substrate starts melting while Subtilisin Carlsberg retained its activity and selectivity.

Calcium ions used in the initial process were found not to be essential for stereoselectivity or the reaction rate but they had a welcome anti-foaming effect. Nevertheless, the original 20 mM calcium hydroxide solution (upper solubility limit for calcium hydroxide) was replaced by more concentrated sodium hydroxide solutions.

The higher substrate concentration also made it necessary to increase the enzyme concentration if the reaction time was not to exceed 2 d. With a substrate concentration of 12% (w/w) at a temperature of 36–40 °C and a pH of 7.5, an s/e of ca. 30 had to be applied [36]. Under these conditions the product was obtained in high enantiomeric excess (>99%) and purity (>99% GLC, silylated).

5.3.1.3 Work-up

Work-up was a major issue in this enzymatic process step. The highly concentrated reaction mixtures favored the formation of persistent emulsions in the extraction steps with ethyl acetate at alkaline and acidic pH. Owing to the tedious phase separation the first extraction steps at both alkaline and acidic pH initially lasted one day and had to be enhanced by the addition of sodium sulfate.

Emulsions are usually formed, for instance, due to over-agitation, interfacial tension, low-density difference, or the inherent nature of the compounds involved. In our case the extraction problems were supposed to stem from the high density of both substrate and product ($\delta > 1$) combined with their high concentrations in the reaction mixture, the emulsifying power of the enzyme, and loose solid fines that floated at the interface. In some very highly concentrated experiments (cf. Section 5.4.2) the organic phase had an even higher density than the aqueous phase.

The emulsion problems could be defused by using dichloromethane as the extractant, because of the larger density difference of the phases and the filtration of the reaction mixture on Dicalite Speedex filter aid prior to the extractions in order to remove solid fines originating from the enzyme preparation [37]. The downstream protocol based on the dichloromethane extraction is described in note [38] and was also carried out on the pilot-scale (cf. Section 5.4.3).

5.3.2
Process Development

The main issues of Process Development were:
i) to reduce the reaction volume further,
ii) to implement a more economic product isolation procedure on the large scale while reducing work load, and
iii) to increase the yield (at equal or better quality).

5.3.2.1 Process Parameters

For the further reduction of the reaction volume we tested even higher substrate concentrations and higher concentrated titrating solutions. As to the former, it turned out that the enzyme worked smoothly even at 20 to 30% (w/v) substrate concentration; and, in addition, at substrate to buffer ratio of 1:1 (w/v) the enzyme still showed a useful activity. In all cases the product was obtained with excellent chemical and enantiomeric purity (both >99%).

As to the concentration of the titrating solution a high NaOH concentration affected the enzyme activity more seriously than a high substrate concentration. Thus, in order to have a high space-time yield it is better to work with a high substrate concentration and titrate with a moderately concentrated sodium hydroxide solution than *vice versa*. However, at pH 9 the enzyme still worked without problems at 33% substrate concentration using 3 M or 4 M sodium hydroxide solution (and even 9 M NaOH worked, however, with a considerably reduced enzyme activity).

The dramatically increased substrate concentration required a concomitant increase of the amount of enzyme to keep the substrate to enzyme ratio constant. In order to tackle the emulsion problems during work-up we tried to reduce the relative enzyme quantity and compensate for this by increasing the enzyme activity (the more so as the high substrate or product concentrations lowered the enzyme activity). Enzyme properties such as the activity depend strongly on the enzyme conformation and therefore can be influenced by all types of chemical and physical reaction parameters, which have a more or less pronounced effect on the enzyme structure. Typical parameters, such as temperature and pH, are often applied in parameter optimization programs (and well-documented in numerous publications). Variation of the pH demonstrated the expected acceleration of the hydrolysis rate with increasing pH. Whereas at values up to pH 9.0 the acid (*S*)-**3** was yielded in excellent quality, at pH 10 a noticeable rate of unspecific hydrolysis occurred. Increasing the reaction temperature revealed a slight decrease in activity even at 45 °C probably owing to minor denaturation or degradation of the enzyme. As a result of substrate precipitation at lower temperatures (m.p. of **2** 36–

39 °C) it was necessary to keep the temperature at a minimum of 37 °C, thus limiting the useful temperature range to 38–43 °C.

Vigorous stirring was crucial for high hydrolysis rates, however, at the same time the formation of foam had to be avoided. Foam inactivated the enzyme in a reversible manner. In one case, on formation of a foam cap the activity was lost almost completely, but was fully restored when the foam was driven back by overpressure onto the reaction vessel and with gentler stirring. Therefore, on a larger scale the reactions were conducted under slight nitrogen overpressure (0.1–0.3 bar) helping to suppress foam formation.

Because at very high substrate and sodium hydroxide concentrations work-up became more tedious due to persistent emulsions, an initial substrate concentration of 20–30% together with 3 M NaOH seemed to be optimal. On the pilot-scale, however, we had to choose a certain filling level of the reaction vessel (a 1000 L fermenter) in order to achieve optimal stirring and, therefore, pilot runs were sometimes conducted at sub-optimal substrate and NaOH concentrations. For the same reasons we defined a broad acceptable range of concentrations.

Based on the above results the following optimal conditions were specified for the enzyme reaction: 25–50 g of **2** emulsified in 100 mL of buffer pH 8.5 at 39–43 °C using a 2–4 M NaOH solution for pH control.

5.3.2.2 Improved Enzyme Preparation

In the course of our development work it turned out that Optimase M 440 was no longer available [39]. Since it was known from another project that the liquid enzyme formulations exhibit a less pronounced emulsifying behavior (cf. [37]), the different commercial bulk preparations, now Alcalases and Solvay Proteases, were re-evaluated. All yielded very good results with respect to chemical purity and enantiomeric excess (>99% each). However, Solvay Protease L 660 (the successor of Optimase L 660) was clearly the most active enzyme, thus requiring the smallest amount of enzyme for the reaction. The clearly higher price of this enzyme [40] should be of no consequence because of the expected simpler work-up and, hence, lower manpower costs.

5.3.2.3 Work-up

The goal of Process Development was to establish a more straightforward and robust work-up procedure. Therefore, it was again mandatory to find a solution for the emulsion problem. In addition, a reliable technique for the separation of emulsions would allow replacing dichloromethane with environmentally more acceptable ethyl acetate. Several technologies are known to be suitable for breaking up or avoiding stable emulsions during liquid–liquid extraction. Among the most common are continuous centrifugation, membrane technologies, the addition of demulsifiers or coagulants, and enzyme immobilization. We decided to introduce continuous centrifugation as a robust and broadly applicable method into our work-up procedure using a disk stack separator. Since we had a Westfalia [41] se-

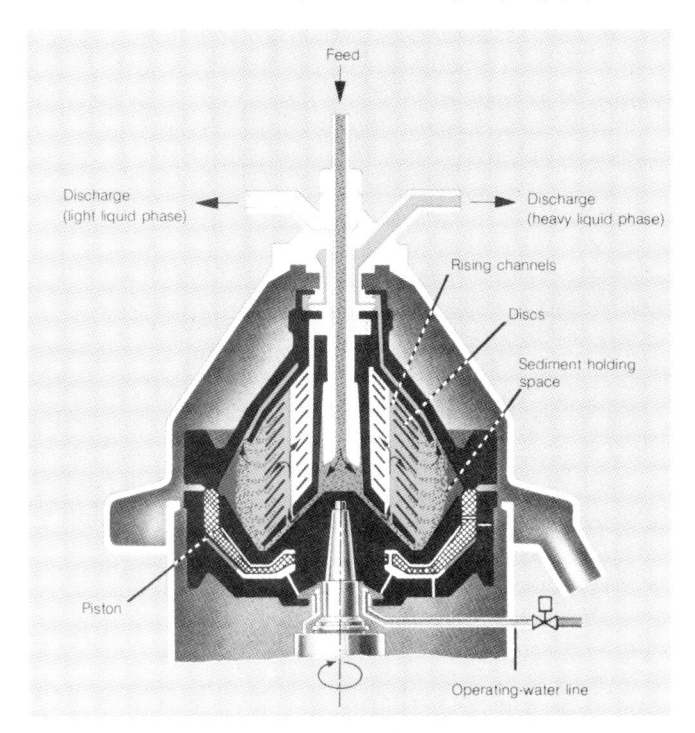

Fig. 5 Bowl of disk stack separator (Westfalia).

parator SA 01-01-175 for the lab-scale and an SA 20-01-576 for the pilot-scale available in our biotechnology pilot plant, this technology could be implemented rapidly. Because of the continuous operation of these separators they are suitable for large-scale operation.

5.3.2.3.1 **Equipment**
Continuous centrifugation is widely used in many different industries, e.g., the dairy, food, beverage, pharmaceutical and biotechnology industry. Sophisticated separators for different processes can be purchased or hired in various sizes from several suppliers. The disk stack separators SA 01-01-175 and SA 20-01-576 from Westfalia Separator AG, Germany, are so called three-phase separators, which are able to separate two liquid phases and a solid phase from each other (Fig. 5).

5.3.2.3.2 **Isolation of Sulfopropionic Acid (S)-3**
After some orientating experiments employing the newly selected liquid enzyme preparation a simple extraction procedure comprising three extractions at pH 8.5 and two at pH 2 could be developed. On comparison of the solid enzyme preparation Optimase M 440 with Solvay Protease L 660, it turned out that the removal of

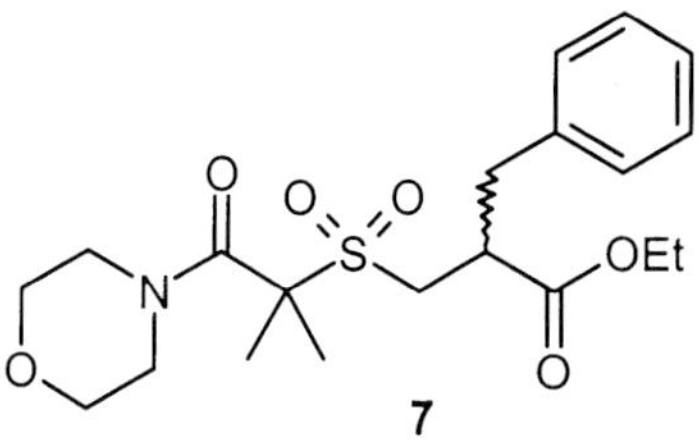

Fig. 6 Structure of compound 7.

solid fines at the phase interface became obsolete. In addition, in the alkaline extraction series the phase separation occurred spontaneously and, thus, the separator was only used in the first acidic extraction step. In the previous procedure (S)-3 was obtained directly from the aqueous phase by acidification [38]. In order to remove traces of insoluble biological material still present at pH 2, a crystallization procedure was developed wherein (S)-3 was crystallized from ethyl acetate by the addition of n-hexane. The procedure is described in note [42].

Owing to an urgent need for material at an early stage of development, the pilot batches had to be carried out according to the not yet optimized procedure described in note [38]. Shortly after that, the remikiren project was abandoned. The newly developed protocol [42], however, could be confirmed in principle in a very similar experimental outline in a follow-up project with the closely related compound 7 (Fig. 6) on the 200 L scale (17 kg 7) using the disk stack separator SA 20 (Westfalia) (cf. [5]).

5.3.3
Pilot Production

As mentioned in Section 5.3.2.3.2, three pilot batches from 120 to 150 kg of (R,S)-2 had to be carried out according to not yet optimized conditions without a phase separator and using the solid enzyme preparation [38]. Nevertheless, the experiments were successful and a total of 160 kg of (S)-3 in high chemical and optical purity were produced.

5.3.3.1 Equipment
The three runs were conducted in analogy with the procedure described in [38] at a concentration of 23–29%. A 1000 L fermenter (Giovanola; stainless steel; pressure-proof) equipped with a blade stirrer, temperature and pH-stat control [43] and pressurized with nitrogen gas was chosen. The dosing system was adjusted to a fast response rate to keep the retardation for the addition of the titrating agent to a minimum. The consumption of the titrating agent was determined by means of weight loss using a mobile vessel on a balance. Work-up and crystallization was conducted in two 1200 L glass-lined reactors equipped with a basket centrifuge and several tanks (1000 L). Each run had to be worked-up in two batches.

Tab. 2 Experimental data and results for the three pilot runs.

Batch no.	Amount of 2	Conc. of 2	Buffer amount	Enzyme amount (% with respect to 2)	Time (h)	Conversion (%)	% ee of (S)-3
2-7-2-92	147.5 kg	29%	520 L	7.08 kg (4.8%)	70	49.4	>99
3-11-2-92	121.8 kg	23%	540 L	7.56 kg (6.2%)	48	46.3	>99
4-14-2-92	153.2 kg	29%	540 L	7.28 kg (4.8%)	70	45.9	>99

Procedure: **2** was emulsified in 2 mM sodium phosphate buffer pH 7.5 under vigorous stirring (100 rpm) at 40°C and ca. 0.2 bar of nitrogen. Optimase M 440 was added and the pH kept constant at 7.5 by the controlled addition of 3.0 M NaOH solution.

5.3.3.2 Enzyme Reaction

The results of the three batches are shown in Tab. 2. The enzymatic hydrolyses went smoothly within the expected time and yielded product of high quality. In spite of the automated pH control the reaction was supervised continuously. The stirrer speed was chosen such (100 rpm) that intensive mixing was secured while the formation of foam avoided (the latter supported by a slight overpressure of 0.2 bar). After termination of the reaction the reaction mixtures were transferred from the fermenter to the extraction plant in a mobile tank and processed as described below.

5.3.3.3 Work-up

Work-up was conducted according to note [38]. It proved to be the rate-limiting step in the process, but nevertheless enabled the desired product to be isolated in high yield [total yield of all three runs 41.6%, 88% of (S)-**3** formed, according to Tab. 2] and excellent quality (>99.5% chem. purity, >99% ee). In particular, the centrifugation (two portions) of the warm reaction mixture and the several extractions/back extractions were lengthy. To produce 1 kg of (S)-**3** required 10.7 man-hours of labor for the work-up. Thus the procedure was suitable for the first pilot campaign, but not for the larger campaigns we had planned. For these we intended to use a Westfalia SA-20 disk separator according to note [42], which would not only cut down the amount of labor [4.4 man-hours per kg (S)-**3**, and at production scale even less] but would allow the use of lower quantities [28 kg per kg (S)-**3** instead of 86 kg] of environmentally more favorable solvents (ethyl acetate and *n*-hexane).

5.4 Discussion

Between March 1989 and February 1992 an enzymatic racemic resolution step was developed from the lab- up to the 150 kg-scale. The choice of the biotransformation in question (Fig. 3) was determined, firstly, by the nature of the substrate, which in a broad sense resembled a phenylalanine derivative and, secondly, by the high selectivity and robustness of the biotransformation itself. Because it was a laundry enzyme, Subtilisin Carlsberg was extraordinarily cheap and stable. While its high enantioselectivity towards the phenylalanine-like substrate was not necessarily surprising, it turned out to be unexpectedly tolerant to extreme concentrations of substrate and salt, to elevated temperature or to a combination of all. It displayed its excellent enantioselectivity and high activity even at 20–30% substrate concentration at 39–40 °C. The elevated temperature allowed the application of the substrate in melted and thus a much more accessible form. In spite of emulsion problems initially encountered during the extractive work-up, at a very early stage of the project we were convinced that we had a cheap, facile, robust, and hence technically feasible reaction to hand, providing us with inexpensive access to the key-building block (S)-**3**.

Because of the initial attractive results of the enzyme reaction, optimization concentrated not so much on an improved performance but on higher substrate concentrations and a straightforward and scalable work-up providing the product in high optical and chemical purity (>99% for both). Once the emulsions that appeared during product isolation could be handled by using a separator or a conventional phase separation [38], the work-up became routine. At the lab-scale kg amounts of the target product could already be prepared. Therefore, and due to its comparative simplicity at later stages of the development, the enzyme reaction was never a limiting factor in development.

A big factor for the success of the process was the close similarity of the substrate to the natural substrates of Subtilisin Carlsberg, an enzyme with extraordinary properties. Since cheap commercial sources for this enzyme were available on the market, the enzyme could be discarded. A major disadvantage of the present chemoenzymatic route was the fact that no satisfactory racemization procedure for the unwanted enantiomer (R)-**2** could be established due to elimination and/or hydrolysis side reactions.

The present enzymatic reaction, however, never entered the production-scale as the target molecule remikiren finally failed in clinical trials.

5.5
Acknowlegements

We would like to express our cordial thanks to our colleagues from the kilo-lab for supply of the pure substrate and to our colleagues from the analytical services for running our samples. We are also grateful for the technical assistance of Gerhard Barwich, Sylvie Biétry, Sophie Brogly, Joseph Flota, François Gantz, Gérard Gasser, Werner Herrmann, Elfriede Kirscher, Markus Kyburz, Beat Schilliger, Evelyne Schumacher, Patrik Stocker, Daniel Strub, Peter Steidle, Barbara Stirnimann and Maurice Wilhelm.

5.6 References

1. W. J. Greenlee, Med. Res. Rev. 1990, 10, 173–236.
2. Q. Branca, W. Neidhart, H. Ramuz, H. Stadler, W. Wostl, Eur. Pat. 0416373A2 1991.
3. Q. Branca, M. P. Heitz, W. Neidhart, H. Stadler, E. Vieira, W. Wostl, Eur. Pat. 0509354B1 1992.
4. B. Wirz, W. Wostl, Eur. Pat. Appl. No. 475255 1992.
5. S. Doswald, H. Estermann, E. Kupfer, H. Stadler, W. Walther, T. Weisbrod, B. Wirz, W. Wostl, Bioorg. Med. Chem. 1994, 2, 403–410.
6. P. Bühlmayer, A. Caselli, W. Fuhrer, R. Göschke, V. Rasetti, H. Rüeger, J. L. Stanton, L. Criscione, J. M. Wood, J. Med. Chem. 1988, 31, 1839–1846.
7. P. Bühlmayer, W. Fuhrer, R. Göschke, V. Rasetti, H. Rüeger, J. L. Stanton, Eur. Pat. Appl. No. 236734 1987.
8. Y. Ito, T. Kamijo, H. Harada, F. Matsuda, S. Terashima, Tetrahedron Lett. 1990, 31, 2731–2734.
9. P. Bühlmayer, W. Fuhrer, R. Göschke, V. Rasetti, J. L. Stanton, Eur. Pat. Appl. No. 184550 1986.
10. M. Nakano, S. Atsuumi, Y. Koike, S. Tanaka, H. Funabashi, J. Hashimoto, M. Ohkubo, H. Morishima, Bull. Chem. Soc. Jpn. 1990, 63, 2224–2232.
11. T. Nishi, M. Sakurai, S. Sato, M. Kataoka, Y. Morisawa, Chem. Pharm. Bull. 1989, 37, 2200–2203.
12. M. Nakano, S. Atsuumi, Y. Koike, S. Tanaka, H. Funabashi, J. Hashimoto, H. Morishima, Tetrahedron Lett. 1990, 31, 1569–1572.
13. S. Atsuumi, M. Nakano, Y. Koike, S. Tanaka, M. Ohkubo, T. Yonezawa, H. Funabashi, J. Hashimoto, H. Morishima, Tetrahedron Lett. 1990, 31, 1601–1604.
14. U. Lerch, J. H. Jendralla, B. Seuring, R. Henning, Eur. Pat. Appl. No. 512415 1992.
15. R. Schmid, E. A. Broger, Proceedings of the Chiral Europe '94 Symposium, September 19–20, 1994, Nice, France.
16. J. B. Jones, Y. Y. Lin, D. N. Palmer, Can. J. Chem. 1974, 52, 469–476.
17. J. F. Beck, J. F. McMullan, Can. J. Chem. 1979, 57, 2516–2517.
18. D. P. Bauer, World Patent No. 8002421 1980.
19. S. T. Chen, K.-T. Wang, C.-H. Wong, J. Chem. Soc., Chem. Commun. 1986, 1514–1616.
20. M. Pugnière, L. G. Barry, A. Previero, Biotechnol. Techn. 1989, 3, 339–344.
21. R. S. Boger, S. R. Crowley, World Patent No. 9116031 1991.
22. S. H. Rosenberg, J. F. Denissen, Eur. Pat. Appl. No. 456185 1991.
23. S. T. Chen, S. C. Hsiao, A. J. Chiou, S. H. Wu, K. T. Wang, J. Chin. Chem. Soc. 1992, 39, 91–99.
24. C. Garbay-Jaureguiberry, I. McCort-Tranchepain, B. Barbe, D. Ficheux, B. P. Roques, Tetrahedron Asymmetry 1992, 3, 637–650.
25. Y. Tomiuchi, K. Oshima, H. Kise, Bull. Chem. Soc. Jpn. 1992, 65, 2599–2603.
26. O. D. Tyagi, P. M. Boll, Indian J. Chem. 1992, 31B, 851–854.
27. S. G. Cohen, A. Milovanovic, J. Am. Chem. Soc. 1968, 90, 3495–3502.
28. T. M. Zydowsky, H. Mazdiyasni, D. B. Konopacki, D. A. Dickman, Tetrahedron Lett. 1993, 34, 935–938.
29. s/e designates the ratio of substrate to enzyme used on a weight basis.
30. S. T. Chen, S. Y. Chen, K. T. Wang, J. Org. Chem. 1992, 57, 6960–6965.
31. B. Poldermans, in Enzymes in Industry (W. Gerhartz, Ed.) (Chapter 4.4 Proteolytic Enzymes), VCH Verlagsgesellschaft GmbH, Germany 1990.
32. Miles Kali-Chemie (MKC), is now Solvay Deutschland GmbH (Hannover, Germany).
33. Novo Nordisk a/s is now Novozymes (Bagsvaerd, Denmark).
34. B. Aleksiev, P. Schamlian, G. Widenov, S. Stoev, S. Zachariev, E. Golovinsky, Hoppe-Seyler's Z. Physiol. Chem. 1981, 362, 1323–1329.
35. C.-S. Chen, Y. Fujimoto, G. Girdaukas, C. J. Sih, J. Am. Chem. Soc. 1982, 104, 7294–7299.

36 Enzymatic hydrolysis: 12.6 g ester **2** was emulsified in 100 mL 70 mM sodium phosphate buffer pH 7.0 at 39 °C. The pH was adjusted to 7.5 (1 M NaOH) and the reaction started by adding 0.4 g Optimase M 440. The pH was maintained by the controlled addition (pH-stat) of 1 M NaOH solution under vigorous stirring at 39 °C. After 47 hours a degree of conversion of 48% was attained.

37 Solid Optimase M 440 contained titanium dioxide particles.

38 Work up procedure: Hydrolysis – 1 kg of **2**, 2 L sodium phosphate buffer, 100 g Optimase M 440, 800 mL 2.0 M NaOH, 39–40 °C. Work-up – the warm reaction mixture was filtered in two portions on a basket centrifuge packed with a Dicalite Speedex layer and the filter cake washed with 2×1 L water and 2×2 L CH_2Cl_2. The combined organic phases were washed with 1 L water. The three aqueous phases were combined and filtered on Dicalite Speedex and the filter cake washed with 2 L water. The filtrate was set to pH 1.5 and the precipitated product (S)-**3** extracted with 4 L CH_2Cl_2. The biphasic mixture containing some foam-like substance was filtered on Dicalite Speedex. The filter cake and the aqueous phase were extracted with 1 and 2 L of CH_2Cl_2, respectively, and the combined organic phases washed with 1 L water, which was extracted with 1 L CH_2Cl_2. The combined organic phases were extracted with 2 L of 1 M NaOH solution and the aqueous phase washed with 1 L CH_2Cl_2. The combined organic phases were re-extracted with 1 L of 1 M NaOH and the combined alkaline phases washed with 1 L of CH_2Cl_2. The product was precipitated by adding 240 mL of 32% HCl, filtered off, washed with 500 mL water and dried (60 °C, 20 mbar, 2–3 d) to give ca. 360 g of (S)-**3** (40–45% yield; content: ca. 99.8%, >99% ee).

39 Miles Kali-Chemie was taken over by Solvay Enzymes.

40 The prices for Solvay Protease L 660 and Alcalase DX 2.5 L were approx. US $24 vs. US $13.

41 Westfalia Separator AG, 59302 Oelde, Germany.

42 Enzymatic hydrolysis using a disk stack separator for work-up: 2 kg (6.40 mol) of ester **2** was emulsified in 4 L of 13 mM borax HCl buffer pH 9 of 40 °C by melting and vigorous stirring. After addition of 200 mL Solvay Protease L 660 the pH-drop was immediately re-adjusted to 8.5 and the pH then kept constant (pH-stat) by the addition of 2 M NaOH under vigorous stirring at 39–40 °C. After completion of the reaction (1586 mL, 49.5% conversion, 42.3 hours) the reaction mixture (8 L) was cooled down to room temperature. (R)-**2** was removed by extraction three times with ethyl acetate (8 L, 4 L, 4 L). Phase separation occurred spontaneously within 10 min. The aqueous solution was then adjusted to pH 2 by the addition of 98% H_2SO_4 (105 mL). After extraction with ethyl acetate (8 L) the emulsion formed was separated on a disk separator (Westfalia SA 1) operated at 7900 rpm (4.9 kg). Complete phase separation was attained at a flux of 22–28 L/h. After the second extraction with ethyl acetate (5 L) the emulsion separated spontaneously. The combined organic phases were washed with deionized water (5 L), concentrated to 2.5 L, and residual water was removed (<1%) by azeotropic distillation with ethyl acetate (5×0.5 L). Crystallization of (S)-**3** was achieved by adding *n*-hexane (4 L) to the 2.5 L of solution under vigorous stirring. After crystallization had started, further *n*-hexane (2 L) was added and the suspension stirred at 4 °C for 1.5 hours. The crystals were filtered off, washed with 1 L of *n*-hexane and dried *in vacuo* at 40 °C to provide 760 g (42%) of (S)-**3** as colorless crystals (content >99%; ee ≥99%).

43 A long pressurized pH-probe was used.

6
Protease-Catalyzed Preparation of Chiral 2-Isobutyl Succinic Acid Derivatives for Collagenase Inhibitor RO0319790

BEAT WIRZ, MILAN SOUKUP, THOMAS WEISBROD, FLORIAN STÄBLER, and ROLF BIRK

Abstract

An efficient enzyme-catalyzed reaction for the large-scale preparation of (R)-2-isobutyl succinic acid 4-ethyl ester (R)-**2a**, a key intermediate in the synthesis of collagenase inhibitor RO0319790 (**1**), is described (Fig. 1). The corresponding racemic diethyl ester substrate **9** is applied at 20% concentration and hydrolyzed regio- and enantioselectively ($E > 100$) at the sterically more hindered secondary ester group using a cheap commercial subtilisin preparation. The desired (R)-monoacid is separated from the remaining antipodal diester (S)-**9** by means of extraction and obtained in >99% ee and >43% yield. The development of the reaction from process research to the pilot-scale is described.

6.1
Introduction

Collagenase enzymes play an important role in the development of rheumatoid arthritis [1, 2]. At Roche Drug Discovery the compound RO0319790 had been identified as an orally active collagenase inhibitor and was tested as a development candidate for the therapy of both rheumatoid arthritis and osteoarthritis. For the preparation of larger amounts of substance the original Research Synthesis from Roche Welwyn [3] had to be re-evaluated. In Process Research an economically more attractive new synthetic concept was developed by Milan Soukup (Fig. 2, route C) [4, 5]. The new Process Research Synthesis contained a combined enantio- and regioselective enzymatic hydrolysis as one of the key steps [5] and the racemization of the unwanted enantiomer. This chapter describes the development of this enzymatic step from the lab- to pilot-scale. The first batches were piloted on the 200 kg-scale and a campaign for the production of 160 kg of RO0319790 was carried out successfully.

Asymmetric Catalysis on Industrial Scale: Challenges, Approaches and Solutions
Edited by H. U. Blaser, E. Schmidt
Copyright © 2004 Wiley-VCH Verlag GmbH & Co. KGaA, Weinheim
ISBN: 3-527-30631-5

Fig. 1 Structures of collagenase inhibitor RO0319790 **1** and 2-isobutyl succinic acid 4-ethyl ester **2**.

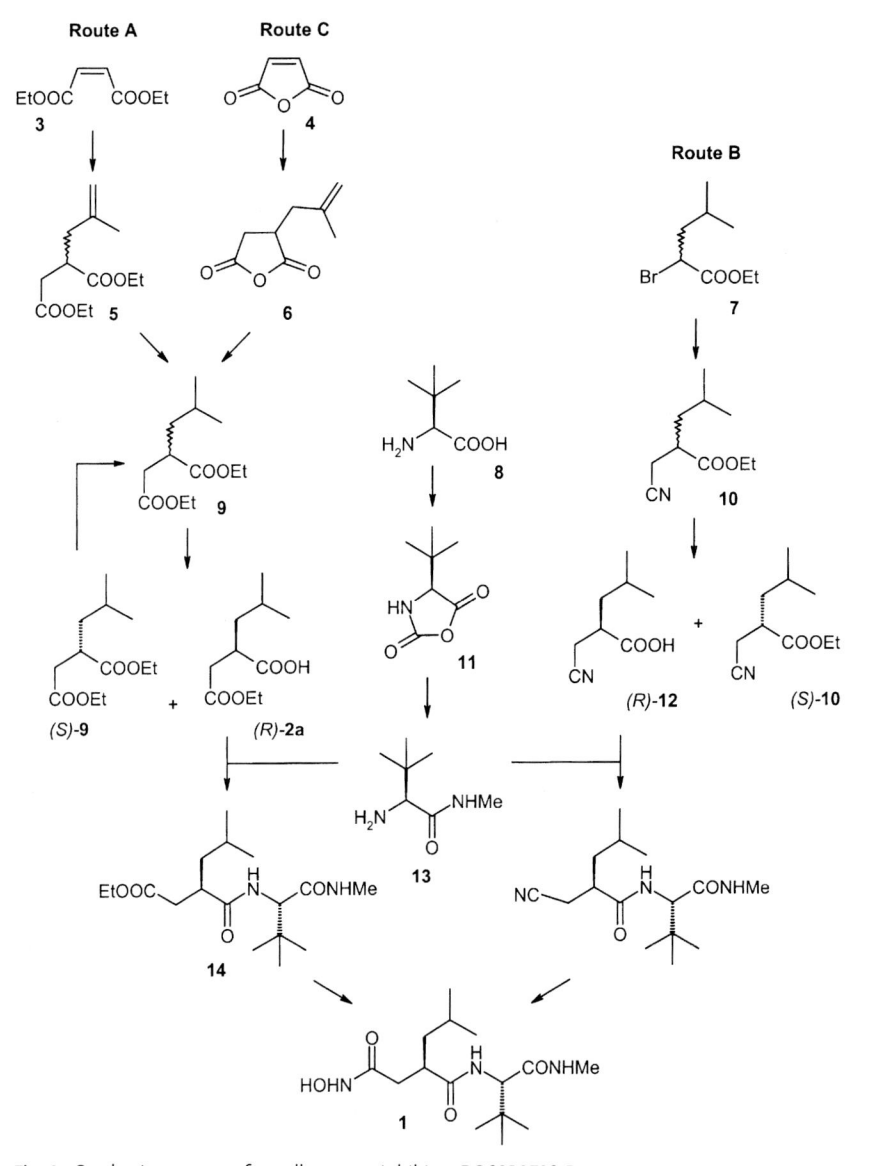

Fig. 2 Synthesis concepts for collagenase inhibitor RO0319790 **1**.

6.2
The Process Research Synthesis

6.2.1
The Synthetic Concept

By retro synthetic analysis collagenase inhibitor RO0319790 (**1**) can be assembled from two chiral building blocks, (*R*)-succinate **2** and (*S*)-*tert*-leucine *N*-methylamide **13**. As the latter can be prepared from commercially available (*S*)-*tert*-leucine **8** our work concentrated in particular on the construction of the first building block **2**. In order to assemble the carbon skeleton of **2** in the most efficient way, extremely cheap maleic anhydride **4** was converted in a known ene reaction with isobutylene to provide the cyclic anhydride **6**. Hydrogenation of the double bond followed by the addition of EtOH/*p*-TsOH yielded the racemic diethyl ester substrate **9** for the enzyme reaction. The enzymatic monohydrolysis of **9** afforded the monoacid (*R*)-**2a**. (*R*)-**2a** was coupled via its acid chloride with leucine amide **13** to ester **14**, which finally was converted into the hydroxamic acid **1**.

6.2.2
Alternative Concepts

In the course of this work Milan Soukup from Process Research also evaluated several other routes to RO0319790.

In one approach (Fig. 2, route C) the resolution of the racemic anhydride **6** was attempted via the regioselective opening with various chiral auxiliaries, however, even with sterically hindered alcohols the diastereoisomeric esters could not be separated on a technical scale by crystallization. In contrast, good selectivity was obtained with (*S*)-phenylethyl hydroxylamine. This pathway, however, failed due to problems at a later stage.

Alternatively (Fig. 2, route B), a nitrile group in principle could serve as a precursor of the final hydroxamic acid function via oxidation to the nitrileoxide and its hydrolysis. This strategy would be based on the nitrile ester **10** as a key intermediate and its racemic resolution. While the resolution step could be achieved enzymatically, the oxidation experiments failed and this approach was finally abandoned.

6.2.3
Enzymatic Approaches

6.2.3.1 Racemic Resolution of Nitrile Ester 10
The first enzymatic reaction investigated in the whole project concerned the introduction of chirality in route B (Fig. 2) by generation of succinic acid mononitrile (*R*)-**12** from its racemic precursor. Since in a broad sense the nitrile ester substrate **10** can be interpreted as an amino acid analogue, proteases recommended themselves as catalysts to be tested. From the literature, 2-substituted succinic and butyric acid esters were known to be resolved by proteases [6, 7]. Proteases are

Tab. 1 Influence of co-solvents, salts, and temperature on the reaction rate and enantioselectivity of Alcalase[a].

No.	Solvent added or salt used (instead of NaCl)	°C	Time for 40% conversion (min)	% ee (R)-12
1	Standard procedure	20	20	94.2
2	5% DMSO	20	13	95.7
3	0.1 M guanidinium chloride	20	14	95.8
4		1–3	29	98.4
5	0.01 M guanidinium chloride	1–2	24	98.5

a) Standard procedure: 100 mg of **10** was emulsified in 15.6 mL 0.1 M NaCl, 4 mM phosphate buffer pH 7.5. The reaction was begun by adding 200 µl Alcalase 2.5 L and the pH maintained under vigorous stirring by adding 0.1 M NaOH (pH-stat).

widely used in bulk amounts in many industrial applications such as laundry, food and leather processing, and others [8]. Hence, many proteases are commercially available at a comparatively low price. In a screening of various commercially available bulk proteases from Solvay Enzymes [9], Novo Nordisk [10], Amano Pharmaceuticals, and Sigma for the enantioselective hydrolysis of the racemic ester, **10** Subtilisin Carlsberg exhibited clearly the best results with respect to both activity and enantioselectivity. To improve the initial result (Tab. 1, entry 1) the influence of a few parameters such as co-solvents, salts and temperature on the enzyme activity and selectivity was investigated (Tab. 1). Among several co-solvents and salt additives tested, DMSO (entry 2) and guanidinium chloride (entry 3), respectively, had a moderately positive effect on both reactivity and selectivity. As DMSO was expected to interfere with the work-up, the use of guanidinium chloride was favored. Low temperature (entry 4) also improved enantioselectivity but at the expense of the reaction rate. A combination of guanidinium chloride (exhibiting its positive influence even at a concentration of 10 mM) at 1 °C was finally chosen as a suitable system (entry 5). At higher, technically more favored substrate concentrations (10–16% w/w), however, the reaction slowed down considerably after 35–40% conversion. Further optimization was abandoned when the more straightforward route C (Fig. 2) via (R)-2-isobutyl succinic acid 4-ethyl ester (R)-**2a** was found.

6.2.3.2 Enantio- and Regioselective Monohydrolysis of Diester 9

Route C (Fig. 2) starts from the extremely cheap bulk agents isobutylene and maleic anhydride **4** and provides diester **9** in two steps. When in the course of process development this attractive access to the potential enzyme substrate **9** could be established, work concentrated immediately on its enantio- and regioselective monohydrolysis. Thus, the sterically more hindered ester group had to be hydrolyzed to the target monoacid (R)-**2a**. Protease-catalyzed reactions with the stated specificity had already been described for several 2-substituted succinate diesters, such as 2-

methyl [6, 11], 2-benzyl [6, 12] or 2-(3,4-dimethoxybenzyl)succinate [13]. In addition, selective lipases/esterases had already been described [14–17].

Thus, an enzymatic reaction with a preference such that the monoacid (R)-**2a** is generated could be expected from the literature. Indeed, the specificity requirements were met surprisingly well again by Subtilisin Carlsberg, which had already been used in the resolution of racemic **10** (cf. Section 6.2.3.1). The development of this reaction is the main topic of this chapter and will be described in Section 6.3.

6.2.3.3 Aminolysis in Organic Solvent Systems

A very straightforward approach in route C (Fig. 2) would have been the direct enzyme-catalyzed peptide formation (cf. Chen et al. [18]) by enantioselective aminolysis of diester **9** with (S)-tert-leucine methylamide **13** or even racemic **13**. This would combine three synthetic objectives: the resolution of (rac)-**9**, the resolution of (rac)-**13** and the coupling step. In orientating experiments monoester **10** was tested as a model substrate. It was contacted with an equal amount of (S)-amine **13** in the presence and absence of an organic solvent. Solid or liquid subtilisin Carlsberg preparations (Alcalase 2.0 T or Alcalase 2.5 L, respectively) were used as the catalyst. Only with the liquid enzyme preparation was the formation of minor amounts of one of two possible diastereoisomeric peptides observed [19], whereas most of the ester was hydrolyzed to the acid. Likewise, a few selected lipases also provided negative results.

6.2.3.4 Coupling of (R)-2a with Amine 13 in Aqueous Milieu

Previously, a series of genetically altered subtilisins were reported [20] which were capable of catalyzing peptide synthesis in water but not the competing hydrolytic reaction. J. Wells and T. Chang from Genentech kindly provided us with two engineered species of subtilisin BPN': subtiligase (C221 A225) and stabiligase (C221 A225 F50 D76 S109 R213 S218). Orientating experiments with ester **9** (2 mM) and (S)-amine **13** (3 mM) carried out under the conditions described in reference [20], however, they did not lead to the corresponding peptidic diastereoisomer(s).

6.3
Enzymatic Resolution of Diethyl 2-Isobutyl Succinate 9

6.3.1
Process Research

6.3.1.1 Selection of the Enzyme
In contrast to most biotransformation projects, the work for the present biotransformation did not start with an enzyme screening. Only Subtilisin Carlsberg, by far the best enzyme obtained in the screening for the racemic resolution of nitrile

ester **10** (cf. Section 6.2.3.1), was tested for the enantio- and regioselective monohydrolysis of diethylester **9** and gave excellent initial results, with respect to both activity and selectivity. Subtilisin Carlsberg exhibited virtually absolute specificity for the 1-carboxyl ester group, which corresponds to the α-carboxy group, in a natural amino acid and provided the desired (R)-configured monoacid **2a** in high enantiomeric and chemical purity. The configuration of (R)-**2a** was assigned by its conversion to the final product.

As Subtilisin Carlsberg was, and still is, one of the cheapest enzymes available on the market, no further proteases were tested. Four Subtilisin Carlsberg preparations from Solvay Enzymes [9] and Novo Nordisk [10] were tested successfully: Solvay-Protease M 440 (solid form) and L 660 (liquid form) as well as the Novo enzymes Alcalase 2.0 T (solid) and Alcalase 2.5 L (liquid).

6.3.1.2 Parameter Optimization

Enzyme properties such as activity or selectivity strongly depend on the enzyme conformation and, therefore, can be influenced by all types of chemical and physical reaction parameters, which have a more or less pronounced effect on the enzyme structure. Typical parameters often applied in parameter optimization programs (and well-documented by numerous publications) are temperature, pH, concentration, and the addition of water-miscible and -immiscible co-solvents, salts or other additives. Our optimization activities were directed towards improving the key factors of the catalyst performance further with regard to: i) activity in order to minimize reaction time and enzyme costs, ii) selectivity, and iii) stability, so that the enzyme can tolerate high substrate concentrations. More generally, overall process costs should be reduced by providing robust process conditions including a facile work-up, which immediately pays back in terms of reduced processing times.

The starting point of the parameter screening was the relatively plain "standard" conditions used in our initial experiments described in note [21]. Since Subtilisin Carlsberg is known to tolerate high substrate concentrations [22], besides the 0.4% concentration, 7.2% was also tested. The initial experiments were already revealing excellent selectivity (>98% ee at 49% conversion (E>100 [23]). Therefore, no further additives such as co-solvents or salts were tested. Instead, the optimization was aimed at a simpler reaction system and technically more attractive conditions.

The nature of the aqueous buffer (0.1 M NaCl, 4 mM sodium phosphate buffer pH 8.5 versus 0.1 M NaCl, 4 mM sodium borate buffer pH 8.5) had no appreciable effect on the enzyme selectivity although the hydrolysis was slightly faster with the phosphate. The buffer composition was simplified by omitting sodium chloride, while at the same time the phosphate concentration was increased to 10 mM in order to compensate for the ionic strength and increase the buffering capacity of the system. The pH value was kept at 8.5 as the optimal compromise between highest enzymatic reaction rate and lowest reaction rate of the unspecific alkaline ester hydrolysis.

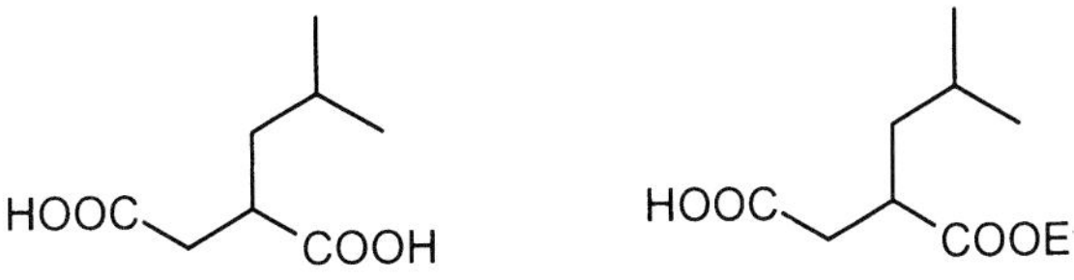

6.3 Enzymatic Resolution of Diethyl 2-Isobutyl Succinate 9

Fig. 3 Compounds **15** and **16**.

Owing to experience previously gained concerning work-up with the solid enzyme formulations [22] only the liquid enzyme preparations of Novo and Solvay, Alkalase 2.5 L and Protease L 660, respectively, were used in the following studies. Both preparations exhibited nearly the same specific activity in the present reaction. An enzyme concentration of <5% (v/w) with respect to **9** ($s/e > 20$ [24]) was used in the subsequent experiments of this section.

Under these newly introduced conditions, and at an increased substrate concentration of 10% (w/w), the product was obtained in high enantiomeric excess (>99%), purity (>98% GLC; silylated) and yield (s [25]). Less than 1% of the undesired diacid **15** was formed [26], and the alternative monoacid **16** could, if at all, be observed only in <0.2% (GLC, silylated) (Fig. 3).

For reasons of technical attractiveness we exploited even higher, physiologically completely unnatural substrate concentrations. As a result, the enzyme retained its excellent stereoselectivity even at 20% (w/w) substrate concentration, however, the reaction time for completion of the reaction exceeded two days.

Usually, in addition to the parameters influencing the enzyme properties, process-engineering parameters such as the degree of conversion (cf. [23]) or re-use of the catalyst also have to be evaluated in enzymatic transformations. Conveniently, the enzyme selectivity was so high that, even at a degree of conversion close to 50%, the product (R)-**2a** was generated in high enantiomeric excess. Re-use of the enzyme by immobilization onto a support or by retention in a membrane reactor was not considered as Subtilisin Carlsberg is an extremely cheap bulk protease [27] and was discarded after use. Assuming an s/e-ratio of ≥ 10 and a yield of 44% the maximum enzyme costs would amount to ca. US $3–5 per 1 kg of (R)-**2a**.

6.3.1.3 Substrate Engineering

In parallel to the diethyl ester **9** the corresponding dimethyl ester was also tested as an alternative substrate for the enzymatic reaction [28]. Compared with **9**, however, the enzymatic hydrolysis slowed down to only 44% conversion after 74 hours and afforded the desired monoacid (R)-**2b** in merely 94% ee along with 6.7% (GC) of the undesired monoacid (corresponding to **16**). Consequently, this approach was no longer considered.

6.3.1.4 Work-up

On the laboratory-scale the reaction products were easily separated by means of conventional extraction with ethyl acetate at neutral and acidic pH. Phase separation in the neutral extraction steps occurred almost instantly. Only in the first of the acidic extraction steps did the aqueous phase remain partly turbid and cloudy. The handling of this problem on the larger scale is described in Sections 6.3.2.2 and 6.3.3.3.

6.3.1.5 Racemization of the Antipodal Diester (S)-9

The present racemic resolution is particularly attractive because it allows easy racemization of the unwanted (S)-diester back to its racemic precursor simply by distillation in the presence of a catalytic amount of NaOEt (1%), which represents a minimum additional effort (see Section 6.3.2.3).

6.3.2 Process Development

The main issues of process development were:
i) the introduction of a conversion control independent of the consumed amount of titrating agent [29],
ii) the reduction of the reaction time to 48 hours,
iii) the further investigation of the effect of the various process parameters, particularly pH and temperature on activity and selectivity as required for process validation, and finally,
iv) the implementation of the product isolation on the larger scale.

6.3.2.1 Process Parameters

Buffer: Again optimization experiments did not reveal an appreciable dependency of the enzyme activity on the nature of the buffer solution (sodium hydrogen phosphate versus sodium hydrogen carbonate). Thus, cheaper sodium hydrogen carbonate was chosen, and, in view of the 20% substrate concentration going to be employed, the buffer concentration was increased to 30 mM in order to prevent local over-basification.

Selection of enzyme: A comparison of the two enzyme preparations Alcalase 2.5 L Typ DX and Protease L 660 revealed a very similar behavior for this biotransformation. With Alcalase the degree of conversion was slightly lower, but product quality was slightly better (Tab. 2). Rating the slightly higher yield as more valuable, further development was continued with Protease L 660. Two different commercial batches of Protease L 660 gave identical results (Tab. 2).

Temperature and pH: As expected, both higher reaction temperature and higher pH led to a higher degree of conversion within 48 hours, but in turn favored the formation of diacid **15** and slightly affected the enantiomeric excess (results not

Tab. 2 Comparison of commercial enzyme preparations.

Parameter	Protease L 660		Alcalase 2.5 L Typ DX
Lot No.	092.897-692	39067	PMN 4536
Conversion (48 h) (%)	45.8	45.8	43.3
Content of (R)-**2a** (%)	97.8	98.6	99.2
Content of **15** (%)	0.67	0.69	0.32
ee (%)	99.4	99.4	99.6
Chem. yield (%)	44.7	45.0	42.9

Conditions: 20% (w/w) **9** in 30 mM $NaHCO_3$, 8.6% (v/w) enzyme solution with respect to **9**, 2.0 M NaOH, pH 8.5, stirrer 440 rpm, EtOAc extraction.

shown). Therefore, the target value for the reaction pH remained between 8.0 and 8.5 at a temperature of 20–25 °C.

Co-solvents: The search for a co-solvent enhancing the enzyme activity was unsuccessful (results not shown).

6.3.2.2 Work-up

For reasons of solvent harmonization with other reactions ethyl acetate was replaced by toluene in the extraction steps (cf. experimental procedure in note [30]). The third washing step to remove the retained diethyl ester (S)-**9** from the hydrolysis mixture appeared unnecessary on the lab-scale, but was recommended as a safety measure against carry-over [content of (S)-**9** in the third extract: 0.05%]. During scale-up, however, it turned out that even four extraction steps resulted in traces of (S)-**9** occurring in the isolated product, due to a microemulsion of toluene droplets, which remained in the aqueous phase (cf. Section 6.3.3.3). The product (R)-**2a** was recovered from the aqueous phase by two toluene extractions at pH 2.0. In the first acidic extraction step the emulsified part of the extraction mixture was centrifuged together with a filter aid (27% Dicalite Speedex with respect to product). The centrifugation was applied to remove a fine, film-like precipitate formed upon acidification (probably denatured enzyme), which prevented complete phase separation. Simpler variants such as pressure filtration or centrifugation of the whole mixture were less reliable in separating the precipitate. Azeotropic removal of water during concentration of the toluene phase allowed the (optional) use of the concentrated product solution directly in the subsequent amidation step. In contrast to the extraction with ethyl acetate the extraction with toluene also allowed to deplete the diacid **15** slightly, which was the only appreciable by-product (0.3% w/w) of the reaction. The results of the modified product isolation procedure are given in note [30].

6.3.2.3 Racemization

Deviating from the original proposal, the racemization of concentrated anhydrous (S)-**9** was conducted in toluene [1:1.5 (w/v)]. Using 5 mol-% of EtONa as a commercially available solution (21% in EtOH), at 50°C the racemization was completed after 60 min. The reaction mixture was washed with aqueous sodium hydrogen carbonate and water. The organic phase, optionally, can be combined with the respective toluene solution of **9** obtained from the preceding hydrogenation/ethanolysis step applied on **6**. Concentration and distillation of the crude product afforded the recycled product in 93% yield and a content of >99%.

6.3.3 Pilot Production

6.3.3.1 Equipment

As outlined in Section 6.3.2.1, temperature, pH range, and reaction time have a strong influence on the outcome of the reaction. At the same time these parameters required a defined profile of the multi-purpose plant to be chosen. Owing to the work-up procedure, it had to be equipped with two reaction vessels and a centrifuge to separate the film-like precipitate, and it had to be comparatively large. As sulfuric acid was not suitable for acidification a steel-enamel version was required. Important specific requirements were a by-pass with a pH electrode and a dosing system with a fast response rate to keep the retardation of or for the addition of the titrating agent minimal.

In this section the results and experience of the first runs on the 200 kg-scale are given. Process Development conducted three runs according to the procedure [30]. The following devices were selected for the campaign: a glass-lined reactor and crystallizer (4000 L each) equipped with a bottom-discharge centrifuge and an automated (and, optional, manual) pH control with dosing from a mobile titrating agent tank.

6.3.3.2 Enzyme Reaction

The enzymatic reaction ran without complications. In spite of the automated control of the pH, the reaction was supervised full time in shift-work. Control of pH and temperature worked smoothly and after 48 hours the three batches showed degrees of conversion as observed in Process Development (Tab. 3).

Tab. 3 Large-scale batches according to the Process Development procedure [30]: hydrolysis for 48 hours at 22–25°C and pH 8.0–8.5.

Run No.	Substrate 9 (kg)	Protease L 660 (kg)	Conversion (%)	Product (kg)	Yield (%)
001	196.6	18.4	44.1	75.0	43.4
002	198.8	18.6	43.2	52.2[a]	30.2[a]
003	198.6	18.4	44.4	73.7	42.6

a) Product loss due to incident.

Tab. 4 Product quality of the large-scale batches.

Batch	Product (R)-2a (%)	GLC/HPLC			KF	
		Diacid 15 (%)	Monoacid 16 (%)	Substrate (S)-9 (%)	Water (%)	ee (%)
001	91.7	0.16	0.05	1.81	0.017	99.3
002	96.0	0.14	0.04	0.28	0.019	99.3
003	92.4	0.17	0.04	0.22	0.018	99.3

6.3.3.3 **Work-up**

The film-like precipitate formed upon acidification (cf. Section 6.3.2.2) could be easily removed by the addition of Dicalite Speedex and centrifugation. The production of one batch required four days. The results from Process Development could be essentially reproduced.

The product quality was assessed by TLC, GLC, and HPLC (Tab. 4). Isolated (R)-2a was obtained as a nearly colorless oil with residual toluene as the main contaminant. Since toluene was used in the next reaction this was not an issue to be concerned about. The main deviation between the 400 g process development batches and the first 200 kg plant batch was the unexpected occurrence of diester (S)-9 in the isolated product (cf. Tab. 4). The high content of (S)-9 in the first batch 001 (Tab. 4) was simply caused by the inefficient stirring device of the vessel used for the extraction. Better results with 0.22% and 0.28% of (S)-9 in the isolated product (Tab. 4, batches 002 and 003) were obtained when the extraction was performed in the reaction vessel with its more efficient stirring device combined with longer stirring and settling times, and the application of four rather than the recommended two to three extraction steps. As the project was discontinued at that time, the influence of trace impurities of (S)-9 in (R)-2a on the purity of the final drug compound **1** had not been investigated and the further optimization of the extraction procedure was abandoned.

Owing to their good degradability, the wastewater fractions containing the enzyme were directed to the sewage works. The Dicalite residue (containing the bulk of denatured enzyme) was dried and delivered for waste disposal.

6.4 Discussion

Between May 1993 and October 1994 an enzymatic racemic resolution step was developed as part of a new technical synthesis from the lab- up to the 200 kg-scale. The choice of the biotransformation in question was predominantly determined by the extremely easy and cheap access to the substrate, and, secondly, by the high selectivity and easy working of the biotransformation itself. The biotransformation was a hydrolysis, the most frequently seen reaction type in the synthetic application of enzymes and technically particularly simple. As the employed enzyme was a laundry enzyme, high substrate concentrations were accepted. From the very start the enzyme exhibited high selectivity and good activity at high substrate concentrations. At a very early stage of the project we were convinced that we had a cheap, facile, robust and hence technically feasible reaction to hand, providing us with an inexpensive access to the key building block (R)-2a. Because of the attractive initial results of the enzyme reaction, optimization work concentrated not so much on an improved performance but on simpler reaction conditions. On the lab-scale, kg amounts of the target product could already be prepared. Therefore, and due to the comparative simplicity of the later development stages, the enzyme reaction was never a limiting factor in the development.

A key factor to the success was that the substrate possessed crucial structural elements in common with the natural substrates of the enzyme, bulky side-chain amino acids. Since cheap commercial sources for the enzyme were available on the market the enzyme could be discarded. Also adding to the process was the fact that the unwanted retained enantiomer could be racemized and recycled by a simple procedure.

The present enzymatic reaction, however, never entered the production-scale as the target molecule, collagenase inhibitor RO0319790, finally failed in clinical evaluation.

6.5
Acknowledgements

We thank Patrik Stocker, A. Cueni, Francois Gantz, Sylvie Courbet, Daniel Strub, Roger Stäubli and Peter Salathé for their skilful technical assistance, Willy Walther and Philippe Schmidlin for determining the ee-values and our colleagues from the analytical services for running numerous samples.

6.6
References

1 K.M. Bottomley, W.H. Johnson, D.S. Walter, J. Enzy. Inhibition 1998, 13, 79–101.
2 M. Brewster, E.J. Lewis, K.L. Wilson, A.K. Greenham, K.M. Bottomley, Arthritis Rheumatism 1998, 41, 1639–1644.
3 P.A. Brown, W.H. Johnson, G. Lawton, EP0497192 A2 1992.
4 M. Soukup, B. Wirz, EP0664284 A1 1995.
5 B. Wirz, M. Soukup, Tetrahedron Asymmetry 1997, 8, 187–189.
6 S.G. Cohen, A. Milovanovic, J. Am. Chem. Soc. 1968, 90, 3495–3502.
7 M. Colombo, M. de Amici, C. de Micheli, D. Pitre, G. Carrea, S. Riva, Tetrahedron Asymmetry 1991, 2, 1021–1030.
8 B. Poldermans, in Enzymes in Industry (W. Gerhartz, Ed.), chapter 4.4, Proteolytic Enzymes, VCH Verlagsgesellschaft GmbH, Germany 1990.
9 Solvay Enzymes GmbH (formerly Miles Kali-Chemie), now Solvay Deutschland GmbH (Hannover, Germany).
10 Novo Nordisk a/s now Novozymes (Bagsvaerd, Denmark).
11 R.L. Gu, C.J. Sih, Tetrahedron Lett. 1990, 31, 3283–3286.
12 R. Conrow, P.S. Portoghese, J. Org. Chem. 1986, 51, 938–940.
13 J.P. Barnier, L. Blanco, E. Guibé-Jampel, G. Rousseau, Tetrahedron 1989, 45, 5051–5058.

14 E. GUIBÉ-JAMPEL, G. ROUSSEAU, J. SALAÜN, J. Chem. Soc., Chem. Commun. 1987, 1080–1081.

15 J. SALAÜN, B. KARKOUR, J. OLLIVIER, Tetrahedron 1989, 45, 3151–3162.

16 E. SANTANIELLO, P. FERRABOSCHI, P. GRISENTI, F. ARAGOZZINI, E. MACONI, J. Chem. Soc., Perkin Trans. I 1991, 601–605.

17 B. WIRZ, P. SPURR, Tetrahedron Asymmetry 1995, 6, 669–670.

18 S. T. CHEN, S. Y. CHEN, K. T. WANG, J. Org. Chem. 1992, 57, 6960–6965.

19 The absolute configuration was not determined.

20 L. ABRAHAMSEN, J. TOM, J. BURNIER, K. A. BUTCHER, A. KOSSIAKOFF, J. A. WELLS, Biochemistry 1991, 30, 4151–4159.

21 Diester **9** (2.00 g) was emulsified in 0.1 M NaCl, 4 mM sodium phosphate buffer pH 8.5 (26 mL) by vigorous stirring. The reaction was begun by adding Alcalase 2.0 T (200 mg). The pH was kept constant by the controlled addition (pH-stat) of 2.0 M NaOH under vigorous stirring. After 49.6% conversion (22 hours) the reaction mixture was washed with ethyl acetate (3×50 mL). The aqueous phase was acidified to pH 2.0 and extracted the same way to give (R)-**2a**.

22 S. DOSWALD, H. ESTERMANN, E. KUPFER, H. STADLER, W. WALTHER, T. WEISBROD, B. WIRZ, W. WOSTL, Bioorg. Med. Chem. 1994, 2, 403–410.

23 C.-S. CHEN, Y. FUJIMOTO, G. GIRDAUKAS, C. J. SIH, J. Am. Chem. Soc. 1982, 104, 7294–7299.

24 s/e designates the ratio of substrate to enzyme used (w/v).

25 Diester **9** (256 g, 1.11 mol) was emulsified in 10 mM sodium phosphate pH 8.5 (2.5 L) under vigorous stirring and the pH adjusted to 8.5. Hydrolysis was started by adding Protease L 660 (12 mL) and the pH kept constant under vigorous stirring by the controlled addition of 2 M NaOH (pH-stat). After 48.7% conversion (47 hours) the reaction mixture was washed with ethyl acetate (3×1.5 L), acidified to pH 2.0 (25% HCl) and extracted with ethyl acetate (2×1.5 L). 105.5 g of monoacid (R)-**2a** (522 mmol, 47%) was obtained as a colorless oil.

26 It could be demonstrated that diacid **15** at a content of 1.95% did not interfere with the subsequent reaction sequence.

27 The price for 1 kg (~1 L) Protease L 660 was ca. US $26 (1994) and for 1 kg (~1 L) Alcalase 2.5 L Typ DX ca. US $14 (1991).

28 Conditions: 9% (w/w) dimethyl ester in 0.1 M NaCl, 4 mM sodium phosphate buffer pH 8.0 with Protease L 660 ($s/e=20$).

29 At higher pH the titrimetric results can be slightly biased by the dissolution of carbon dioxide into the reaction mixture, leading to a higher apparent (gross) degree of conversion.

30 Diester **9** (400 g, 1.719 mol, content 99.0%) was emulsified in 30 mM sodium hydrogen carbonate pH 8.7 (1550 g) under vigorous stirring and the pH adjusted to 8.5. Hydrolysis was begun by adding Protease L 660 (37.13 g; flask rinsed with 50 g of the above buffer) and the pH kept constant under vigorous stirring (ca. 350 rpm) at 21 °C by the controlled addition of 2.0 M NaOH (pH-stat). After 48 hours and a consumption of 403.35 mL NaOH solution (45.9% conversion, HPLC) the reaction mixture was washed with toluene (900, 200, and 200 mL), acidified to pH 2.0 (89.47 g of 11.65 M HCl) and extracted with toluene (600 mL). The upper turbid toluene phase (after 30 min; still containing water) was homogenized with a suspension of Dicalite Speedex (42.0 g) in toluene (320 mL) and the mixture centrifuged (220 cm^2 filter cloth Dacron DA 50, 28 min). The combined aqueous phases were extracted with toluene (200 mL) and the toluene phase centrifuged through the Dicalite residue. The residue and the devices were rinsed with toluene (in total 100 mL) and the combined toluene phases (comprising 62 mL aqueous phase of the second centrifugation) filtered through a filter plate (9 cm diam.; FibraFix AF6, Filtrox-Werk AG). The filter was washed with toluene (50 mL). The combined toluene phases were washed with water (180 mL) and evaporated (19–58 mbar, bath 55 °C). The residue was dried (14 mbar/23 °C) for 8 hours under stirring to give 157.91 g (45.4%) of (R)-**2a** in 99.4% ee as a yellowish oil (content 98.7%, 0.27% **15**, 0.14% toluene, 0.12% water).

7
An Innovative Asymmetric Sulfide Oxidation: The Process Development History Behind the New Antiulcer Agent Esomeprazole[1]

Hans-Jürgen Federsel and Magnus Larsson

Abstract

The proton pump inhibitor Losec®/Prilosec®, which uses the racemic sulfoxide omeprazole as the active ingredient, has been an extraordinary success story ever since the first introduction into the market in 1988. However, there was room for improvement with respect to its clinical profile, for example to reduce inter-patient variation by increasing the bioavailability and this led, after conducting a thorough screening investigation, to the identification of the (S)-enantiomer of omeprazole as a compound fulfilling these goals. In order to conduct more comprehensive studies on the properties of this molecule, later to be named esomeprazole, larger quantities of material were needed and so the first attempt was to separate the stereoisomers in omeprazole using preparative chromatography. Albeit this approach was rather labor intensive, it afforded the required amounts of several hundred grams of each enantiomer in satisfactory purity. For further scale-up this methodology had some clear drawbacks so the focus was aimed at developing an asymmetric synthesis. When applying the procedure reported by Kagan using titanium as the catalyst together with a chiral auxiliary, such as tartaric acid esters, in the presence of H_2O to our prochiral sulfide pyrmetazole, the outcome was disappointing as the isolated product had an optical purity of only 5% ee. Nonetheless our endeavors continued and, very unexpectedly, we found that by adding a base such as diisopropylethylamine to the reaction mixture the ee increased immediately to around 60%. Further work resulted in a full-scale catalytic process transferred into commercial production operating on the multi-ton volume per annum and providing the desired esomeprazole product in chemical as well as optical yields well above 90%. The following describes the details behind this achievement, starting with the background and proceeding via the earliest attempts onto the details of how the work in the area of enantioselective catalysis was conducted and its further progress to the final unique method of manufacture.

[1] Partly presented in lecture format (orally) at the following conferences and symposia: ChiraSource 2000 in Lisbon, Portugal, October 2–4, 2000; Chiral Europe 2000 in Malta, October 26–27, 2000; 18th SCI Process Development Symposium, Cambridge, UK, December 11–13, 2000; ACS ProSpectives Conference Process Chemistry in the Pharmaceutical Industry, Barcelona, Spain, February 24–27, 2002; 14th International Symposium on Chirality, Hamburg, Germany, September 8–12, 2002.

Asymmetric Catalysis on Industrial Scale: Challenges, Approaches and Solutions
Edited by H. U. Blaser, E. Schmidt
Copyright © 2004 Wiley-VCH Verlag GmbH & Co. KGaA, Weinheim
ISBN: 3-527-30631-5

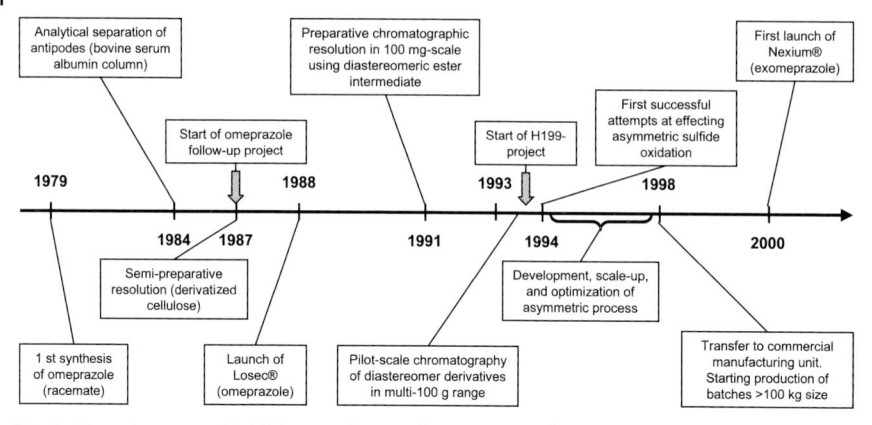

Fig. 1 Some important highlights on the way from omeprazole to esomeprazole.

7.1
Major Events at a Glance: An Overview of Achievements from 1979–2000

The story about to be told is essentially concentrated in the 1990s but its origins go back further, by more than two decades. At this point a group of researchers at Hässle in Mölndal (an R&D company belonging to the former Astra and localized outside Gothenburg on the west coast of Sweden) started work on new ways to treat acid-related disorders in the gastro-intestinal tract. The crowning of their endeavors was the first synthesis of omeprazole (a racemic sulfoxide), which was followed more than 20 years later by the launch of esomeprazole, its (S)-enantiomer, in the form of a new product named Nexium® (see Fig. 1). The following is an account of how the synthesis of the latter was developed and transferred into a manufacturing process for commercial production (under the project acronym H199).

7.2
Introduction: Omeprazole (Losec®) as the Starting Point for a Challenging Project

With the long-awaited launch of Losec®, after receiving the first market approval in 1988, it is no exaggeration to state that the treatment of acid-related disorders (e.g., peptic ulcers) underwent a dramatic change and provided the patient with a healing opportunity of unsurpassed efficiency. This year also marked the starting point of a success story for a product that within a number of years climbed to the very top of the world-ranking list of most sold prescription drugs displaying peak sales figures of a stunning $6000 million per annum in the late 1990s! The Active Pharmaceutical Ingredient (API) omeprazole, a moderately complex molecule first synthesized in 1979, represents the culmination of a project that was initiated back in the 1960s at the research company Hässle in Mölndal. A sophisticated *in vivo* screening in dogs was applied in the search for potent compounds,

Fig. 2. Structural formula of omeprazole.

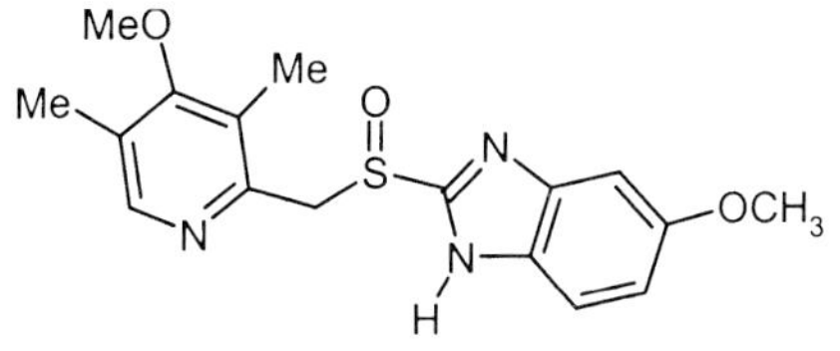

Omeprazole

and it was not until later (1977) when the key biological target was unraveled as being the unique H^+-, K^+-ATPase, which is an enzyme residing in the parietal cells of the gut and is solely responsible for eliciting protons (acid) into the lumen. Compounds were screened for their ability and potency to inhibit gastric acid secretion in the dog model, work that resulted in the selection of omeprazole (Fig. 2) as the ultimate candidate drug (CD). Because of the fact that the compounds inhibit this particular enzyme the whole class of products originating from these research endeavors has consequently been given the illustrative name proton pump inhibitors (PPIs), and indeed constitute a very significant advancement in the GI (gastro-intestinal) area, allowing severe medical conditions to be treated appropriately [1]. Their mode of action at a molecular level has been established in detail and displays a fascinating series of molecular transformations before the active species interacts with and binds to specific mercapto groups and thereby inhibit the enzyme [2].

Once the outstanding pharmacological properties of omeprazole were fully understood and the drug product had started its victorious journey into the global marketplace, it became apparent that a strategy addressing the need for follow-up compounds was required. It was vital to address weaknesses that had been identified and provide the public with a product that would offer noticeable benefits over the already existing highly efficacious ones delivered by Losec®/Prilosec® (trademark in USA). Thus in 1987 Hässle (today AstraZeneca R&D Mölndal) finally defined the goals for such a new acid inhibitor research program and which was to reduce the variation in acid inhibition between patients. This was likely to be achieved by a compound with increased bioavailability. The chemical approach was to change the substitution pattern on the pyridine and benzimidazole rings. In this search for new chemical entities (NCEs) that could match the requirements, a whole range of molecules that were analogues of the parent structure were synthesized. This again clearly showed the rather unique features of omeprazole and underpinned the fact that finding a follow-up would indeed be an extremely tough challenge. Nonetheless a number of compounds managed to pass the initial pre-clinical tests to allow them to be nominated as new CDs, two examples of which are shown in Fig. 3 [3].

Later on, one of the single enantiomers of omeprazole, esomeprazole, was selected as a CD. This may seem strange and poses the question as to why had the single enantiomers not been investigated at an earlier stage? Omeprazole displays

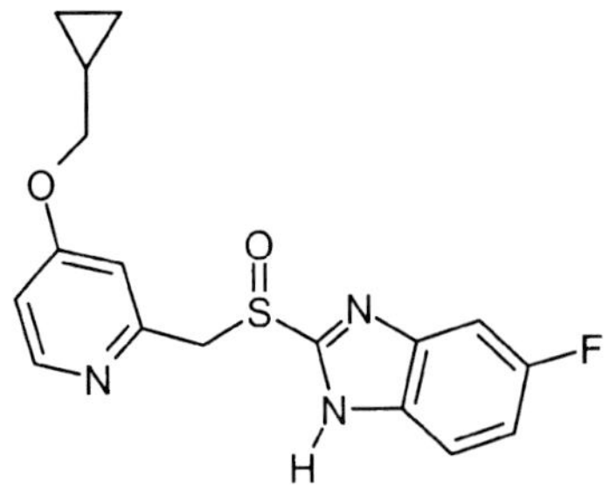

Fig. 3 Examples of post-omeprazole CDs.

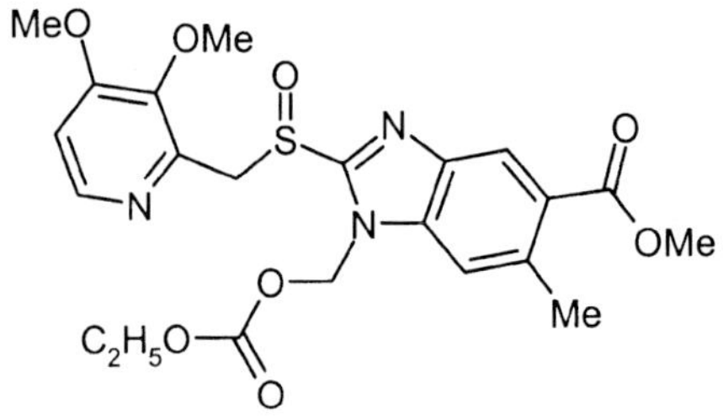

chirality by virtue of its sulfoxide functionality and a couple of analytical separations of the two isomers were performed during the 1980s. However, early on a single isomer was not considered for development. The reason being that the accepted mechanism of acid inhibition led to the prediction that both isomers of omeprazole would have exactly the same effect at the site of action because acid-catalyzed conversion of either isomer to the same *non*-chiral sulfenamide species would occur at the same rate for both isomers. Thus while it was known that the isomers of chiral drugs often have different activities [4], the logical assumption was that this was not the case here. Furthermore, before 1990 only milligram amounts of the single isomers had been prepared, which was not enough for *in vivo* testing. For these reasons the "single-isomer concept" did not look very attractive. However the Hässle scientists could not turn away from the idea of a possible difference in metabolism between the two isomers.

Initially the efforts were merely focused on *structural* analogues of the parent molecule but gradually shifted over to *stereoisomers*. Once esomeprazole, the (S)-isomer, had been selected for development a close and intensive collaboration was established between Medicinal Chemistry in Mölndal and Process R&D Södertälje, in order to be able to meet the expectations from the business to more or less guarantee the successful and timely delivery of a robust and commercially viable manufacturing process. Suffice to say that the launch of Nexium® in 2000 as the first PPI in enantiomerically pure form is testimony to the achievement of this goal, as well as that of the entire global project team succeeding in bringing a new and medically important product to the market [5].

7.3
Early Attempts: Turning Unfavorable Odds into Success at Last

Switching back to the early days at the beginning of the 1990s when the hunt for a suitable omeprazole-successor was still vigorously being pursued, the pre-clinical evaluation including *in vivo* testing, required access to at least 100 mg of each investigated compound. In the case of the stereoisomers the task was addressed in a way that must be regarded as most straightforward under the circumstances, since the racemate was abundantly available (commercial production on a multi-ton scale), namely by applying a conventional resolution procedure. However, the rather extreme acid sensitivity of omeprazole, $t_{1/2} \approx 2$ min at pH 1–3 [1, 5] (a quite intriguing property of a compound displaying such outstanding inhibition of proton release in the gut) prevents a simple precipitation of a diastereomeric acid-addition salt. Instead a circumventory derivatization procedure was developed enabling a chiral handle to be covalently attached before performing the actual separation. Thus following the very first approach (in 1984) where the enantiomers had been directly resolved analytically [6] and later on also semi-preparatively on a milligram scale [7] after chromatography on a chiral stationary phase, this derivatization procedure provided the required gram quantities needed to identify the important biological difference between the two enantiomers. This procedure also offered a considerably more attractive alternative from a scale-up perspective, since a diastereomeric mixture could be successfully separated on a chromatographic system operating in an achiral mode [8]. As the delivery of the first large scale batch to satisfy the requested amount of 500 g of each enantiomer was critical and it was judged that developing an entirely chemistry-based process (not including chromatography) would be too time consuming to be seriously considered, the initial approach had to be based entirely on this separation methodology that had proven to be operable in the low gram range. How this challenging task, a solution to which was far from obvious, was accomplished will be the theme of the next section.

Derivatizing any given racemic substrate should preferably be characterized by as few chemical transformations as possible and the steps involved need to operate in very high yields (quantitative if possible) to avoid unnecessary losses of valuable material. After conclusion of the successful separation, the reverse order manipulations required to lead back to the target (now in an optically enriched form) should be simple and selective, so that no degradation occurs leaving a virtually pure material after work-up not contaminated by residues of reagents and other chemicals utilized. In the present case this meant running a three-step synthesis where the benzimidazole nitrogen was substituted with a hydroxymethyl group via reaction with formaldehyde followed by a chloro-dehydroxylation ($CH_2OH \rightarrow CH_2Cl$) and a final O-alkylation (yielding an ester) with L-mandelic acid to produce a mixture of two diastereomers as shown in Fig. 4.

There are two key features in the sequence leading up to the diastereomeric mixture: (i) the equilibrium in the first step between the starting material and N-hydroxymethyl adduct and (ii) a problem of regioisomers caused by the presence of two nitrogen atoms (in the benzimidazole moiety) of similar basicity. The former phenomenon has the effect that when applying a one-pot procedure with no work-up and isolation of the $N\text{-}CH_2OH$ species before adding thionyl chloride, then the overall yield will be extremely poor at a level of <15% for the first two steps. For an explanation of this a reference back to the previously mentioned acid sensitivity of the omeprazole molecule is appropriate, since the equilibrium mixture after the formaldehyde reaction contains considerable quantities of starting material, which is then rapidly degraded as a consequence of the low apparent pH in the solution due to the

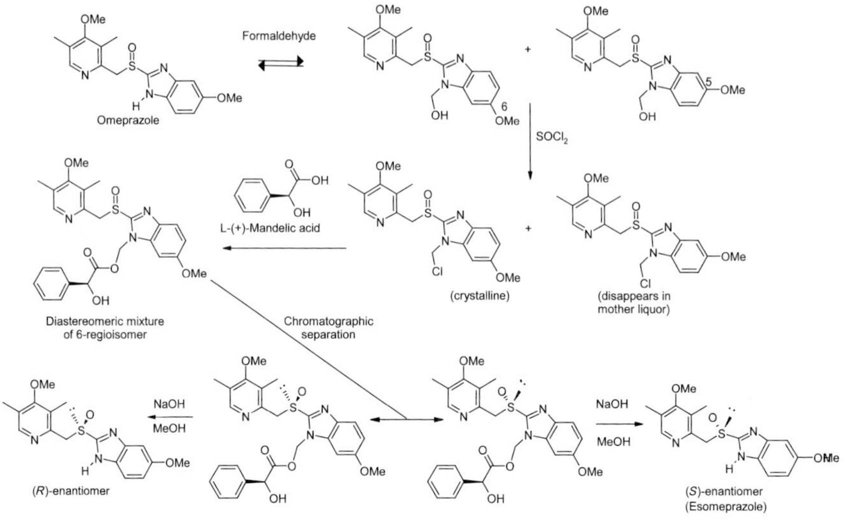

Fig. 4 Diastereomeric resolution path via chromatographic separation to omeprazole enantiomers.

generation of strongly acidic byproducts (HCl, SO$_2$) from the use of SOCl$_2$. The second problem stems from the fact that both imidazole-N^1 and -N^3 display reactivity against formaldehyde that only partly discriminates between them and hence a mixture of 5- and 6-methoxy regioisomers (the nitrogen atom thus substituted will always be denoted as number 1 which causes the OMe-group to be in either position 5 or 6) is formed in roughly a 1:3 ratio [9]. Both are equally valuable in the further processing but as it turned out only the 6-isomer was obtained as a crystalline material whilst the 5-analogue remained in the mother liquor. Fortunately this does not lead to any other major problems besides having to accept a rather severe yield penalty, which in our first scale up (performed according to a scheme of two identical runs/step) resulted in the isolation of an average yield of not more than 13.6% (19.2 and 8.0%, respectively, indicating a non-robust process) at 95 area-% purity (HPLC) of the 6-Ome-N^1-CH$_2$Cl intermediate. In the final step the crude product was used as an alkylating agent in order to convert enantiomerically defined mandelic acid (L-configuration; >99% ee) to afford a diastereomer mixture in an acceptable and reproducible 70% yield. Overall, starting from 40 kg omeprazole the three-step procedure thus described was operated at a 250–500 L pilot scale and rendered some 5.3 kg (~9% yield, 93% pure) in total of the mandeloyl-derivative with an invested effort of almost 300 man hours (total cycle time) spread over a period of 40 days! This was a considerable resource input just up to this early stage but clearly justifiable under the given circumstances where time to delivery was by far the most important success factor.

Having access to the targeted amount of diastereomers it was now time to separate them chromatographically. Small scale-preparative experiments had shown that reverse phase conditions on an ordinary C$_{18}$-derivatized silica stationary phase would be suitable for this purpose using CH_3CN-ammonium acetate buffer pH 7 (60:40) as eluant. Thus with an in-house medium-sized HPLC column (Ø 15 cm, length 1 m) at our disposal we set out to separate the crude product into its stereoisomeric components. All in all 430 (!) injections onto the column loaded with 3 kg of solid support were required, each amounting to about 12 g and the run time per cycle was roughly 2 hours. The total eluant volume thus collected was about 10 m^3, but one slightly complicating factor was the relatively low solubility of the starting diastereomer mixture and this feature forced us to use a more CH_3CN-rich solvent-buffer system (70:30) just to enable a high enough concentration to be injected (2 m^3 were used for this purpose). Furthermore, the instability of the fractions collected under the operating conditions (NH_4OAc buffer at pH 7) towards hydrolysis (10% degradation within 7 hours in the current eluant) and, more seriously, sulfoxide stereomutation (loss of 15–20% ee in only 16 hours), a known property that is explained by reversible rearrangement to sulfenic acid [10], required an expedient work-up as soon as proper analyses showed which fractions to combine. This was affected by adjusting to pH 11 using NaOH and then allowing the mixture to stir for a few hours at room temperature in order to cleave off the appendage in the form of a mandelic acid residue and formaldehyde to liberate the two omeprazole enantiomers. Acidification to transform the product into its neutral form and extraction into an organic solvent (CH_2Cl_2), followed by a final evaporation to dryness finished off this extremely tedious and labor-intensive exercise. Based on the crude mixture that was actually submitted to chromatography the yield of each enantiomer was about 65%.

To summarize, the endeavor just described afforded about 900 g each of the two stereoisomers as oils at an optical purity of around 96% ee, an achievement that at the time was felt to be absolutely fabulous. In terms of isolated yield, however, the outcome was not that brilliant as only 4.5% (!) based on the theoretical content of the respective isomer in the 40 kg of racemic starting material was able to be retrieved. This was not the end, as because the product was needed in an H_2O-soluble form it had to be converted into its Na-salt, a step that at this stage functioned rather poorly. Hence the amount shrunk to almost half so that it was barely possible to meet the demand for 500 g of (S)- and (R)-enantiomer, respectively. The good thing however was that the oily neutral forms were converted into crystalline salts, which in addition also improved their optical quality. Being absolutely delighted that these precious materials were now ready for delivery and further use in other departments, an important question remained to be addressed, that of whether this procedure, ignoring its premature developmental state, had the potential to be scaled up to a reasonably efficient and robust production process for commercial purposes. So whilst letting the staff unwind, after the pressures they had been under to prepare the desired quantities in a timely fashion and compile loads of data and accumulated experience to be able to draw the right conclusions for the continuation of this project, it was also necessary to start thinking in other directions aimed at identifying other potential approaches. This inasmuch as a new order was already on its way for 5 kg of pure enantiomer, ten times the quantity that had just been delivered.

7.4
The Way Forward: Options and Development Strategies

As mentioned in the previous paragraph the chromatography-based resolution had considerable weaknesses, no matter how attractive it might seem in the light of a virtually unlimited in-house access to racemic omeprazole. Yield was the obvious problem as a process of this sort could never offer more than the theoretical 50% (unless recycling of the wrong antipode followed by inversion to the desired stereoisomer would be included), but also the derivatization that had to be performed, the instability of the eluant fractions and the copious volumes to be handled (~ 12 m^3/kg enantiomer) were factors to be considered and evaluated. A challenging question to ask at this point in time was: would this process, even after development and optimization, suffice to produce 50 kg, 1 ton, or as much as 50 tons per annum? Based on the above it is fairly simple to imagine that we felt quite uneasy about the likelihood of succeeding. In order to drive the development forward it required some imaginative thinking.

The first step in this direction was to do the following two things:

- Starting from the target compound, perform a retrosynthetic analysis in order to identify possible new ways of constructing the molecule.
- Figure out a strategy that building on accumulated knowledge in the field would allow us to concentrate efforts on the most promising approaches.

The first item was really to find out if "new" chemistry was available that we did not have access to when developing the commercial omeprazole process in the 1980s. As it turned out no matter how unprejudiced our approach to the problem, really tackling it from all possible angles, we could not identify any other route that would be better suited for the synthesis of the molecule than the current one for omeprazole (see Fig. 5). In one sense this was pleasing as it showed that a good job had been done in the first place!

Focusing now on strategic options, again some intensive discussions led to the identification of three principally different approaches that needed to be carefully evaluated and compared on their respective merits and weaknesses:

- Asymmetric oxidation of the prochiral sulfide (pyrmetazol), the penultimate intermediate in the manufacture of omeprazole (see Fig. 5).
- Direct resolution of omeprazole without prior derivatization ("classical" salt formation).
- Resolution of omeprazole via covalent diastereomeric derivatives (in principle following a procedure described in Section 7.3 or variations thereof).

As the third option was one where considerable "hands-on" experience was available, our chromatography-based procedure definitely had to be kept "alive" in the sense that development resources were justified for improving the performance. However, a forecast of 5 kg production simply by linearly extrapolating data from the ½ kg batch, resulted in a really frightening scenario where considerable manpower (peak values of 20–30 people during certain stages) had to be allocated over

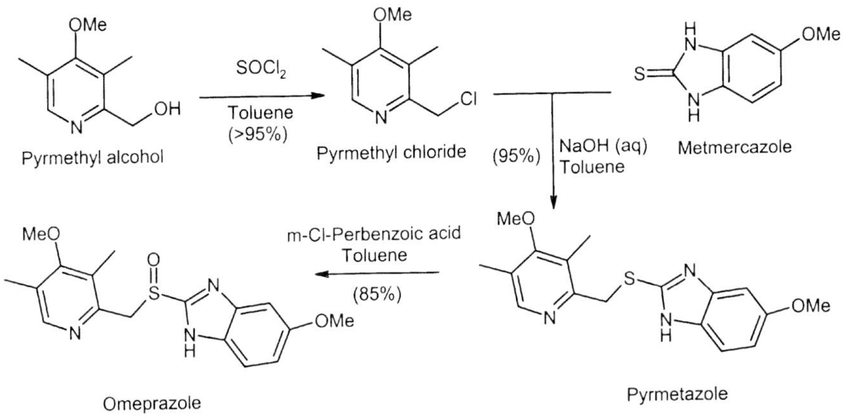

Fig. 5 The synthetic pathway to omeprazole.

a period as long as half a year. This was not a very pleasing outlook but to be prepared for this "worst case" at least the optimization of the two chemical steps up to the CH_2Cl-intermediate were seen as prioritized, inasmuch as the likelihood of improving on the low yields obtained in the just completed campaign (13% average yield in two runs), was regarded as very high. As it turned out this directed effort really paid off in the sense that by isolating the CH_2OH-adduct (which drives the equilibrium to the right) and then subjecting this crystalline material to chloro-dehydroxylation resulted in a considerable yield increase for the chloromethyl compound at 60–70%. In parallel, some speculative discussions on the direct resolution route were conducted and some interesting ideas presented, but given the unfavorable properties of the omeprazole molecule for successfully conducting such a process, it was decided unanimously to invest all efforts possible in the remaining alternative, namely to try to design an enantioselective sulfide oxidation. Running alongside each other, two sidetracks had been set up at a contractor where the possibility to conduct either a selective bioreduction of the racemate to leave behind the desired isomer [11] or to bio-oxidize the sulfide using various microorganisms [12] were evaluated. Although both routes were shown to be workable, the decision was taken not to develop them any further.

Over that past 30 years or so an enormous body of literature has grown up on all sorts of transformations presenting a truly variable outcome with regard to their capability to deliver stereochemically defined products. This is particularly so in the highly significant area of asymmetric oxidations [13], a subset of which is the conversion of prochiral sulfides, that is unsymmetrically substituted sulfur compounds, to the corresponding sulfoxides. Whilst it is rather fascinating to realize that the first ever stereocontrolled oxidation (diastereoselective fashion) was reported as early as in 1907 [14], an equally significant contribution to the field was presented in 1960, when two independent groups demonstrated that an enantioenriched sulfoxide could be prepared from an achiral aryl methyl sulfide, albeit in only 4.3% ee [15]. Shortly after, a preparatively feasible procedure was pub-

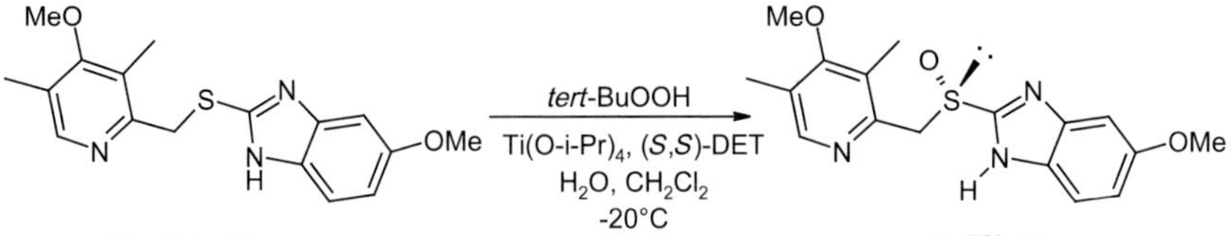

Fig. 6 Attempted enantioselective sulfide oxidation following the Kagan protocol.

lished by Andersen [16] built on a derivatization (via a Grignard reaction) and resolution protocol that allowed sterically defined sulfoxides to be synthesized. A large number of reports have been accumulated representing both chemo- and biocatalytic approaches, which are testimony to the tremendous advances in this area from then on [17]. Here we specifically mention the seminal work by the Kagan [18] and Modena [19] groups, who in the first half of the 1980s, building on the famous titanium-based Sharpless system [20], managed to obtain breakthrough results that really made the chiral sulfoxide area flourish.

Contrary to the Sharpless conditions these systems require either one crucial equivalent of H_2O (Kagan) or an extra excess of the chiral tartaric acid auxiliary (Modena), but their performance was really outstanding and provided the sulfoxide product in stereochemical purity ≥90% ee, e.g., using methyl p-tolyl sulfide as substrate (careful preparation of the catalyst even yielded stunning purities >98% [21]). However, when the steric bias becomes less pronounced, i.e., when going to larger and larger alkyl substituents (from methyl via ethyl to benzyl in the case just mentioned) the purities quickly drop off to very poor levels (from 90 via 74 to 7% ee). Evidently this reaction shows a pronounced sensitivity towards the steric bulk of the groups attached to the sulfur atom, with the big loss in ee appearing when these are of relatively similar size (p-tolyl and benzyl). As this molecular architecture resembles the one in the omeprazole sulfide, not much hope was put into achieving a successful outcome, and indeed when applying literature procedures [18] for experiments that had been already conducted in 1990, a disappointing result of only ~5% ee was obtained (see Fig. 6).

However, when we explored the Davis' chiral oxaziridines [22], another well-documented class of reagents to be used in asymmetric oxidation procedures, the situation improved dramatically. Thus the desired sulfoxide was obtained in as high a quality as 40% ee, which was seen as extremely encouraging especially since we had found that crystallization (of the racemate) increased the value to >94% ee (the desired product could be retrieved from the mother liquor) [23]. The inherent drawback of this transformation is that it operates in a stoichiometric way with regard to the oxidant which, given the high expenditure (even considering a possible recycling loop), precludes its usage for the current purpose (the issue is the cost of the goods).

7.5
The Breakthrough: What a Difference a Day Made!

There was a strong feeling that the ultimate aim for the development work should preferably be a catalytic (or semi-catalytic) process. In spite of the poor prospects of success when applying the procedures based on the Sharpless/Kagan methodology (Fig. 6), it was decided to continue for at least some time before abandoning this avenue. Thus a period of roughly three months was allocated to this investigation, then the progress was to be evaluated and a decision made on future directions. During this phase a group of four people (synthetic and analytical chemists) were devoted full time to the work, to ensure the best prospects for success.

Our efforts were amply rewarded and a real breakthrough was achieved when, very unexpectedly, finding that the addition of a base had a tremendous impact on the transformation to the desired sulfoxide enantiomer [24]. We now knew that our sulfide could in fact be oxidized in an asymmetric fashion so we were on the right track! Some facts for this crucial progress will be described in more detail below.

After a furious collection of primary laboratory data analyzed predominantly on chiral HPLC to determine the optical purity of the desired product, some very interesting and surprising results appeared that caused tremendous excitement and confirmed that our original hopes could indeed be satisfied.

- Addition of base had, as already mentioned, a pronounced effect and promoted higher ee values.
- The amount of H_2O was directly coupled to the stereochemical outcome.
- The choice of solvent has a notable impact on ee value.
- The ratio between $Ti(O\text{-}iPr)_4$/tartaric acid ester should be 1:2.
- Reaction time and temperature were significant but of less magnitude.

Some data and facts behind each of these main conclusions are briefly presented in consecutive order.

7.5.1
The Base

Running the oxidation in CH_2Cl_2 results in 23% ee in the absence of Hünigs base (after 200 min at 20 °C and a conversion ~70%) whereas addition of 0.29 equiv. renders a product with as high as 72% ee (180 min at 20 °C and almost 90% conversion). A further increase in the amount of base by a factor of about 3 (to 1.0 equiv.) only has a slight effect on the optical purity (78% ee). Under these conditions (i.e., not using the superior conditions discovered later regarding addition order and reaction time/temperature) other bases, such as for example triethylamine, diisopropylamine, piperidine, DMAP, 2,6-lutidine, and *N,N*-dimethylaniline, all gave a product with higher stereochemical purity compared with a reference reaction where the base was missing. Shifting to chiral bases, such as (R)- and (S)-N,N-dimethyl-1-phenylethylamine, did not offer any added value and use of extra strong bases, e.g., DBU and 1,1,3,3-tetramethyl-guanidine, instead gave a dramatic decrease in ee [25]. Later on a similar finding on the beneficial effect of amines in promoting enantioselective sulfide oxidations were reported by Modena et al. who used chiral trialkanolamines [26].

7.5.2
The Solvent

The Kagan methodology can only be operated practically in a small range of solvents, higher dielectricity constants leading to increased ee values [27]. Generally this is a severe and complicating limitation when designing a production process. However, much to our surprise we found that the oxidation now proceeded well in a wide range of solvents. Thus apart from chlorinated solvents used in the Kagan procedure, a series of solvents could be applied which included aromatics such as toluene, ketones, e.g., methyl isobutyl ketone, esters, e.g., ethyl acetate, and ethers, e.g., tetrahydrofuran (THF). Strongly polar solvents, for example DMF, on the other hand caused a sharp decrease in the stereochemical induction and so only 6% ee was noted in this particular case, which virtually brought us back to the starting point. One feasible explanation for this phenomenon could be that the solvation power is indirectly exerting a strong and pronounced effect on the catalytic Ti-complex with a concomitant dramatic loss in stereoselectivity.

7.5.3
The Chiral Auxiliary

Varying the ratio between Ti(O-iPr)$_4$ and the diethyl tartrate auxiliary using EtOAc as solvent from 1:1 to 1:4 did not have any impact on the ee value, but the kinetics of the reaction changed tremendously. This means that in the former case (at 20 °C) only 2.5% of the starting prochiral sulfide remains after 140 min, whereas in the latter this figure is 24.5%. Furthermore, comparing different chiral diols (e.g., other tartrates, (S)-1,2,4-butanetriol, 2,3-O-isopropylidene-D-threitol) actually showed that a range of structurally rather dissimilar auxiliaries could also be used. Because of market availability and other process-dependent considerations diethyl tartrate was finally chosen for the current application.

7.5.4
The Amount of H_2O

The crucial importance of water was already evident in the early Kagan investigations and is well documented in the literature [18]. The profound influence of this component was further corroborated in our case and depending on the specific design of the reaction system (type of solvent, catalyst loading, ratio Ti/auxiliary) the best results obtained were somewhere in the range between 0.3 and 1.0 equiv. H_2O/equiv. Ti.

7.5.5
The Effect of Oxidation Time

From early on the reaction time for the oxidation did not appear to have a particularly pronounced impact on the ee values and this was also confirmed. The total yield is of course closely coupled to the reaction time but this parameter was more conveniently studied at a later stage.

Now that a set of clear trends had been established, some major conclusions could be drawn from the data gathered by using the % ee purity of the product as the key reporter signal, but also including for example unreacted sulfide, overoxidized sulfone, and unwanted antipode:

- In CH_2Cl_2 using 1 equiv. Ti-catalyst, ee values increase by increasing the concentration (100–500 mM range) and to a lesser extent by higher temperatures (from –30 to 20 °C).
- Experiments in methyl isobutyl ketone with 1 equiv. Ti-catalyst indicated that the amount of H_2O, the ratio Ti/tartrate ester ligand, and the reaction temperature had a strong influence. The best relation between wanted and unwanted sulfoxide stereoisomers was obtained at 0.5 equiv. H_2O and with Ti/diethyl tartrate equal to 1:2 at 40 °C.
- Using toluene under the same conditions a more or less similar picture emerged. Interestingly the formation of sulfone is supported by higher amounts of H_2O and higher concentrations and reached >9% when these parameters were set at 1.4 equiv. and 900 mM, respectively.
- Reducing the Ti-catalyst to 0.5 equiv. and maintaining toluene as the solvent further confirmed the process set points as mentioned above (0.5 equiv. H_2O, Ti/tartrate 1:2 at 40 °C).

This then concluded our efforts aimed at designing an oxidative asymmetric transformation applied to our sulfide substrate. The understanding of the reaction had reached a much more advanced level during this investigation, and given the extraordinary positive results the decision was an easy one to make: to go ahead with all available resources devoted to this approach!

7.6
Going from a Bench-scale Synthesis to a Fully Fledged Process

A platform had been created from which it was felt that the enantioselective sulfoxide synthesis could be developed into a process that would match logistical as well as quality requirements. However, it is important to stress that other variations of the titanium-catalyzed approach had been concomitantly identified but the decision was taken to focus solely on the present option. One of the most important and challenging aspects in this regard was to demonstrate how the oxidation step could be fitted into the existing manufacture of the prochiral sulfide (pyrmetazole) and, if not, what changes or modifications would be needed. This proved to be a major undertaking requiring development in the laboratory while running pilot trial campaigns in parallel to gain scale-up experience. Hence our resources were devoted intensely to these studies, a summary of which are presented here, covering a time span of >1½ years, leading up to a point when the tolerance limits were ready to be established.

7.6.1
Solvent

The literature [18, 19, 21] states that halogenated solvents work best (e.g., CH_2Cl_2, $ClCH_2CH_2Cl$) but our previous findings clearly show that the oxidation actually runs well in a rather broad range of solvents, with the best options from our in-house production point of view being EtOAc, methyl isobutyl ketone and toluene. Using the last, the stereoselectivity was independent of the concentration in the interval 0.5–1.0 M. One key factor in deciding to choose this solvent was the good opportunity this opened up to attach the omeprazole process, which affords the sulfide as a toluene solution to be fed directly into the subsequent asymmetric conversion into the sulfoxide.

7.6.2
Base

A number of different parameters were considered when selecting which base to be used in the process. Thus cost, availability, environmental load and effluent handling were assessed alongside of course the ease of operating the process in practice. At the end of this exercise, diisopropylethylamine (Hünigs base) stood out as the preferred one.

7.6.3
Composition of Titanium Complex

The interplay between Ti, tartrate ester, and H_2O was of crucial importance to the outcome, a feature well documented in the literature [18]. According to our results, varying the last two components while keeping the former constant (at 1 equiv.) showed that the best results (>75% ee) were obtained using around 0.5 equiv. H_2O and 2 equiv. diethyl tartrate. In order to achieve the best stereochemical outcome it was recommended to prepare the Ti-complex separately under carefully controlled conditions and only thereafter add it to the substrate [13, 21]. Much to our surprise, when first adding the three components forming the complex to a toluene solution of pyrmetazole, an increase in ee from 74 to 82% was achieved. Further studies revealed that the presence of amine at this stage of pre-formation of the catalytic complex had a negative impact.

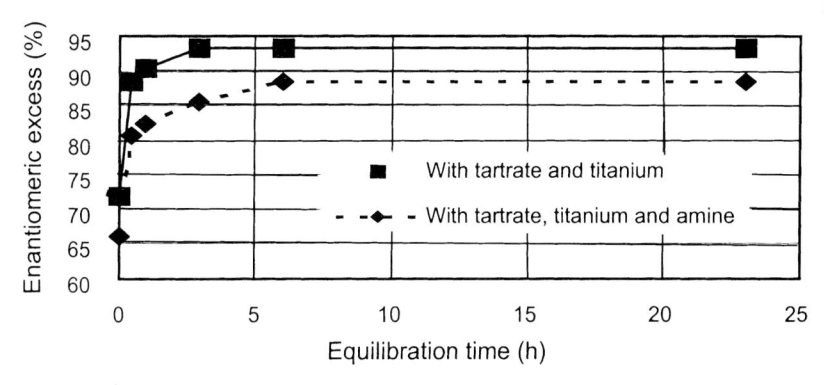

Fig. 7 Performance (expressed as % ee) of asymmetric sulfide oxidation as function of time and presence/absence of base during the equilibration.

7.6.4
Equilibration of Titanium Complex

It is not unusual that catalytic systems show sensitivity towards elevated temperatures especially when exposed over prolonged periods of time. Surprisingly we discovered that in our case the contrary is valid, and it is in fact beneficial to allow the catalytic complex to age or equilibrate at a higher temperature before carrying out the oxidation. A more in-depth investigation of the role played by the time when creating the equilibrated Ti-complex gave very interesting results, as the sulfoxide obtained after 3 hours at 30 °C had, under the circumstances, an extremely good quality of 94% ee. However, when trying to reach equilibrium in the presence of Hünigs base the selectivity was markedly poorer and it took 6 hours to reach 89% ee (see Fig. 7).

Further studies also revealed that the temperature during this equilibration had an effect and could be used to speed up the process, but when operating at too high a level the complex obviously started to degrade with a concomitant loss in ee. The optimum for our purposes seemed to be around 50 °C where an approximately 1 hour time slot for reaching equilibrium was sufficient without running any appreciable risk of degradation.

7.6.5
Equivalents of Ti-Complex

Generally the literature [18] states that 0.5–1.0 equiv. of the Ti-complex should be applied when attempting an oxidation of sulfides. However, when using our new conditions a high selectivity was achieved (91% ee) with as little as 0.04 equiv. When adapting the process to the simulated conditions and restraints in our current omeprazole production it was discovered that the method was non-robust. It was, however, found that robustness could be achieved by simply compensating with larger amounts of Ti-complex (relative to sulfide) and thus the following se-

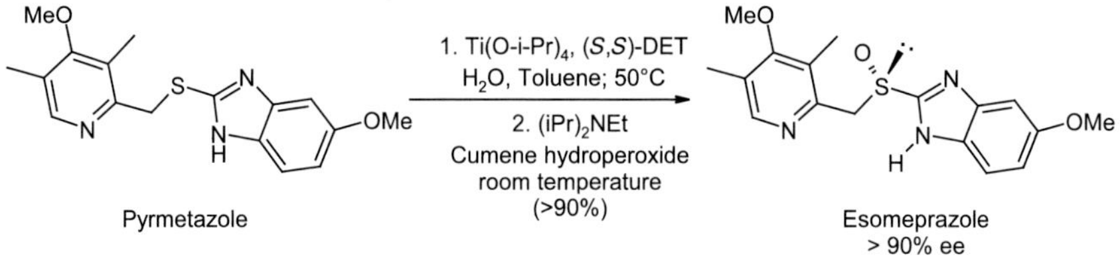

Fig. 8 Asymmetric oxidation of pyrmetazole to esomeprazole using Ti-catalysis.

ries of experiments, intended to mimic a large scale situation, clearly indicated a positive trend:

Ti-complex (equiv.)	% ee
0.1	85.4
0.2	92.6
0.5	93.5

This then provided a tool to increase the operability and performance in a production environment if there should be a need to do so.

7.6.6
Layout of Synthesis

With this wealth of information on our hands a clear picture was emerging showing how to conduct the synthesis of our sulfoxide. Thus starting with the pyrmetazole sulfide (assembled according to the standard omeprazole route as shown previously in Fig. 5) the synthesis is as follows (see Fig. 8).

7.7
The Big Test: Going to Plant Scale

Chemical synthesis can work beautifully on a small scale in the laboratory where the conditions can be meticulously controlled and the course of a reaction monitored in great detail, but it is not until a scale-up to the plant has been conducted that data on its robustness and sensitivity to various parameters will become obvious. Thus in parallel with the efforts put in to developing and optimizing the asymmetric step, comprehensive resources were invested in conducting trial runs on various scales in the pilot plant. A compilation of significant batches is shown in Tab. 1 together with conditions of operation and comments on the outcome. It is worth noting the increase in process performance quality over time as a result of development and optimizations exemplified, e.g., by a steady improvement in yield and % ee.

Tab. 1 Data from selected pilot plant batches to illustrate the development of the oxidation.

	Spring 1994	Spring 1994	Autumn 1994	Autumn 1995	Comments
Charged Sulfide	1.6 kg	1.6 kg	6.7 kg	104 kg	Pyrmetazole (sulfide substrate used as base for calculation)
Reaction					
Solvent for the oxidation	EtOAc	EtOAc	Toluene	Toluene	
Mol% of water	100	50	36	Optimized	
Mol% of diethyl tartrate	195	100	150	Optimized	
Mol% of Ti(OiPr)$_4$	100	50	60	Optimized	
Equilibration	No	No	Yes	Yes	Approx. 1 h at 50 °C
Mol% of diisopropylethylamine	50	50	52	Optimized	
Mol% cumene hydroperoxide	84	92	99	103	Oxidation at 30 °C, 1–1.5 h
Enantiomeric excess after oxidation (HPLC)	79.9%	80.5%	93.9%	92.7%	
Unreacted sulfide (HPLC)	28.4%	19%	6.5%	2.5%	
Sulfone (HPLC)	3%	3.8%	1.6%	2.4%	The only byproduct formed in significant amounts
"Yield"	62%	70%	89%	92%	Calculated as 100% (S)
Workup					
Precipitation of the racemate to upgrade the material	Yes	Yes	No	No	The crude product is dissolved in acetone from which the racemate precipitates
Enantiomeric excess after precipitation of the racemate (HPLC)	99.7%	95.2%			

Tab. 1 (continued)

	Spring 1994	Spring 1994	Autumn 1994	Autumn 1995	Comments
Yield after workup	38.4%	59.6%	66.8%	74%	Yield for the product [as 100 % (S)], measured by assay/volume of the solution that goes into the salt forming step
Comments	First pilot-plant batch. Extractive work up with NH$_3$ and methyl isobutyl ketone	Extractive work up with NH$_3$ and methyl isobutyl ketone	Equilibration implemented. Extractive work up with NH$_3$ and methyl isobutyl ketone	Ratio (relative amounts) between H$_2$O/DET/Ti = approx. 0.5 : 2 : 1	

7.8
Reaching the Final Target: A Robust Commercial Process

With the wealth of data and experience accumulated from laboratory studies and pilot scale-up trials, the time had come to bring all the bits and pieces together in a consolidated final process. A key activity in this respect was to establish tolerance limits, especially for the critical parameters identified; another aspect was to ensure that mutual amounts of reagents and substrate were all operating in the right direction to reach a global optimum in yield and quality with regard to the product. As it had been demonstrated in pilot batches that the catalysis worked well in the range 0.04–0.5 equiv. Ti (relative to sulfide), this parameter was kept at this range. The relative default ratio between Ti/tartrate/H_2O of approximately 1:2:0.5 was used as the base case throughout this series of experiments (amounts given in equiv. always expressed relative to Ti(O-iso-Pr)$_4$ except for when discussing cumene hydroperoxide).

7.8.1
Water

Running at higher amounts (0.9 equiv.) did not show any negative impact on conversion or ee, but as the levels increased beyond this (1.5 and 2.1 equiv.) both of these values were sharply reduced. Also the amount of sulfone had a tendency to increase at the highest H_2O content.

7.8.2
Diethyl Tartrate

At lower levels (<2 equiv.) there was a clear trend towards forming increased amounts of sulfone (8.3% at 0.5 equiv.) whereas the sulfoxide stayed relatively constant. This was also true for the ee except for the lowest value (0.5 equiv.) where a meager 12.6% ee was obtained (see Fig. 9).

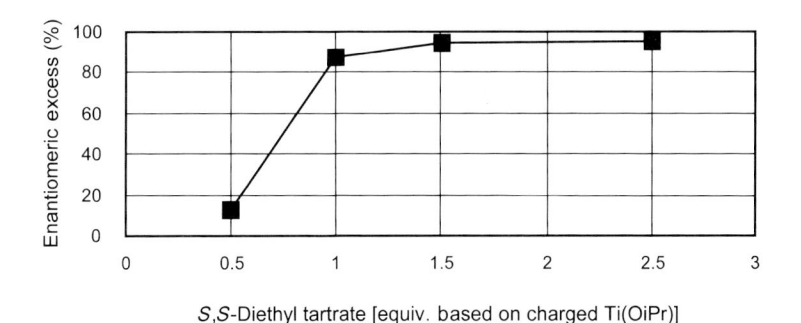

Fig. 9 Enantiomeric excess as a function of added equiv. (S,S)-diethyl tartrate.

7.8.3
Hünigs Base

On investigating the range between 0.5 to 3.0 equiv. the process operated in a very stable mode when checking a number of key parameters (sulfoxide content, ee, sulfone, conversion).

7.8.4
Cumene Hydroperoxide

The interval between 0.9 and 3.0 equiv. (relative to sulfide substrate) was studied and the general trend showed that the sulfide conversion went up and reached completion at >1.3 equiv. Both sulfoxide content and ee remain relatively stable (much greater than 90%) and it is only at the highest peroxide levels (≥2 equiv.) when the former falls off due to a considerable sulfone formation (14–19%) (see Fig. 10). This tendency was strengthened when reaction times were prolonged, and so on leaving the mixture with 14% sulfone overnight the content had increased to over 42%.

7.8.5
Stability of the Catalytic Complex

In order to determine the stability of the catalytic species all reagents including sulfide were added and then kept for varying lengths of time (from instantaneously up 43 hours) before adding the oxidant. Thus the data showed that storage for shorter periods of time, being mindful of the fact that problems do occur in commercial manufacturing which require some time to have them sorted out, did not have any measurable impact on the outcome (high sulfoxide content and ee), whereas at longer times the rate of conversion decreased and a gradual reduction in ee was obtained.

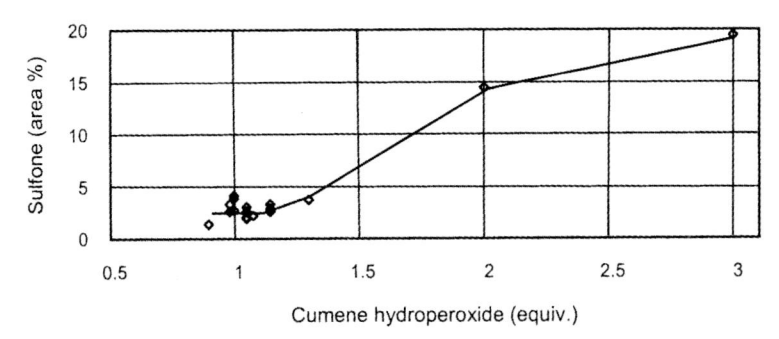

Fig. 10 Formation of sulfone byproduct as function of equiv. cumene hydroperoxide added (relative to sulfide).

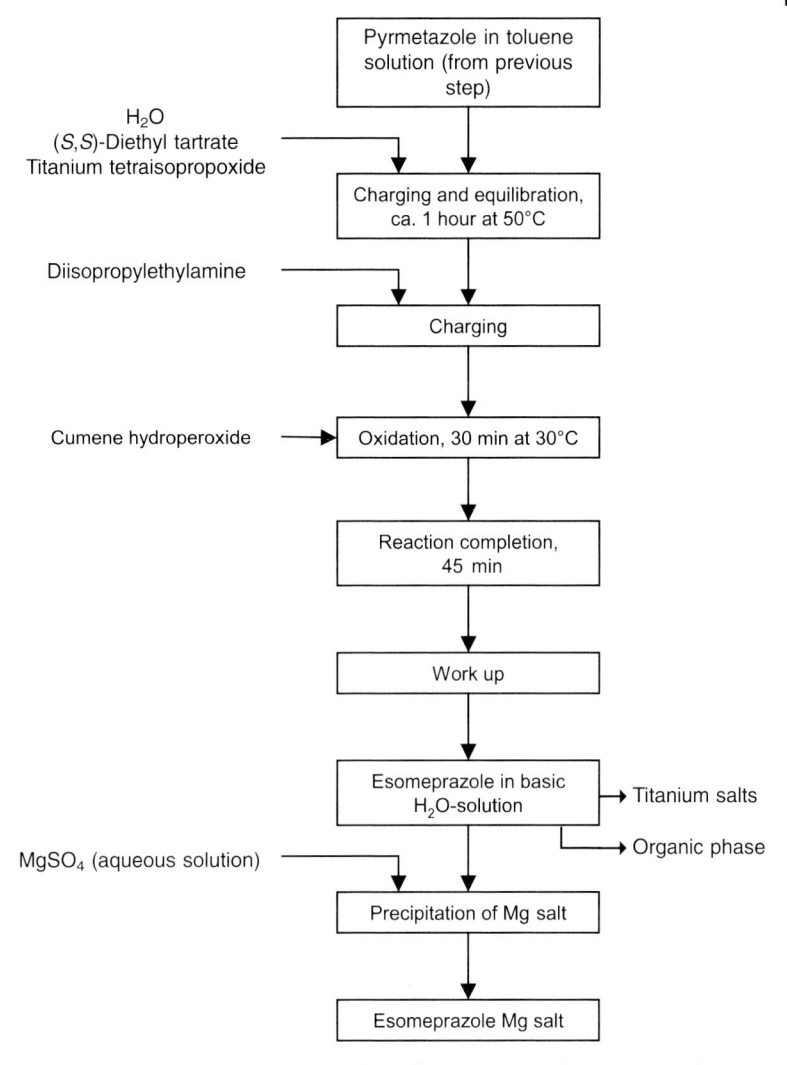

Fig. 11 Process flow of asymmetric oxidation reaction from pyrmetazole → esomeprazole.

7.8.6
Reaching the Goal – A Robust Process

When combining the data and experience described above with that accumulated over the years, which due to space limitations are not described in this chapter, a fully fledged process emerged [25, including experimental details]. Its outstanding performance and robustness with regard to yield, quality and operability has now been proven at the plant scale over some time and a simplistic flow chart (see Fig. 11) gives a good impression of how the production is conducted.

7.9
Points to Learn and Conclusions

At the end of this long and effort-consuming journey aimed at designing a full-scale process for esomeprazole, it is definitely the right time to reflect on what has been achieved and what learning experience this has brought about. Before going into more details, however, it is important to realize that with a good collaborative climate and team spirit in a cross-functional environment coupled with a clear vision of what needs to be done, challenges of this sort can be successfully managed. The focus was always directed towards the target to identify a route that could be scaled up. Of foremost importance to this whole project was the fact that a novel asymmetric process showing some unique features had been discovered and developed into a full commercial scale operation. Initial setbacks when applying literature procedures have not scared off those involved and instead a surprising breakthrough changed the situation into something much more favorable.

The process described herein has a proven track record over some years of reliable and robust commercial manufacturing at a batch size of >100 kg. Its scope is far wider than being restricted only to the current sulfide pyrmetazole, as has been demonstrated in-house in a number of additional examples [24, 25]. In fact others have explored this process further on their own substrate sulfides and obtained equally outstanding results [28]. Compared with previously published methods our work has established a new way of effecting an enantioselective sulfide oxidation with the following characteristics:

- Addition of a base to directly promote a high ee.
- Change in the order of adding the reagents allowing the Ti-catalyst, (S,S)-diethyl tartrate, and water (in approximately a 1:2:0.5 ratio) to equilibrate in the presence of the substrate at elevated temperatures for a short period of time.
- A considerably increased choice of suitable solvents (toluene, EtOAc, various ketones), not limited to environmentally unfriendly chlorinated ones (e.g., CH_2Cl_2).
- Performing the oxidation at higher and technically more convenient temperatures than before ($\geq 20\,°C$ vs. $-20\,°C$).

Also favorable from an atom economy/efficiency perspective [29] the process operates in a really catalytic mode with a turnover number (ton) ~4–16 and turnover frequency (tof) ~3–12/h. From a logistical point of view the new oxidation was adapted so that it could be linked to an already existing method of preparing the sulfide starting material without requiring any appreciable modifications or re-designs.

7.10
Acknowledgements

The achievements described herein have only been possible thanks to the wholehearted dedication and professional commitment to the task by numerous co-workers and colleagues in Process R&D Södertälje and Medicinal Chemistry Mölndal, respectively. Worthy of special mention in this context due to their seminal contribution to the successful outcome of this work are from the former site Hanna Cotton and from the latter Dr. Thomas Elebring, Dr. Lanna Li, Gunnel Sundén, Dr. Henrik Sörensen, and Sverker von Unge. For highly valuable comments on the manuscript special recognition is due to Christina Fregler, Dr. Per Lindberg and Sverker von Unge all residing at the Mölndal site and Gabrielle Lindquist and Bo Lindqvist of Bulk Production in Operations Södertälje.

7.11
References

1 P. Lindberg, A. Brändström, B. Wallmark, H. Mattsson, L. Rikner, K.-J. Hoffman, Med. Res. Rev. 1990, 10, 1–54.

2 P. Lindberg, P. Nordberg, T. Alminger, A. Brändström, B. Wallmark, J. Med. Chem. 1986, 29, 1327–1329.

3 For a) H259/31, see: A. Brändström, P. Lindberg, G. E. Sundén, US Patent 5039808; b) H326/07, see: A. Brändström, P. Lindberg, G. E. Sundén, Patent Appl. WO 91/19711.

4 H.-J. Federsel, Endeavour 1994, 18, 164–172; b) I. Szelenyi, G. Geisslinger, E. Polymeropoulos, W. Paul, M. Herbst, K. Brune, Drug News Perspect. 1998, 11, 139–160 and references cited therein; c) I. Agranat, H. Caner, J. Caldwell, Nature Rev. Drug Discov. 2002, 1, 753–768 where also regulatory and patentability issues are addressed.

5 For excellent and brief overviews describing the "Gastrin project" from the late 1960s to date see: a) E. Carlsson, P. Lindberg, S. von Unge, Chem. Brit. 2002, 38(5), 42–45; b) L. Olbe, E. Carlsson, P. Lindberg, Nature Rev. Drug Discov. 2003, 2, 132–139.

6 S. Allenmark, B. Bomgren, H. Borén, P.-O. Lagerström, Anal. Biochem. 1984, 136, 293–297.

7 P. Erlandsson, R. Isaksson, P. Lorentzon, P. Lindberg, J. Chromatogr. 1990, 532, 305–319.

8 P. Lindberg, S. von Unge, Patent Appl. WO 94/27988.

9 T. Alminger, R. A. Bergman, H. Bundgaard, P. Lindberg, G. E. Sunden, Patent Appl. WO 88/03921.

10 A. Brändström, P. Lindberg, N.-Å. Bergman, L. Tekenbergs-Hjelte, K. Ohlson, I. Grundevik, P. Nordberg, T. Alminger, Acta Chem. Scand. 1989, 43, 587–594.

11 D. Graham, R. Holt, P. Lindberg, S. Taylor, Patent Appl. WO 96/17077.

12 R. Holt, P. Lindberg, C. Reeve, S. Taylor, Patent Appl. WO 96/17076.

13 An excellent account of the entire field is given in "Asymmetric Oxidation Reactions" T. Katsuki, Ed., Oxford University Press, Oxford, 2001. Special focus on oxidation of hetero atoms, e.g., sulfur, is devoted to chapter 4 by H.B. Kagan, pp. 153–70.

14 A. McKenzie, H. Wren, J. Chem. Soc. 1907, 91, 1215–1228.

15 a) K. Balenovic, N. Bregant, D. Francetic, Tetrahedron Lett. 1960, (6), 20–22; b) A. Mayr, F. Montanari, M. Tramontini, Gazz. Chim. Ital. 1960, 90, 739–749.

16 a) K. K. Andersen, Tetrahedron Lett. 1962, 93–95; b) K. K. Andersen, W. Gaffield, N. E. Papanikolaou, J. W. Foley, R. I. Perkins, J. Am. Chem. Soc. 1964, 86, 5637–5646.

17 a) H. B. Kagan, P. Diter, Organosulfur Chem. 1998, 2, 1–39; b) P. G. Potvin, B. G. Fieldhouse, Tetrahedron: Asymmetry 1999, 10, 1661–1672; c) F. van de Velde, I. W. C. E. Arends, R. A. Sheldon, Topics Catal. 2000, 13, 259–65; d) F. Fache, E. Schulz, M. L. Tommasino, M. Lemaire, Chem. Rev. 2000, 100, 2159–2231; e) W. Adam, C. R. Saha-Möller, P. A. Ganeshpure, Chem. Rev. 2001, 101, 3499–3548; f) M. Matsugi, N. Fukuda, Y. Muguruma, T. Yamaguchi, J. Minamikawa, S. Otsuka, Tetrahedron 2001, 57, 2739–2744; g) B. Saito, T. Katsuki, Tetrahedron Lett. 2001, 42, 3873–3876; h) A. Massa, A. Lattanzi, F. R. Siniscalchi, A. Scettri, Tetrahedron: Asymmetry 2001, 12, 2775–2777; i) A. Massa, F. R. Siniscalchi, V. Bugatti, A. Lattanzi, A. Scettri, Tetrahedron: Asymmetry 2002, 13, 1277–1283; j) Z. Han, D. Krishnamurthy, P. Grover, Q. K. Fang, C. H. Senanayake, J. Am. Chem. Soc. 2002, 124, 7880–7881; k) P. J. Hogan, P. A. Hopes, W. O. Moss, G. E. Robinson, I. Patel, Org. Process Res. Dev. 2002, 6, 225–229; l) Z. Han, D. Krishnamurthy, P. Grover, H. S. Wilkinson, Q. K. Fang, X. Su, Z.-H. Lu, D. Magiera, C. H. Senanayake, Angew. Chem. Int. Ed. 2003, 42, 2032–2035.

18 a) P. Pitchen, H. B. Kagan, Tetrahedron Lett. 1984, 25, 1049–1052; b) P. Pitchen, E. Duñach, M. N. Deshmukh, H. B. Kagan, J. Am. Chem. Soc. 1984, 106, 8188–8193; c) S. H. Zaho, O. Samuel, H. B. Kagan, Tetrahedron 1987, 43, 5135–5144; d) H. B. Kagan, F. Rebiere, Synlett 1990, 643–650.

19 F. DiFuria, G. Modena, R. Seraglia, Synthesis 1984, 325–326.

20 A few representative citations from this vast area: a) T. Katsuki, K. B. Sharpless, J. Am. Chem. Soc. 1980, 102, 5974–5976; b) R. M. Hanson, K. B. Sharpless, J. Org. Chem. 1986, 51, 1922–1925; c) A. Pfenninger, Synthesis 1986, 2, 89–116; d) Y. Gao, R. M. Hanson, J. M. Klunder, S. Y. Ko, H. Masamune, K. B. Sharpless, J. Am. Chem. Soc. 1987, 109, 5765–5780; e) S. S. Woodard, M. G. Finn, K. B. Sharpless, J. Am. Chem. Soc. 1991, 113, 106–113; f) M. G. Finn, K. B. Sharpless, J. Am. Chem. Soc. 1991, 113, 113–126.

21 J. M. Brunel, P. Diter, M. Deutsch, H. B. Kagan, J. Org. Chem. 1995, 60, 8086–8088.

22 a) F. A. Davis, M. C. Weismiller, C. K. Murphy, R. T. Reddy, B. C. Chen, J. Org. Chem. 1992, 57, 7274–7285; b) F. A. Davis, R. T. Reddy, W. Han, P. J. Carroll, J. Am. Chem. Soc. 1992, 114, 1428–1437.

23 S. von Unge, Patent Appl. WO 97/02261.

24 a) M. Larsson, U. Stenhede, H. Sörensen, S. von Unge, H. Cotton, Patent Appl. WO 96/02535; H.-J. Federsel, Chirality 2003, 15, S128–S142; c) H.-J. Federsel, Nature Rev. Drug Discov. 2003, 2, 654–664.

25 H. Cotton, T. Elebring, M. Larsson, L. Li, H. Sörensen, S. von Unge, Tetrahedron: Asymmetry 2000, 11, 3819–3825.

26 F. DiFuria, G. Licini, G. Modena, R. Motterle, J. Org. Chem. 1996, 61, 5175–5177.

27 H. B. Kagan, E. Duñach, C. Nemecek, P. Pitchen, O. Samuel, S. H. Zhao, Pure Appl. Chem. 1985, 57, 1911–1916.

28 H. Hashimoto, T. Urai, Patent Appl. WO 01/83473.

29 For more details see: a) B. M. Trost, Angew. Chem. Int. Ed. Engl. 1995, 34, 259–281; b) R. A. Sheldon, Pure Appl. Chem. 2000, 72, 1233–1246.

8
Development of a Biocatalytic Process for the Resolution of (R)- and (S)-Ethyl 3-amino-4-pentynoate Isomers Using Enzyme Penicillin G Amidohydrolase

Ravindra S. Topgi

Abstract

The beta-amino acid, (S)-ethyl 3-amino-4-pentynoate, a chiral synthon, was an intermediate in the synthesis of an anti-platelet agent, xemilofiban. A biocatalytic process using the enzyme Penicillin G amidohydrolase was developed to resolve (R)- and (S)-enantiomers of ethyl 3-amino-4-pentynoate in enantiomerically pure form. The design of experimental approach used was one to optimize the critical reaction parameters to control the stereoselectivity of the enzyme Penicillin G amidohydrolase.

8.1
Introduction

As a part of ongoing efforts to synthesize a potent, orally active anti-platelet agent, xemilofiban **1** [1], development of an efficient chemoenzymatic process for **2**, the chiral β-amino acid ester synthon (Fig. 1) was proposed. The scheme emphasized the creation of the stereogenic center as the key step. In parallel with the enzymatic approach, chemical synthesis of the β-amino acid ester synthon emphasized formation of a chiral imine, nucleophilic addition of the Reformatsky reagent, and oxidative removal of the chiral auxiliary. This chapter describes a selective amidation/amide hydrolysis using the enzyme Penicillin G amidohydrolase from *E. coli* to synthesize (R)- and (S)-enantiomers of ethyl 3-amino-5-(trimethylsilyl)-4-pentynoate in an optically pure form. The design of the experimental approach was applied in order to optimize the critical reaction parameters to control the stereoselectivity of the enzyme Penicillin G amidohydrolase.

8.2
Enzymatic Approach

Penicillin G amidohydrolase (PGA) (E.C. 3.5.1.11) from *E. coli* ATCC 9637 is a unique enzyme that exhibits exceptionally high affinity to phenylacetic acid and its derivatives [2]. This capability has been used advantageously to achieve moder-

Asymmetric Catalysis on Industrial Scale: Challenges, Approaches and Solutions
Edited by H. U. Blaser, E. Schmidt
Copyright © 2004 Wiley-VCH Verlag GmbH & Co. KGaA, Weinheim
ISBN: 3-527-30631-5

Xemilofiban, **1**   β-amino acid ester synthon, **2**

Fig. 1 Anti-platelet agent xemilofiban and β-amino acid ester synthon.

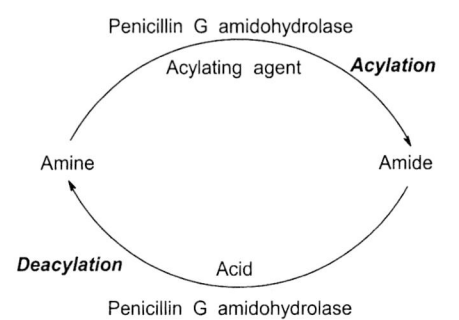

Fig. 2 Acylation and deacylation by enzyme Penicillin G amidohydrolase.

ate to excellent stereochemical discrimination between corresponding enantiomers in the hydrolytic cleavage of the phenylacetyl group from α-aminoalkylphosphonic acids [3], α-, β- and γ-amino carboxylic acids [4], sugars [5], amines [6], peptides [7], and esters of phenylacetic acid [8].

The catalytic activity of the enzyme PGA is pH dependent. In the pH range 7–8.5, PGA exhibits deacylation activity, whereas at acidic pH 5–6, acylation is observed [9]. This pH-dependent enzyme activity offers the possibility of exploring selective amidation/amide hydrolysis to synthesize (R)- and (S)-enantiomers of ethyl 3-amino-5-(trimethylsilyl)-4-pentynoate in optically pure form (Fig. 2).

8.2.1
Deacylation

In the presence of triethylamine, phenylacetyl chloride reacts with a racemic amine to yield a phenylacetyl derivative **4**. Heptane or ethyl acetate can be used as the reaction medium. Reaction at room temperature for 2 hours gave the desired product **4** (Fig. 3). As the final product xemilofiban **1** does not contain the trimethylsilyl group at the alkyne carbon, it was preferable to perform desilylation prior to the enzymatic deacylation. Desilylation of the phenylacetyl derivative **4** was achieved by treatment with 0.2 equivalents of sodium ethoxide in ethanol at room temperature. Desilylation to provide **5** was complete in 1 hour.

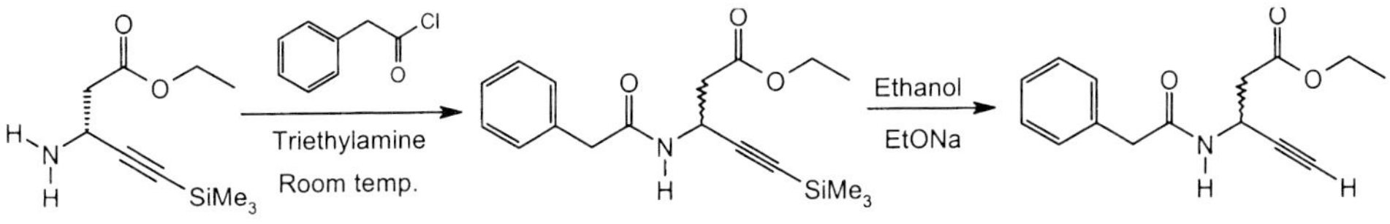

3 4 5

Fig. 3 Synthesis of desilylated ethyl 3-(*N*-phenylacetyl)-4-pentynoate.

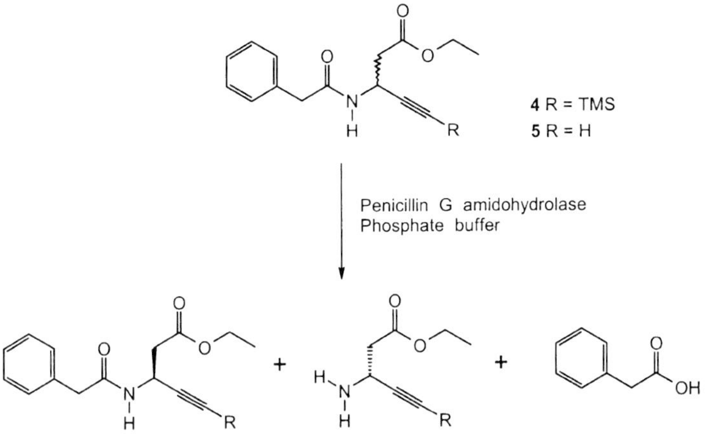

Phenylacetamide (*S*)-isomer — Amine (*R*)-isomer — Phenylacetic acid

6a R = TMS; **6b** R = H
Yield: 96% 95%

7a R = TMS; **7b** R = H
Yield: 91% 92%

8

Fig. 4 Hydrolysis of phenylacetamides using Penicillin G amidohydrolase.

Racemic amides **4** and **5** were then independently subjected to the PGA-catalyzed deacylation reaction. Because of the selective preference of PGA towards the (*R*)-enantiomer, the rate of deacylation of the (*R*)-enantiomer is significantly higher than the (*S*)-enantiomer for both **4** and **5**. The result is the selective deacylation of the (*R*)-enantiomers. The racemic ethyl 3-(*N*-phenylacetyl)-5-(trimethylsilyl)-4-pentynoate **4** was placed in the phosphate buffer (pH 7.4). The pH of the medium was adjusted to ~7.5. At room temperature, immobilized enzyme PGA was added to the reaction medium and the pH of this medium was maintained at approximately 7.4 using dilute potassium bicarbonate solution. The reaction was usually complete in approximately 20 hours. After completion of the reaction, the reaction mixture was extracted with ethyl acetate. The ethyl acetate extract was then extracted with dilute hydrochloric acid. The organic phase contained the (*S*)-amide **6a** (96% yield). The pH of the aqueous phase was adjusted to ~8 using potassium carbonate and the aqueous phase was extracted with ethyl acetate to obtain (*R*)-amine **7a** (91% yield). Similarly, starting with the racemic ethyl 3-(*N*-phe-

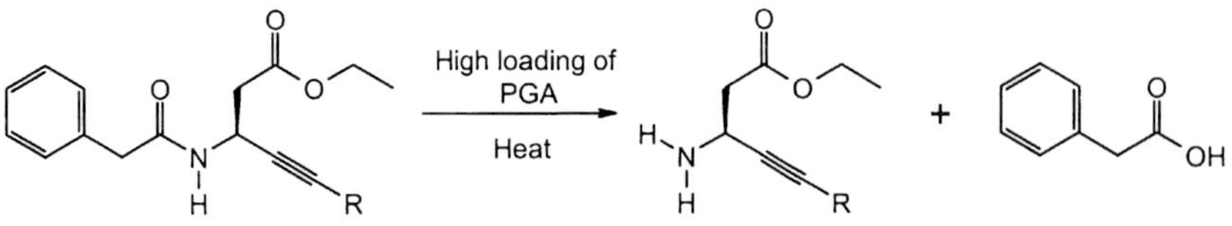

(S)-Phenylacetamide | (S)-Amine | Phenylacetic acid

6a R = TMS; **6b** R = H | **9a** R = TMS; **9b** R = H | **8**

Fig. 5 Hydrolysis of (S)-phenylacetamide using Penicillin G amidohydrolase.

nylacetyl)-4-pentynoate **5**, (*S*)-amide **6b** (95% yield) and (*R*)-amine **7b** (92% yield) were obtained (Fig. 4).

The (*R*)-amine **7** is a liquid, whereas the (*S*)-amide **6** is a solid. This difference in physical properties in effect enabled easy separation of the amine antipodes by filtration. The desired (*S*)-amide **6** needs to be deacylated prior to the synthesis of the final compound, xemilofiban **1**. The slow rate of deacylation of the (*S*)-amide by the enzyme PGA can be enhanced significantly by using a high concentration of the enzyme. The temperature can also be raised to increase the rate of enzymatic hydrolysis. The (*S*)-amide **6** obtained was deacylated to liberate the free (*S*)-amine **9** using excess PGA (Fig. 5).

8.2.2 Acylation

The equilibrium of the enzyme acylation reaction can be shifted towards the synthesis of the amide by precipitation of the acylated product formed (Fig. 6). The racemic ethyl 3-amino-5-(trimethylsilyl)-4-pentynoate **3** is an insoluble liquid, whereas the (*R*)-phenylacetamide **10** is an insoluble solid. The racemic ethyl 3-amino-5-(trimethylsilyl)-4-pentynoate **3** was added to dilute hydrochloric acid. The pH of the reaction medium was then adjusted to 6. Phenylacetic acid (2 equiv.) was added and the pH of the medium was readjusted to 6. Soluble PGA (50 units/100 mg of racemic amine) was added, and the reaction was stirred at room temperature. After completion of the reaction, the pH of the reaction mixture was adjusted to 4. Filtration of the reaction mixture gave (*R*)-amide **10** in quantitative yield. Chiral HPLC analysis of this isolated amide showed the absence of (*S*)-amide. The pH of the filtrate was raised to ~8, and the filtrate was extracted with ethyl acetate to obtain (*S*)-amine **11** (yield: 90%) (Fig. 6). The chiral HPLC analysis indicated an *R*:*S* ratio of 2:98.

Compared with the deacylation approach, the acylation methodology does not require prior chemical phenylacylation and also eliminates the final deacylation of the (*S*)-amide, which requires an excess of PGA.

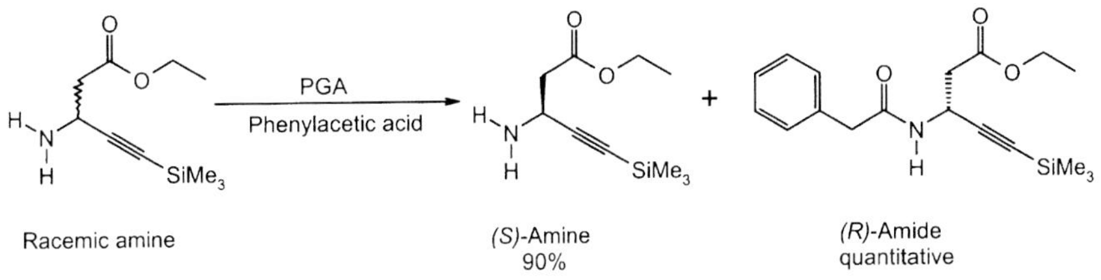

Fig. 6 Acylation using Penicillin G amidohydrolase.

8.3
Optimization of Reaction Conditions

8.3.1
The Enzyme Penicillin G Amidohydrolase

The commercial supply is primarily derived from expressing the enzyme Penicillin G amidohydrolase (E.C. 3.5.1.11) ATCC 9637 in *E. coli*. Soluble enzyme was obtained from Calbiochem and Boehringer-Mannheim, and immobilized enzyme was obtained from Sigma, Röhm-Pharma and Boehringer-Mannheim. The recovery of the immobilized enzyme made it cost effective compared with the soluble enzyme.

8.3.2
Reaction Monitoring

In deacylation, as the enzyme cleaved the phenylacyl group, phenylacetic acid was formed, which lowered the pH of the reaction medium. Base was added to maintain the starting pH. (Note: Use of ammonium hydroxide led to the formation of desilylated byproducts; desilylation was eliminated when bicarbonates were used.) This approach was not required in the acylation reaction. At pH above 7.5 the (R)- and (S)-amines are practically insoluble in water. Organic solvents were used to extract the free amines from the aqueous reaction medium at pH ~8.0. *p*-Fluorobenzoyl, 1-naphthoyl, and phenylacetyl derivatives of the racemic amine were prepared and their behavior on the chiral HPLC column was studied. Based on ease of preparation and HPLC analysis, the 1-naphthoyl derivatives (Fig. 7) were preferred. Reversed phase HPLC analysis on a Vydac-C18 analytical column used a gradient of acetonitrile (0.1% triethylamine) in water (0.05% phosphoric acid) to quantify the total amide in the reaction mixture. Chiral HPLC analysis on (*S,S*) Whelk-O Chiral column used isopropanol:hexane (30:70) as a solvent system to separate and quantify the (*R*)- and (*S*)-enantiomers.

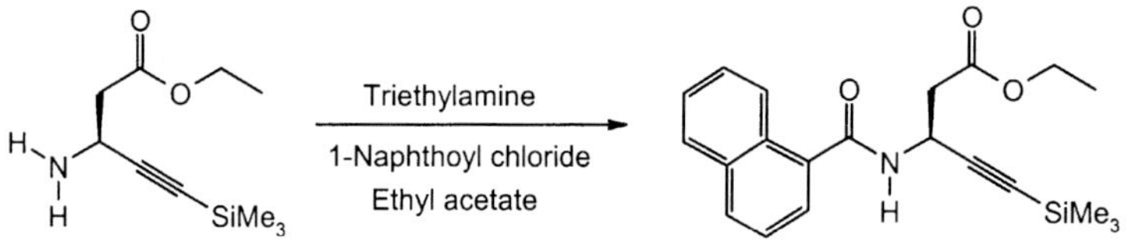

Fig. 7 Synthesis of 1-naphthoyl derivative.

8.3.3
Acyl Group Specificities

The enzyme active site has a well defined structural requirement for the acyl moiety. PGA shows a rather broad substrate specificity [10]. In the case of substituted phenylacetamides, PGA displays higher specificity to the phenylacetyl group. Based on bulk availability, the phenylacetyl group was preferred to *p*-hydroxyphenylacetyl and phenoxyacetyl groups. It has been reported that the esters of phenylacetic acid have enhanced reactivity in PGA enzyme reactions [11]. Commercially available acylating agents such as methyl phenoxyacetate, methyl phenylacetate, ethyl phenylacetate, and phenylacetic acid were used. All of these reagents were compatible with the conditions used in the enzyme reaction. Methyl phenoxyacetate gave higher yields of amide than methyl phenylacetate. Phenylacetic acid proved fairly reactive under enzymatic reaction conditions. However, based on ease of handling and the possibility of recyclization without derivatization, phenylacetic acid was the preferred acylating agent.

8.3.4
Organic Co-solvent

The interaction between the enzyme reaction center and acyl moiety of the substrate is a prerequisite for the product formation. Therefore, solubility of the substrate in the reaction medium is important. Although the enzyme PGA prefers an aqueous environment, reactions can be run in the presence of a low percentage of organic co-solvent [12]. In order to solubilize the substrates in water, a minimum amount of organic co-solvent was used. Experiments were performed using acetone, tetrahydrofuran, acetonitrile, isopropanol, methanol, and toluene. Methanol and acetonitrile showed compatibility in the PGA-catalyzed deacylation, whereas acetone was a more acceptable solvent in acylation reactions. In the absence of an organic co-solvent, the enzyme performance was equally good even though the reaction was a heterogeneous mixture. The enzymatic reaction carried out in a medium without organic co-solvent gave one enantiomer in the solid form and the other in the liquid form. This allowed easy separation of the enantiomers.

8.3.5
Phosphate Buffer

Phosphate buffer is known to enhance the activity of the enzyme PGA [13]. In the present studies, the enzyme PGA displayed satisfactory activity in water without phosphate buffer.

8.3.6
pH of the Reaction Medium

As the pH of the reaction medium dictates the activity of the enzyme PGA, pH plays an important role. The acylation reaction was studied at different pH values ranging from 4.5 to 7.5. The enzyme showed maximum acylation activity around pH 5.3. However, acylation activity declined sharply at pH values below 5.2. The pH range from 5.25 to 6.25 (Fig. 8) seems to be optimum for phenylacetamide formation.

8.3.7
Optimum Amount of Acylating Agent

In order to determine the optimum concentration of the acylating agent, phenylacetic acid, the amount of amine (0.47 mmol) and enzyme PGA (3.16 units, Röhm-Pharma) were kept constant and the acylating agent was varied from 0.125 to 1.5 mmol (17 mg to 204 mg) (Fig. 9).

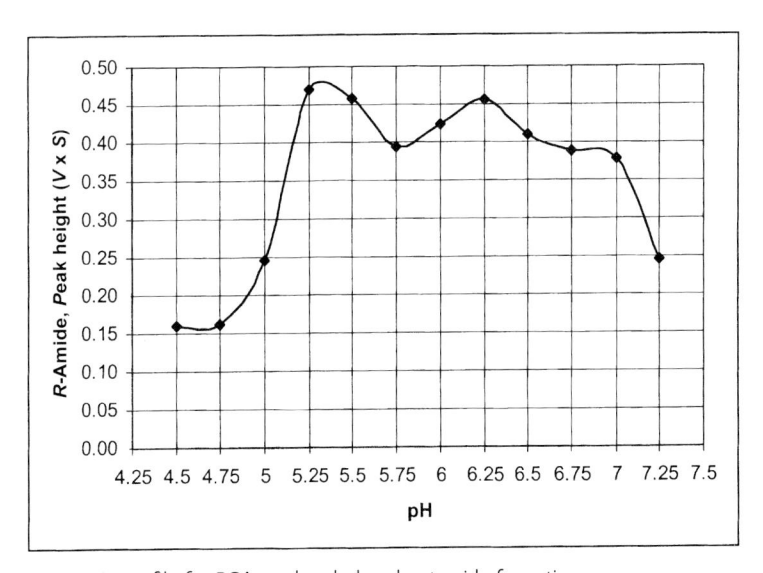

Fig. 8 pH profile for PGA-catalyzed phenylacetamide formation.

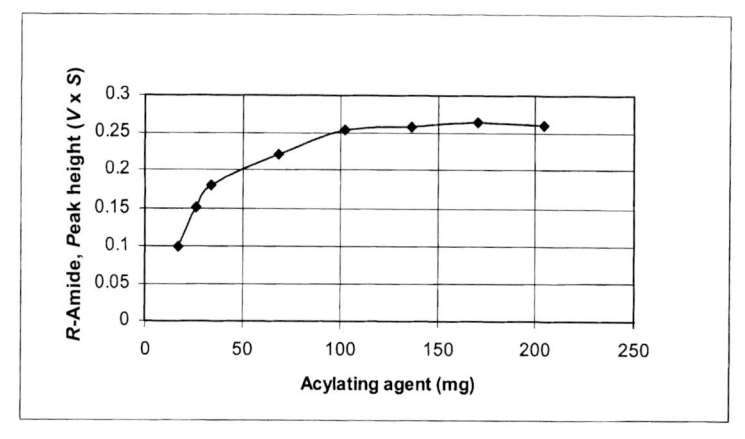

Fig. 9 Variation of amount of phenylacetic acid.

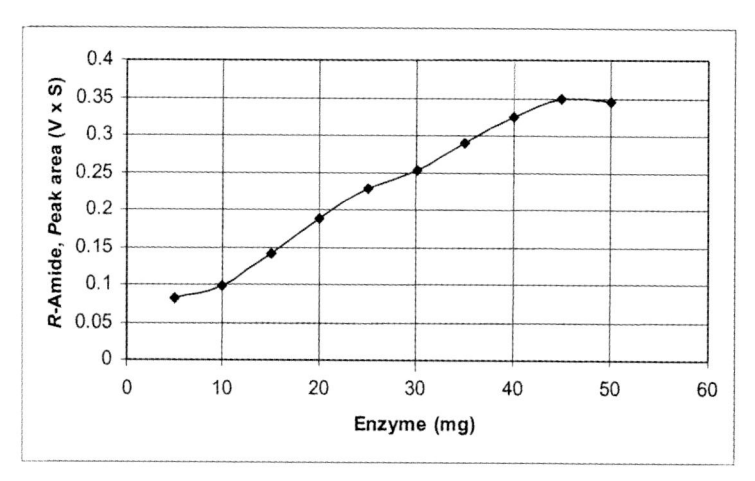

Fig. 10 Variation of amount of PGA.

8.3.8
Optimum Amount of Enzyme

A minimum amount of enzyme and high turnover are necessary for a cost-effective operation. The amount of the racemic amine ($\sim$100 mg), phenylacetic acid (193 mg) and reaction volume (10 mL) were kept constant at pH 6. Water was used as the reaction medium. The amount of immobilized enzyme was varied from 0.8 to 80 units (5 mg to 50 mg) (Röhm-Pharma). The reaction was carried out for 16 hours at room temperature (Fig. 10).

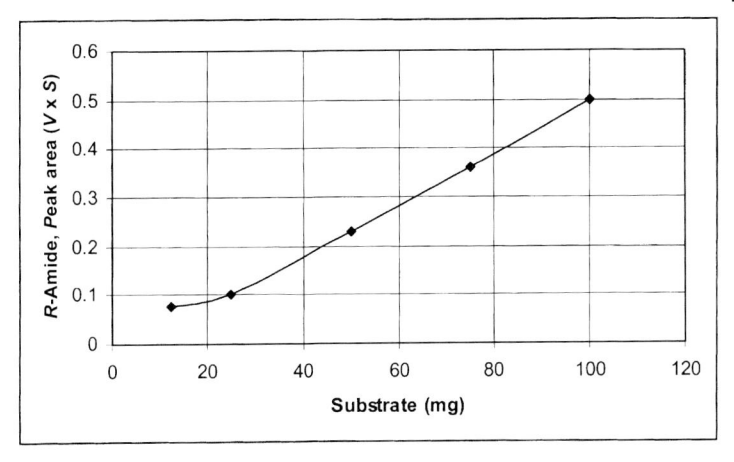

Fig. 11 Variation of amount of substrate, racemic amine.

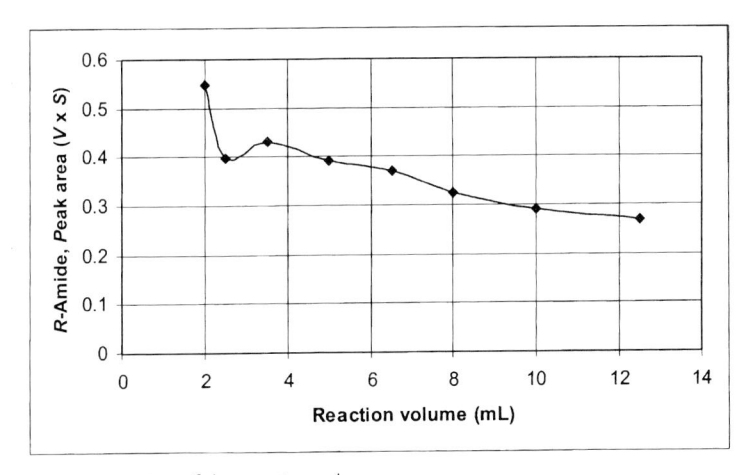

Fig. 12 Variation of the reaction volume.

8.3.9
Optimum Amount of Substrate

The amount of substrate, racemic amine, was varied from 0.06 to 0.47 mmol (13 mg to 100 mg) while phenylacetic acid (2 mmol), immobilized enzyme (30 units), reaction medium (10 mL) and pH 6 were kept constant (Fig. 11).

8.3.10
Optimum Reaction Volume

The reaction volume appeared to be critical in the acylation reaction as it governs the interaction between the amine, phenylacetic acid, and the enzyme-active site.

Keeping the amounts of the reactants as well as other variables constant, the reaction volume was varied. The results (Fig. 12) indicate that the smaller reaction volumes give higher conversion of the amine to the corresponding amide [14].

8.3.11
Enzyme Activity

The reactions were carried at a racemic amine concentration of 0.25 M. The activity of the enzyme was reduced from 387 units/g dry to 253 units/g dry over 15 runs. Based on the standard activities, half-activity was extrapolated for 22 cycles. The results over 14 cycles indicated 99.7% purity, 98.3% ee and 41.7% yield of the desired (S)-amine.

8.3.12
Design of Experiment Study

Based on the results obtained from the single factor variation, optimum reaction conditions were approximated. Under these conditions, enzymatic resolution was quite remarkable. The further optimization of the reaction conditions involved a design of experiment (DOE) study. Based on the values obtained from single factor variation limits for each factor, constant factors and responses measured, a 2^5 DOE study was undertaken (see Tabs. 1–4).

The study emphasized the enantiomeric excess (ee) as well as the molar yield of (S)-amine. The statistical analysis provided an insight into the various interac-

Tab. 1 Values for DOE study.

Factor	Units	Low value	Median value	High value
Amine	g	25	37.5	50
Enzyme	units/g amine	60	100	140
Phenylacetic acid	g/g amine	0.55	0.65	0.75
pH	–	5.2	5.6	6.0
Temperature	°C	28	33	38

Tab. 2 Constant factors.

Factors	Values/description
Reaction volume	500 mL
Enzyme source	Immobilized PGA enzyme from Boehringer-Mannheim
Amine	From pilot plant
Reactor	2 L with water jacket, and thermometer and mech. stirrer inlets
Type of agitation	Mechanical
Rate of agitation	200 rpm

Tab. 3 Responses to be measured.

Response	Method of measurement	Type of measurement
(R)-amide yield	Reversed phase HPLC	Quantification
	Chiral HPLC	Quantification
(S)-amine yield	Chiral HPLC of 1-naphthoyl derivative	Quantification
(R)-amide ee	Chiral HPLC	ee
(S)-amine ee	Chiral HPLC of 1-naphthoyl derivative	ee
Desilylation	Chiral HPLC	% and quantification
Reaction completion	Reversed phase HPLC	Formed (R)-amide
	Chiral HPLC	ee of (S)-amine

Tab. 4 Optimum reaction conditions.

Factor	Description	Value
Amine concentration	High	100 g/L
Enzyme stoichiometry	Low	60 units/g amine
Phenylacetic acid stoichiometry	Low	0.55 g/g amine
pH	Higher than median	5.6–5.7
Temperature	Low	28 °C

tions. The single factors identified as influencing the formation of the (R)-amide were the amount of enzyme, amount of phenylacetic acid, and pH. The combined interactions influencing the formation of the (R)-amide observed were pH/temperature and enzyme/phenylacetic acid ratio. While lower pH and lower temperature were preferred, high temperature, and low pH were unfavorable. The interaction between enzyme and phenylacetic acid indicated that the high enzyme concentration was compatible with either high or low phenylacetic acid concentration. However, low enzyme concentration required low phenylacetic acid concentration. In addition to the single factors (temperature, amine to enzyme ratio, and pH) the combined interactions such as phenylacetic acid/pH, amine/phenylacetic acid, enzyme/pH, and phenylacetic acid/temperature influenced the ee of the (S)-amine.

8.4 Conclusion

Penicillin G amidohydrolase proved to be an effective and efficient catalyst for enantioselective acylation as well as deacylation. Synthesis of the desired antipode [(R)- or (S)-enantiomer] can be achieved by selecting either amidation or amide hydrolysis, as demonstrated in the synthesis of (R)- and (S)-enantiomers of ethyl 3-amino-5-(trimethylsilyl)-4-pentynoate. Statistical analysis of single and multi-factor variations provide insight into the various interactions. Optimization of enzymatic reaction conditions to achieve higher ee values involves the identification of the reaction conditions that expand the energy difference between the transition states of competing diastereomeric pathways. The present biocatalytic approach, which offers the possibility of synthesizing separately both enantiomers of β-amino acids/esters, could serve as a method of choice, especially from the points of view of generality, economy, and simplicity of experimental procedure.

8.5
Acknowledgments

Dr. John Ng for support and editing; Dr. Bryan Landis, Dr. Ping Wang and their bioprocess groups at St. Louis; Larry Miller for providing analytical methodology; and Kathy McLaughlin, Dr. John Dygos and Dr. Russell Linderman for valuable discussions and manuscript preparation are gratefully acknowledged.

Figs. 1 to 6 and 8 are reprinted from reference 1(f) with permission from Elsevier Science.

8.6
References

1 a) J. A. Zablocki, J. G. Rico, R. B. Garland, T. E. Rogers, K. Williams, L. A. Schretzman, S. A. Rao, P. R. Bovy, F. S. Tjoeng, R. J. Lindmark, M. Toth, M. Zupec, D. E. McMackins, S. P. Adams, M. Miyano, C. S. Markos, M. N. Milton, S. Paulson, M. Herin, P. Jacquin, N. Nicholson, S. G. Panzer-Knodle, N. F. Haas, G. D. Page, J. A. Szalony, B. B. Taite, A. K. Salyers, L. W. King, J. G. Campion, L. P. Feigen, J. Med. Chem. 1995, 38, 2378; b) K. Babiak, S. Babu, J. R. Behling, M. L. Boys, K. J. Cain-Janicki, W. W. Doubleday, P. Farid, T. J. Hagen, E. A. Hallinan, D. W. Hansen Jr., D. E. Korte, K. T. McLaughlin, J. R. Medich, S. T. Nugent, V. Orlovski, J. M. Park, K. B. Peterson, D. R. Pilipauskas, B. S. Pitzele, S. Tsymbalov, G. L. Stahl, preparation of ethyl 3(S)-{4-[4-(aminoiminomethyl) phenyl]amino-1,4-dioxobutyl]amino}-4-pentynoate. US Patent: 5536869, July 16, 1996; c) J. Cossy, A. Schmitt, C. Cinquin, D. Buisson, D. Belotti, Bioorg. Med. Chem. Lett. 1997, 7, 1699; d) M. L. Boys, Tetrahedron Lett. 1998, 39, 3449; e) J. Ng, R. S. Topgi, Curr. Opin. Drug Discov. Develop. 1998, 1(3), 314 and references cited therein; f) R. S. Topgi, J. S. Ng, B. Landis, P. Wang, J. R. Behling, Bioorgan. Med. Chem. 1999, 7(10), 2221, and references cited therein.

2 a) R. Didziapetris, B. Drabnig, V. Sehellenberger, H.-D. Jakubke, V. Svedas, FEBS Lett. 1991, 287, 31; b) I. B. Stoineva, B. P. Galunsky, V. S. Lazanov, I. P. Ivanov, D. D. Petkov, Tetrahedron 1992, 48, 1115.

3 a) V. A. Solodenko, T. N. Kasheva, V. P. Kukhar, E. V. Kozlova, D. A. Mironenko, V. K. Svedas, Tetrahedron 1991, 47, 3989; b) V. A. Solodenko, M. Y. Belik, S. V. Hgalushko, V. P. Kukhar, E. V. Kozlova, D. A. Mironenko, V. K. Svedas, Tetrahedron Asymmetry 1993, 4, 1965.

4 a) A. L. Margolin, V. K. Svedas, I. V. Berezin, Biochim. Biophys. Acta 1980, 616, 283; b) E. Anderson, B. Mattiasson, B. Hahn-Hagerdal, Enz. Microb. Technol. 1984, 6, 301; c) A. L. Margolin, Tetrahedron Lett. 1993, 34, 1239.

5 H. Waldmann, Kontakte 1991, 2, 33.

6 A. Romeo, G. Lucente, D. Rossi, G. Zanotti, Tetrahedron Lett. 1971, 21, 1799.

7 V. K. Svedas, I. Yu. Galaev, Yu. A. Semiletov, G. A. Korshunova, Bioorgan. Khim. 1983, 9, 1139.

8 a) C. Fuganti, C. M. Rosell, S. Servi, A. Tagliani, M. Terreni, Tetrahedron Asymmetry 1992, 3, 383; b) E. Baldaro, P. D'Arrigo, G. Pedrocchi-Fantoni, C. M. Rosell, S. Servi, A. Tagliani, M. Terreni, Tetrahedron Asymmetry 1993, 4, 1031; (c) R. S. Topgi, in Applied Biocatalysis in Specialty Chemicals and Pharmaceuticals (B. C. Saha, D. C. Demirjian, Ed.), ACS Symposium Series 776, Washington, 2001, p. 200.

9 J. G. Shewale, B. S. Deshpande, V. K. Sudhakaran, S. S. Ambedkar, Process Biochem. Intl. 1990, 97.

10 a) P. Hermann, Biomed. Biochim. Acta, 1991, 50, 19; b) I. B. Stoineva, B. P. Galunsky, S. V. Lozanov, I. P. Ivanov, D. D. Petkov, Tetrahedron 1992, 48, 115.

11 a) C. Fuganti, P. Grasselli, S. Servi, A. Lazzarini, P. Casati, Tetrahedron 1988, 44, 2575; b) C. Fuganti, C. M. Rosell, R. Rigoni, S. Servi, A. Tagliani, M. Terreni, Biotech. Lett. 1992, 14, 543.

12 E. Andersson, B. Mattiasson, B. Hahn-Hagerdal, Enzyme Microb. Technol. 1984, 6, 301.

13 C. Ebert, L. Gardossi, P. Linda, Tetrahedron Lett. 1996, 37, 9377.

14 M. J. Zmijewski, B. S. Briggs, A. R. Thompson, I. G. Wright, Tetrahedron Lett. 1991, 32, 1621.

Subject Index

a

7-ACA (7-aminocephalosporanic acid) 117
– chemical route 120
– comparison of chemical and two-step biocatalytic process 127
– single step biocatalytic synthesis 121
– two-step biocatalytic synthesis 122
Amano PS-30 368, 374
amidase from *Klebsiella oxytoca* 109
amidase from *Pseudomonas fluorescens* 136
α-amino acids: large-scale manufacture 271
D-amino acid oxidase (DAO) 122
L-amino acids
– processes using L-amidases 135
– processes using L-amino acylases 133
– processes using L-hydantoinases 137
– via addition of ammonia to α,β-unsaturated acids 143
– via lactam hydrolysis 139
– via reductive amination 140
– via transamination 142
amino acylases from *Aspergillus oryzae* 134
7-aminocephalosporanic acid (7-ACA) 117
(1R,4S)-1-amino-4-hydroxymethyl-cyclopent-2-ene 114
aminopeptidase 135
L-aspartate ammonia lyase from *Escherichia coli* 143
asymmetric hydrogenation 23, 39, 55, 71, 91, 259, 269, 283
asymmetric cyclopropanation 343
asymmetric reduction 107, 323
asymmetric sulfide oxidation 413
asymmetric transfer hydrogenation 201

b

BIPHEMP 79
– synthesis 81
Brevibacterium flavum whole-cell catalysts 143
N-boc-(S)-3-fluorophenylalanine 272
4-boc-piperazine-2(S)-N-tert-butylcarboxamide 299
(R)-1,3-butanediol
– large-scale preparation 223
(R)-1,3-butanediol by stereospecific oxidoreduction 217
(R)-1,3-butanediol dehydrogenase from *Candida parapsilosis* IFO 1396 225
N-tert-butoxycarbonyl-(S)-3-fluorophenylalanine 273
(S)-2-[(tert-butylsulfonyl)-methyl]-hydrocinnamic acid 385

c

Candida parapsilosis IFO 1396 223
L-Carnitine via asymmetric biocatalysis 106
CATHy™ catalysts (Catalytic Asymmetric Transfer Hydrogenation) 202
– industrial process 210
– hydrogen donor 206
CBZ-D-proline [(R)-N-CBZ-proline] 113
chiral catalysts: overview 10
chiral ligands 79
– BIPHEMP 79
– DIPAMP 29
– DuPHOS 269
– josiphos 289
– methylcyclohexyl-o-anisylphosphane 26
– Ph-β-glup 41
– salen 166
– xyliphos 60
chiral phosphane ligands 23

Asymmetric Catalysis on Industrial Scale: Challenges, Approaches and Solutions
Edited by H. U. Blaser, E. Schmidt
Copyright © 2004 Wiley-VCH Verlag GmbH & Co. KGaA, Weinheim
ISBN: 3-527-30631-5

4-chloro-3-acetoxybutyronitrile 246
(R)-4-chloro-3-hydroxy-butyrate 247, 251
(R)- and (S)-3-chloro-1,2-propanediol (CPD) 238
Comomonas acidivorans A:18 111
(CPD) assimilating bacteria 238
crotonobetainyl-CoA-hydrolase 106
cyanohydrin
– large-scale production 158
– via chiral metal catalysts 154
– via cyclic dipeptides 154
– via lipases 156
trans cyclopropane carboxylic acid 335
– comparison of four routes 346
– via asymmetric cyclopropanation 340
– via Jacobsen epoxidation 341

d
DCP (2,3-dichloro-1-propanol) assimilating bacterium 235
(R)- and (S)-2,3-dichloro-1-propanol (DCP) 235
– large-scale trials using immobilized cells 236
– control logic system 240
diethyl 2-isobutyl succinate
– pilot production 408
– process research 403
– process development 406
(S)-2,2-dimethylcyclopropane carboxamide 111
DIPAMP 29
– synthesis 31, 265
(R)-N-diphenylphosphinyl-1-methylnaphthyl-amine 213
L-dopa process
– effect of solvent and anions 46
– immobilization of the catalyst 49
– VEB ISIS variant 39
DuPHOS rhodium(I) catalysts 269
dynamic kinetic resolution 372

e
EATOS (environmental assessment tool for organic synthesis) 99
enantioselective heterogeneous hydrogenation
– keto ester 95
– diketo ester 95
enantioselective reduction of keto acid
– with immobilized *Proteus vulgaris* 94
– with D-LDH 94
Enterobacter sp. DS-S-75 248
– cultivation and preparation 248

(S)-epichlorohydrin (EP) production 236
Escherichia coli 107, 110
2-ethoxycarbonyl-3,6-dihydro-2H-pyrane 352
– optimization of the resolution 355
– protease screen 354
ethyl (R)-2-hydroxy-4-phenylbutyrate (HPB ester) 91
ethyl (R)-4-chloro-3-hydroxybutanoate 230

f
FDH (formate dehydrogenase) from *Candida boidinii* 141, 328
(S)-4-fluorophenylethanol 213
(R)-3-(4-fluorophenyl)-2-hydroxy propionic acid 323, 331
formate dehydrogenase (FDH) 326

g
glossary 16
glutaryl-7-ACA acylase (GA) 123
(R)- and (S)-glycidol 238

h
hetero Diels-Alder-biocatalysis approach 349
HKR (hydrolytic kinetic resolution) reaction 165
– isolation of resolved epichlorohydrin 196
– process optimization and scale-up 190
– (salen)-Co complex 166
(S)-HNL (hydroxy nitrile lyases) from *Hevea brasiliensis* 162
homogeneous catalyst
– catalyst charging 278
– catalyst removal 279
(R)-HPB ester (ethyl 2-hydroxy-4-phenylbutyrate) 91
– comparison of routes 96
– environmental assessment 100
– health/toxicology 100
– mass balance 97
– safety aspects 99
– substrate 209
– synthetic pathways 92
2-hydroxy acids
– enzyme loading and replenishment 333
– reactor set-up and preparation 332
(S)-3-hydroxy-lactone 247
– ton-scale production 251

i

industrial manufacture
– choice of precatalysts 275
– general considerations 9
– practical and economic considerations 274
– substrate synthesis, purity and catalyst loading 276
2-isobutyl succinic acid 399

j

Jacobsen hydrolytic kinetic resolution (HKR) 165
josiphos ligands 286
– preparation of josiphos (R)-(S)-PPF-P(^{t}Bu)$_2$ on a technical scale 291
– xyliphos 60

k

kinetic resolution 349, 365, 385, 399, 437
kinetic modelling and simulation of the HKR reaction 176
Knowles' catalyst 259

l

D-lactate dehydrogenase (D LDH) 326
D-LDHs from *Leuconostoc mesenteroides* 328
L-leucine dehydrogenase from *Bacillus sphaericus* 141
Lonza biotin process 284
– laboratory screening experiments 286
– parameter screening and optimization 290
– substrate screening 287
Lonza dextrometorphan process 293
– economic process 296
– scale-up 297

m

(R)- and (S)-mandelic acid derivatives 163
MEA imine hydrogenation 57
mechanism of isoxazoline ester racemization 370
MeOBIPHEP 79
– synthesis 82
(S)-3-[2-{(methylsulfonyl)oxy}ethoxy]-4-(triphenylmethoxy)-1-butanol methanesulfonate 349
(S)-metolachlor
– ligand fine tuning 61
– chiral switch 55
microbial resolution of 2,3-dichloro-1-propanol, 3-chlor-1,2-propanediol 233

Monsanto L-dopa process 23
Mycobacterium neoaurum 135

n

Nishiyama catalyst 343

o

(R)- and (S)-2-octanol 316
2-oxabicyclo[3.3.0]-oct-6-en-3-one 316

p

patent problems 14
Penicillin G amidohydrolase 437
(S)-3-phenoxybenzaldehyde cyanohydrin 162
processes
– overview 5
prochiral allylic alcohols
– asymmetric hydrogenation 78
– synthesis 76
proline acylase in a strain of *Arthrobacter* sp. 113
Pseudomonas putida 135
Pseudomonas sp. DS-K-19 246

r

resolution of diethyl 2-isobutyl succinate 403
resolution of 2-ethoxycarbonyl-3,6-dihydropyran 360
resolution of ethyl-3-amino-4-pentynoate 437
– optimization of reaction conditions 441
resolution of ethyl sulfopropionate 385
– pilot production 394
– process development 391
– process research 389
Rh-DIPAMP 259
– ligand synthesis 31, 265
– scope 260
Rhodococcus erythropolis CB101 114

s

(salen)-Co complex 166
– activation 169
– preparation 167
– scale-up considerations 174
screening of microorganisms 219, 221
secondary alcohol dehydrogenase from *Candida parapsilosis* IFO 1396 226
sodium 3-(4-fluorophenyl)-2-oxo-propionate (6) 331
Solvay Protease L 660 392
Subtilisin Carlsberg 389, 403

sulfide oxidation
– chiral auxiliary 424
– effect of base 423
– effect of solvent 424
– effect of oxidation time 425
– equilibration of titanium complex 427
– process flow 433
– scale-up 428

t

technical problems 14
(R)-1-tetratol 212
time lines 13
(R,R,R)-α-tocopherol 71
– total synthesis 74
– stereochemical analysis 84
transformations
– overview 10

(R)- and (S)-3,3,3-trifluoro-2-hydroxy-2-methyl-propionic acid 108
Trigonopsis variabilis 123

u

unnatural amino acids 259
– preparation of enamides 263

v

VEB ISIS L-dopa process 39

w

whole-cells from *Brevibacterium flavum* 143

x

xyliphos
– scale-up of the ligand synthesis 63